HEAT PUMPS

Prospects in Heat Pump Technology and Marketing

Proceedings of the
1987 International Energy Agency Heat Pump Conference
Prospects in Heat Pump Technology and Marketing

Orlando, Florida
April 28–30, 1987

HEAT PUMPS

Prospects in Heat Pump Technology and Marketing

HEAT PUMPS

Prospects in Heat Pump Technology and Marketing

Proceedings of the
1987 International Energy Agency Heat Pump Conference
Prospects in Heat Pump Technology and Marketing

Orlando, Florida
April 28–30, 1987

Editor
Kay H. Zimmerman
Oak Ridge National Laboratory
Bldg. 9102-2, P.O. Box Y
Oak Ridge, TN 37831

Technical Editor
Raleigh H. Powell, Jr.
Oak Ridge National Laboratory

LEWIS PUBLISHERS, INC.

Library of Congress Cataloging-in-Publication Data

International Energy Agency Heat Pump Conference
(1987 : Orlando, Fla.)
Heat pumps.

Bibliography: p.
Includes index.
1. Heat pumps—Congresses. I. Zimmerman, Kay H.
II. Powell, Raleigh H. III. International Energy
Agency. IV. Title.
TJ262.I6 1987 621.402'5 87-3677
ISBN 0-87371-107-6

LEWIS PUBLISHERS, INC.
121 South Main Street, P.O. Drawer 519, Chelsea, Michigan 48118

PRINTED IN THE UNITED STATES OF AMERICA

Sponsoring Organizations

Electric Power Research Institute
Palo Alto, California

Federal Ministry for Research and Technology
Bonn, Federal Republic of Germany

Gas Research Institute
Chicago, Illinois

Heat Pump Technology Center of Japan
Tokyo, Japan

Ministry of International Trade and Industry
Tokyo, Japan

Oak Ridge National Laboratory
Oak Ridge, Tennessee

Swedish Council for Building Research
Stockholm, Sweden

Swedish National Board for Technical Development
Stockholm, Sweden

U.S. Department of Energy
Washington, D.C.

Organizing Committee

F. A. Creswick (Chairman)	Oak Ridge National Laboratory, Oak Ridge, Tennessee
J. Berghmans	Katholieke Universiteit Leuven, Heverlee, Belgium
T. Bostroem	Swedish Council for Building Research, Stockholm, Sweden
R. J. Fiskum	U.S. Department of Energy, Washington, D.C.
J. F. Harris	Groupe MAC, Paris, France
W. Hochegger	Energiesparhaus Forschungsges, Graz, Austria
K. O. Holzapfel	IEA Heat Pump Center, Karlsruhe, Federal Republic of Germany
P. A. Joyner	Electric Power Research Institute, Palo Alto, California
A. R. Maret	Gas Research Institute, Chicago, Illinois
E. Miura	Heat Pump Technology Center of Japan, Tokyo, Japan
H. Miyama	Ministry of International Trade and Industry, Tokyo, Japan

Oak Ridge National Laboratory Conference Staff

Frederick A. Creswick	Organizing Committee Chairman
Kay H. Zimmerman	Conference Administration
Norma F. Cardwell	Conference Coordinator
Vivian A. Jacobs	Publications
Gail Anderson	Editing
Sharon H. McConathy	Editing
Raleigh H. Powell, Jr.	Editing
Pamela J. Lewis	Registration
Joseph A. Pietsch	Preconference Site Visits
Kathi H. Vaughan	Local Tours and Social Events

Preface

Whether we call it energy conservation, rational energy use, or energy efficiency, heat pumps make obvious good sense. Electric heat pumps provide an alternative to oil or gas energy for space heating or a doubling or tripling in efficiency over resistance heating. Thermally activated heat pumps promise an improvement in the efficiency of the utilization of natural gas and oil by 50 to 100%, as compared with conventional furnaces and boilers. In industrial applications, substantial reductions in energy use are possible by using heat pumps to upgrade waste energy streams.

This was the motivation in 1978 when 13 International Energy Agency (IEA) member countries entered in the Implementing Agreement for a Programme of Research and Development on Advanced Heat Pump Systems. At that time, heat pumps were obviously an important technology to implement a stated IEA goal of "cooperation to reduce excessive dependence on oil through energy conservation, development of alternative energy sources, and energy research, development, and demonstration." It is difficult to imagine that heat pumps would not also be an important part of future energy technologies, because it seems reasonable to assume that future energy prices and supplies will warrant rational energy use.

It is also clear that the use of heat pumps is now a problem in many parts of the world. Falling oil prices have drastically affected the economic viability of currently available heat pumps for space conditioning, except in those areas where a strong cooling requirement exists. Research and development budgets have been reduced sharply, and many manufacturers have withdrawn from the field. Consequently, what can we say about the future for heat pumps in view of the present situation?

The 1987 IEA Heat Pump Conference has been structured to address these current issues and potential solutions to present problems.

What works? What does it take to do it correctly? What can be learned from those countries where vigorous heat pump markets exist? Will technological advances improve the situation? How should equipment manufacturers, energy companies, and governments provide support?

One thing we have learned from the past decade of energy-related activities is that energy conservation works! Substantial reductions in energy use per capita, per household, or per unit of gross national product have been accomplished throughout the world. In the buildings sector, this has been accomplished primarily by improved building thermal envelopes and application of existing technology to improving equipment efficiency. Even so, in both the buildings and industrial sectors, we have barely scratched the surface of possible future improvements in the efficiency of energy utilization. Advanced heat pumps will be one of the key technologies involved in achieving the ultimate in the rational use of energy in the future.

The members of the Organizing Committee are hopeful that this international conference will be mutually beneficial to all participants, and that the interchange of ideas and information made here will contribute in some measure toward making future heat pump technologies available in a timely manner when they are needed.

Fred Creswick, Organizing Committee Chairman
Oak Ridge National Laboratory

Table of Contents

Section IV. Heat Pumps for Industrial Applications

Section VIII. Government's Role in Energy Conservation

Section IX. Additions

Section I

Opening Plenary Session

Chapter 1

Heat Pump Research Carried Out by the European Communities

Peter Zegers

ABSTRACT

CEC heat pump research is executed in a four year Non-Nuclear Energy R and D Programme running from 1985 to 1988; at present 22 CEC funded heat pump projects are being carried out by different research organizations in Europe. This programme has two main objectives :

- Reduce the cost and improve efficiency of heat pumps for domestic and commercial heating applications in order to achieve economic feasibility and bring about a breakthrough on this market will consumes 26 % of the primary energy in the E.C.
- Develop industrial heat pumps which can produce heat up to 300°C.

Research to achieve the first objective for compressor heat pumps is aiming at the use of fluid mixtures, development of more efficient and cheap compressors, of oil free compressors, better control and development of efficient low pollution combustion engines. Work on absorption heat pumps is mainly focussed on fluid pairs and on the development of cheap and efficient heat exchangers.
Research on high temperature industrial heat pumps is carried out along several lines. For compressor heat pumps, fluid mixtures for high temperatures (180°C) are being investigated. For high temperature absorption heat pumps, work is focussed on new fluid pairs and on the study, construction and testing of different cycles (absorption heat pumps, heat transformers and other cycles). Finally, the possibility to develop solid / fluid absorption heat pumps up to 300°C (solid/fluid combinations, reactor, heat exchangers) is systematically explored.

HEAT PUMPS: Prospects in Heat Pump Technology and Marketing (Proceedings of the 1987 IEA Heat Pump Conference), Edited by Kay H. Zimmerman

INTRODUCTION

The European Community is funding heat pump research in its Non-Nuclear Energy R and D Programme. 22 R and D projects are being carried out by different organizations in Europe on compression and absorption heat pumps in the framework of this programme which runs from 1985 to the end of 1988. In total 6 million ECU (*) is being spent on this research of which the Commission is paying 60 %.

IMPACT OF HEAT PUMPS ON ENERGY SAVINGS, OIL SUBSTITUTION AND POLLUTION ABATEMENT

R and D on heat pumps forms part of a broader energy policy of the Commission which aims at :

- Energy saving
- Substitution of oil by coal, nuclear or renewable energy
- Pollution abatement.

Energy saving

For the provision of energy in the European Community about 30 billion ECU per year is spent on energy investment, this is about 35 % of the total industrial investments. In addition over 100 billion ECU per year is spent on fuels. Energy savings of only a few percent will lead to the saving of billions of ECU.
The primary energy consumption in the European Community in 1985 was 940 million tonne oil equivalent. This energy was distributed over the different demand sectors as shown in Table 1.

Table 1. Primary energy use in the E.C.

Buildings 41 %		Industry 41 %			Transport 18 %
Heating 26 %	Power 15 %	Power 13 %	Non energy use 7 %	Process heat 21 %	18 %

* 1 ECU (European Currency Unit) = 1,1 $

Heat pumps may contribute to energy savings in two sectors : heating of buildings and industrial process heat.
Heat pumps can provide heat up to temperatures around 120°C. For heating in buildings heat is needed at temperatures of around 70 - 90 °C, which can easily be provided by heat pumps. The potential market is thus the whole building stock and if one assumes that heat pumps use 30 - 50 % less energy than conventional heating systems, the potential energy savings will lie between 8 and 13 % of the overall primary energy use.
Heat pumps in industry mainly serve as a tool to transform waste heat into heat at a higher temperature level where it may be used. Process heat in industry is required in a temperature range from 100°C to 1500°C and the amount of heat in industry as a function of temperature is roughly as indicated in Fig. 1. The use of heat pumps for industrial process heat is limited to applications which use temperatures below 120 °C. This limits the potential market to about 10 % of the overall industrial process heat consumption. The potential energy savings lie around 1 % of the overall primary energy consumption. If heat pumps could be developed which produce heat at temperatures up to 400 °C the energy saving potential could be increased to around 3 - 4 % (see Fig. 1). Development of high temperature heat pumps is therefore an important objective of this programme.

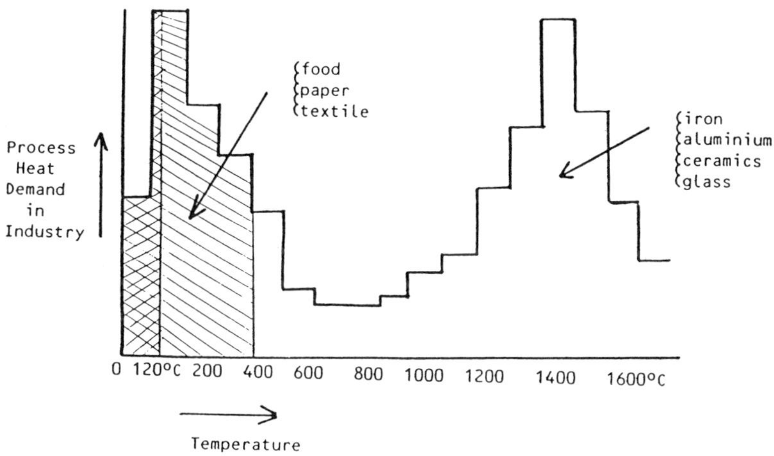

Fig. 1 : Typical process heat demand curve as a function of temperature (for Germany)

As for the size of the heat pumps one may expect that for the case of a large scale introduction of heat pumps about 50 % of this heat will be provided by small heat pumps (typically 10 kW) for individual heating of dwellings. Large heat pumps of over 100 kW will be needed for the remaining 50 % (heating of offices, industrial buildings, apartment buildings, hospitals, schools etc.).

Substitution of oil

The use of oil in the EC presently amounts to 42 % of the primary energy use (down from 63 % in 1973). It is the objective of the Commission to further reduce this percentage for two reasons :

- In the long term oil will be in short supply
- Around 80 % of the oil is imported; mainly from politically unstable areas

Table 2. Oil consumption in the different demand sectors

	Heating in Buildings	Electricity	Industrial Process Heat	Transport
Primary en. use	26 %	28 %	21 %	18 %
Oil use	55 %	16 %	37 %	85 %

Of the different demand sectors (see Table 2) the transport sector has the highest percentage of oil consumption. It is however very difficult to replace oil in this sector by other fuels in the short and medium term. In the long term electrical vehicles may possibly take a share of the market.
Other sectors which have a large oil consumption are : heating of buildings and industrial process heat. In both sectors heat pumps and in particular compressor heat pumps could contribute to substitution of oil as they provide heat by using electricity which can be produced from nuclear energy, coal or renewable energy. Absorption heat pumps could replace oil by gas, coal or wood.

Pollution

The conversion and use of energy has always been an important cause of environmental pollution; e.g. exhaust gases of cars, of power plants of domestic heating systems and of industrial boilers. Recently environmental pollution has taken a dramatic dimension as it is more and more clear that a large part of the forests in Europe are threatened by acid rain caused by such a pollution. The cost for society of environmental abatement is presently estimated to be 3 - 5 % of the GDP and is likely to increase due to an increasingly severe legislation.

Potentially heat pumps could make a large contribution to pollution abatement, in particular in densily populated areas. Heat pumps produce heat 30 - 50 % more efficient than conventional systems and give per unit of heat produced less pollution. If used on a large scale this heat pump could make a large contribution to the pollution abatement; about 28 % of the primary energy consumption could be provided by heat pumps.
In addition the major part of the heat pumps will be electrically driven heat pumps which would replace coal, oil or gas fired conventional systems by a completely clean system. The additional power required would result in a somewhat increased pollution at the electricity plant but this pollution could be more easily taken care of.
For fossil fuel fired absorption heat pumps the reduction of the pollution level would mainly be brought about by lower energy use, which could be 50 %.
Another aspect of heat pumps could have a negative impact on the environment. In particular, some fluorochemicals used in heat pumps can attack the ozon layer. In view of the present worry on keeping the ozon layer intact a ban on the use of these refrigerants in the USA and Europe may very well be possible. This would have a strong impact on heat pump R & D and the consequences should be investigated.

OBJECTIVES FOR EC HEAT PUMP R AND D

Work carried out in the E.C. heat pump programme is aiming at two objectives :

- Reduction of the cost and improvement of the efficiency of compressor heat pumps by 50 %
  Although heat pumps save energy and reduce pollution levels theay are in Europe economically only marginally attractive. At present in Europe around 200.000 heat pumps have been installed and a breakthrough can certainly not be expected in the near future. The main reason is that heat pumps are still too expensive and the performance is too low. However, both for cost and performance there is scope for improvement. Large scale production could lead to cost reductions of 30 - 50 %; research could also lead to cost reductions. In particular for compressor heat pumps the practical seasonally averaged COP values are a factor three lower than theoretically can be expected. R and D should therefore be able to increase performance levels by 50 %.
  The second objective is the development of high temperature heat pumps for industrial applications. The potential for application of heat pumps in industry could strongly increase if the temperature range of heat pumps could be extended for present values of around 120°C to 300°C. To that end new refrigerants, working fluid paris, solid gas pairs and new cycles are being investigated.

RESEARCH PROGRAMME

Heat pump research in the European Community is carried out in its third Non-Nuclear Energy R and D Programme (1985 - 1988). At present 22 projects have been approved for a total cost of 6 million ECU of which the Commission pays 50 %.

Compressor heat pumps

Compressor heat pumps are commercially available but their economic feasibility is marginal. Moreover their seasonally averaged efficiencies (from primary fuels) of 90 - 100 % are considerably lower than for other heat pump types such as absorption heat pumps or internal combustion engine driven heat pumps. On the other hand, electrical heat pumps could make a major contribution to pollution abatement and have the additional advantages to allow the replacement of precious premium fuels such as oil and gas by coal or nuclear energy. R and D on compressor heat pumps was therefore continued, the more so as there is still scope for considerable (50 %) improvement of the efficiency and a reduction in cost.
The second Non-Nuclear Energy R and D Programme (1979 - 1983) established that the compressor heat pump could be improved in the following way :

- Improved compressor design leading to higher efficiency and lower cost.
- Dissolution of lubrication oil of the compressor in the refrigerant leads to a decreased performance; ways should be found to avoid this.
- Use of non-azeotropric mixtures can lead to increased performance and lower cost.
- Fluctuation in the refrigerant flow should be avoided.

Four projects are aiming at the improvement of compressor heat pumps :
GENERAL SUPPLY, Greece is developing a simple oil free highly efficient and cheap compressor. It is a rotating compressor with a maximum speed of 15 000 rpm, the volume swept per rotation is 2 x 100 cm^3.
UNIVERSITY OF ULSTER, UK is studying a number of ways to improve the compressor heat pump. This group has studied the influence of dissolution of compressor lubricant oil in the refrigerant and found that this could decrease the performance by as much as 30%. Two options exist to avoid this problem. The first one is the development of a lubricant oil free compressor; this way is followed by GENERAL SUPPLY and FICHT. A second possibility is to develop lubricants which are dissolving less in the refrigerant; as a first step this project is studying dissolution properties of a wide range of lubricants. Experiments with simple rolling compressors are also being carried out and first experiments indicate that they give a better performance than reciprocating or screw compressors. Other ways to improve the compressor heat

pump are the development of methods to control evaporator superheat with a microcomputer based expansion system and the development of a cycle in which an active process is used for flashing the refrigerant to evaporating pressure and which according to Granryd should lead to significant performance gains.

R and D on the use of non-azeotropic mixtures in the previous E.C. Non-Nuclear Energy Programme (1979 - 1983) showed that these mixtures could improve the performance by 10 - 20 % and reduce the cost. In this programme ELF and IFP France are developing non-azeotropic mixtures for industrial heat pumps which produce heat above 110°C.

It is known that for on/off operation of a heat pump the duration of the operation interval has a strong influence on the overall performance and it is generally assumed that operation of at least 15 - 20 minutes is required to optimize heat pump performance. Shorter time intervals could reduce the performance by 10 - 15 %. On the other hand it has also been established that very short operation intervals can improve operation performance of heat pumps as compared to continuous operation. This is being investigated by FORDSMAND, Denmark. First experiments showed improvements of 20 %, with cycle times of 20 - 60 seconds and with a ratio of on and off time operation of 0,25.

Contrary to electrically driven heat pumps, internal combustion engine driven heat pumps allow the recovery of waste heat of the engine, which is added to the heat produced by the heat pump. This leads to considerably higher efficiencies. Also the heat output can be varied more easily by regulating the speed of the motor and switching off one or possibly more cylinders of the compressor. The heat output can thus be matched to the heat requirements of the house which leads to further energy savings.

Two projects are concerned with internal combustion engine (ICE) driven compressor heat pumps.

The first project by FICHTEL AND SACHS, Germany deals with a 20 kW ICE driven heat pump developed in the second E.C. Non-Nuclear Energy Programme. A barrier for introduction on the market was the high pollution caused by the engine. In this project a lean burn gas engine and a low polluting Diesel engine are being developed, suitable for operation with a 20 kW heat pump. The following objectives have been set :

	Oil	Gas
NO_x	800 (2650) mg/m^3	400 (2500) mg/m^3
CO	600 (650)	310 (1250)
CmHn	510 (680)	65 (360)
Soot	0,1 - 0,2 (1,0 - 14) in Bosch number	

The data in brackets give the present values.

The second project by FICHT, Germany aims at the construction of a combustion engine - compressor unit where the crankslot has two piston rods one for the piston of the combustion engine and one for the piston of the compressor; in this way transmission losses are minimized. The conical rotary slide engine is a lean burn engine with very low CO and NO_x levels, far below the levels legally required. Due to the low number of components the cost is expected to be low and the reliability high.

Absorption heat pumps

Absorption heat pumps are expected to have a much higher efficiency (PER = 1.3 - 1.5) than electric heat pumps and they may use a variety of fuels such as coal, oil, gas, wood or even waste heat. Absorption heat pumps do cause more pollution than electric heat pumps which are intrinsically clean. However as compared to conventional oil or coal combustion, absorption heat pumps cause a much lower pollution in particular due to the much higher energy conversion efficiency.

A large part of the research work in this programme on absorption heat pumps is aiming at the development of high temperature heat pumps for applications in industry.

For absorption heat pumps chemical decomposition of the working fluid pairs is often the bottleneck for achieving a high temperature heat production. Research on working fluids therefore has a high priority. In the second Non-Nuclear Energy R and D Programme (1979 - 1983) a systematic search for new working fluid pairs has been carried out, which resulted in a number of promising pairs which will be further studied by the UNIVERSITY OF ESSEN, Germany in the ongoing programme. Stability, compatibility with materials used in heat pumps and also heat and mass transfer problems will be investigated.

For high temperature operation the absorption heat pump is at a disadvantage as the temperature of the produced heat at the absorber and condenser, is considerably lower than the temperature of the generator which has the highest temperature in the circuit; here decomposition is likely to occur first. This is not the case for a heat transformer where medium temperature heat is given to the generator and evaporator to produce high temperature heat at the absorber. In a heat transformer heat is thus produced in the highest temperature point of the circuit. This makes the heat transformer intrinsically more suitable for high temperature operation than absorption heat pumps. It has the additional advantage that, apart from waste heat, no extra energy input is needed.

Heat transformers are therefore being developed by GEA, Germany in particular for operation at high temperatures. Working fluid pairs identified by the UNIVERSITY OF ESSEN, suitable for high temperature operation, will be tested in a heat transformer producing 20 kW heat at 170°C from 50 kW waste heat at 100°C. In addition also an absorption heat pump will be built, with the new working fluids. Different combinations of the heat transformer and the absorption heat pump will be tested and of the most promising combination a 100 kW installation will be built.

A similar project will be carried out by the UNIVERSITY OF MUNCHEN, Germany. Here a combination of a heat transformer for the low temperature part and an absorption heat pump for the high temperature part is envisaged. As a working pair LiBr/H_2O will be used. This combination is interesting as it combines the advantages of providing sufficient heat at high temperatures with the ability to also use all waste heat from the process at lower temperatures. The absorption heat pump alone generally produces sufficient heat but can not recover all the waste heat of an industrial process; on the other hand a heat transformer can generally not cover the heat demand but can use all the waste heat. There is no risk for decomposition of LiF as this component is stable up to high temperatures.

A third two-stage heat pump type is constructed by BATTELLE, Germany. This is a two-stage LiBr/H_2O absorption heat pump with which 100 kW of heat at 130°C will be produced from waste heat at 105°C.

A periodically operating absorption heat pump is being developed by the UNIVERSITY OF AACHEN, Germany. This heat pump consists of two reservoirs where liquids are separated, but where vapors can freely move from one reservoir to the another. They are used as a generator and condensor for half a period and as an absorber and evaporator for the other half period. During the generator/condensor phase, heat is given to the generator and vapour is condensed in the condensor where heat is extracted. During the evaporator/absorber phase low grade heat evaporates the refrigerant which is absorbed in the absorber where released heat is extracted. The cycle time is 1 - 1,1/2 hour. An advantage of this heat pump type is that no solution pump is required. The working fluids are CH_3OH/H_2O-LiBr. In the second programme a 10 kW unit has been built which produced heat at 50°C (with the generator at 120°C and the heat source at 0-10°) with a PER value of 1,23. In the ongoing programme, research is aiming at improving this absorption heat pump and at the construction and testing of a second pilot plant.

A second periodic absorption heat pump is developed by the CNRS, France. The main feature of this heat pump is a new concept for the generator and condensor and the possibility for storage of the working fluids. A 30 kW unit is expected to be in operation in 1988.

Reduction of cost is an important objective in this research programme also for absorption heat pumps. As a large part of the heat pump is related to heat exchangers a cost reduction and improved performance of heat exchangers can make a large contribution to economic feasibility of absorption heat pumps, heat transformers or other cycles. To that end, DUINTJER, Netherlands is developing a platefin heat exchanger which has the promise of good performance and cheap mass production. This type of heat exchangers will be tested in heat pumps for heat and mass transfer and flow resistance. If satisfactory the manufacturing of the heat exchangers will be developed.

An interesting option to develop high temperature heat pumps is the use of solid and gas as working media. It is believed that, with suitable solids and gases, heat pump operation up to 300°C is possible. The fact that such absorption heat pumps always operate periodically could be a disadvantage for industrial applications; however solutions for this problem may be found. Other problems are the development of suitable heat exchangers and reactors. This topic is studied by three heat pump projects.
A group of one Italian (CNR) and four French (CNRS, UTC, CETIAT, BLM) organizations is carrying out a systematic search for solid/gas combinations (with a emphasis on zeolite/water and active coal/methanol) suitable for operation up to 250°C. Other important R and D topics are the design of the reactors and heat exchangers in the different components of the heat pump, the optimization of the components, and the operation of cycles which consist of different stages.
The TECHNICAL UNIVERSITY OF DENMARK will systematically investigate NH_3 (refrigerant) and different metal halides (absorbants) in a number of cycles. In this project an attempt will be made to design a system which operates in a quasi-continuous way in order to circumvent the main draw back of solid absorption heat pumps : periodic operation. Two small systems will be constructed : a air to air heat pump for energy recovery from a drying process and a 2,3 kW two-stage heat transformer producing heat up to 300°C.
INPG, France investigates the possibility to develop a solid absorption heat pump with NH_3 (refrigerant) and graphite containing metallic salts. It is believed that this system, which operates via insertion of metallic salts in graphite, will have a much better performance than metal halides and pure NH_3.

ACKNOWLEDGEMENT

The author wishes to thank Ir. J.A. Knobbout for his help and valuable advice, as expert of the Commission.

Section II

Operating Experience with Heat Pumps for Residential and Commercial Space-Conditioning Applications

Chapter 2

Operating Experiences with Heat Pumps in North America

Carl C. Hiller

ABSTRACT

Central air conditioning systems are used mostly in general. However they are not energy-efficient since they require the large power of fans and pumps, not featuring the good partial load characteristics.

Instead of central system, multi-zone air source heat pump air conditioners (MZHP) have been increasingly used for air conditioning residences and office buildings in Japan.

MZHP system consists of an outdoor unit and plural indoor units. Wall suspended type indoor units are available for residential use, and ceiling mounted cassette type for commercial use. Each indoor unit is independently operative by each remote controller to control room temperature for each zone. Capacity-controllable compressors are provided in outdoor units.

The operation data of MZHP system compared with central system will be explained.

INTRODUCTION

There is a variety of so called central air conditioning systems. Typical of these systems are combination of chilled and hot water chiller with fan coil units and duct type heat pump air conditioning system for residence. For commercial use, there are combination of chilled and hot water chiller with fan coil units or air handling units and duct type heat pump air conditioning system.

HEAT PUMPS: Prospects in Heat Pump Technology and Marketing (Proceedings of the 1987 IEA Heat Pump Conference), Edited by Kay H. Zimmerman

below 50 tons. Some heat reclaim systems for water heating are larger than 50 tons, and can be classified as heat pumps, although they are generally associated with some cooling application. Industrial heat pump applications are currently rare in the U.S., with the exception of vapor recompression applications.

Information on heat pump operation in the U.S. has been gathered from informal contacts with users and installers, more rigorous structured surveys of users, and field tests. Information from informal discussions with users and installers is useful for structuring formal research efforts, but is not sufficiently quantitative to allow conclusions to be drawn without additional information.

Several structured surveys of heat pump users have been conducted under sponsorship of the Electric Power Research Institute (EPRI). These surveys have produced the first quantitatively verifiable information on heat pump service life in the U.S. Additional information has been gathered on heat pump repairs, proficiency of repair personnel, and customer satisfaction.

Electric utilities, EPRI, manufacturers, universities, Oak Ridge National Laboratory, and various other state and federal agencies have sponsored numerous field tests of a variety of heat pumps in diverse applications. Such tests are normally conducted to determine performance and efficiency characteristics in different climates, to study impacts on electric utility peak demand, and/or to evaluate design improvements.

This paper discusses findings from user surveys, general observations from field testing efforts in the U.S., and specific results from selected tests.

USER SURVEYS

Customer Satisfaction

Surveys of customer satisfaction with heat pumps [1,2] have found that the majority of users are satisfied or very satisfied with all aspects of the heat pump, including quality of comfort delivered, cost of operation, and reliability. This is especially true for newer vintage heat pumps, installed since 1980. There have been selected brands which were more prone to operational problems than the average. Market forces have caused such brands to either improve product quality, or cease to have significant market share.

While not quoted by the majority of heat pump users, there are several types of complaints that occur frequently enough to be of interest. These complaints include cool air drafts during heating operation, noise, and repair frequency. Investigation of the "cool blow" complaint indicates that in many cases this is due to improper duct and register design. The result is direct high velocity discharge of outlet air onto persons in the room, instead of proper low velocity air distribution. Register modifications or energizing the first stage of resistance heat below a given ambient air temperature are simple fixes to this complaint.

Noise complaints are usually related to either the indoor or outdoor fans. Modest air flow rate reductions via motor speed adjustment can often reduce this complaint. Installation of noise barriers around the outdoor unit of air-to-air heat pumps can also reduce this complaint. Noise levels vary considerably between brands and models.

A small number of users have complained about excessive repairs, but these complaints were for the most part associated with older vintage heat pumps. Newer vintage units, especially those installed since 1980, experience far fewer repairs or complaints of any kind. Research has shown, moreover, that a significant number of the complaints of frequent repairs were related to inadequate proficiency of maintenance personnel. This phenomenon was especially noticeable in regions and time periods when a rapid increase in the number of heat pump installations occurred. In such cases, installers who were not properly trained in heat pump installation and maintenance often began installing heat pumps in attempts to capitalize on the growing market. This situation is normally self correcting to some extent as users learn what to expect from their heat pumps and from maintenance personnel. Unfortunately, training and experience only with air conditioning has been shown to not be sufficient for installing and maintaining heat pumps. Special training specific to heat pumps is generally advisable. Education and training of service personnel must remain a high priority due to continued evolution of equipment designs and personnel turn-over.

Heat Pump Life

Surveys of heat pump service life revealed, as shown in Figure 1, that the median service life (age at which 50% of the units had been removed from service and 50% remain in service) of heat pumps in northern climates is between 15 and 16 years, while it is slightly greater than 20 years in southern climates [3,4]. Nearly half of the heat pumps removed from service were fully operational when replaced, and most were replaced with new heat pumps.

Compressor Life

Service life of heat pump compressors was found on average to be greater than 13.5 years, with some brands exhibiting compressor lives of more than 16 years [3,5,6]. Newer units should have even longer compressor lives if observed trends continue. An increasing number of units are experiencing no compressor problems during their normal total system lifetimes, having reached 40% at last available count. Typically, less than 3-4% of all heat pumps have experienced multiple compressor replacements. Most heat pump owners can therefore expect to replace the compressor no more than once, if ever, during the normal total system lifetime.

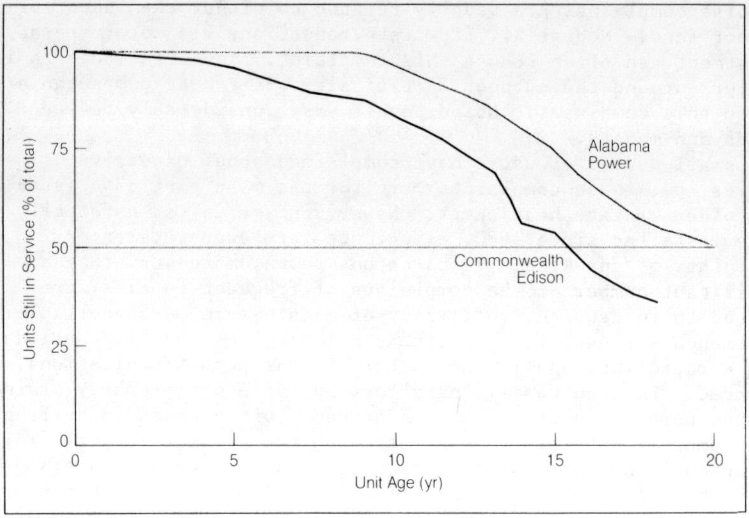

Figure 1

Surveys Conducted by Alabama Power Company (APCo) and Commonwealth Edison Company (Edison) and Sponsored by EPRI Indicate Heat Pump Service Life Ranges From 15 Years to More Than 20 Years

HEAT PUMP FIELD PERFORMANCE TESTS

Table 1 gives a partial list of the types of heat pumps that have been field tested by various organizations in the U.S. It can be seen that numerous tests have been performed on air-source units. Many of these tests are evaluations of new features, ranging from new compressors to better controls. New applications of existing equipment are also included. Other types of heat pumps tested include an assortment of ground source, solar assisted, and storage coupled units.

Data Interpretation

Due to differences in data collection techniques and data analysis procedures, comparison of results from different field tests must be done with caution. A thorough understanding of procedures and assumptions used for data collection, analysis, recovery from data loss, and normalization of results is advised before drawing conclusions from such comparisons.

Type		Configuration
Air Source		
		Single Package
		Dual Split
		Tripple Split
		Dual Compressor
		Two Speed
		Variable Speed (inverter drive)
		Room Units
		Package Terminal Units
		Multizone Units
	With Water Heating	
		Desuperheating
		Full Condensing Dedicated Stand Alone
		Full Condensing Integrated With Space Conditioning
	Dual-Fuel	
		Forced Air
		Hydronic
		Oil Backup
		Propane Backup
		Natural Gas Backup
Ground Source		
		Ground Water
		Vertical Closed Loop
		Horizontal Closed Loop
		Crawl Space Coupled
Storage Coupled		
Solar Assisted		

Table 1
Heat Pump Equipment Types Evaluated

FIELD TEST OBSERVATIONS

Despite differences in data collection and analysis procedures used, numerous general observations can be drawn from the various heat pump field tests that have been performed [7]. These observations are supported with examples from specific field tests sponsored by the Electric Power Research Institute (EPRI), described in Table 2.

Air-to-Air Heat Pumps

Much information is known about the efficiency, electrical demand, and reliability of air-to-air heat pumps:

1. Air-to-air heat pumps can perform reliably in both northern and southern U.S. climates.

2. Air-to-air heat pumps can save significant amounts of operating energy compared to electrical resistance heating in all climates that have significant hours of heating operation.

3. If controls are properly adjusted and functioning, conventional air-to-air heat pumps can reduce peak electrical demand compared to electric resistance heat. Care must be taken, however, to avoid poor operating strategies, such as improperly controlled night setback, in order to attain the reductions in peak demand.

4. Improper refrigerant charge can significantly reduce seasonal efficiency and negatively impact user comfort and satisfaction. It can also adversely affect life of critical components in the system, such as the compressor.

5. High efficiency units are more susceptible to reductions in overall efficiency due to controls maladjustments or other problems than are lower efficiency units. This is because the low energy consumption of high efficiency units causes problems to have a greater percentage impact.

6. Defrost controls are getting better, but there is still room for improvement.

7. Thermostat placement, method of mounting, and control features can significantly impact comfort and energy consumption, through impacts on cycling rate and effective setpoint.

Tables 3 and 4 summarize results of two related heat pump field test efforts [8,9] performed in the 1980-83 time period.

The multi-capacity units (dual-speed, Batchtown Ill. and dual-compressor, Manlius N.Y. and Minneapolis, Minnesota.) performed well. The dual speed unit performed especially well in the cooling mode. One potential drawback was noted with these units,

Project No.	Title/Description	Report No.	Date
RP789-3	Monitoring of Improved Air-Source Heat Pumps	EM-3978	1986
RP2033-9	Monitoring of Central and Room Heat Pumps	EM-4674	1986
RP1201-14	Monitoring of Residential Ground-water Source Heat Pumps In The Northeast	EM-4372	1986
RP2033-14	Heat Pump Experiments Using Crawl Spaces as Ground-to-Air Heat Exchangers	EM-4908	1986
RP2033-19	Evaluation of a Novel Dual Fuel Heat Pump	EM-4479	1986
RP1495-1	Performance Analysis of Air-to-Air Heat Pumps	EM-4226	1985
RP1191-6	Performance Comparison of Air- and Ground-Coupled Heat Pump Systems	EM-3408	1984
RP2033-5	Heat Pump Water Heaters	EM-3582	1984
RP789-1	Northern Climate Heat Pump Field Performance Evaluation	EM-2319	1982
RP1191-6	Performance Monitoring of Ground-Coupled Solar-Assisted Heat Pump Systems	EM-1697	1981
RP432-1	Load and Use Characteristics of Electric Heat Pumps in Single-Family Residences	EA-793	1978
RP385	Solar Energy and the Heat Pump in a Northern Climate	EA-407	1977
RP1940-5	Survey of Heat Pump Field Performance Data	None	N/A
RP1201-15	Monitoring of Multizone Heat Pumps	None	N/A

Table 2
EPRI Heat Pump Test Projects

Table 3
Heat Pump Performance Summary

Site Location	Heat Pump System Description	Measured			Normalized		Expected		Peak Power Demand Reduction Compared to		Comments
		HPF	HSPF	SEER	HSPF	SEER	HSPF	SEER	I^2R Furnace or Boiler (kW) (%)	Single Capacity Heat Pump (kW) (%)	
Batchtown, Illinois	Triple-split, Two Speed-Single Compressor Heat Pump	6.3	--	8.3	5.8	8.4	6.4	7.3	3.3 (32%)	0.6 (8%)	Measured HPF = 3/81-12/81. Heat pump did not operate 12/80-2/81.
Central Square, New York	Air-Water Hydronic System With Oil Boiler	6.1	--	NA(2)	6.1	NA(2)	NA(3)	NA(2)	10.8 (94%)	NA(1)	Heating-only system, no ARI ratings available. Prototype refrigerant-to-water heat exchanger. Measured HPF from 11/80-5/81.
Dallas, Texas	Double-Split Heat Pump	--	6.4	6.4	7.1	6.5	6.6	6.0	2.8 (43%)	NA(1)	Heat pump oversized.
Kingsport, Tennessee	Triple-Split Heat Pump With Domestic Water Heater	--	5.7	6.3	5.5	6.4	5.4	6.8	1.6 (22%)	NA(1)	Expected HSPF/SEER do not include effects of desuperheater used to heat domestic water.
Liverpool, New York	Double-Split Heat Pump With Gas Furnace	7.4	--	7.6	7.3	7.7	6.4	6.9	13.2 (96%)	NA(1)	Measured HPF from calendar year 1981.
Manlius, New York	Triple-Split, Two-Compressor Heat Pump	--	5.6	NA(4)	5.7	NA(4)	NA(5)	NA(5)	2.3 (19%)	1.6 (14%)	Cooling load too small to determine SEER.
Minneapolis, Minnesota	Triple-Split Two-Compressor Heat Pump	5.5	--	NA(4)	5.5	NA(4)	5.7	NA(4)	1.0 (6%)	1.0 (6%)	Cooling load too small to determine SEER. Measured HPF from 1981 calendar year.

Note: (1) Not applicable. These heat pump systems use one single speed compressor.
(2) Not applicable. Heat pump not used for cooling.
(3) Not applicable. Prototype refrigerant-to-water heat exchanger in system with no ARI ratings.
(4) Not applicable. Cooling load too small to determine cooling performance.
(5) Not applicable. No ARI ratings for this system.

Table 4
Heat Pump Performance Summary - Central Units

Site Location	Heat Pump System Description	Measured HPF	Measured HSPF	Measured SEER	Normalized HSPF	Normalized SEER	Expected HSPF	Expected SEER	Power Demand Reduction Compared to I^2R Furnace or Boiler (kW) (%)	Power Demand Reduction Compared to Single Capacity Heat Pump (kW) (%)	Comments
Batchtown, Illinois	Triple-Split, Two Speed - Single Compressor Heat Pump		6.6	8.9	6.3	9.0	6.4	7.5	1.7 (12%)	1.3 (10%)	Measured performance from 10/82 - 9/83.
Birmingham, Alabama	Double-Split Heat Pump with Gas Furnace	7.3		5.9	7.3	5.9	6.7	6.7	1.4 (55%)	NA(1)	Measured HPF and SEER from 12/82-9/83.
Cazenovia, New York	Double-Split Heat Pump with Oil Furnace		7.7	6.8	7.7	6.8	7.7	6.8	17.7 (95%)	NA(1)	Indoor temperature kept at 71F for cooling. Heat pump oversized for cooling, large cycling losses. Measured performance from 9/82 - 8/83.
Central Square, New York	Air-Water Hydronic System with Oil Boiler		7.0	NA(2)	7.1	NA(2)	NA(3)	NA(3)	13.3 (98%)	NA(1)	Heating - only system, no ARI ratings available. Measured performance from 9/81 - 4/82.
Dallas, Texas	Double-Split Heat Pump		4.8	6.1	4.9	6.1	6.3	5.7	1.9 (34%)	NA(1)	Heat pump oversized. Measured and normalized performance during heating lowered due to refrigerant leaks. Measured performance from 10/82 - 9/83.
Kingsport, Tennessee	Triple-Split Heat Pump with Domestic Water Heater		4.5 5.3	4.7 6.5	4.5 5.0	4.9 6.7	5.1 NA(3)	6.5 NA(3)	0.7 (8%) 0.2 (3%)	NA(1) NA(1)	38HQ127 heat pump with separate desuperheater from 10/81 - 9/82, 38QW027 heat pump with integral desuperheater from 10/82 - 9/83. Expected values do not include effects of desuperheater used to heat domestic water. Indoor temperature kept at 69F for cooling
Liverpool, New York	Double-Split Heat Pump with Gas Furnace		7.0 7.8	6.6 8.3	7.0 7.7	NA(4) 8.4	6.5 6.9	NA(4) 7.2	16.9 (97%) 12.8 (95%)	NA(1) NA(1)	38RQ034 heat pump from 9/81 - 8/82, 38QF036 heat pump from 9/82 - 8/83.
Long Island, New York	Air-Water Hydronic System with Oil Boiler	6.8		NA(2)	6.8	NA(2)	NA(3)	NA(3)	15.6 (99%)	NA(1)	
Minneapolis, Minnesota	Triple-Split Two-Compressor Heat Pump	4.9		5.4	5.2	5.5	5.4	6.8	1.9 (13%)	1.2 (6%)	Data from 9/81 - 8/82 for heating and 7/83 - 8/83 for cooling.

Note: (1) Not Applicable. These heat pump systems used one single-speed compressor.
(2) Not Applicable. Heat pump not used for cooling.
(3) Not Applicable. No ARI ratings available for this system.
(4) Not Applicable. Cooling load too small to determine cooling performance.

however. In all cases, the units were designed to stop the compressor when changing the speed or number of compressors in use. This caused a much larger number of on-off cycles than are typically observed in single capacity units, resulting in increased cycling losses and unknown effects on reliability. This phenomenon was less evident in the cooling mode at the Batchtown site because the majority of the cooling could be satisfied in the low capacity mode, and the thermostat was designed for a much lower cycling rate in air conditioning mode, compared to the heating mode.

The deliberately oversized unit at Dallas, Texas exhibited good efficiency during the first test, but would have performed even better if a smaller unit were utilized, with reduced cycling losses. In the second test, efficiency of the deliberately oversized unit was lower than observed earlier because a refrigerant leak developed that went undetected for a considerable length of time. The oversizing of the unit masked the performance degradation until efficiency calculations were performed. Once the leak was repaired the unit again functioned well.

A rigorous program of air-to-air heat pump laboratory tests followed by field tests of 9 identical heat pumps showed the importance of proper application and adjustment of controls [10]. This information was used to develop improved simulation models of heat pump performance. Thermostat mounting and adjustment was found to have a major impact on cycling rates and therefore on cycling losses. Heating seasonal Coefficients of Performance (COP) found during these tests ranged from 1.35 to 1.75. Thermostat mounting and adjustment, and usage patterns were found to be more important than climatic differences in determining energy efficiency.

Table 5 summarizes results of field tests on several room and package terminal heat pumps, which are single package units designed for thru-the-window or thru-the-wall mounting [9]. These units were equipped with passive defrost, and were designed to switch from heat pump to pure electric resistance heating below an ambient temperature of 32 F. The units performed well as heat pumps above 32 F, but due to their nature of being zonal heating devices, they saw much of their operation only when the outdoor ambient was below 32 F. This reduced their overall efficiency. The units were also somewhat noisy, resulting in limited use in bedroom applications, except for morning warm-up.

Water Heating Heat Pumps

Water heating can be accomplished either by utilizing the superheat in the refrigerant before condensation occurs (desuperheating), or by utilizing both the heat of desuperheating and of fully condensing the refrigerant. The former, known as desuperheater water heating, has the advantage that it can be acheived without special system design, and can therefore be retrofit to existing equipment. The latter, known as full

Table 5
Heat Pump Performance Summary - Room Units

SITE LOCATION	INSTALLATION	HEAT PUMP	MEASURED PERFORMANCE HPF	MEASURED PERFORMANCE EER	CAPACITY SUPPLIED BY HEAT PUMP (2) (% OF TOTAL)	ENERGY (KWH)	SAVINGS (3) (% OF TOTAL)	COMMENTS
Atlanta, Georgia	Motel Room	Carrier 52AQA312311AA Packaged Terminal Heat Pump	6.2	7.6	75%	1211.0	46%	Heat pump only HPF was 8.6. (1)
Baltimore, Maryland	Kitchen/Dining Room Window	Carrier 51QF315311 Room Heat Pump	4.2	7.4	36%	122.3	20%	Heat pump only HPF was 7.6. (1)
Belleville, Illinois	Office	Carrier 52AQA312311AA Packaged Terminal Heat Pump	4.2	7.2	41%	71.7	21%	Heat pump only HPF was 7.0. (1)
Boston, Massachusetts	Bedroom Window	Carrier 51QF211311 Room Heat Pump	--	8.7	--	--	--	Very little heating season operation.
Liverpool, New York	Living Room	Carrier 51QF315311 Room Heat Pump	4.7	5.6	50%	701.0	29%	Heating performance only for October 1982 through January 1983 due to instrumentation problem. Heat pump only HPF was 8.0. (1)

(1) "Heat pump only HPF" is the heating performance factor of the unit neglecting the resistance heat output capacity and input energy consumption.

(2) Percent of the total heating capacity supplied by the heat pump in the reverse cycle heating mode of the compressor.

(3) Calculated energy savings of the monitored heat pump as compared to an electric resistance unit.

condensing water heating, can only be achieved in systems that have been designed for such operation, but provides more water heating ability than desuperheaters.

Tests of desuperheater water heaters have shown that significant amounts of hot water can be delivered as a byproduct during summer air conditioning operation. However, since desuperheater water heating can only be provided when there is a need for space heating or cooling, limited amounts of hot water are available during spring and fall months when there is neither a heating nor a cooling load. Moreover, in winter, desuperheater water heating subtracts from heat available for space heating at ambient temperatures below the balance point.

Full condensing water heating is most frequently provided by dedicated heat pump water heaters. These units provide cooling and dehumidifying as a free byproduct. Another device which is expected to begin appearing in the U.S. market soon is the integrated space conditioning heat pump with full condensing water heating ability. This unit can provide much more hot water than desuperheater water heaters because it can operate in dedicated heat pump water heater mode during periods when there is no space heating or cooling demand.

Water heating from desuperheater water heaters or full condensing heat pump water heating operation can significantly reduce water heating energy consumption, with no adverse impact on either the equipment or the quantity and temperature of hot water delivered.

Tables 3 and 4 summarize the performance of two versions of one manufacturer's desuperheater water heater on an air-to-air heat pump. The units performed well, and saved considerable hot water energy consumption during the summer months, as shown in Figure 2. This savings is not reflected in the performance results shown in Tables 3 and 4. Little or no water heating energy savings were noted in the winter, spring, and fall months.

A survey of field tests [11] of first generation dedicated heat pump water heaters, conducted by electric utilities and others, showed general agreement that properly installed, those early units had seasonal COP's of around 2.0, regardless of climate. There was little climatic dependency because these units are not installed outside, and do not use outside air as the heat source. Improperly installed heat pump water heaters uniformly exhibited seasonal COP's of around 1.5. The installation error common to these units was failure to disconnect the lower electric heating element in the existing water heater. Another common problem was wiring errors during installation. Both the wiring errors and the failure to disconnect the lower heating element were symptoms of a more general problem. In almost every heat pump water heater field test, the installers (normally plumbers) initially failed to read and follow the installation instructions. Correct installations were obtained normally only after measures, such as installation inspections, were taken to ensure that the installer learned how to correctly install the units.

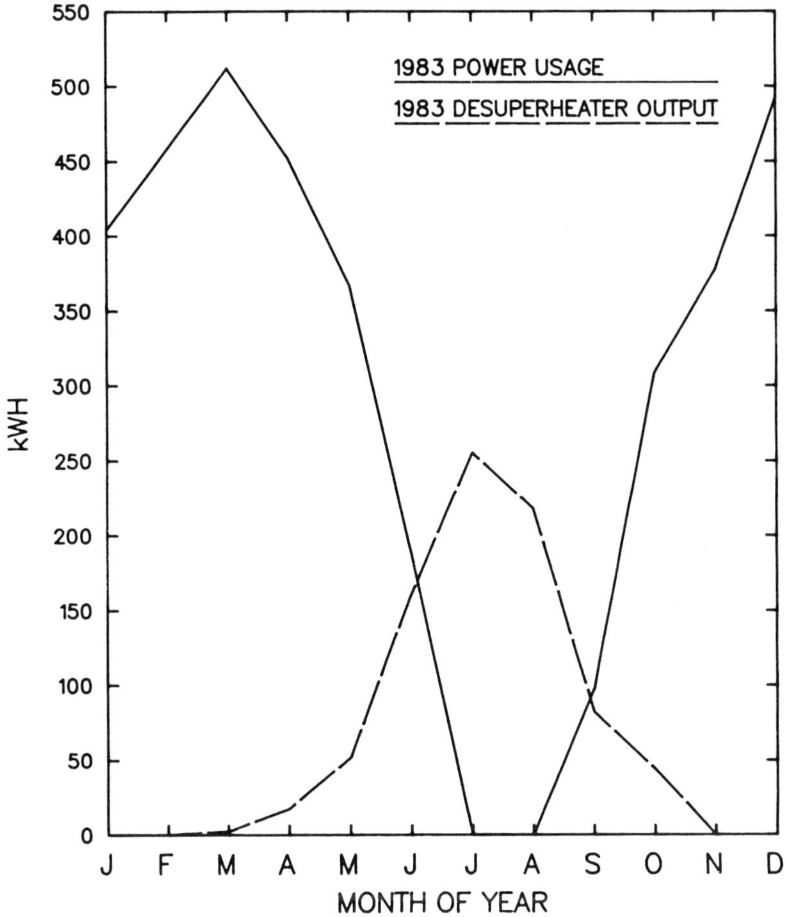

Figure 2
Domestic Water Heater Energy Usage
and Desuperheater Outputfor Kingsport, Tennessee

Dual-Fuel Heat Pumps

Heat pumps which utilize some form of fossil fuel as the supplemental heat source are referred to as dual-fuel units. The fossil fuel source is normally natural gas, propane or heating oil. Such units have the advantages of both high efficiency during heat pump operation, and high heating capacity at very cold ambient temperatures without high electrical energy demand.

Heat pumps can save significant amounts of seasonal energy compared to fossil fired heating systems, when used in conjunction with such systems as dual-fuel heating systems. Energy and operating cost savings will always be positive if an intelligent operating strategy is used to ensure that the mode (i.e. heat pump or fossil furnace) with the lowest operating cost at any given ambient temperature is utilized at that temperature. Peak heating electrical demand is entirely eliminated with such dual-fuel systems.

As seen in Tables 3 and 4, all of the dual-fuel heat pumps operated well. The hydronic units, which utilize baseboard hot water convectors to heat the structure, had somewhat lower seasonal efficiency than the units that were added on to forced-air furnaces because a higher condensing temperature was required to deliver the heat to the conditioned space through the water loop. Both types of units exhibited attractive seasonal efficiencies because the heat pumps were operated only at relatively warm outdoor ambient temperatures, with the fossil fired auxiliary heat utilized at lower temperatures. Large reductions in peak electrical demand, greater than 10kW, were observed with these units because there was no heat pump operation at all during the cold winter peak demand periods.

Ground Source Heat Pumps

Several approaches to utilizing the ground as the heat source/sink have been tried experimentally. These include ground water source heat pumps, closed loop ground coupled with heat exchangers in the ground, and crawl space earth coupled where the crawl space was utilized as a ground-to-air heat exchanger.

Ground water source heat pumps utilize wells, lakes, or other sources of water that is heated and cooled by the ground. Pumping power must be included in calculations of overall efficiency. Large amounts of water are required, necessitating some form of water disposal or direct use. Lakes, ponds, or reinjection wells are typically used for water disposal. First cost of such systems are currently high, unless an adequate well or other water source already exists or the well can be justified for other purposes.

Closed loop ground coupled heat pumps utilize either vertical or horizontal loops of piping inserted into wells or trenches dug specifically for that purpose. Again, first cost of such systems is currently high because of high costs of installing the ground heat exchanger. Vertical loop systems in general operate somewhat better than horizontal loop systems, especially in cooling mode. This is because heat transfer to/from the ground is highly dependent on the amount of soil moisture present, and horizontal loops drive soil moisture away from the piping more readily than do vertical loop systems during cooling operation [12].

Use of the crawl space beneath a house to preheat air before passing it through the outdoor unit of a conventional air-to-air heat pump allows a significant improvement in the efficiency of the heat pump, but much of the gain is lost because of increased heat loss through the floor of the house, even with high levels of insulation. One approach which utilized a closed-loop air flow circuit in the crawl space reduced peak electrical demand of the heat pump by as much as 30% compared to a conventional air-to-air unit, through reduction in number and duration of defrost cycles and reductions in backup electrical resistance heat use during short periods of very cold weather [13]. However, yearly energy savings were minimal or negative.

In general, ground water source and closed loop ground coupled heat pumps can operate at significantly greater seasonal efficiency and with significantly reduced peak electrical demand, compared to conventional air-to-air heat pumps.

Tests on ground water source heat pumps [14] yielded seasonal heating Coefficients of Performance of 1.8 to 3.0, even though these units were retrofitted into homes which had smaller than recommended ducting and air flow rates because they were initially equipped with fossil furnaces. The units exhibited sensitivity to variations of ground water source temperature which sometimes caused nuisance trips on low evaporator pressure protection devices. This problem was caused to some extent by heat exchanger fouling due to microorganisms from the well and other contaminants. Chemical treatments of the well helped control these problems.

Tests on vertical closed-loop ground coupled heat pumps [15] showed that the yearly heating and cooling electrical energy consumption of the ground coupled system was around 30% less than that for an air-to-air system. Similarly, peak electrical power demand for the ground coupled system was around 70% less than for the air-to-air heat pump. This large reduction of electrical demand resulted from complete elimination of electric resistance backup heat use, saving approximately 3.5 to 4.0 kW of demand. Proper sizing and installation of the ground coil was confirmed experimentally to be essential to acceptable operation of closed-loop ground coupled heat pumps.

Storage and Solar Coupled Heat Pumps

Storage coupled heat pumps are primarily associated with heat reclaim units for domestic water heating, and sometimes are used for heat storage as part of large commercial cool storage installations. Storage works well in such applications. Use of storage has also been investigated for residential applications. There, however, performance benefits have generally not justified the added costs and complexity [16].]

Solar collectors have been investigated for boosting the source temperature of heat pumps, particularly those equipped with some type of energy storage. Again, performance benefits have generally not justified the added costs and complexity.

LOOKING TO THE FUTURE

U.S. heat pump experiences to date have identified the need for improvements in the following areas; field test and data analysis techniques, training of maintenance personnel, and selected hardware to continue improvements in efficiency, electrical demand reduction, and reliability.

Standardized normalization techniques should be developed that allow comparison of results from different field tests. These techniques should include adjustment to average weather conditions, and to predetermined usage profiles. For example, standardized sets of house sizes, indoor temperature settings, set back strategies, hot water usage and temperature settings, and other factors should be developed to represent "average", "low" and "high" energy users. The attractiveness of different heating and cooling options can then be meaningfully compared for different climates and lifestyles.

Newer heat pumps have more features than ever before, and are becoming progressively less similar to conventional air conditioning units. Therefore, more emphasis must be placed on maintaining a supply of properly trained service personnel through training and certification programs. Development of better diagnostic devices and easier to service units is also warranted.

Continued efforts to develop higher efficiency heat pumps with greater low temperature heating capacity are also justified. Development of units with improved features should also continue. Development of such improved units should be accompanied by applications research activities aimed at expanding their use in new and better ways. More attention needs to be devoted to the development of reliable, foolproof controls that avoid excessive cycling of units.

Tests are currently underway on next generation heat pumps, including units with capacity control using continuously variable compressor and fan speed modulation, full condensing water heating capability, and zonal heating capability. Preliminary results indicate that the full condensing water heating feature saves considerable amounts of water heating energy in all seasons compared to both conventional water heating and desuperheating water heating. Hot water recovery rates with units that have integrated space conditioning and full condensing water heating capability are in some cases greater than with fossil fired water heaters.

Continuously variable speed modulation of compressors and fans promises to reduce cycling losses, allow use of larger compressors for increased low temperature capacity, and improve efficiency through more efficient heat exchange at low load conditions. These same features also allow the ability to selectively heat or cool only portions of a structure, thereby allowing reductions in peak electrical demand and further reductions in energy consumption without significantly impacting user comfort.

SUMMARY

Experiences with heat pumps in the U.S. have been generally favorable. They perform efficiently and reliably in most climates. Problems experienced with early vintage units have for the most part been eliminated. In order for this high level of satisfaction to persist, however, efforts to improve product quality and repair proficiency must continue.

REFERENCES

1. "Heat Pump Market Analysis: Heat Pump Owner Study", Consumers Power Company, Market Research and Pricing Department Study, Michigan, March 1984.

2. Electric Power Research Institute funded survey of heat pump owners in northern Illinois, 1985, formal report not yet published on these findings.

3. "Heat Pump Life and Compressor Survival In a Northern Climate", Electric Power Research Institute, Report No. EM-4660, July 1986.

4. "A Survey of Heat Pump Service Life", Electric Power Research Institute, Report No. EM-4163, July 1985.

5. "An Analysis of Heat Pump Compressor Life", Electric Power Research Institute, Report No., EM-4659, July 1986.

6. N. C. Lovvorn and C. C. Hiller, "Heat Pump Compressor Life in Alabama", paper presented at the January 1987 ASHRAE Winter Annual Meeting, New York City.

7. "Bibliography of the DOE Building Equipment Research Program", DOE/CE-0037/1, September 1985.

8. "Monitoring of Improved Air-Source Heat Pumps", Electric Power Research Institute, Report No. EM-3978, June 1986.

9. "Monitoring of Central and Room Heat Pumps" Electric Power Research Institute, Report No. EM-4674, July 1986.

10. "Performance of Air-Source Heat Pumps", Electric Power Research Institute, Report No. EM-4226, November 1985.

11. "Heat Pump Water Heaters", Electric Power Research Institute, Report No. EM-3582, May 1984.

12. "Tech House I Horizontal Coil Ground Coupled Heat Pump 1983-84 Annual Performance", ORNL/Sub/81-7685/3& 92, June 1985.

13. "Heat Pump Experiments Using Crawl Spaces As Ground-to-Air Heat Exchangers", Electric Power Research Institute, Report No. EM-4908, December 1986.

14. "Monitoring of Residential Ground Water-Source Heat Pumps In The Northeast", Electric Power Research Institute, Report No. EM-4372, January 1986.

15. "Performance Comparison of Air- and Ground-Coupled Heat Pump Systems", Electric Power Research Institute, Report No. EM-3408, January 1984.

16. "ACES Tests at the TECH Site: 1981", ORNL/CON-96, August 1983.

Chapter 3

Operating Experience with Heat-Pump-Type Room Air Conditioners for Cold Districts

Tsunehiko Minagawa and Zenkichi Yamaguchi

ABSTRACT: Air conditioning in Japanese houses in general is mainly made by independent and restricted systems by reason of the living custom, building construction, etc. and central air-conditioning systems are not in common use.

Conventional heat pump type room air conditioners using air heat source have been problematic in insufficient capacity due to lowering of the outside air temperature, defrosting operation, running costs and the like. Jointly with Hitachi, Mitsubishi, Matsushita and Toshiba, Tohoku Electric Power Company has developed, to solve such problems, heaterless inverter air conditioners which can be practicably used even in districts where the yearly average of daily minimum temperature is -7°C, and made them available to the market.

This paper describes the field test results such as heating capacity, rise time, defrosting time, energy consumption efficiency, running costs, etc. and summary of the development of subject technology.

INTRODUCTION

It is said that about 52 per cent of the national land of Japan is heavy snowfall areas. Especially, the meteorological condition in Hokkaido and Tohoku is cold and snowy, and some areas of the districts where the minimum temperature in winter may become under -20°C. Accordingly, air conditioning is indispensable to all buildings such as houses in general, offices, factories, educational institutions.

By this reason, large quantities of energy is consumed for

HEAT PUMPS: Prospects in Heat Pump Technology and Marketing (Proceedings of the 1987 IEA Heat Pump Conference), Edited by Kay H. Zimmerman Printed in U.S.A.

heating. The ratio of energy for heating to the total energy consumption in an average household is, as shown in Table 1, about 66 per cent for Hokkaido and about 50 per cent for Tohoku compared with about 30 per cent for all districts of Japan [1]. Seeing from the quantity of heat consumption, Hokkaido is about 3.7 times as much as the national average and Tohoku about 2.2 times.

Table 1. Composition of energy consumption by household purpose in 1984.

	Heating	Cooling	Water heating	Lighting	Total	MJ/year
Hokkaido	65.6	-	19.0	15.4	100%	62,753
Tohoku	49.9	0.4	26.5	23.2	100%	49,358
Kanto	29.2	1.2	38.5	31.1	100%	38,641
Hokuriku	43.9	1.3	29.0	25.8	100%	45,596
Tokai	25.8	2.0	40.1	32.1	100%	35,255
Kinki	23.5	3.1	39.8	33.6	100%	34,795
Chugoku	28.9	2.3	34.9	33.9	100%	33,547
Shikoku	22.2	2.1	31.7	39.0	100%	30,253
Kyushu	25.2	2.0	34.4	38.3	100%	30,043
Japan	29.7	1.5	37.8	31.0	100%	38,001

About 90 per cent of energy for heating is, as shown in Table 2, obtained from kerosine [1]. This Table also suggests that the spread of electoric heating including the heat pump system is narrow.

Table 2. Composition of energy consumption for house heating in 1984.

	Electricity	Kerosine	Coal	Gas & other	Total
Hokkaido	1.7	86.9	11.3	0.1	100%
Tohoku	3.6	89.6	0.1	6.7	100%
Japan	6.8	72.3	1.9	18.0	100%

Such heavy burden of energy for heating in the family budget in cold and snowy districts magnifies economic and psychological pressure. Since combustion energy obtained from pertroleum, coal, etc. involves many problems in environmental pollution, convenience, economy, etc., many people desire, if economically advantageous heat source is available, that the former will be converted into heat pump type room air conditioners.

Heat pump type room air conditioners have many merits in energy efficiency, safety, cleanness and convenience which other heating machines cannot provide. Through further enhancement of heating capacity and improvement of efficiency, each electric manufacturer has promoted the rapid spread of heat pump type room air conditioners as the main item of heating sppliances.

As shown in Figure 1, national demand for air conditioners in 1985 marks the maximum record of 3.35 million sets (118% of the previous year), of which demand for heat pump type machines that had won high appreciation as heating appliance shows high growth to

3.35 million sets (119% of the previous year, 64% of the total air conditioners) [2].

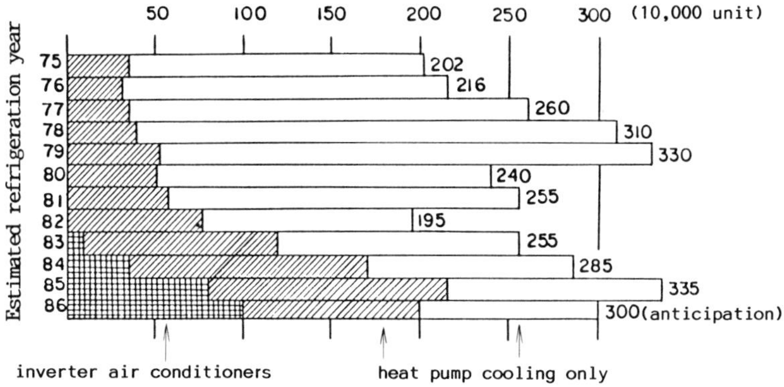

Fig.1. Transition in demand for room air conditioners.

Since, however, heat pump type room air conditioners use the open air as the heat source, they have shortcomings such as insufficient heating capacity, slow rise of room temperature and frequent defrosting operations, when the open air temperature falls and much quantity of heat is required in cold districts. Such shortcomings have hindered the full-scale spread of heat pump type machines.

In consideration of such circumstances, this Company has, since 1982, taken part in development of heat pump type room air conditioners well usable even in cold districts jointly with three electric manufacturers. Based on the results of field tests of the trial products in seven Prefectures of Tohoku District, we have put to practical use heat pump type air conditioners, although with auxiliary heaters, which can be used in all areas of Tohoku District. Such machines have been marketed since the autumn of 1983 by the manufacturers participating in the joint development project.

In order to make the developed machines more efficient, this Company continually developed the element technology from 1983 jointly with four manufactures and proved performance, efficiency and the like of the trial products incorporating such element technology. Based on such proof, the manufacturers participating in the joint development project put the new model to practical use and has marketed it with good reputation since April 1985.

Here will be described the results of field tests in the northern Tohoku District of the newly developed heat pump type room air conditioners for cold districts and summary of the developed element thechnology.

RESULTS OF FIELD TESTS

Appreciation of performance of heat pump type room air conditioners in cold districts is centered on feeling of heating, rise time, defrosting operation, periodic energy consumption efficiency and running costs. For the purpose of such appreciation, field tests of the trial products incorporating the element technology developed during this project were performed in 12 major cities of three prefectures of the northern District (Aomori, Iwate and Akita) from December 1984 through March 1985.

Heating Capacity at Low Open Air Temperature

While output of the compressors used in these tests was 0.7~0.9 kw, heating capacity at the outside air temperature of -7°C was 12.6~15.9 MJ/h (3,000~3,800 kcal/h) without auxiliary heaters with slight difference by types of developed products. This is nearly equal to 12.6~16.3 MJ/h (3,080~3,900 kcal/h) of the capacity of the machines with auxiliary heaters (about 1 kW) developed last time.

As shown in Figure 2, the ratio of beating capacity at -7°C of the outside air temperature compared to 7°C of the standard temperature was 64% for the machines developed last time and became 76~88% by the development of the element technology at this time, achieving remarkable improvement of drop of heating capacity in low temperature areas.

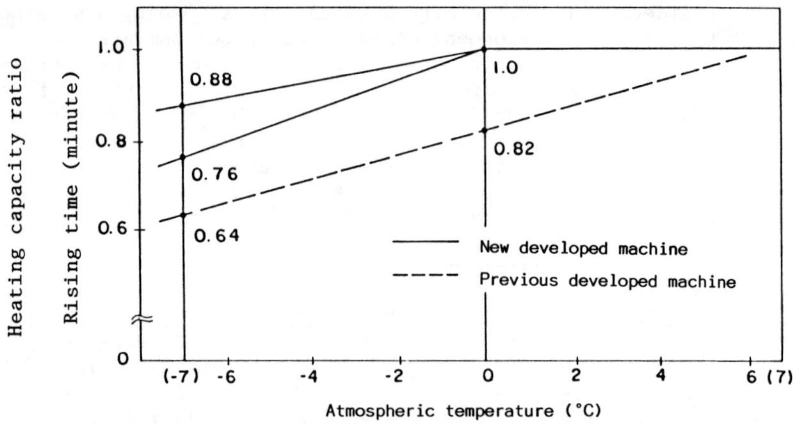

Fig.2. Atmospheric temperature and heating capacity ratio.

Rise Time of Room Temperature

As shown in Figure 3, time to raise room temperature from 0°C to 18°C at the outside air temperature of 0°C was about 15 minutes, that is, one third of 40~45 minutes in case of the machines developed last time.

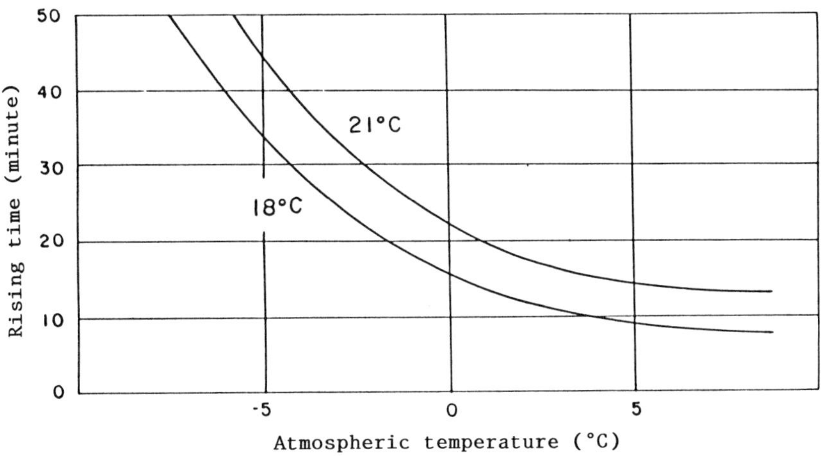

Fig.3. Example of rising characteristics.

Defrosting Time and Quantity of Frost

As shown in Figure 4, time of defrosting from heat exchangers became about 2~4 minutes, that is, nearly half of about 7 minutes in case of the machines developed last time. +4~-4°C of maximum frosting temperature zone of the machines developed last time could be reduced to +2~-2°C and prevention of ineffective defrosting operation and remarkable decrease in quanty of frost were intended.

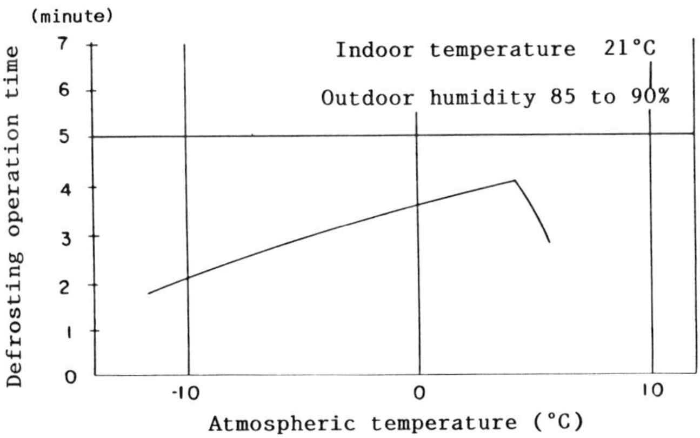

Fig.4. Example of defrosting operation characteristics.

Energy Consumption Efficiency During Heating

Energy consumption efficiency during heating periods was 10.34~11.55 KJ/h·W (2.47~2.76 kcl/h·W) in the average for the north Tohoku three prefectures, achieving improvement by 30~37% against 7.79~8.87 KJ/h·W (1.86~2.12 kcal/h·W) in case of the machines developed last time.

Appreciation by Field Test Monitors

On feeling of heating

Monitors using the machines in living rooms nearly between 13.2 m^2 and 16.5 m^2 were satisfied with softer feeling of heating than conventional warm air circulators.

On rise

On rise in the morning, most of the monitors appreciated that rooms became enough warm at the hour of rising. Such appreciation might be caused by the fact that rise time of the newly developed machines was reduced and preliminary heating operations were made for half or one hour before the hour of rising.

On defrosting operation

While defrosting operation is influenced by meteorological conditions such as outside air temperature, humidity and snowfall, it may happen once per 50~90 minutes and last only for 2~3 minutes even at the open air temperature of +2~-2°C which is the zone of maximum frosting temperature. Therefore, most of the monitors appreciated defrosting operation did not get on their nerves.

On running costs

Since the living custom and operation hour of each monitor differed and a part of test patterns was specified, consumed electric energy was 400~1,000 kWh/month, showing considerable variance between monitors. However, monitors estimate that, if the mode of operation suitable to the normal living condition would be adopted, 300~500 kWh/month will do.

The latest field tests were performed in 9.9~19.8 m^2 big rooms where the trial products were installed, while the product is designed to be used in an about 13.2 m^2 big room. Simulation based on the tests in case of using in a 13.2 m^2 big room gives the running casts of 9.400 yen/month ($59/month, one US dollor is equivalent to 160 Japanese yen) in the average value for the northern Tohoku three prefectures.

The simulation also gives the energy consumption efficiency of 11.76 KJ/h·W (2.81 kcal/h·W). The energy unit price obtained from this value is 50.57~56.85 yen/MJ (0.32~0.36 $/MJ, 12.08~13.58 yen/1,000 kcal), which is nearly equal to the estimated value of 57.52 yen/MJ (0.36 $/MJ, 13.74 yen/1,000 kcal) for the Forced Flue type kerosine stove.

MAJOR ELEMENT OF TECHNOLOGY DEVELOPED

Development of High Efficiency Compressors for Inverters Used in Cold Districts

Mainly as a result of development and application of a high efficiency compressor in which pressure loss in the suction process and exhaust process at the refrigeration cycle during high speed and high power operation of the compressor is decreased and the mechanism of separation of refrigerator oil from refrigerant is improved, energy consumption efficiency has been elevated as shown in Figure 5.

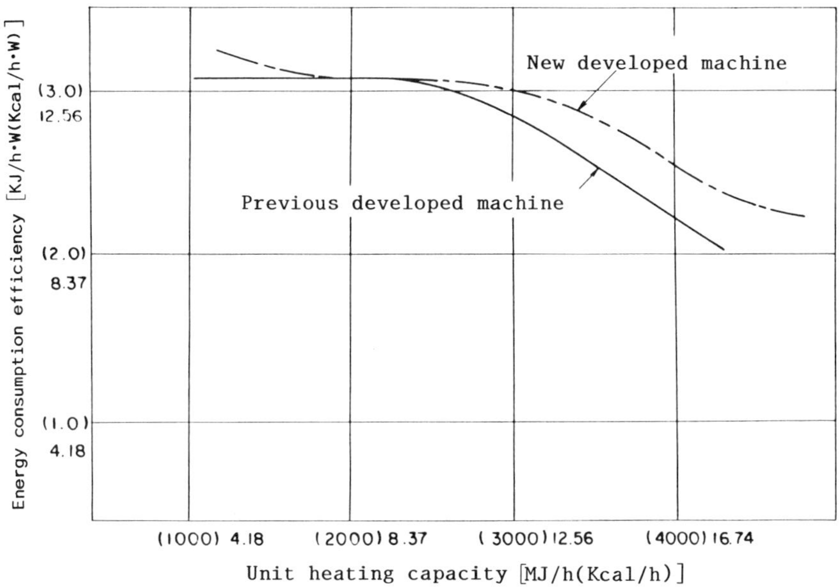

Fig.5. Example of unit performance comparison (compressor).

Adoption of Refrigerator Oil Adapting to Low Outside Air Temperature

Refrigerator oil with low pour point and floc point and excellent cold-proof characteristic has been selected, and operation reliability of the compressor in case of low outside air temperature secured.

Development of Compressor Preheating System

Since about 14.5% of the total quantity of heat is required to heat the compressor during rise for heating, the compressor preheating system has been developed. This system has made the rise time shorter by lessening quantity of heat necessary for heating the compressor during rise and increasing heating capacity.

Adoption of New Mixed Working Fruids and Development of Optium Control System for Refrigeration Cycles

By adopting new mixed working fluids which operates most effectively in the entire outside air temperature zone and developing the optimum control system for refrigeration cycles, enhancement of heating capacity and improvement of the starting characteristic especially at the time of low outside air temperature have been realized.

Development of New Frost Detecting System

By developing the detecting argorithm of air temperature and heat exchanger temperature and of air temperature and humidity and heat exchanger temperature as shown in the example of Figure 6, prevention of ineffective defrosting operation and reduction of defrosting time have been realized.

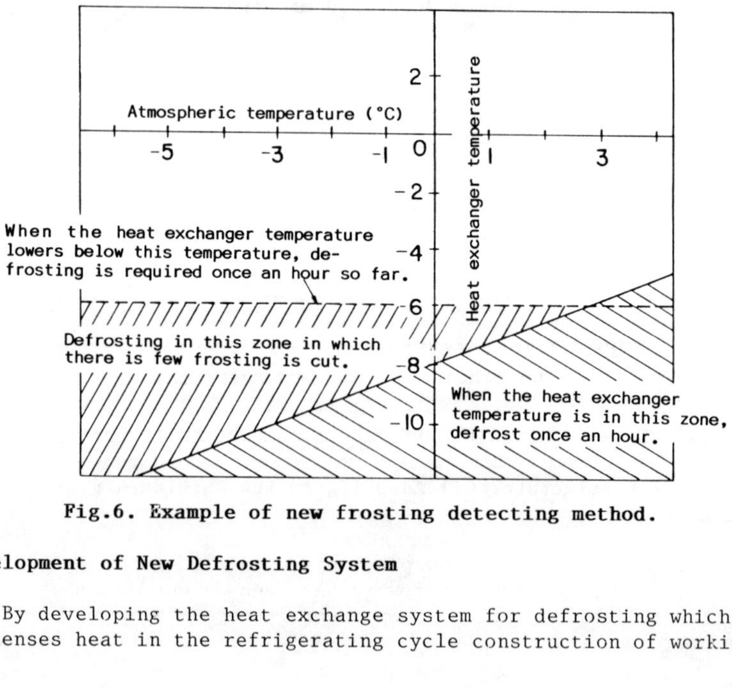

Fig.6. Example of new frosting detecting method.

Development of New Defrosting System

By developing the heat exchange system for defrosting which condenses heat in the refrigerating cycle construction of working fluid unified with the compressor, reduction of defrosting time has been realized.

Development of Frost Decreasing System

By developing the frost decreasing system involving improvement of shape of heat exchangers and surface treatment of heat exchangers as shown in the example of Figure 7, elevation of efficiency in defrosting operation has been realized.

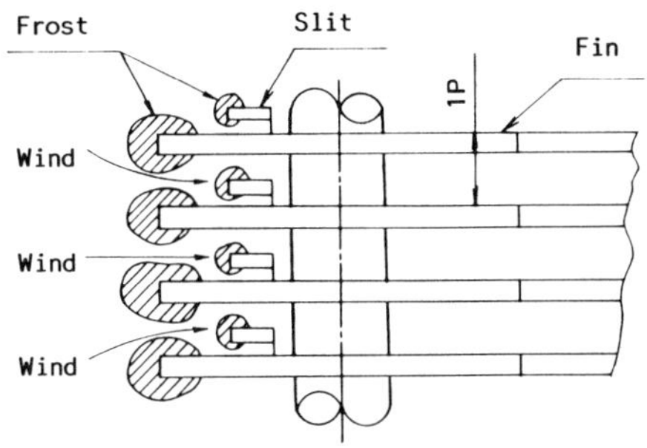

Conventional fins

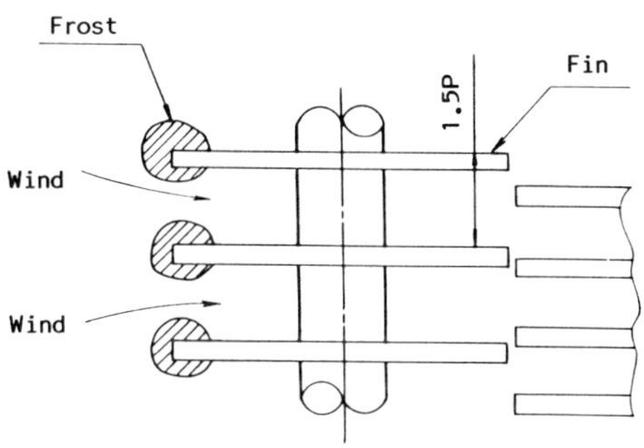

Rough and fine fin pitch

Fig.7. Example of improved fin shape.

Improvement of Cooling Characteristic

By adoption of new type propeller fans and improvement of shape of suction and supply openings of heat exchangers, increase of draft air and lessening of noise of the unit has been realized.

APPLICABILITY

As shown in the example of occurence of specific outside air temperature in cold districts during the heating period illustrated in Figure 8, hours when outside air temperature is below -7°C is only a few, i.e., 1.9% in Morioka and 0.1% in Akita.

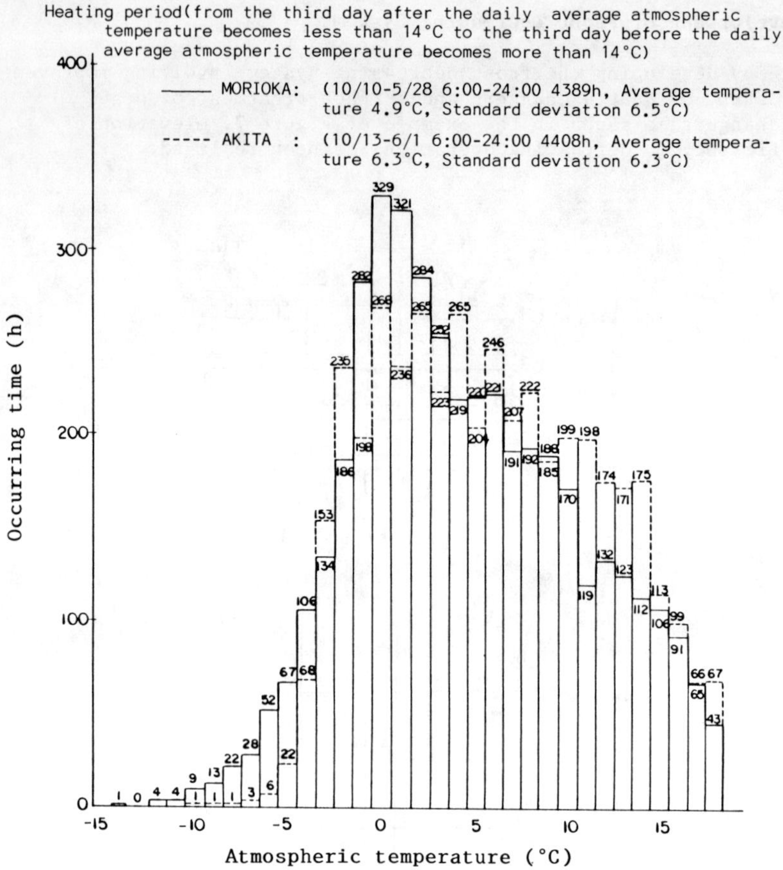

Fig.8. Occurring time of atmospheric temperature (HASP data).

The standard of heating for heat pump type room air conditioners is, as shown in Figure 9, 11.28 MJ/h (2,695 kcal/h) for a Japanese-style room of a wooden house of 13.2 m^2 and 9.40 MJ/h (2,246 kcal/h) for a Western-style room of ferroconcrete apartment of 13.2 m^2 in the example of Morioka where monthly average of daily minimum temperature is lowest in Tohoku District.

Therefore, we have reached a prospect that, if the heating capacity of heat pump type room air conditioner is ensured to be more than 3000 kcal/h at -7°C of outside air temperature, it can be used well even in cold areas such as Tohoku District.

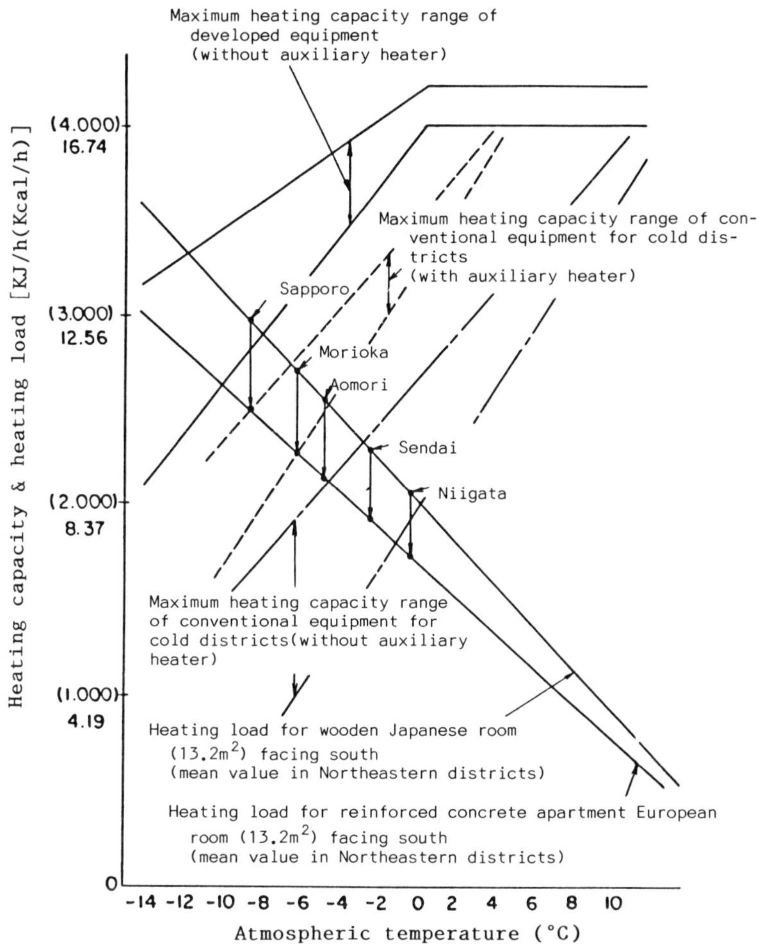

Fig.9. Atomospheric temperature and heating capacity heating load characteristics.

CONCLUSION

Thanks to sales of the new type products incorporating the element technology, the number of units held and used in the seven Tohoku prefectures (including Niigata) for 1985 has reached 104 thousand units (130.4% of the previous year). These products are spread not only in Tohoku District but also in cold and snowy districts throughout Japan and their characteristics are highly appreciated.

We believe that the new type of products developed this time are nearly satisfactory in both performance and energy consumption efficiency and we are confident that they will contribute much more to comfortable living in cold districts.

REFERENCES

1. Living Environment Planning Institute: *Statistical Year Book on Energy for Household Use,* 1984.

2. *Dempa Shimbun* (The Electric and Electronic News), Oct. 29, 1985.

Chapter 4

Operating Experience with Multizone Air-Source Heat Pumps in Residential and Commercial Applications

Katsushige Kawahara and Kazuo Tanaka

ABSTRACT

Heat pumps first began appearing in North America, primarily the United States, in significant numbers in the 1950's. Reliability problems with these early units lead to system improvements and a reemergence beginning in the mid-1960's. Since that time many surveys and tests have been performed to determine efficiency and reliability of the ever improving new units. Some poorly performing heat pump brands persisted through the mid-1970's, but most units installed since that time have performed well. The current paper presents general conclusions based on the many surveys and field tests that have been performed, and includes example results from selected research efforts.

INTRODUCTION

Heat pump use in the United States has increased steadily since the mid 1960's. More than one million new units of all types were installed in 1985. Air-to-air heat pumps dominate the U.S. market, but significant numbers of water source units are also sold, primarily for commercial applications in buildings with central circulating water piping systems. Variants of air and water source heat pumps have been utilized in a small, but increasing number of applications. Examples include ground water and closed loop ground coupled heat pumps, hydronic heat pumps, heat pump water heaters, dual-fuel heat pumps, and solar assisted heat pumps. Heat pump use in the U.S. is largely limited to the residential and small commercial sizes having heating outputs

HEAT PUMPS: Prospects in Heat Pump Technology and Marketing (Proceedings of the 1987 IEA Heat Pump Conference), Edited by Kay H. Zimmerman Chelsea, Michigan 48118. Printed in U.S.A.

Central air conditioning system circulates chilled and hot water with pump or cool and warm air with fan to convey heat between air conditioning zone and heat source unit. Power to convey this heat is not generally altered even under light load operation, unless VWV or VAV system is employed. Power required for heat conveying in central air conditioning system may sometimes exceed 50% of the annual total power required for air conditioning system.

MZHP system, on the other hand, utilizes the latent heat of refrigerant to convey the required amount of heat when necessary. Compared with central air conditioning system conveying heat with chilled and hot water or cool and warm air, therefore, MZHP system can remarkably reduce heat conveying power.

In addition, when air conditioning is not required, its operation can be suspended for each indoor unit thus resulting in energy saving when partially loaded.

We will choose duct type air conditioner as an example of central air conditioning system for residence and explain the simulated example of power consumption to compare with MZHP system for residence.

As to those for commercial use, we will compare the actually measured values of power consumption during the summer season for the building and evaluate these systems. This building was air conditioned by combination of centrifugal water chiller, hot water boiler, induction units and air handling units and then replaced by MZHP system.

MZHP SYSTEM FOR RESIDENTIAL USE

Outline of MZHP System

The basic concept of the MZHP system for residential use is to get cooling or heating only when and where it is needed. The system consists of an outdoor unit (compressor and condenser section), connected by refrigerant piping, and power and control lines to multiple indoor fan coil units, which can be operated independently and individually.

The MZHP system is not limited to cooling in summer, but has an additional function of heating by adopting a heat pump system. The heat pump is well accepted as an efficient way of cooling/heating using a reverse cycle refrigeration.

Capacities of the outdoor units range from about 4 to 6.5 kW/h in both cooling and heating modes; and those of the indoor fan coil units are from about 2 to 4 kW/h each. The indoor fan coil units are two types; wall and floor mounted, and can be fit with supplementary electric heaters, if necessary.

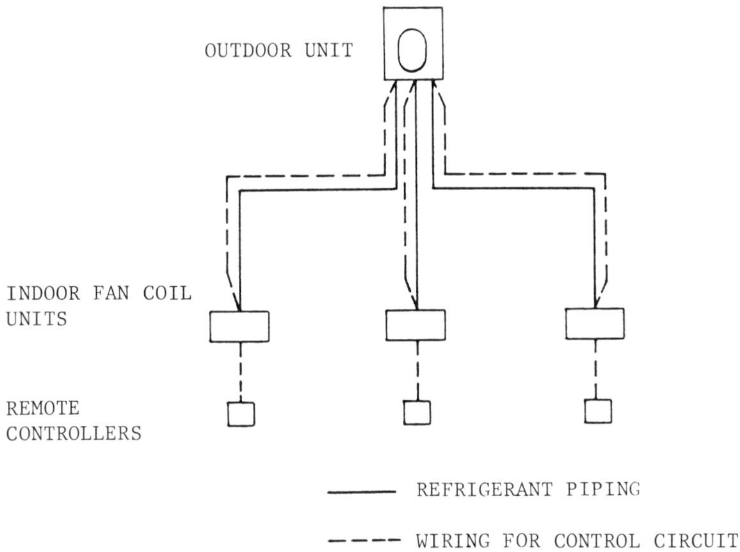

Fig. 1. An example of the system.

Comparison of MZHP System with Ducted Air Conditioning System

There are various air conditioning systems for residence. Here, air conditioning system by duct type heat pump air conditioner was chosen and its power consumption was compared with same by multi-zone air conditioning system for evaluation of energy saving characteristic. Assuming the air conditioning hours and its operating ratio for each room when both systems are adopted for the same residence, we calculated the power consumption for 3 months during the summer by simulation.

Building and Air Conditioning System

The residence chosen for this evaluation has 5 rooms and approx. 140 m^2 of floor space.

Figure 2. shows installation example of multi-zone air conditioner. 2 sets of multi-zone air conditioners each for 2 or 3 rooms are installed and their operation is switched over depending on application hours. No room is continuously occupied by day and night. On the other hand, duct type heat pump air conditioning system is assumed to be operated simultaneously for each room.

Table 1 shows the specifications and performance of each air conditioning equipment compared here. Because of considerable difference in COP of duct type heat pump air conditioners depending each model, 3 levels of COP were established. COP in this table shows the value including conveying power of fan.

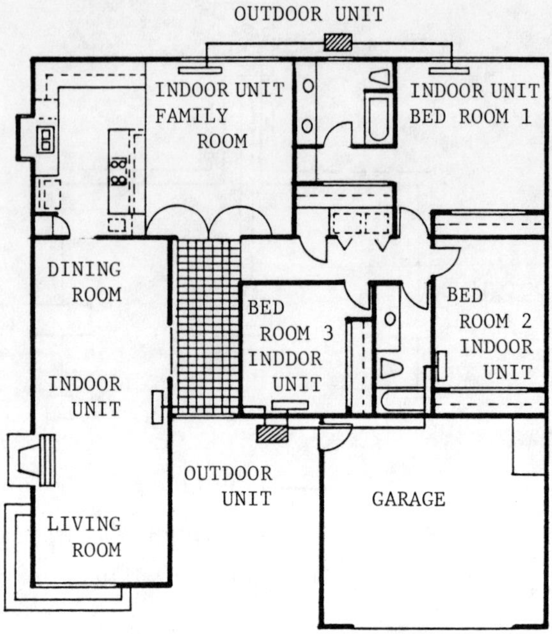

Fig. 2. Installation example.

Table 1. Performance of air conditioning equipment

		Room used	Air conditioning capacity (kW/h)	Power consumption (kW)	COP
Multi-zone heat pumps	A	Dining room	4.80	2.06	2.33
		Bed room 2 + 3	4.94	2.15	2.30
		Dining room + bed room 2 + 3	6.34	2.33	2.72
	B	Family room or bed room 1	3.14	1.45	2.17
		Family room + bed room 1	4.41	1.65	2.67
Ducted heat pumps	C	All rooms	11.6	3.64	3.19
	D	All rooms	11.6	4.55	2.55
	E	All rooms	11.6	5.46	2.12

Note: Outdoor temp. 35°CDB Indoor temp. 27°CDB, 19.5°CWB

Establishment of conditions for simulation

The most important condition to be established for this comparative simulation is the occupied hours for each room or operating hours of air conditioners.

As shown in Figure 3. bed rooms are set to be occupied from 18 to 9 and dining room and family room from 6 to 21. Since no family member are usually at home between 9 and 18, the occupied ratio during this time zone was set at 0.7 for more realistic condition. This value cannot be uniformly set, but it is considered not seriously affect the results.

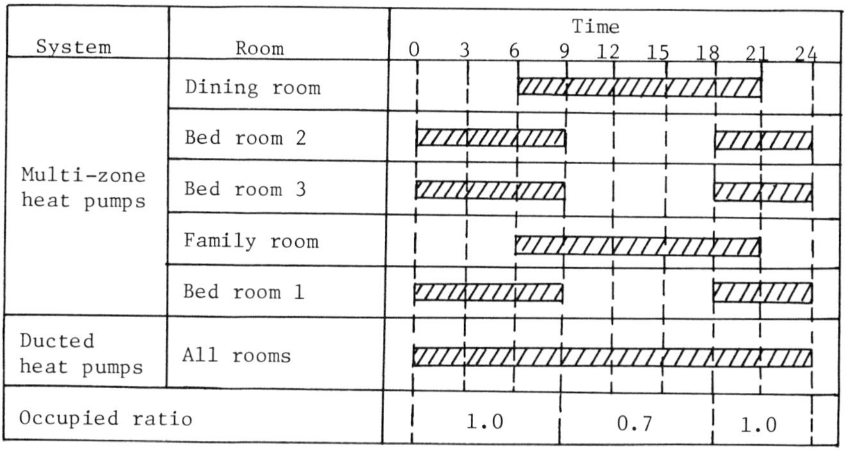

Fig. 3. Establishment of occupied hours.

Outdoor temperature in Osaka during 3 months of July - Sept. was set for our purpose. Appearance time of outside temperature range for each 1°C over 4 time zones (6-9, 9-18, 18-21 and 21-6 o'clock) was calculated.

Assuming the variation in cooling load in room and cooling capacity of air conditioner depending on outside temperature as its parameter, we made out cooling capacity diagram and load diagram for each room.

Results of simulation and evaluation

Based on these set conditions, we obtained the load factor for each outside temperature and power consumption by air conditioner and calculated the power consumption by air conditioner by integration with the appearance time.

Figure 4. and Table 2 show the total of power consumption during 3 months of July - Sept. separately calculated for indoor unit (fan) and outdoor unit.

In comparison of duct type air conditioner, D-unit, with multizone air conditioner, power consumption resulted in reduction by approx. 60% for indoor unit (fan) and approx. 33% for outdoor unit, thus totalling to approx. 36%.

Because of low duct static pressure set for duct type air conditioner, power consumption by indoor fan remained at approx. 13%.

It can be concluded from the results of this simulation that superiority of multizone air conditioner due to its zone control is evident.

Insufficient capacity of multizone air conditioner in case of its full simultaneous operation was a matter of concern. Because of low outisde temperature during the set time zone (6-9, 18-21), however, no problem was found in the course of our calculation.

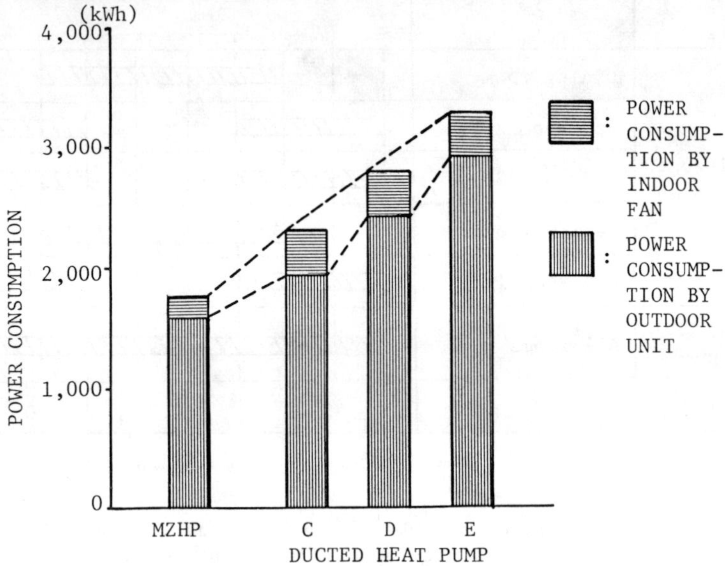

Fig. 4. Comparison of power consumption.

Table 2. Power consumption

		Power consumption (kWh)		
		Indoor fan	Outdoor unit	Total
Multi-zone heat pumps		148	1,627	1,885
Ducted heat pumps	C	364	1,932	2,296
	D	364	2,413	2,778
	E	364	2,896	3,260

Now that we have finished explaining the most important merit, energy saving of the MZHP system.

MZHP SYSTEM FOR COMMERCIAL USE

Outline of MZHP System

This system is mainly intended to air conditioned office buildings with floor space of 2,000 m^2 and less.

It consists of outdoor units equipped with plural of capacity control compressors of 5 HP, plural of indoor units and refrigerant piping connecting these units. Each indoor unit can set indoor temperature and control its operation with remote controllers separately operated. Its operation can be also remotely controlled and supervised with central control board.

Figure 5. shows an example of this system.

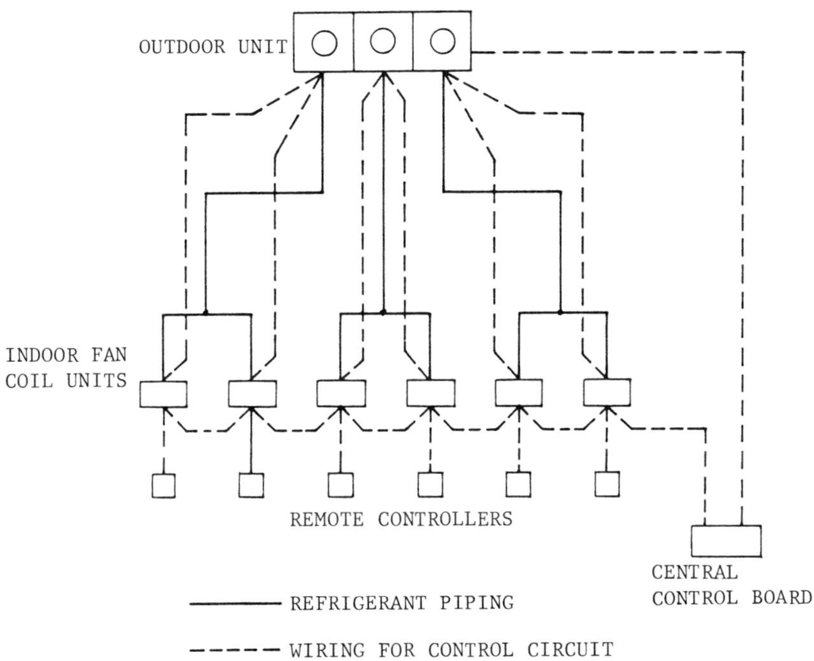

Fig. 5. An example of the system.

Figure 6. shows refrigerant piping diagram for a part of this system. Outdoor unit contains oil separator to control amount of refrigeration oil in compressor during light load operation and refrigerant regulator to regulate amount of refrigerant, etc., as well as compressor and cross fin coil. Indoor units contain electronic control valve to control the flow of refrigerant depending on operation mode as well as cross fin coil and fan.

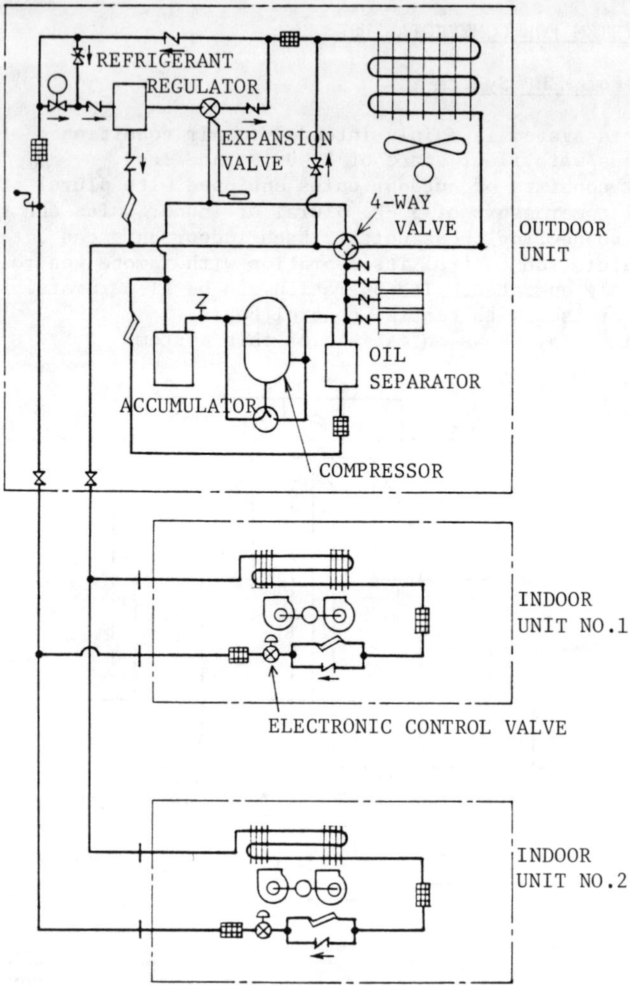

Fig. 6. Refrigerant piping diagram.

Building and Air Conditioning System

This building was used as a general office building for about 20 years. Because of increasing cooling load and deterioration of its air conditioning equipment, however, central air conditioning system consisting of a centrifugal water chiller and a hot water boiler so far employed was replaced by MZHP system. This building has 5 stories and total floor space of about 6,700 m^2.

Old air conditioning system

For air conditioning purpose, chilled water from a centrifugal water chiller during summer and hot water from hot water boiler during winter were sent to air handling units and induction units. Induction units were mainly set in perimeter zone and interior zone was air conditioned by air handling units set on the roof. Figure 7. shows the layout of old air conditioning system and Table 3 shows its specifications.

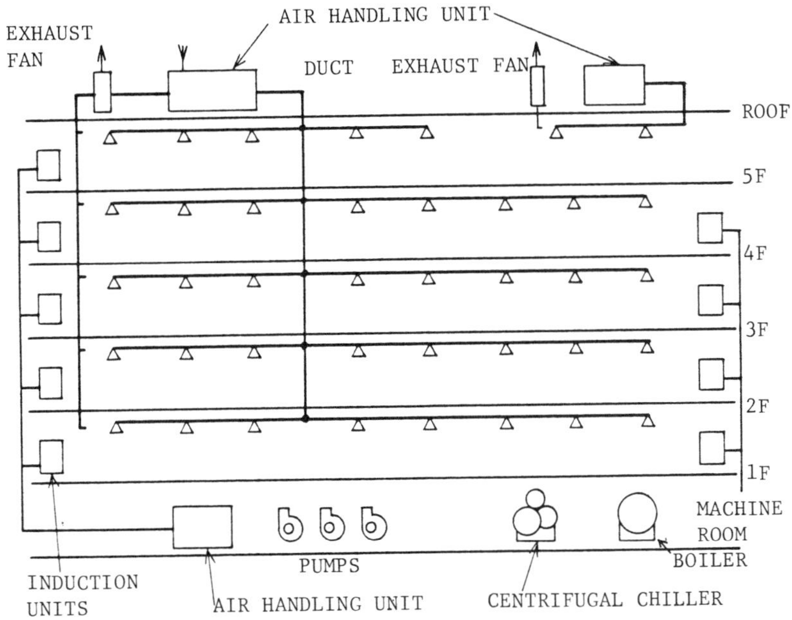

Fig. 7. Old air conditioning system.

Table 3. Specifications of old system

Units		Capacity and number of units
Heat source units	Centrifugal chiller	170RT x 1
	Cooling tower	170RT x 1
	Hot water boiler	942 kW x 1
Air conditioners	Air handling units	66,000 M^3/h (3 units in total)
	Induction units	123 units
Auxiliary units	Pumps	3.7 kW x 3, 5.5 kW x 1 11 kW x 1, 15 kW x 1

New air conditioning system

Both of perimeter and interior of office sections except meeting rooms are air conditioned by MZHP system. Fresh air is treated by heat pump air conditioners and total enthalpy heat exchangers and sent to each floor.

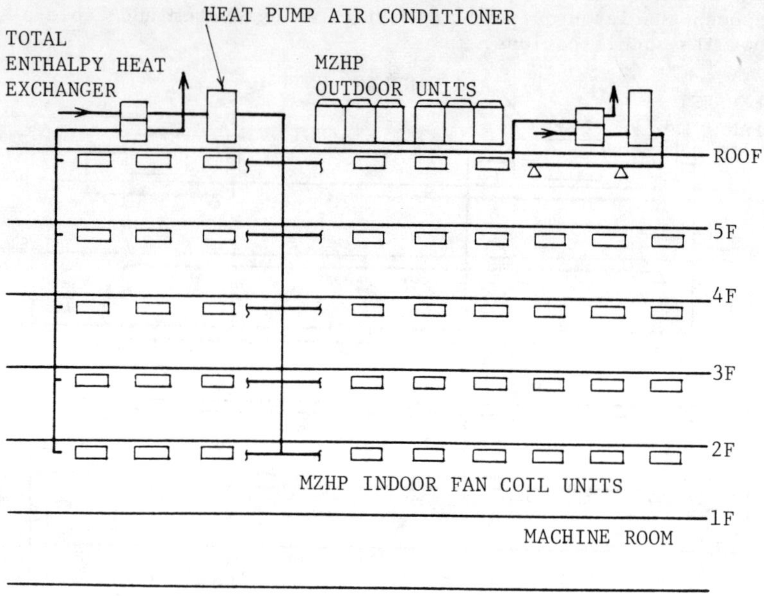

Fig. 8. New air conditioning system.

Table 4. Specifications of new system

Units		Capacity and number of units
Multi-zone heat pumps	Outdoor units	1,000 kW (16 HP x 8 units, 20 HP x 5 units, 25 HP x 5 units)
	Indoor units	7.32 kW x 125 units 14.53 kW x 1 unit
For treatment of fresh air and meeting room	Air source heat pump air conditioners	174 kW (10 HP x 1 unit, 20 HP x 1 unit, 30 HP x 1 unit)
Auxiliary units	Total enthalpy heat exchangers	3 units

Figure 8. shows the layout of new air conditioning system and Table 4 shows the specifications of air conditioning units.

Comparison of Operation Data and Evaluation

Power consumption for air conditioning by the old system was actually measured for 3 months during July - September 1983. Power consumption for 3 months during July - September 1984 after renewal of air conditioning equipments to MZHP was also actually measured. It is impossible to make to accurate comparison of these data for these two seasons because of inconformity in outdoor air conditions or load conditions. These data, however, are sufficient enough to evaluate the tendency of these two systems.

Power consumption for air conditioning was reduced by approx. 33% as a result of renewal of the system. Power consumption by heat source units was reduced by approx. 18%, while that by conveying units was reduced by approx. 56%.

This example has, therefore, proved that MZHP system for commercial use, typical of decentralized air conditioning system, is distinctly superior to central air conditioning system in respect of energy saving.

It is added that the summer of 1984, when the above measurement was conducted for MZHP system, was extraordinarily hot.

Table 5. Reduction of power consumption for air conditioning

	Power consumption (kWh)		
	Conveying units	Heat source units	Total
Old system	48,850	75,400	124,250
New system	21,450	61,550	83,000
Reduction	27,400 (56%)	13,850 (18%)	41,250 (33%)

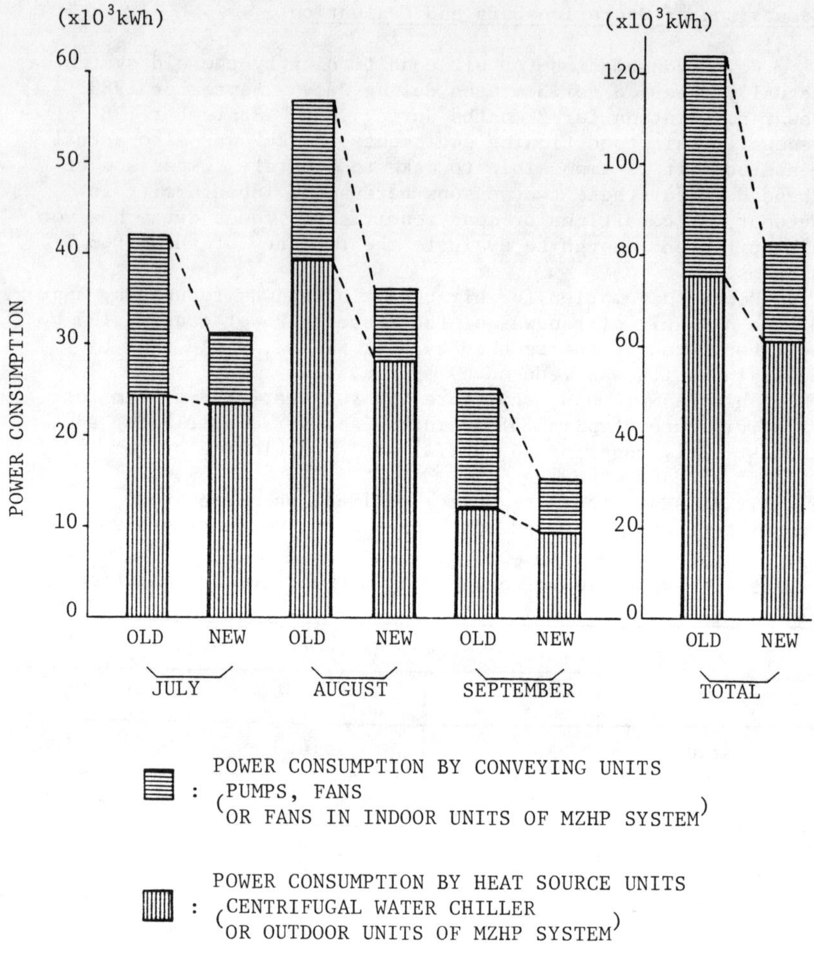

Fig. 9. Comparison in power consumption by air conditioning.

POSTSCRIPT

Superiority of MZHP system in energy saving is believed to be fully recognized so far.

It can be also presumed that the data for heating operation not presented in this paper will show the same tendency.

Compared with central air conditioning system, however, there remain some other problems to be solved.

First, fan coil unit can evidently control the capacity more easily incase of chilled and hot water compared with flow control of refrigerant.

For realization of more comfortable air conditioning by MZHP system, therefore, it is necessary to develop high reliable method to control the flow rate of refrigerant.

In addition, total capacity of refrigeration units especially in MZHP system for commercial use tends to be larger than central air conditioning system. In general, it is 20 - 30% larger compared with heat pump water chiller system.

In the next place, there is the following problem in regard to the heating capacity at low temperature, that is basic characteristic of heat pump air conditioner.

In the heat pump, the heating capacity as well as the COP decreases, when the outdoor temperature goes down, despite the fact that the heating load increases. Therefore, in order to compensate for the imbalance between the cooling and heating capacity, back-up heaters such as electric heaters must be used.

The frost on the coil of the outdoor unit remarkably deteriorates the heating function when the outdoor temperature goes down to some level. Therefore, it is necessary to develop an efficient defrosting method so that the MZHP system is enough strong for the colder regions.

We are confident, however, that MZHP system will further develop as a superior air conditioning system to central air conditioning system after solution of each of these problems by way of new controlling methods. (such as inverter controlled compressor, electronic expansion valve, etc.)

Chapter 5

Operating Experience with Air-Source Heat Pumps in Europe

Christian A. Vidal

ABSTRACT : The European market for heat pumps is essentially different from those in the United States and Japan, which tend to emphasize air-conditioning single room units and low-cost air to air equipment. Europe is mainly a heating only market and the common distribution system is the hydronic system. Within the European scene, France constitutes a significant example of experience in the development of heat pumps. Most of them are air to water heat pumps working with oil boilers and operating in «bivalent» mode. Also air to air systems, are now developing in the commercial field. Technically speaking, the equipment has attained a high degree of reliability and consumer satisfaction is high, as it is shown by the survey and statistical measurement campaigns carried out by official bodies. Further technical improvements and the new marketing policy for air source heat pumps in France are presented in this paper.

APPROACHES TO THE EUROPEAN HEAT PUMP MARKET

Heat pumps for heating and air-conditioning in residential or commercial premises offer a considerable advantage for users and the community : they permit a large reduction in heating costs, save energy and limit the atmospheric emission from combustion of fossil fuels.

The European market for heat pumps is essentially different from those in the United States and Japan : firstly, many Japanese and US installations meet an air conditioning demand which does not exist in Europe. Japanese and American heat pumps are usually air to air, and therefore of a lower unit cost than the air to water heat pumps needed to fit hydronic systems in Europe. Futhermore, the two millions

HEAT PUMPS: Prospects in Heat Pump Technology and Marketing (Proceedings of the 1987 IEA Heat Pump Conference), Edited by Kay H. Zimmerman Chelsea, Michigan 48118. Printed in U.S.A.

annual sales of heat pump units in Japan refers in general to single room units used mainly for air conditioning during summer, whereas European market is typically counting larger output single households units.

It is clear that factors affecting the heat pump market differ considerably across European national markets. Those factors include climate, oil/gas/electricity price ratios, electricity generation source, existing mix of installed heating systems, government and power utility policies. Europe is mainly a heating only market and the total heat pump systems has to be amortized by energy cost savings alone. The common distribution system is the hydronic one, integration of heat pumps and averall system control are more expensive.

This means that very differential succes has been achieved with heat pump sales across the European territories. The first start of this market was introduced by the second oil price shock in 1979. At that time, heat pumps seemed to be the ideal instrument to beat the oil producing countries and to reduce heating costs to a low level. And of course every one wanted to participate such as power utilities, heat pump manufacturers, installers and consumers. But it follows a big decline in European market over the last three years (1983 - 1986) as shown on figure 1 by the evolution of sales in France.

Fig. 1. Evolution of sales by type of system in France.

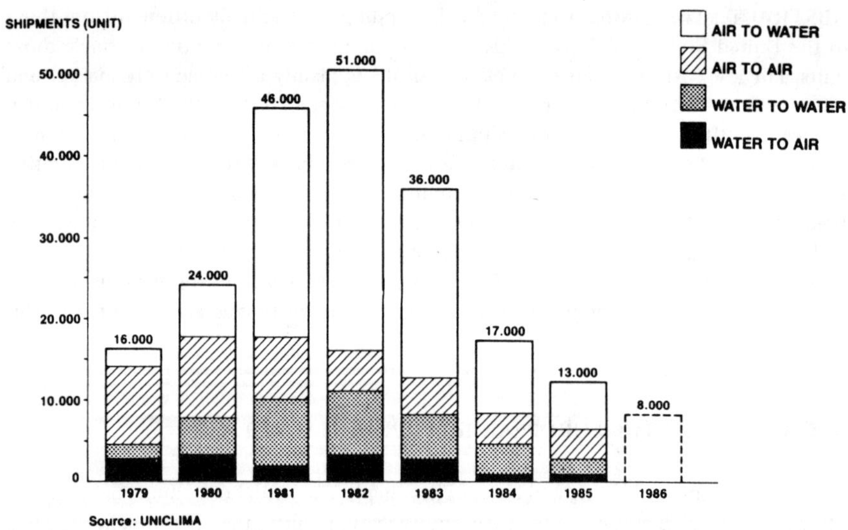

This drop of heat pump market is attribuable to several reasons :

— a low quality installation at the beginning, giving the market a reputation of low reliability and achieving diminished energy savings : heat pumps not suitable for heating purposes, oversized heat pump units, incorrect integrated into heating system and more cycling than energy saving,

— the second reduction in the oil price initially doubling the consumer's payback period and then further extending this period,

— the high cost of installation and maintenance of heat pumps,

— the small scale of heat pumps programs, losing the benefits of scale economics in manufacture, focused consumer marketing and simplification of expert installer training,

— the lack of a pan-European coordinated effort across an identifiable and viable mass market,

— and more recently the reactor accident in Chernobyl.

In the same time, policy for developing heat pumps has been different across the European countries. Thus, Sweden has led in terms of heat pump installations, resulting from its low electricity/oil price ratio, its extensive district heating, and the substantial support given by government and power utilities. In France, heat pumps programs have suffered from government price ceilings which have deterred installers. In Germany, a variety of local initiatives have been taken but the oil price fall has dramatically reduced heat pumps demand. In Austria, installations have still succeeded due to government and power utility support mainly around Linz country. Whereas the heat pump market in the United Kingdom, Holland and Belgium has been stunted by the availability of cheap nationally generated natural gaz.

Nevertheless, the European market for heat pumps remains very large and a second start may occur in the next years. Heat pump is a very attractive heating and air conditioning device. Both energy crisis during the seventies gave it the opportunity to have a large market to substitute to existing oil boilers. The market potential may be evaluated at several millions installations in residential, commercial and industrial sectors mainly in the residential retrofit market. The principal mass market lead to retrofit installations of heat pumps to oil fired hydronic domestic heating systems. Air conditioning and space heating equipment for commercial premises, also constitute a promising market especially in the south of Europe mainly in the form of low power reversible heat pumps.

All this constitute the basis for a considerable development of heat pumps.

OPERATING EXPERIENCE WITH HEAT PUMPS IN FRANCE

Within the European scene, France constitutes a significant example of experience in the development of heat pumps and particularly of air source systems. Considerable efforts have been made, both in technical and commercial terms, with mixed success : more than 60.000 air to air and more than 120.000 air to water heat pumps have been installed, mostly in replacement of oil-burning boilers. France has adopted a deliberate policy of developing heat pumps under the initiative of the national electricity utility «Electricité de France» with the cooperation of a limited number of efficient manufacturers.

Figures 2 and 3 show the evolution of air source heat pump sales in France from 1979 to 1985 classified by thermal power. Figure 2 reports to air to water heat pumps and we may see the big drop of sales of 1983. Figure 3 reports to air to air systems and shows a newly growth of less than 6 kW power devices due to the development of air conditioning and space heating single room units in the commercial sector.

Fig. 2. Air to water HP : Sales by thermal power in France.

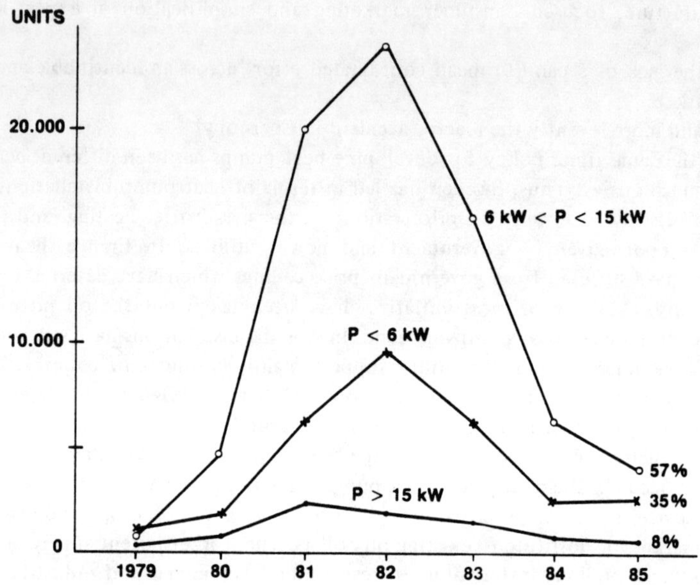

Fig. 3. Air to air HP : Sales by thermal power in France.

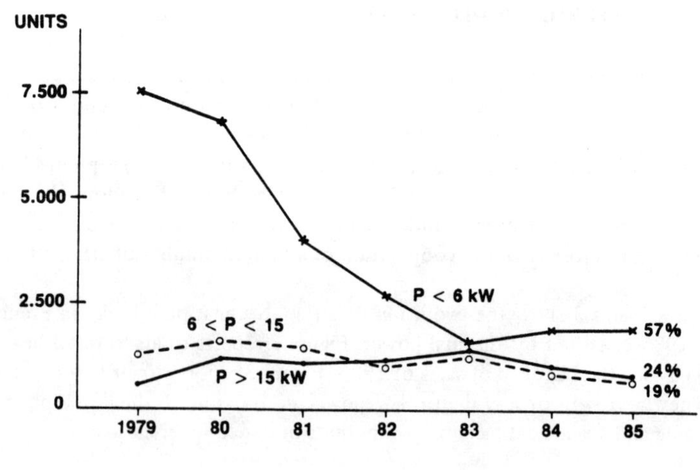

Exhaust air source heat pumps

In the 1975-1980 period before the second energy crisis, a big effort was led for new buildings by «Électricité de France», the French electricity utility. In order to improve the economical performances and the primary energy balance of direct electric heating, small air to air heat pumps using exhaust air of ventilation as could source were designed and developed.

These small heat pumps (one electrical kW for three thermal kW produced) were able to supply the basic heating demand in mid season. The supplementary heat demand, during the coldest days was made by direct heating.

A relative development of this devices was observed. However, to day the market of such heat pumps remains small (some thousands units annually sold, eight thousands sold in 1980 as shown on figure 3). The economical performance of these heating systems is not attractive enough, compare to direct electric heating to fight against it efficiently.

Air to water heat pumps fo retrofit market

After the second energy crisis, the strong increasing of oil price gave the heat pump a new chance to compete with existing oil boilers for heating hot water in hydronic systems. From 1980 was developed in France the PERCHE system, that means «Heat Pump Added to Existing Oil Boilers», working on a bivalent mode for single family houses, block centrals of dwellings and commercial buildings.

Heat pumps working with oil boilers generally use outside air source. The heat pump only heats the water of the heating system above a temperature (generally between 4 and 8°C) which corresponds to its capacity ; by lower temperature, the heat pump is stopped and the boiler takes over on its own ; during intermediate periods, when the outside temperature is between approximately – 5°C and 5°C, the boiler and the heat pump operate simultaneously. The stopping temperature of heat pump depends of its eveporator low limit temperature and of the network distribution temperature. Heat pump generally produce hot water at a maximum temperature of 55 to 60°C. It is calculated to provide about 50 % of the heating needs. Figure 4 shows the principle of an outside air to water heat pump.

Fig. 4. Principle of an outside air to water heat pump.

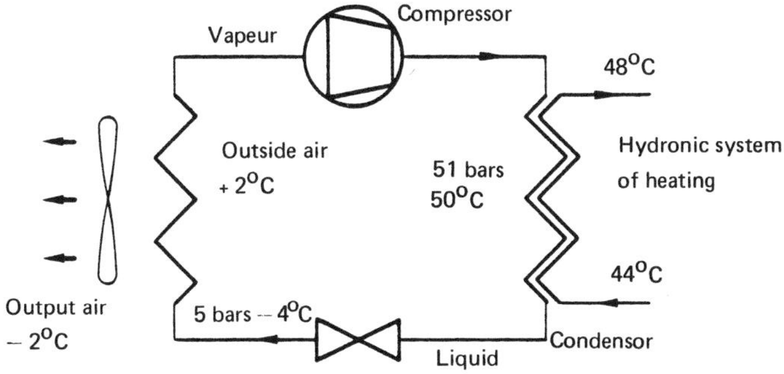

For individual houses, the standard unit mainly proposed by French manufacturers is a single outside unit with compressor and heat exchangers in the same box located outside the house. The connection with the existing water heating system is made with only two pipes of water to reduce the installation cost (figure 5).

Fig. 5. Principle of heat pump installation in individual house.

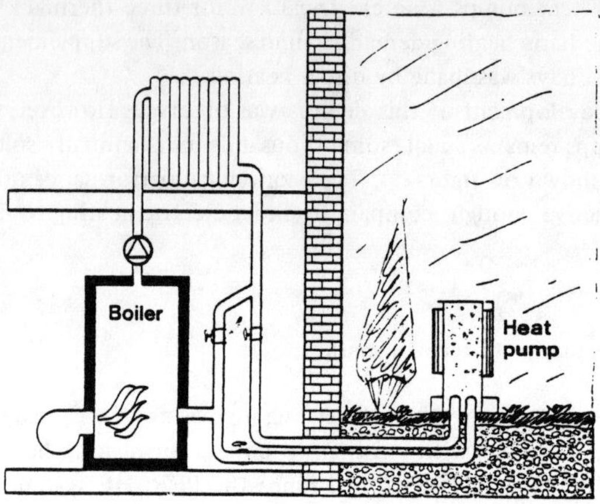

The principle of heat pump used as backup of a boiler in collective dwellings or commercial buildings are the same as small units explained before. Monoblock outside units are proposed until more than 150 kW. The condensor of heat pumps is situated in series with the return water pipes to the boiler as shown on figure 6.

Fig. 6. Principle of heat pump installation in collective housing.

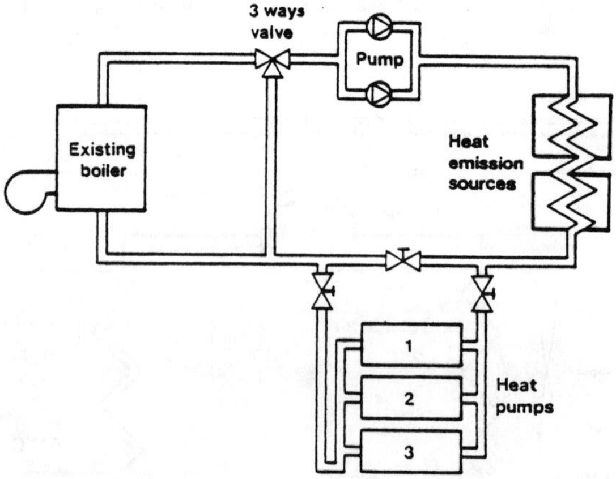

Marketing experience in air to water heat pumps

The market potential for retrofit installations of heat pumps to oil fired hydronic heating systems in France is about six millions of individual and collective dwellings corresponding to 17 millions tons of fuel oil yearly consumed. Many buildings of commercial sector with permanent occupancy are also involved (hospitals, hotels...).

Figure 7 shows the variation of the relative price of electricity and domestic oil between 1973 and 1985 with an oil boiler efficiency of 0.75 and a heat pump average COP of 2.5.

From 1984, Électricité de France provides consumers using bivalent systems of heating a particularly attrative new tariff option called EJP or «disconnection on peak days». This option includes a very high price for electricity during the peak period, the duration of which is fixed at 400 hours in the year (22 days of 18 hours, the choice of which is left to the utility) and a much lower than the average price for the rest of the year. Users who can accomodate disconnection from the network by using oil other than electricity throughout the peak period choose this tariff option : this is the case of bivalent systems users.

Fig. 7. Thermal useful kWh cost versus time electricity fuel (individual house) in constant CF 1985.

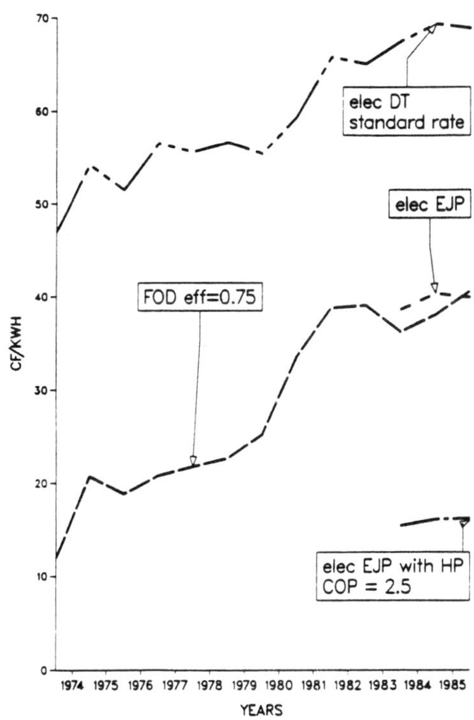

In this condition, the heat pump use is always economically justified : the money saving are between 40 and 50 % and the payback time about 4 to 6 years. The initial oil consumption is reduced about 80 %.

From 1982, a big technical and marketing effort has been undertaken to develop outside air to water heat pumps by the French government authorities and the French utility Électricité de France.

Single family existing houses

In 1982, a very ambitious program of heat pump development was decided for small units (electrical power between 1.2 kW and 5 kW) standards have been elaborated and fixed prices of installations were given : for example 25.000 FF for a 2.5 electric kW heat pump. However, in spite of the economical interest for customers, the market dropped dramatically after 1983 to achieve in 1985 the same volume as 1980 (see figure 2). The main reasons of this drop are the following :

— the first houses equipped with heat pumps were big ones with high oil consumption (up to 5000 liters/year). People living in such houses had large facilities to invest in energy savings. In fact, the oil consumption of most French single family houses does not exceed 3000 liters/year ; in that case, the economical interest is lower,

— after the second energy crisis, the fuel oil cost decreased in constant money on the contrary of the electricity cost,

— the marketing policy, has been very active but also coercitive with fixed investment prices not accepted by installers who rejected this device,

— a marketing inquiry showed the potential market was smaller than expected previously ; 80 % of consumers are waiting for the existing boiler out of order before deciding to change it. In such a case, they need a compact device including heat pump and boiler but the capital cost is, then, very high,

— finally, a payback time under five years seems to be a upper limit to convince customers to buy a heat pump . This pay back time is now hard to achieve for small or medium houses.

Collective dwellings and commercial sectors

The theorical market is also important but the ways to decide to install heat pump are quite different as individual houses. In collective private dwellings the decision upon heating systems need a majority vote of the owners that always take a long time. For private commercial sector or social housing, the decision is faster because there is only one owner of the building.

Marketing policy of Électricité de France in that field is to sign agreements with social housing owner and heating supply companies which assure maintenance of large heating systems in order to incite them to install heat pumps.

Several hundred installations of middle size (50 to 500 thermal power heat pumps) are now in operation and show, for most of them, energy savings with payback time roughly 5 years sometimes less.

New marketing developments

For the next future, marketing action are principally directed towards :

— standardizing the equipment and promoting the quality label,

— consumer advertizing based an high brand image and the financial advantage at short and medium term offered by heat pumps : loan repayments costing less than the operating saving,

— bonus and incentive subsidies during the development phase of the product,

— promotion of the new electricity off peak tariff which, in return for power cut-off on peak days, offers electricity at a price per kWh 40 % less than the usual tariff,

— training and selection of a network of specialized and skilled installers,

— long term garantee with a maintenance contract for users who are worried about product reliability.

Technical experience

Most of heat pumps now commercialized in France have to receive qualification label called NF (French Norm) which is issued by the French Standardization Association. It garantees equipment quality, design level of thermal performances, noise limitations and electricity safety :

— the minimum COP must be 2.5 for an outside temperature of 7°C, a hot water distribution temperature of 50°C and an air hygrometry of 80 %,

— the power noise level must be limited at 70 decibels.

The heat pump standardization concern units up to 60 kW of thermal power. Test methods for COP and acoustic level measurements are strictly defined. Standards for higher size units (thermal power up to 150 kW) have been undertaken.

Figure 8 shows the variation of COP versus outside temperature of a standard air to water heat pump.

Fig. 8. Variation of performances versus outside temperature.

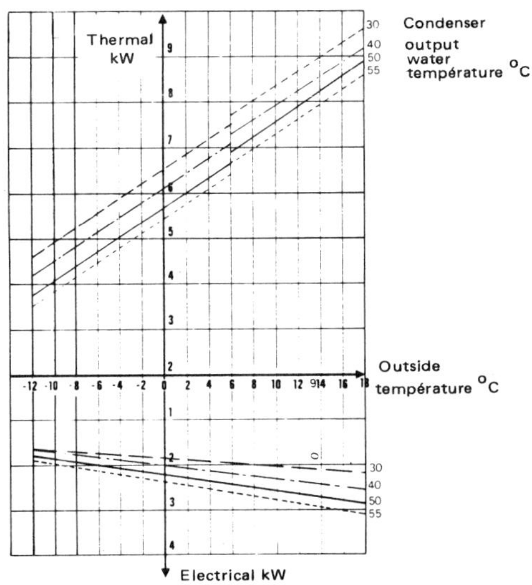

Experimental survey

A large program of technical and statistical survey has been conducted for last three years by the French National Scientific and technical Centre for Building Research (C.S.T.B.) sponsored by Électricité de France. More than 400 small units in single family house and several larger installations of outside air to water heat pumps working as backup of oil boilers have been monitored. Seasonal efficiency and annual consumption of energies before and after the heat pump installation were measured in operating conditions.

The results confirmed the equipment has attained a high degree of reliability with average seasonal efficiency exceeding 2.5. The initial oil consumption has been confirmed to a reduction of 70 to 80 %. The breakdown frequency has attained the lower level of less than 5 %.

Endurance tests

In addition of the ground experimental survey, comparative tests of endurance and reliability under severe conditions have been made in 1985 by official laboratories on 8 outside air to water heat pumps of 2 to 3 electrical kW. These tests were to simulate five years of real operating :

— high stop and start frequencies of compressor even at very low temperature (– 15°C to – 20°C during several following days),

— cyclic test of defrost system under the same conditions,

— vibration test simulating road transportation in the most severe conditions,

— salt mist test for corrosion resistance measurement,

— variation of acoustic and thermic performances compare to the standards.

The cyclic performance tests were compared to the standard performance of heat pumps given by the manufactures. For most heat pumps there were all confirmed. Very few incidents have been observed at the issue of the tests. No refrigerant leakage and mechanical problems have been noticed. Iron - sheets resist well at the corrosion test. Snow did not disturb the heat pumps.

All the test results have been published in a French consumer magazine with the name and brand of heat pumps.

The following tables 1 and 2 show examples of thermal and acoustic performances measured before and after the tests. Heat pumps were tested using the standardized test method (norm NF E 38-101) :

— condensor output water temperature : 50°C,

— condensor input - output difference of temperature : 5°C,

— global COP is the ratio of energy measured at the condensor divided by total electrical energy consumed during the heating period,

— integrated COP is the same ratio including the electrical energy consumed during the reverse defrosting cycle.

Table 1. Thermal performances before and after tests

	Cond. input water temp. °C	Evap. input air Temp. °C	Evap. input air Humidity %	Thermal Power kW	COP Global	COP Integrated
Before	45.1	7.0	86	7.4	2.6	–
Endurance	45.6	2.9	87	6.5	2.4	2.4
Tests	47.0	– 6.7	85	4.5	1.9	1.8
After	45.0	6.9	87	7.6	2.7	–
Endurance	45.6	3.0	86	6.7	2.5	2.4
Tests	46.8	– 7.0	86	4.8	2.0	1.9

Table 2. Acoustic performances before and after tests

Heat pump	Acoustic power level in dBA Before tests	After tests
N°1	73.0	72.2
N°2	65.2	65.7
N°3	68.6	69.4
N°4	70.1	73.0
N°5	68.9	70.4
N°6	68.5	–
N°7	69.4	70.6
N°8	72.7	72.8

Further technical improvements

Further improvements are studing by manufacturers and laboratories in the area of :

— compressor performance,
— non azeotropic refrigerant,
— automatic defrosting systems,
— regulation and management of the bivalent systems,
— machine reliability,
— thermal and acoustic optimization of the constituant parts, putting the accent on reducing compressor noise and developing flexible links between the compressor and heat exchangers, as well as static evaporators.

Heat pump/atmospheric exchanger associations are also beginning to develop and offer certain advantages ; they promise improved performance that may lead to an integral heating system that can cover all needs no matter what the climatic conditions.

Air to air reversible heat pumps

Small split of 1 to 3 electric kW air to air reversible heat pumps for room air-conditioning and space heating are now developing on the French market. These systems are well known in Japan and US where several millions are sold. They can be installed in the main room of a single family house or in commercial premises (stores, retail outlets, small offices, hotels, restaurants...) and constitute a promising market specially in the south of Europe.

On a marketing point of view, the interest of them are :

—possibility of cooling during summer and heating during winter with a high efficiency,

—low investment cost because of large development of standard units by manufacturers.

A large range of split-system reversible air conditioners offers a luxurous individual environment control for low cost installation. New design, very low sound level of the indoor air treatment part and the outdoor condensing unit, the heating and cooling performances and often an infrared remote control without wire are the basic features of these systems (fig. 3).

Technically speaking, these air to air reversible heat pumps are more and more improved under the impulse of Japanese manufacturers : lower noise level, more compact system, use of rotative compressors.

Table 3. Some characteristics of units developed by the French Manufacturer AIRWELL.

Power supply (fan included) kW	Nominal capacity Cooling (1) kW	Nominal capacity Heating (2) kW	COP	Sound level (3) Outside dBA	Sound level (3) Inside dBA
1.030	2.380	2.460	2.72	42	33/37 (4)
1.305	3.050	3.275	2.81	42	34/42 (4)

(1) International standards type A : 27°C/19°C wet bulb - outside air : 35°C/24°C wet bulb.

(2) International standards inside 19°C - outside 6°C wet bulb.

(3) Overal acoustic pressure in dBA at 4 m under nominal conditions.

(4) Low/normal speed.

CONCLUSION

In the present context of lively competition among the different forms of energy, the air source heat pumps have every chance of satisfactory development. However, in Europe, the market is presently rather depressed mainly because of the drop of oil and gas energy prices.

The equipments have now attained a high degree of performance and reliability. Beyond some marketing errors that have been made before, the future of heat pumps on the European market will depend of :

— the reduce of capital cost compare to ordinary heating systems,

— the evolution of energy prices,

— the development of air-conditioning systems associated with reversible air to air heat pumps mainly in commercial building appliances,

— the support of governments and power utilities.

A coordinated Pan-European effort perhaps remains to be undertaken to develop a mass market.

REFERENCES

1. International Heat Pump Promotion Group, The Addison Consulting Group, Swedish Council for Building Research. Document of June 1986 : «A pan European market implementation program for the small residential heat pump».

2. P. Berhondo, «Heat pumps for heating of building and industrial processes : the French situation», 13th Congress of the World Energy Conference, Cannes, 5/11 October 1986.

3. Centre Scientifique et Technique du Bâtiment, Essai comparatifs de huit pompes à chaleur air extérieur/eau, Déc. 1985.

Chapter 6

Exhaust Air as a Heat Source for Heat Pumps

Henrik Enstroem and Lars-Olof Glas

Abstract

Exhaust air from different types of buildings is a frequently used heat source for heat pumps in Sweden. The special properties of and practically observed problems associated with the use of exhaust air as a heat source are described and proposals for technical solutions are presented. Operating results from one large installation (3.4 MW) are reported. A description is given of a couple of main installation types for combined heat pump and cooling operation. A fundamental but simple formula for calculation of the economics is presented with examples.

Introduction

Exhaust air from housing accomodation, offices and other buildings is a popular source of heat in Sweden. There are more than 30 000 installations throughout the country, most of them in owner-occupied private houses. More than every other private house built in Sweden today is equipped with an exhaust air heat pump, normally with a heat output of about 1 kW. Many exhaust air heat pump installations are much larger, with outputs up to about 3 MW.

Financial considerations in the form of lower heating cost are the prime incentive behind the installation of exhaust air heat pumps. There are, however, several interesting side effects. For purposes of energy conservation, preference is given to keeping

HEAT PUMPS: Prospects in Heat Pump Technology and Marketing (Proceedings of the 1987 IEA Heat Pump Conference), Edited by Kay H. Zimmerman

ventilation at a minimum. In the case of an exhaust air heat pump, an increase in the ventilation flow rate has only a marginal effect. Let us take the case of a normal apartment: in the absence of a heat pump, a 50 per cent increase in the exhaust air flow will add 70 US$ to the annual heating bill. If a heat pump is installed, this figure drops to 20 US$ per year. In the latter case it is possible ventilate more frequently for added protection against damp and mildew, both of which can lead to very expensive repairs. In houses suffering from a radon gas problem, there is less contention between the need to ventilate and the desire to conserve energy.

A two-fold benefit can be derived from cooling plant and heat pump operation in buildings with air conditioning and heating systems. Very elegant solutions embodying cooling and exhaust air heat pump systems can be installed in buildings that utilize air as the medium for heating and cooling and where a mechanical ventilation system is needed. The office-block building is a typical case in point.

Special properties of exhaust air

The heat source is the limiting factor in the output of exhaust air heat pumps. The exhaust air flow is, after all, determined by the size of the building and the purposes for which it is used. It is not a variable that can be controlled at discretion to suit the requirements of the heat pump. By the same token, the temperature of the exhaust air is a fixed parameter. Admittedly measures can be taken to reduce leakage or evacuation from cold areas, but the scope for improvement is strictly limited.

For this reason the only way to change the output of the heat pump is to control the drop in temperature of the exhaust air. Cost-benefit studies have shown (1) that a lowering of the temperature in the neighbourhood of 20 °C is usually the best for living accomodation. This is not a general rule of thumb, but merely an example: conditions are never identical in any two buildings.

The coefficient of performance of the heat pump will deteriorate in direct proportion to an increase in the reduction of temperature for which it is rated. This is primarily attributable to lower evaporation temperature, although the fact that the condensing temperature will increase for a higher rated output is a contributory factor. A lower balance temperature also leads to a risk of a longer operating time with a capacity-controlled installation. On the other hand, the relative effect of auxiliary outputs will be less in larger installations. Nevertheless it is normally more profitable to rate the capacity of the installation for a relatively large reduction of the temperature of the exhaust air. Reducing this reasoning to its essentials, it is fair to say that in a "reasonably" rated equipment installation, where the limiting factor is the flow of the heat source, the profitability of the heat pump system will be in inverse proportion to its coefficient of performance.

In discussing the profitability of heat pumps, the coefficient of performance is often assigned an intrinsic value all its own. The facts as quoted above afford an excellent example of the fact that the coeffient of performance is only one of several factors affecting profitability and by no means always the most important. Factors that are equally important to the economic outcome, and this does not apply specifically to the EHP, include availability, adjustment, monitoring of operation, heat output, credit terms and grants, profitability criterion, energy tariffs and investment, operating and maintenance costs. The careful purchaser will need to prepare a detailed cost estimate for each installation.

The salient characteristics of exhaust air as an operational source of heat are:

- Relatively stable temperature throughout the year.
- Sharp daily, weekly and annual variations in humidity.
- Relatively stable flow determined by the the building and the purpose for which it is used.
- Condensation is formed at fin temperatures the same as or below the dewpoint.
- Frost formation occurs at fin temperatures below 0 °C.
- Major installation work is performed in the building(s).
- Filters will normally be required upstream of the cooling coils/evaporators.

Problems observed in practice

Confirmed malfunctions directly attributable to exhaust air are:

- Clogged filters upstream of the coils throttle flow. This impairs comfort and on occasions causes malfunctions and a deterioration in performance. This is because a reduction in flow rate lowers the evaporation temperature and, where relevant, the temperature of the brine.
- Substantial reductions in exhaust air temperature caused by duct runs on the roof or through unheated attics and leakage or extraction through cold premises cause symptoms similar to a low flow rate.
- Problems have been encountered with direct expansion systems when the evaporator is installed on the roof and other refrigerating equipment in the cellar. The considerable difference in height causes a pressure drop in the refrigerant that could lead to the formation of gas bubbles. The increase in volume caused by the formation of gas interferes with the operation of the expansion valve preventing it from porting an adequate mass flow. The formation of bubbles can be prevented by ensuring that the refrigerant is adequately subcooled.
- Condensing water leakage.
- Poor or non-existent defrosting.

Although the simplest method of defrosting is to use a timer, an event-control system may also be installed. Normal practice is to use passive defrosting whereby heat extraction is reduced or discontinued allowing the heat in the exhaust air to melt the frost. In such cases it is essential to prevent the formation of frost icing up the spaces between the fins anywhere on the coils.

The following four defrosting methods are feasible in practice.

1. Switching off the heat pump and the brine pump.
2. Stopping the brine in sections of the system. This will entail the installation of several parallel coils.
3. Reducing the capacity of the heat pump to lower the cooling effect sufficiently to enable the exhaust air to defrost the fins. When the brine/evaporation temperature rises above 0 °C, the heat in the pipes can also be used for defrosting.
4. Shutting off individual coils or groups of coils from the main flow. An internal pump is then used to circulate the brine through the coils. In this case the brine is heated by electricity or similar source.

It is a wise policy to plan for defrosting when the system installation uses the brine/evaporation temperatures close to 0 °C. It does not take much for the temperature to drop below 0 °C and for frost to form on the coils. A decrease in exhaust air temperature and/or lower flow rate (typically caused by a clogged filter) outside the design envelope will lead to an immediate reduction in fin temperature. If this occurs in a facility that lacks a defrosting device, one of the following two sequences will occur:

- The coils will ice up completely. Ventilation will obviously suffer and the circuit breaker on the fan motor may open. The evaporation temperature in the heat pump will fall rapidly until the low pressostat trips. If the fan continues to operate defrosting will occur and the heat pump will restart and repeat the cycle.
- The heat pump is controlled by the temperature of the outgoing brine to prevent the formation of frost. Output and performance factor will fall without triggering any alarm and this could continue for quite some time.

The largest exhaust air heat pump in the world

The performance factors that can be attained will depend on the reduction in exhaust air temperature and the temperature of the heat sink (apart from all the other factors associated with a carefully installed and adjusted facility). For obvious reasons the best possible performance factor will not necessarily be synonymous with optimum cost efficiency. On the contrary, the fact of the matter is that the rated heat extraction from the exhaust air suggests an inverse ratio for these factors within reasonable limits. The performance factor may be very low when preference is given to a central installation for reasons of service and maintenance and, as the case may be, lower installation costs. The world's largest operational exhaust air heat pump installed in the town of Täby to the north of Stockholm has a thermal output of 3.4 MW and a total seasonal performance factor of about 2.3. It should be borne in mind that the heat emission temperature is frequently between 70 and 80 °C.

The Täby facility is interesting in several respects in that it demonstrates the feasibility of using exhaust air as a heat source for local district heating networks. A description of the facility together with operating experience and observed performance are given below.

The City Boiler Station facility in Täby supplies three housing estates with a total of about 2 500 apartments. A school, day-care centre, hotel and several shops and office premises are connected to the system. The boiler station was previously exclusively oil-fired with an annual consumption of 5 500 m^3 of low-sulphur fuel oil 4, equivalent to 50 000 GWh of delivered heat. Today the facility has a 4 MW electric boiler and a 3.4 MW heat pump facility in addition to the oil burners.

The heat pump recovers heat in the exhaust air from about 2 000 of the apartments in the area. Heat is recovered indirectly using the brine to carry the heat from the exhaust air to the evaporator in the heat pump unit. The brine, consisting of more than 100 m^3 of a 20 per cent ethanol-water solution, circulates in a system of pipes, partly laid underground, from the boiler station where the heat pump is located to the attics or roofs of the apartment blocks where the cooling coils exchange heat between the exhaust air and the brine. There is a total of 51 sites where heat is recovered from the exhaust air in the area. The brine system is divided into three main loops each equipped with motorized gate valves. Motorized valves are used, together with check valves, to contain the brine to minimize discharge in the event of a leakage. They can also be used for automatic defrosting of the cooling coils.

A 9 MW oil-fired boiler was removed from the boiler station to make room for two STAL VSP73EC screw compressors powered by 10 kV electric motors. R12 is used as refrigerant. Pumps, refill tanks, expansion tanks, monitoring equipment, automatic control cubicles, high-tension switchgear and other equipment are located in the boiler station.

The use of a central facility offers several advantages over a system incorporating several local units in the buildings in terms of operation and maintenance.

The return circuit in the heating culvert is run to the heat pump condensers connected in series in the water circuit. Water temperature is raised by up to 15 °C, depending on operating conditions at the time. The heat recovered by the heat pump is sufficient to cover all requirements for about 2 000 hours per year at ambient temperatures in excess of 13-15 °C. At lower temperatures, auxiliary heat is provided by the electric or oil-fired boilers.

The heat pump is estimated to supply more than 27 GWh, or 55 per cent of total annual heating requirements of 50 GWh. If heat pump input energy is included, the annual energy savings are more than 15 GWh, or 30 per cent. The heat pump saves 3 000 m^3 of fuel oil per year with a reduction of about 55 tonnes in sulphur pollution alone. Rated performance has been achieved during the six months that the facility has been in operation so far. The few malfunctions that have occurred are associated with the adjustment and calibration of the equipment.

Operating results when using the heat pump

The operating results from the seven first months of commercial operation of the heat pump are evident from Fig. 1. The lower portion of the columns show the total electrical input to the heat pump facility. Since the full column height represents the heat emission, the upper portion corresponds to the amount of energy saved by the heat pump. All in all, 9.7 GWh was delivered during the period, with an input of 4.2 GWh, giving a COP of 2.3. This figure may seem low, but shows good agreement with calculated performance data. A reason is the high heat emission temperature, frequently in the vicinity of 80 °C. The temperature can nevertheless be lowered with relatively simple means and give room for more profitable operation.

During the month of May in particular, the operation of the heat pump was limited by requirements for a high pipeline temperature and, at times, by too low a water flow in the piping network. These problems could be remedied with technically simple actions. In this context there is reason to stress the importance of regarding the entire heating system as an integrated system, including heat pump, electric and oil-fired boilers, pipelines, substations and the distribution network and heat technical status of the property.

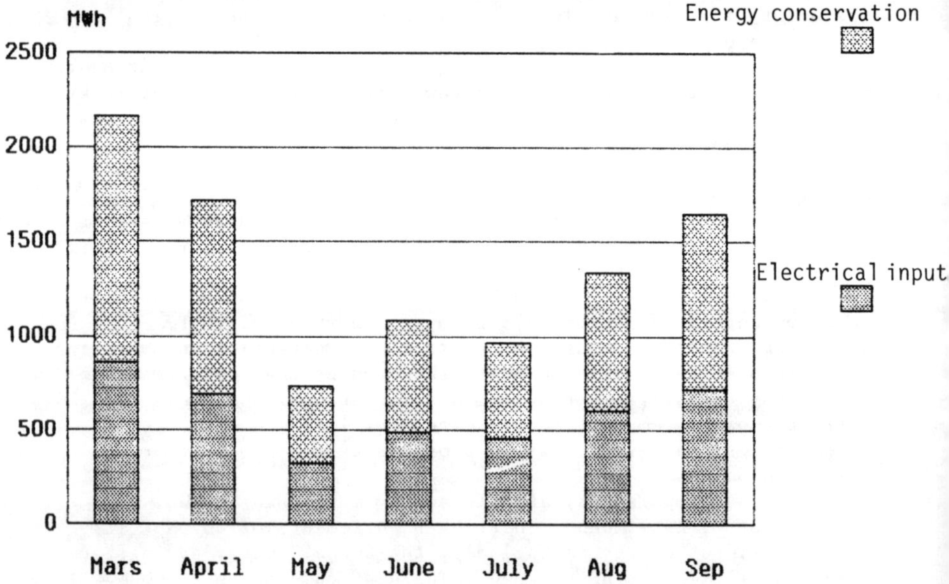

Figure 1. Main operating data 1986

Using Exhaust Air Heat Pumps for Cooling Purposes

If a cooling facility is required for air conditioning or some other purpose, supplementary equipment for heat pump operation can frequently be added at relatively low cost. The overall operating cost of the heat pump will improve substantially. Figure 2 illustrates an example of an installation of this type. It can be used simultaneously to cool and heat or for either one of these purposes. Figure 3 illustrates the performance of an installation normally designed either for cooling or heating. This combined cooling-heat pump equipment was installed between 1970 and 1974 in 20 or so department stores and office blocks in Sweden. Most of the installations have an excellent record of reliability and operating economy. For cooling purposes in the summer, the supply air coils are cooled and in one case, a Frenger-type suspended ceiling. In winter the heat pump supplies hot water to the same coils to heat the supply air. Ambient air can also be used as a heat source in the absence of exhaust air, typically at night.

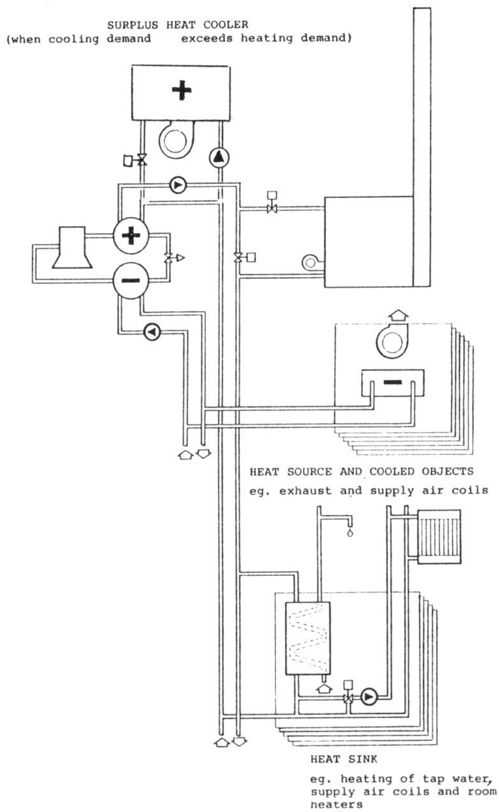

Fig 2. Combined heat pump and cooling plant

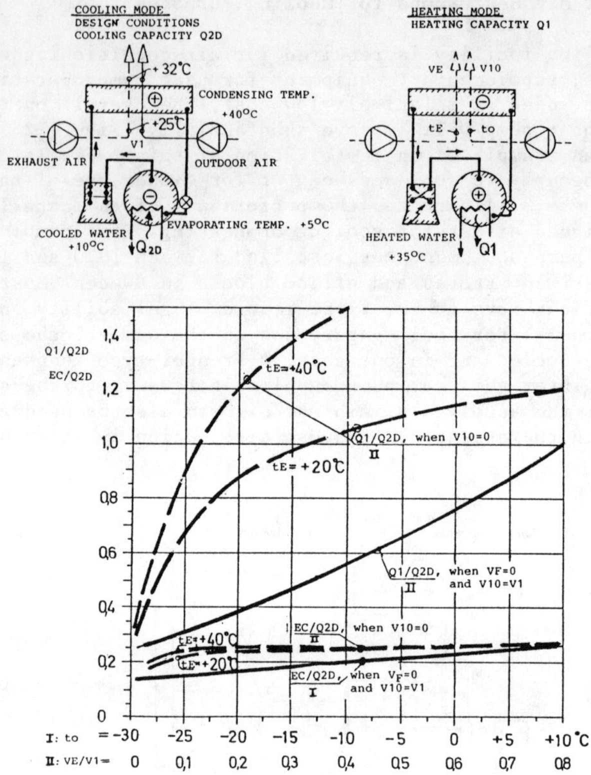

Fig3 . Combined heat pump and cooling plant, capacity in heating and cooling mode

The diagram in Figure 3 relates to the reciprocating compressors with a power consumption of above 50 kW and with Q2D (kW) as cooling capacity at design conditions:

- evaporating temperature +5 °C at outlet water temperature +10 °C.
- Condensing temperature +40 °C at an air flow of V1 to the condenser coils for cooling operation with +25 and +32 °C inlet and outlet temperatures. V1   1.2 · Q2D/1.2 · (32-25) = = 0.14 · Q2D (m^3/s).

Figure 3 illustrates the heating capacity Q1 that is obtained and the compressor power consumption EC needed for the rated cooling performance Q2D, firstly for an exhaust air heat source with +20 and +40 °C inlet temperature with different relative exhaust air flows VE/V1, where V10 = 0, and secondly with ambient air as the heat source, VE = 0, with different ambient temperatures t0. Figure 3 can, for example, be used to determine when the ambient air will be a more efficient source of heat than the exhaust air. This will be evident from the following examples:

- VE/V1 = 0.2 and tE = +25 °C interpolated from the diagram will give:

Q1/Q2D = 0.95

With V10 = V1, VE = 0, i.e. with ambient air as the heat source will give:

Q1/Q2D > 0.95

for

tOI > +8 °C

Obviously ambient air and exhaust air may be mixed for heat pump operation. In such cases the relative output can be read off from the unbroken lines marked:

"Q1/Q2D for V10 = V1, VE = 0" and "EC for V10 = V1, VE = 0" with t0 (in diagram) = tE · VE/(VE + V10) + tOR · V10/(VE + V10). tOR for this calculation is the temperature of the air from the outdoor air fan.

Heat Pump Operating Economy

The profitability of a heat pump installation may be calculated from the Equation (1):

$$S = T \cdot (HC - HH/COPT) - I \cdot (a + i + m) \quad (1)$$

where

- S = total savings US$/year and kW Q1
- T = equivalent heat pump operation time at full capacity, hours/year
- HC = cost of heat saved by heat pump (Q1) US$/kWh
- HH = cost of energy to operate heat pump (EC + E), US$/kWh
- a = amortization factor for I, 1/year
- i = interest for I, 1/year
- m = maintenance factor for I, m · I = total maintenance cost of the heat pump installation, m = 0.01 to 0.02 per year.

To get a positive saving:

$$T \cdot (HC - HH/COPT) > I \cdot (a + i + m)$$

or

$$T \cdot (HC - HH/COPT)/I > a + i + m$$

Example: air conditioning plant also operated as heat pump with condensing temperature t1 = +40 °C and evaporation temperature t2 = -5 °C.

T = 3000 HC = 200 US$/$m^3$ oil)/(9000 kWh/m^3)
= 0.022 (US$/kWh)

HH = 0.05

Referring to Figure 4, t1 = +40°C and t2 = -5 °C will give EC/Q2 = (1/0.50) · 45/268 = 0.34

With auxiliary pumps and fans connected to the heat pump operation of $E = 0.05 \cdot EC$ power demand. Figure 3 will give $Q1/Q2 = 1 + 0.95 \cdot 0.34 = 1.32$ and $COPT = 1.32/(1.05 \cdot 0.34) = = 3.7$. Normal value of I is assumed at 100 US$ per kW Q1.

$T \cdot (HC - HH/COPT)/I = 3000 \cdot (0.022 - 0.05/3.7)/100 = 0.25$

S is positive, if $a + i + m < 0.25$.

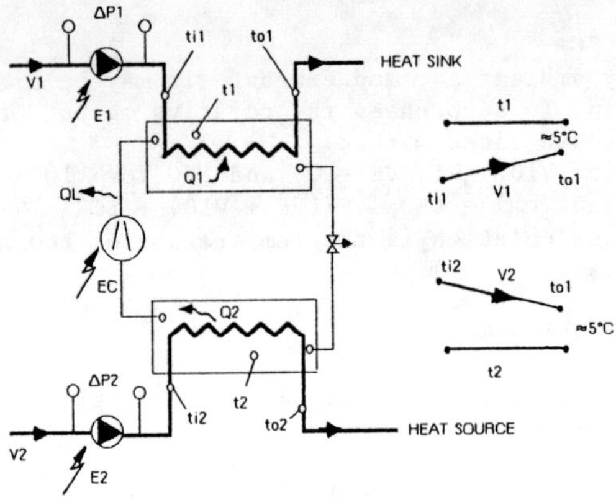

$$EC = \frac{Q2}{et} \cdot \frac{t1-t2}{t2+273}$$

– $et \approx 0{,}60 - 0{,}0025 \cdot (t1-t2)$
total efficiency incl motor
EC >100kW

$E = V \cdot \Delta P/e$
(W)= m³/s) (N/m²)

– for both fans and pumps
e = total efficiency incl motor

$Q1 = EC + Q2 - QL$

$QL \approx 0{,}05 \cdot EC$
incl heat losses from motor

$COPT = Q1/(EC + \sum E)$

$\sum E \approx 0{,}05$ to $0{,}10 \cdot EC$
incl fans , pumps, compressor oil heating, controlsystem etc.

Fig4. Heat pump efficiency

REFERENCE

1. H. Enström, "Sizing of Large Exhaust-Air Heat Pumps", IEA Newsletter, vol 3, No 4, 6-12, December 1985.

Chapter 7

Development of a System for Hot Water Supply by a Heat Pump That Recovers Heat from Daily Wastewater

Masaaki Ukaji

ABSTRACT: At present, hot waste water from the bathroom and cookery of the hotels and residential houses is discharged to the sewerage and at the same time, a large amount of heat is discarded uselessly. This kind of waste heat is abundant, but the waste water is seldom used for any purpose because it is dirty and the temperature level is low. This thesis states an effective energy saving method which uses waste heat for heating water. We use the water source heat pump with the automatic cleaning mechanism of the evaporator for this aim.

This system is operated satisfactorily in a city hotel in Tokyo and achieved good results that the daily average COP is 4.5 to 5.5. At present, we have only an experience of using waste water from the bathroom of hotels for this development, however this method will be applied for heat recovery from the toilet and other sewage in near future. As a result, the application will be widened over the housing complex, hospital, and other buildings.

INTRODUCTION

The idea has been in Europe and America for about 20 years that daily waste water from baths and kitchens can be utilized for providing the heat source for the heat pump, though there has been little information on actually realized examples of such idea. The possibly reasons why there were few actual practices and experiments in this field can be considered to consist in a lowered heat-exchange performance due to both the contamination of the evaporator and mechanical failures caused by foreign bodies in waste water. Hotels and housing complexes still discharge hot waste water from bath and kitchen directly into the sewer. This is

HEAT PUMPS: Prospects in Heat Pump Technology and Marketing (Proceedings of the 1987 IEA Heat Pump Conference), Edited by Kay H. Zimmerman

waste of energy resources, and in Japan, a country poor in energy resources, more importance should be set on development of the effective technique of recovering waste heat. This report describes effectiveness of the Hot water-supplying system by Heat pump Recovering heat in daily waste water (HHR) as a future effective energy saving technology, mainly based on the data from the experiments conducted with the real-scale experiment system set up in a certain hotel in Tokyo.

CONCEPT OF THE HEAT RECOVERY SYSTEM

Figure 1 shows the basic configuration of the heat recovery system for daily waste water using the heat pump. Hot waste water from bath and/or kitchen is saved in a sump pit for the later use as the heat energy source for the heat pump. Such waste water is sent to the evaporator in the heat pump, where the hot waste water is deprived of its heat and then sent back to the sump pit as cold water. On the other hand, feed water is fed to the condenser and heated there with heat recovered from waste water in order to be supplied as hot water.

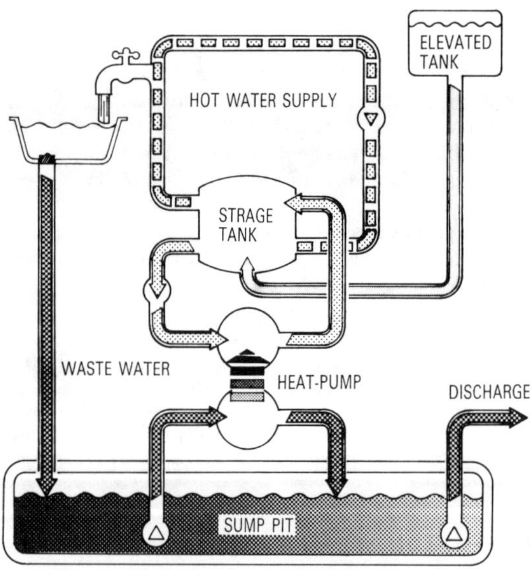

Fig. 1. Outlines of hot water-supplying system by heat pump recovering heat in daily waste water.

APPLICATION ON HOTEL

In Japan hotels are known to be buildings where a great deal of energy is spent. Especially energy for hot water supply is calculated to account for approx. 60% of all the energy consumption quantity in a hotel. Therefore, recovering and reutilization of heat generated in supplying a great deal of hot water to be consumed is one of the effective means to realize energy saving.

This document reports on the hot water supply system operating in a certain hotel in Tokyo as an example of application of HHR whose concept was described in Section 2. Please note that the experiment was conducted in the real-scale as the final stage of the HHR development project.

Now, we are going to explain, for the purpose of carrying out this project, the actual conditions of waste water, the outlines of the system used, its operation records as well as economic assessment results.

Researches on the Actual Conditions of Waste Water

For the purpose of obtaining basic data required for the designing of system, researches were made, before execution of this project, on a variety of points about actual state of waste water in a hotel, such as quantity, quality, and temperature of the supplied/drained water and service hot water in the hotel. The research was made in a city hotel in Tokyo (with 686 beds on the upper floors and 1099 beds on lower floors).

Quantity and temperature of waters for various uses

Figure 2 shows measurements which are made for the water supply and draining system in lower floors. Hot water-supplying flow rate, as having its close reaction with the temperature at outlets of supplied hot water, is calculated as water flow of 43°C (equivalent to bathing water), assuming that, before used, hot water is mixed with feed water. It was found that temperature of waste water was 30°C or more all the year round probably because such water is waste largely from bath and has little seasonal variations.

Table 1 lists measurements of quantity and temperature and calorie of different types of water related to the guest rooms. It was found that most calorie (70 - 90%) spent for heating water remains in waste water.

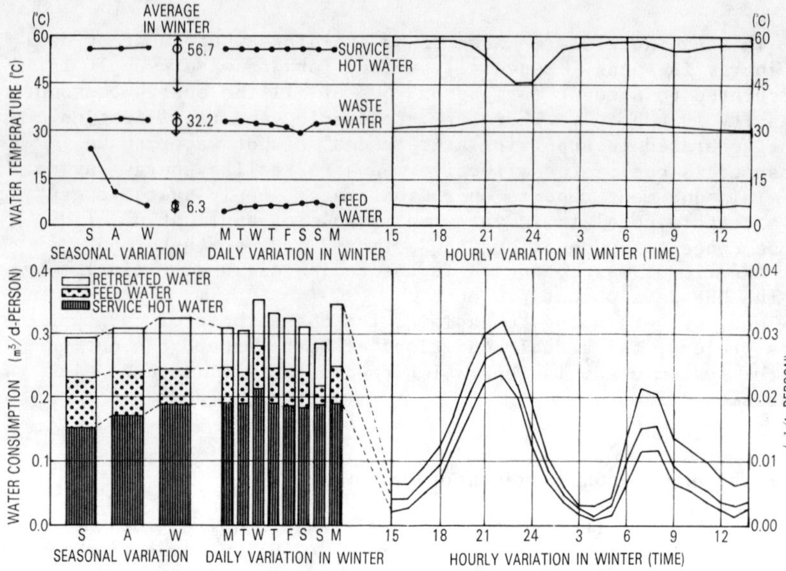

Fig. 2. Quantity and temperature of each kind of water for guest rooms on lower floors.

Table 1. Quantity, temperature, calorie of each kind of water from guest rooms

		Lower floors			Upper floors		
Season of research (Total days)		Summer (9)	Autumn (9)	Winter (7)	Summer (8)	Autumn (9)	Winter (8)
Water quantity measured (m^3/d-person)	Feed water	0.210	0.213	0.233	0.260	0.204	0.245
	Service hot water included in the above item	0.087	0.108	0.141	0.095	0.105	0.133
Measured tempera-ture (°C)	Feed water	23.5	13.9	6.3	24.4	14.6	5.9
	Service hot water	55.2	56.0	55.7	55.2	57.7	54.3
	Waste water	32.5	31.1	32.2	32.7	33.2	32.4
	Kind of waste water	Waste water of bath room			Ditto		
Calorie (MJ/d-person)	Service hot water	11.7	19.1	27.7	12.3	18.5	26.1
	Waste water	7.8	15.3	25.6	8.9	16.5	25.9

Water quality analysis

Table 2 shows the results of quality analysis of waste water from the guest rooms in winter. Foreign bodies of 1 mm or more in size contained in such water were hairs, tea leaves, etc. N-hexane extracts were quite little, only 3 g/m^3 in 14:00 and 23:00 when quantity of waste usually reaches peak. As no oil film was found in sample, it was assumed that waste from guest rooms mostly contained interfacial active agent such as soap and shampoo.

Table 2. Quality of waste water from guest rooms (g/m^3)

Sampling time 1985 (2/18)		7:30	9:30	11:30	14:30	17:30	20:00	22:00	23:00
SS	10mm on	<0.2	<0.1	<0.1	<0.1	<0.1	<0.05	<0.01	<0.01
	1mm on	1.6	0.4	0.5	0.6	0.8	0.9	0.1	0.1
	1 mm pass	13	20	13	10	8	15	24	21
Residue on evaporation		230	240	260	230	220	250	240	230
n-hexan ex-tracts		<1	2	1	3	<1	1	2	3
BOD		20	21	20	18	9	9	24	18
COD		15	12	19	22	11	10	16	12

Aptitude as heat source water

The above results show that the most of calorie used for hot water supply, remaining in various waste waters from each of guest rooms, can be recovered successfully as a heat source for the heat pump, which will be able to satisfy the thermal conditions required.

As to quality, such waste contains relatively large quantity of hairs. This can cause mechanical failures in the system. Appropriate precautions against possible occurrences of malfunction of the machine shall be considered necessary. The nature of waste in Japanese hotels may be slightly different from that in western hotels, due to differences in the climates, cultures, and ways of living.

System Applied

Principles for development and outlines of the system

We have developed the system by taking into consideration the following targets:

- Reduce the cost as low as possible, with its possible spread.
- Attain a higher COP.
- The system must be of maintenance free type, in principle.

In order to attain the above targets, we have selected a shell and tube type among many of evaporators for heat pump, because of its long experience for practical use and of its good durability attested. To resolve the problem of the contamination of the tube inner wall, we have developed a new automatic cleaning equipment. Table 3 lists the outlines of building and equipment information of the hotel. The configuration of the system is as shown in Figure 3. The overall major equipment is shown in Figure 4.

Waste water from guest rooms is saved in the sump pit and pumped to the evaporator of the heat pump by the heat source water pump. Here waste water is deprived of its heat and then sent back to the sump pit. Slime produced inside the tube of the evaporator as waste water passes through the tube will be removed by the automatic cleaning equipment.

In the hot water system, water saved in the storage tank is pumped out to the condenser by the hot water circulating pump and heated there. Such heated water is supplied to guest rooms in the lower floors in the building via the existing hot water supply system of the hotel.

Table 3. Outlines of building & equipment

Building	Equipment (sanitary)
Location: Shinjuku-ku, Tokyo	Feed water equipment: Gravity water supply system
Main application: Hotel, shop	Retreatment equipment: Gravity water supply system
Building area: 681.91 [m^2]	Hot water supply equipment: Central supplying system
Total floor area: 13,906.61 [m^2]	Waste water equipment: Sanitary drain, waste water, storm drain separate system
Floor area of typical floor 651.16 [m^2]	(Heat recovery system)
Floors: 18 stories above and two under the ground, pH 2 stories	Heat recovery heat pump unit: 335,000 [kJ/h],(93 kW)
Main structure: 2nd basement to 4F/SRC, 5F-18F/S	Storage tank: 4 [m^3]
Number of guest rooms: 337 (676 beds)	Automatic cleaning equipment: 1

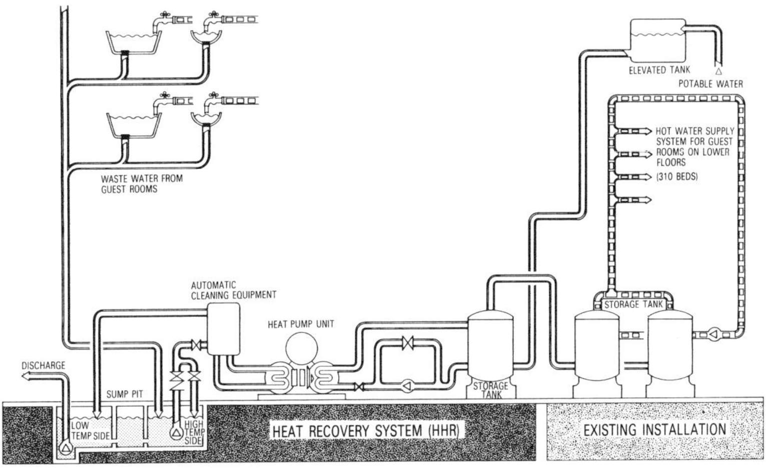

Fig. 3. Hot water-supplying system by heat pump recovering heat in daily waste water.

Fig. 4. General view of main equipment.

Automatic cleaning equipment

Figure 5 is the flow chart of the automatic cleaning equipment. The equipment is intended to automatically clean the tube of the evaporator with the following operation:

- The ball circulating pump is activated periodically once a day and circulates sponge balls (dia. approx. 20 mm) through the closed circuit including the evaporator so that circulating balls remove slime and other contaminants attached the inlet port and inner wall of the tube. The balls circulate per cycle for 16 minutes.
- Back washing is conducted for approx. two minutes before starting of the ball circulating pump by switching the three-way valve to clean the tube inlet port where hairs and other foreign bodies are most easily caught.

Figure 6 shows how sponge balls move under water, cleaning the tube of the evaporator. The cleaning efficiency by means of balls is shown on Figure 7.

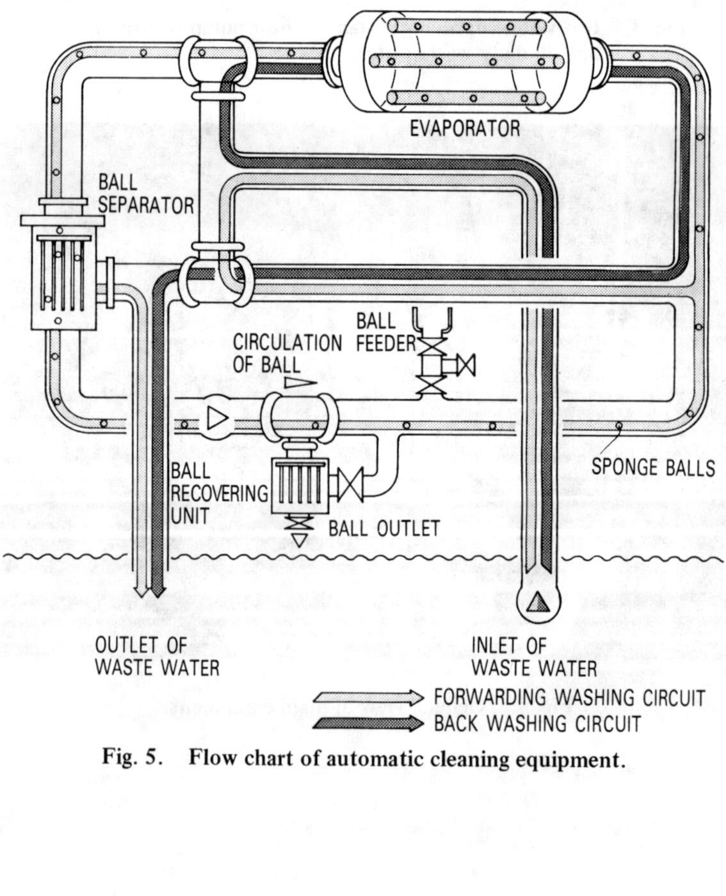

Fig. 5. Flow chart of automatic cleaning equipment.

Fig. 6. Sponge balls circulating through the evaporator.

Fig. 7. Efficiency actually realized by automatic cleaning equipment.

Operation Record

The system has been successfully operated since April, 1986. Figure 8 indicates the operation record of the system for June to September. COP of the heat pump, which depends on temperature at the outlet port of the condenser, has kept at a high level of ave. 4.5 - 5.5 per day. Recently in Japan COP ≒ 2.5 represents a running cost equal to that of kerosene. Thus COP of our system can be evaluated as quite advantageous. According to our calculation based on the operation record, the system provides the annual energy saving of 50% or more in the primary energy conversion (see Table 4).

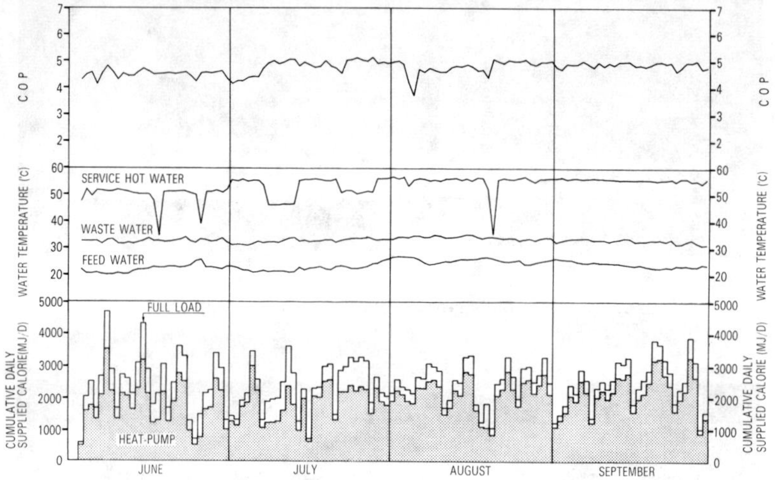

Fig. 8. Operation record.

Table 4. Comparison of consumed energies (converted in primary energy)

(GJ/year)

Head \ System	H H R *	Conventional ** system
Oil-fired boiler	–	2,037.2
Heat pump	800.5	–
Heat source water pump	66.6	–
Total	867.1 (43%)	2,037.2(100%)

*1 kwh— Primary energy: 10,256 kJ (2450 kcal)
Secondary energy: 3,600 kJ (860 kcal)

**Kerosene calorific power
— Primary energy: 37.3 kJ/m^3 (8900 kcal/lit)
Secondary energy: 34.3 kJ/m^3 (8200 kcal/lit)

Economical Evaluation

In determining whether the system is to be installed or not, the relation between the initial cost and the running cost will be critical parameters. In this study, we made an economical evaluation on the annual running cost calculated based on data on the past operation and the initial cost invested to construction of the experimental system.

Table 5 indicates the initial costs and annual running costs of our experimental system and conventional system (oil-fired boiler). The annual running cost is ¥1.43 million (US$9,200*), i.e. approx. 45%, less than that of the conventional system, while the initial cost is increased by ¥9.0 million (US$58,000*). The increment in the initial cost will be paid off in approx. 6.3 years by the saved amount of the annual running cost. Please note that the rather long period for paying off is due to the fact that our experimental system had to contain various experimental elements, causing a higher initial cost. Our case study showed that the system if applied to a middle- or large-scale hotel with 300 beds or more can be paid off in four years or so.

According to the evaluation using the LCC (Life cycle cost), the system has the LCC ratio of 0.78 in 15 years (0.73 for a practical system) (see Figure 9). It is expected that the system will be more attractive as the cost will be reduced by the future mass production.

Table 5. Economic assessment (1)

Head \ System	HHR	Conventional system	Difference in cost
Initial cost (ten thousand yen) (US $)	2,500 (≒160,000)	1,600 (103,000)	A : 900 (58,000)
Running cost (ten thousand yen) (US $)	163 (10,500)	306 (19,700)	B : 143 (9,200)
Difference recuperation cost	A/B ≒ 6.3 years		

[Calculation condition]

1) Hot water supply load/year: 1405 (GJ/year)
   (Assumed on the basis of actually measured data upto now)
2) Electric power cost: 19.30 (yen/kwh), 12.45 (cents/kwh)
3) Kerosene cost : 56 (yen/lit), 36.1 (cents/lit)

* Calculated on the exchange rate of ¥155 per US$1.00.

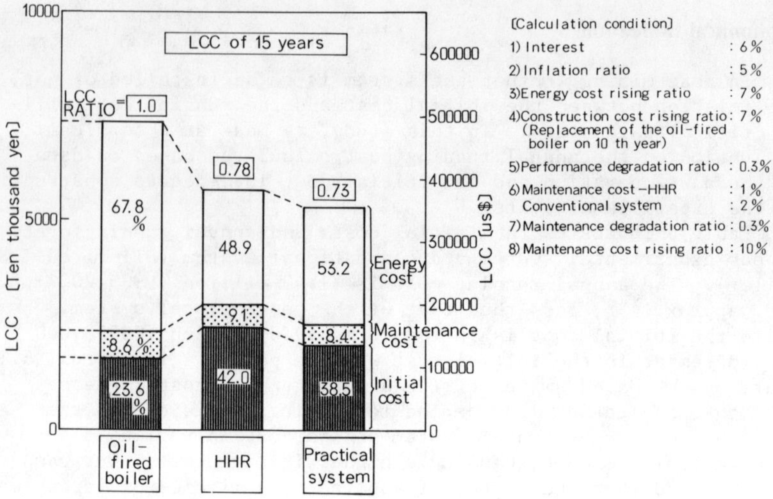

Fig. 9. Economic assessment (2).

CONCLUSION

This document delivered the outlines of the system and its economical features mainly looking at the HHR applied to a hotel in Tokyo. This can be summarized as follows:

(1) As waste water from guest rooms of the hotel is 30°C or more and the most of calorie spent for heating hot supply water was conserved in waste water, it can be used successfully as a heat source for the heat pump (if contaminants are successfully removed).

(2) The system has the high COP of daily average 4.5 - 5.5 and 40% of the annual running cost reduction rate will be obtained compared with that of oil-fueled unit. (Energy saving rate is 50% or more in the primary energy conversion.)

(3) Difference in the initial costs between this system and the conventional system can be paid off in four years or so. The LCC rate is 0.73 which is regarded as economically excellent.

(4) Cooling hot waste water in the heat recovery stage may lead to prevention of heat contamination of the big city.

Though the present study considers, for heat source, waste water from guest rooms of a hotel only, we are further developing a technology to establish an advanced technology for recovering heat from more contaminated waste (sanitary and other sewage) based on the results obtained from this experiment. When such technology is established, the hot water-supplying system by heat pump recovering heat in daily waste water will be applied to a wide range of users, such as the housing complex and hospital, and the other buildings.

REFERENCES

1. SAT. Sugiyama et al., "Installation Example (3), *Shinjuku Washington Hotel Annex (The Heat Recycle System Using Hot Waste Water)", The Society of Heating, Air-Conditioning and Sanitary Engineers of Japan, Journal of Society Vol. 60, No.9, 1986.*

2. HAJ. Yoshida et al., "Development of Heat Recycle System Using Hot Waste Water", *Architectural Institute of Japan, Summaries of Technical Papers of Annual Meeting Architectural Institute of Japan (Environmental Engineering), pp. 487-494, 1986.*

3. MIN. Kojima et al., "Research of Heat Recycle System Using Hot Waste Water by Heat Pump", *The Society of Heating, Air-Conditioning and Sanitary Engineers of Japan, Summaries of Technical Papers of Annual Meeting,The Society of Heating, Air-Conditioning and Sanitary Engineers of Japan, pp. 9-16, 1986.*

Section III

Advances in Electric Heat Pumps

Section III

Advances in Electric Heat Pumps

Chapter 8

Technology, Economics, and Future Potential of Variable-Capacity Heat-Pump Room Air Conditioners

Sei Suma

ABSTRACT: The air conditioning industry is showing great interest in the application of inverter technology to small and medium-sized air conditioners. In Japan, about one million units of variable capacity heat pump room air conditioners were sold in 1985, and the market is expected to expand in the coming years. This paper discusses the technologies applied in the variable capacity (inverter) heat pump room air conditioner. As examples of such technologies, 1)the control system of inverter heat pumps, 2)the non-reverse defrosting refrigeration cycle, 3)research in rotary compressors, and 4)application of the inveter system are discussed. The economics of the inverter heat pump is also compared with that of the conventional heat pump. Futhermore, future technical trends and future potential are described.

INTRODUCTION

In order to increase heating capacity and to improve energy savings, various kinds of capacity control systems have been developed. Before 1981, the methods of capacity control in room air conditioners in Japan depended on mainly refrigeration cycle techniques such as liquid injection, gas injection and gas release cycle and electric heaters. In 1981, the variable capacity heat pump with inverter system came on the market. Figure 1 shows the sales of room air conditioners in Japan. After the inverter heat pump came on the market, the occupation rate of heat pump for all room air conditioners rose gradually, and it has reached about 65% according to 1985 statistics. The ratio of the inverter heat pump

HEAT PUMPS: Prospects in Heat Pump Technology and Marketing (Proceedings of the 1987 IEA Heat Pump Conference), Edited by Kay H. Zimmerman Chelsea, Michigan 48118. Printed in U.S.A.

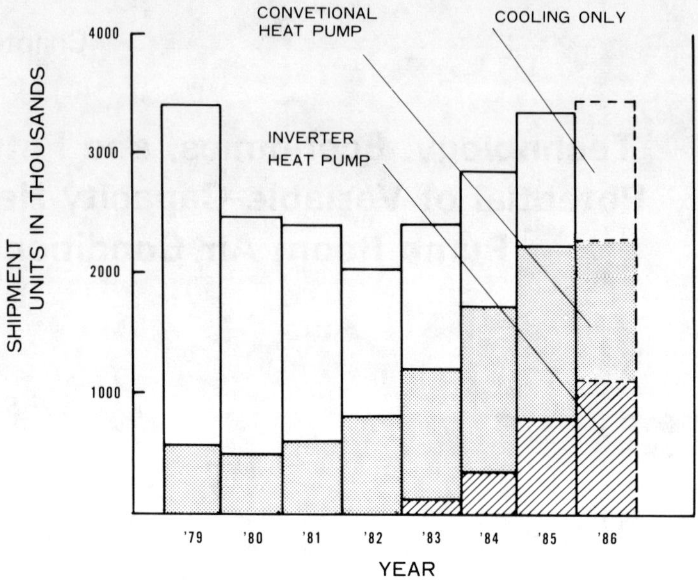

Figure 1. Sales of room air conditioners in Japan.

among all the heat pump room air conditioners has risen rapidly and at present about 50% of all heat pumps are the inverter aided type. In Japan, the term "INVERTER" has became very popular and the "INVERTER" is being applied in many electric home appliances fields such as refrigerator,ice stocker,flourescent lamp, oil heating devices, etc.

In this report, the technology and economical advantage of the inverter heat pump and future technical trends of the variable capacity heat pump are discussed.

TECHNOLOGY OF INVERTER HEAT PUMP

Characteristics and Techniques of Inverter System

An example of basic characteristics of inverter heat pump are shown in Figure 2. Figure 2.(a) shows the variation of both heating capacity and input power as a function of frequency. Figure 2(b) shows the variation of heating capacity almost linearly as a function of outdoor temperature. Heating capacity and input power increase as the frequency increases. EER increases as the frequency decreases, because the heat exchanger capacity relatively increases in the low-frequency region and consequently low compression operation is obtained. At outdoor temperatures above 0°C , the conventional air conditioners are ON-OFF controlled because heating capacity is larger than the load, with a consequent energy loss.

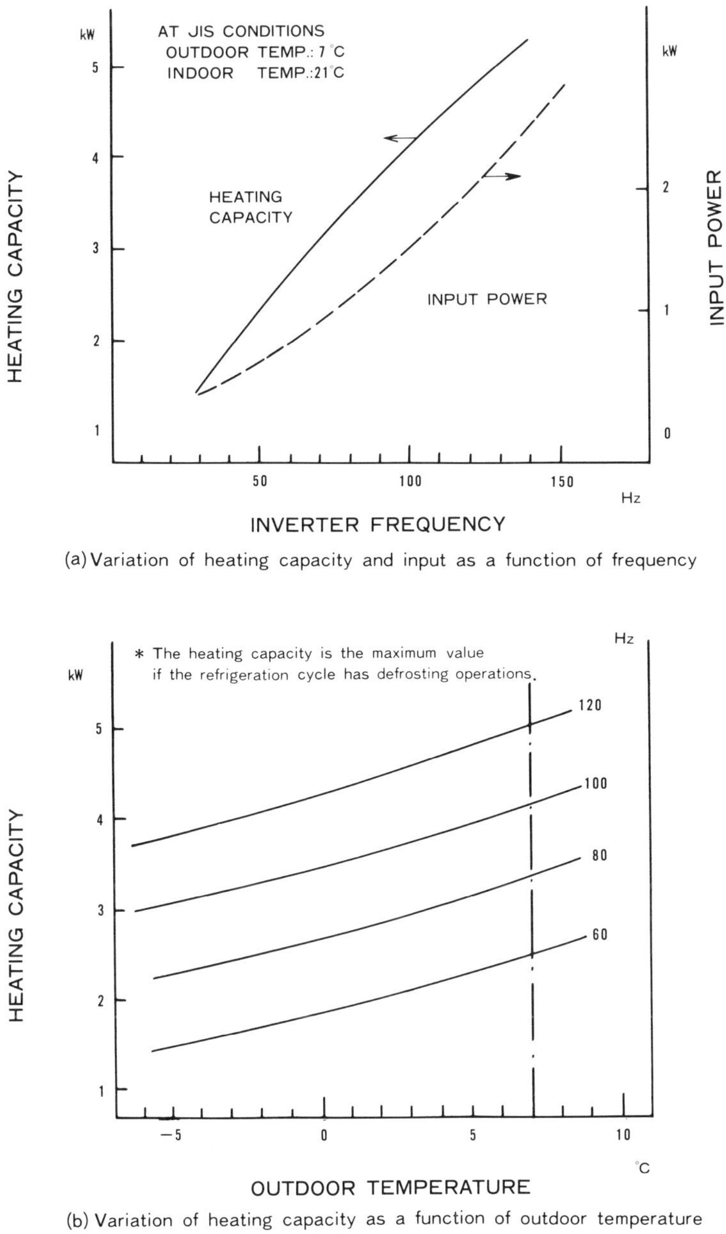

(a) Variation of heating capacity and input as a function of frequency

(b) Variation of heating capacity as a function of outdoor temperature

Figure 2. Basic characteristics of inverter heat pump.

However, in the case of the inverter air conditioner, the unit stays in continuous operation in the low speed zone of 30-60Hz for most loads. As a result, because the operating efficiency is high at low rotational speeds of the compressor and ON-OFF cycling loss is so much less, energy can be saved to a great extent. The economics of an inverter heat pump will be discussed later.

Control System [1]

An example of inverter heat pump system control is shown in Figure 3. Since the inverter air conditioners most available in the market are split type, the control sections of both the indoor and outdoor units have been equipped with a respective micro computer. The indoor controller section receives an instruction of operation mode and set temperature from the remote controller and computes the optimum operating frequency of the compressor with the input values from the room temperature sensor and heat exchanger sensor. Then the signal is transmitted to the outdoor controller section. The outdoor controller section instracts the inverter section to apply an AC power of specified frequency to the compressor. The inverter uses the pulse width modulation(PWM) type which is widely used for the reason of simple configuration of the main circuit and good response to load change.

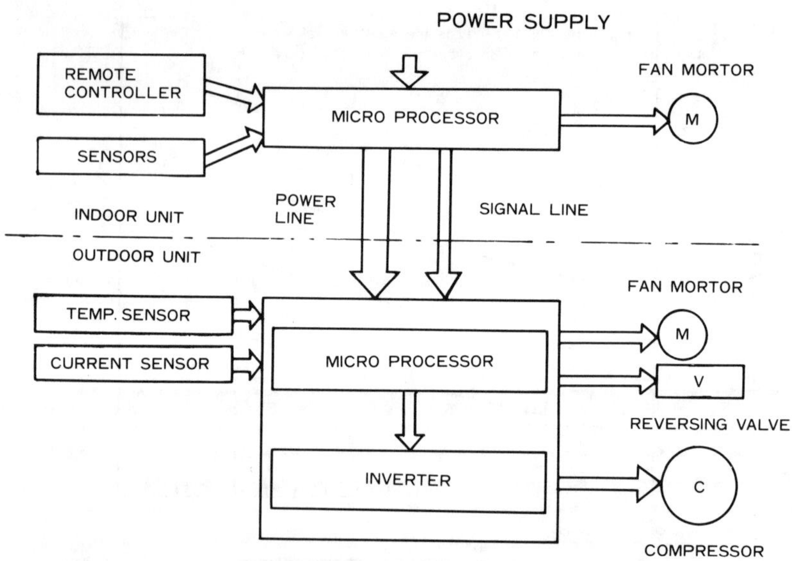

Figure 3. Control system of inverter heat pump room air conditioner.

Several controlling systems have been adopted in order to make the best use of the merits of a variable capacity air conditioner. For designing the control system and for selecting the operation frequency patterns, the following essential factors and points should be taken into consideration.

(1) Comfort (room temperature, air draft, air temperature)
(2) Energy savings (heating capacity and compressor size)
(3) Other designing criteria (total electric current, etc)
(4) Additional new merits whenever required.

Figure 4 shows the operation frequency control pattern under heating mode and the automatic indoor fan control. When starting operation, the room temperature quickly rises to the approximate level as set, since the operation frequency is quite high, as determined by the indoor controller monitoring difference ΔT between room temperature (Tr) and the set temperature (Ts) shown in Figure 4. After the room temperature reaches the set temperature, the frequency is set to the values which give enough heating capacity corresponding to the load. At the automatic indoor fan control mode, the fan speed decreases as $|\Delta T|$ decreases, and after the room temperature reaches the set temperature, air draft and noise are minimized. In addition to the frequency control mentioned above, other main frequency control examples are shown in Table 1.

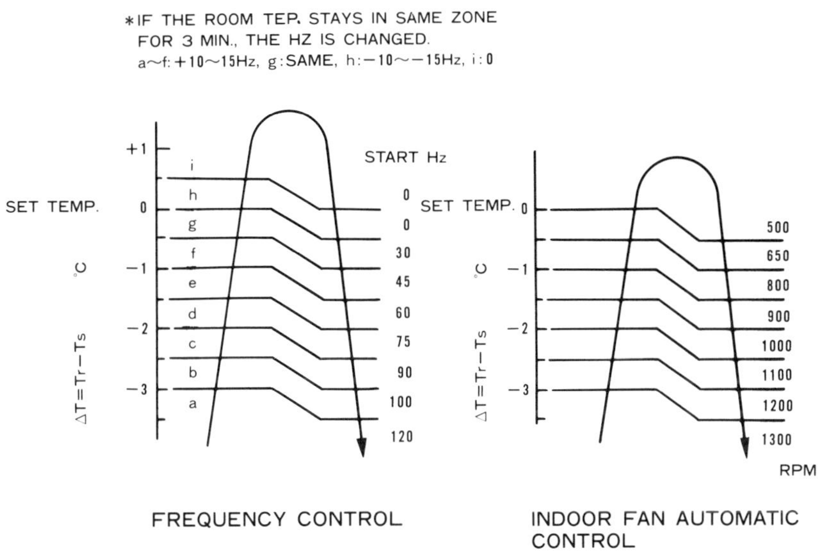

Figure 4. Examples of frequency and indoor fan control.

Non Reverse Defrosting Cycle [2]

The defrosting operation of a heat pump is generally a reverse cycle operation. During the defrosting operation, the room temperature sharply drops because the heating capacity is zero or minus. Furthermore, the inversion noise of the reversing valve is by no means low. Improving these problems, a new non reverse defrosting refrigeration cycle with inverter system began to be utilized. Figure 5 shows an example of non reverse refrigeration

Table 1. Other control examples

Control	Control content
Current release control	The outdoor current sensor detects input control current to the compressor. The compressor speed is controlled to maintain a current of a set value or less.
High temperature release control	When the indoor heat-exchanger temperature detected by the indoor heat-exchanger sensor is a set value or above, the compressor speed is reduced to prevent excessive increase of condensation pressure.
Sleep operation control	When sleep operation mode is selected, the low capacity operation range increases with time according to the set temperature.
Radiation sensor control	The set temperature is adjusted using a correction value according to the difference between the temperature detected by the radiation sensor and room temperature. The radiation sensor detects surrounding surface temperature.

cycle with use of the electronic control expansion valve. When the preset quantity of accumulated frost is detected in the heating operation, the compressor is driven at maximum speed and the opening of the expansion valve is reduced in order to raise the compressor temperature. Then the expansion valve is fully opened together with the bypass two-way valve of the defrosting hot gas and the compressor speed is kept at maximum, which pervents refrigerant stagnation in the indoor coil, and which raises the suction pressure. In this way, the defrosting work is completed quickly using the compressor's high input and the sensible heat of the compressor. The defrosting time is shortend to half the average length previously required , and the drop in the room temperature during the defrosting operation is decreased because the indoor coil is never cooled. In this method, there is no conversion noise of the reversing valve.

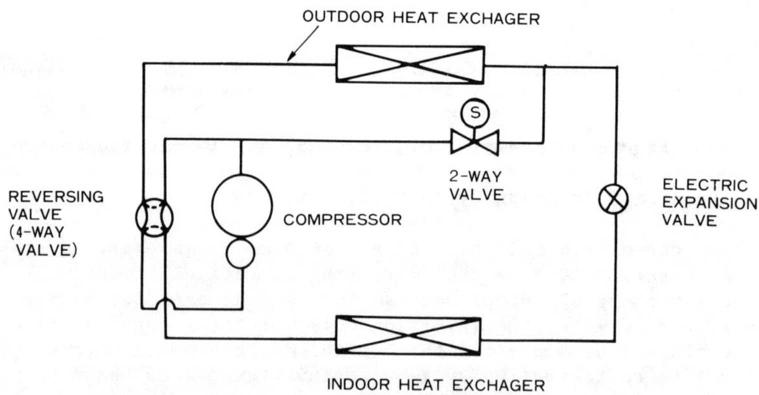

Figure 5 Non-reverse defrosting cycle.

Research in Rotary Compressors [3][4][5]

Rotary compressors of rolling piston type have many advantageous characteristics such as high efficiency, reliability, small size, light weight. The rotary compressors, therefore, are most popularly used in inverter heat pump room air conditioners, and much research is performed in order to improve their performances and to maintain their reliability. The characteristics of the typical capacity control rotary compressor are shown in Figure 6.

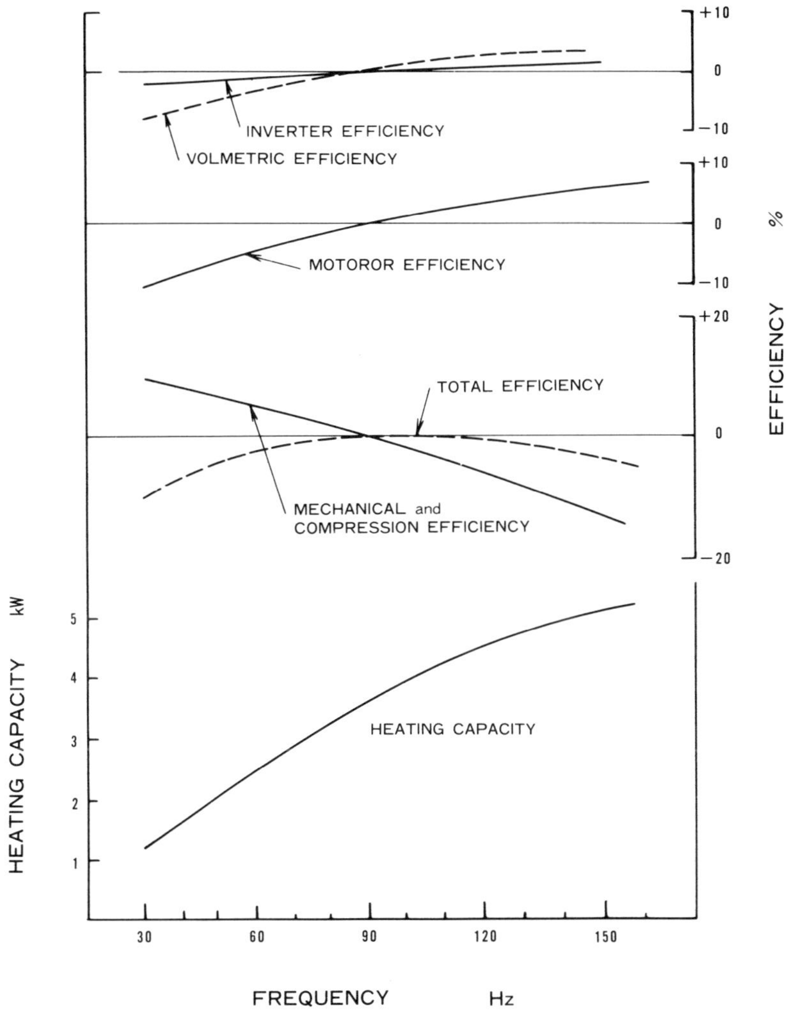

Figure 6. Characteristics of rotary compressor.

The inverter control system is by PWM control of the ratio in relation to voltage and frequency being constant, and this system posseses high efficiency suitable for the inverter motor control system. The efficiency of the motor rises as the frequency goes up. Regarding the inverter efficiency, it rises slightly as the frequency increases.

The volumetric efficiency is determined primarily according to the leakage from the sliding surface. Since the leakage decreases as the rotational speed increases, it will result in the volumetric efficiency rises as the operating frequency increases. The compression efficiency is significantly related to work loss in gas suction and discharge. This work loss is determined by the gas flow area and the piston speed. Therfore, the compressor efficiency goes down as the operating frequency increases. The mechanical efficiency decreases as the operating frequency increases because the friction losses increase as the sliding speed as well as load increases. Total efficiency (COP) is determined from the efficiencies mentioned above, and it has a maximum point for frequencies like a gentle curve of second degree.

The total system efficiency depends not only on the compressor efficiency but also on the whole refrigeration cycle such as heat exchangers, control of an expansion valve, fans and system control. The improvement of compressor performance, however, is a basic technique essential for improving the total system efficiency.

Many compressor companies have made extensive efforts to study and develop a wide variety of innovative and effective compressors. The main subjects of research and development for rotary compressors are design improvement,simulation, performance improvement, noise reduction and component life time improvement. The trends of research work for improving each efficiency of rotary compressor may be clasified as follow.

- Designing component dimensions for reduction of gas leakage and friction losses
- Suction and discharge passage shape for reducing compression losses
- Component materials and their surface treatment for reducing friction losses and improving their reliability

Application of Inverter System

The inverters are used not only for heat pump room air conditioners but also for package (commercial) heat pump air conditioners. Where the heating capacity is large, a large and expensive inverter and compressor are required. If a large compressor is used, its minimum capacity is limited and capacity control range is narrow, since the inverter minimum frequency is limited. In this case, system energy efficiency such as Heating Seasonal Performance Factor(HSPF) will be lower. In order to solve this problem, a two compressor-one inverter system has been developed of which refrigeration cycle is shown in Figure 7. [6]

This system consists of two completely separated refrigeration cycles. Only the fins of the indoor heat exchanger are shared. Therefore, there is no fear that the compressor oil might accumulate in one compressor. Figure 8 explains the function of the capacity control system. One refrigeration cycle is operated with an inverter("inverter cycle") and the other is operated with the commercial power at constant frequency("base cycle"). When a high heat load is required, both cycles operate, while only the inverter

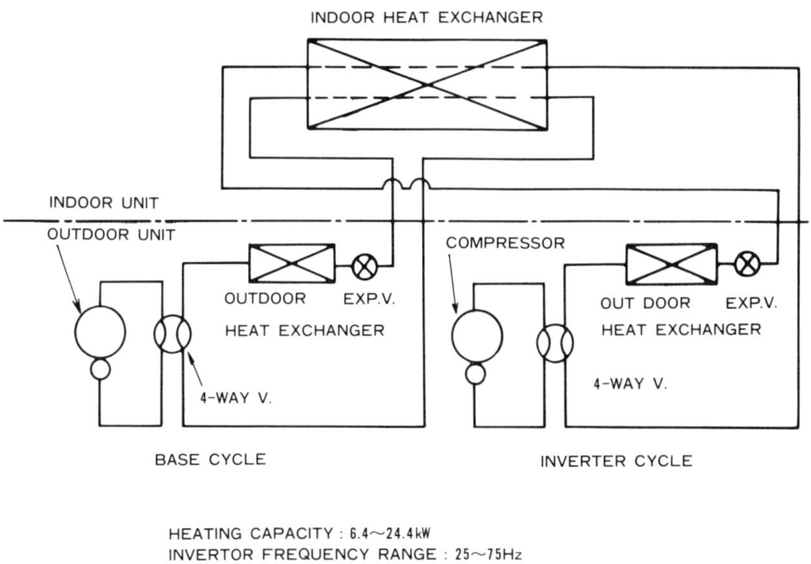

Figure 7. Two-compressors and one inverter refrigeration cycle

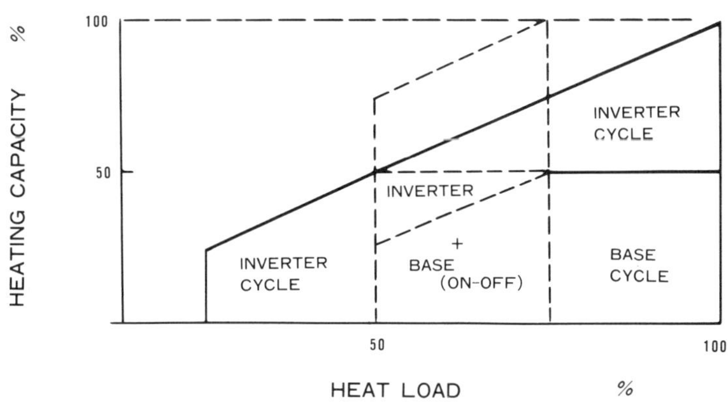

Figure 8. Capacity control of two compressor and one inverter system.

cycle operates at low heat load. For a medium load, the inverter cycle operates all the time, though the base cycle operates intermittently. In this two compressors-one inverter system, the heating capacity range is 100-25 %. If one-large size compressor and one corresponding inverter were to be used, the capacity range might be only 100-50 %. Since this system works effectively for high and low heat loads by using an inverter system, the energy saving will be realized up to about 30 % comparing with a conventional heat pump air conditioner installed in the Tokyo area, Japan.

ECONOMICS OF INVERTER HEAT PUMP ROOM AIR CONDITIONER

The economics of the inveter heat pump room air conditioner should be evaluated from a view point of average annual energy efficiency such as HSPF. However, the calculation method of HSPF for inverter heat pump room air conditioners has not yet agreed upon formally.

Figure 9 shows an example of characteristics of input power of both the inverter and conventional heat pump room air conditioners as a function of outdoor temperature. In this figure, the heating capacity range of a inverter heat pump is 1.45-4.88 KW and the heating capacity of a conventional unit is 4.2KW. The input of inverter heat pump is less than that of a conventional heat pump in the figure because the inverter heat pump is operated continuously, responding to its heat load. On the contrary, the conventional type uses inefficient ON-OFF cycling. Total seasonal input power can be calculated using these input power data and the temperature distribution. The total calculated input power of inverter and conventional heat pumps is 860 KWH and 1150 KWH respectively. The inverter heat pump has achieved about 25% of energy saving comparing with the conventional heat pump. Of course, this energy saving rate will change according to the temperature distribution shapes and heating capacity. When an adequate heating capacity and control system is selected, more than 30% energy saving can be expected.

FUTURE TECHNICAL TRENDS AND POTENTIAL OF CAPACITY CONTROL

The inverter heat pump room air conditioners have already gained their position in the market as mentioned before. Of course, their cost is higher than that of the conventional heat pump. However, users and traders have recognized their merits. At the present time, the heat pump room air conditioners are sold in mild climate areas mainly, and the diffusion rate of heat pumps in cold regions is still low. In the future, one of the important determinants for wider applications of the heat pump room air conditioner in Japan is how promptly and effectively the manufacturers can develop an advanced heat pump which will be accepted in cold climate areas. Recently, some models for cold climate areas have been introduced on a commercial basis. Some models have improved defrosting characteristics and the others have

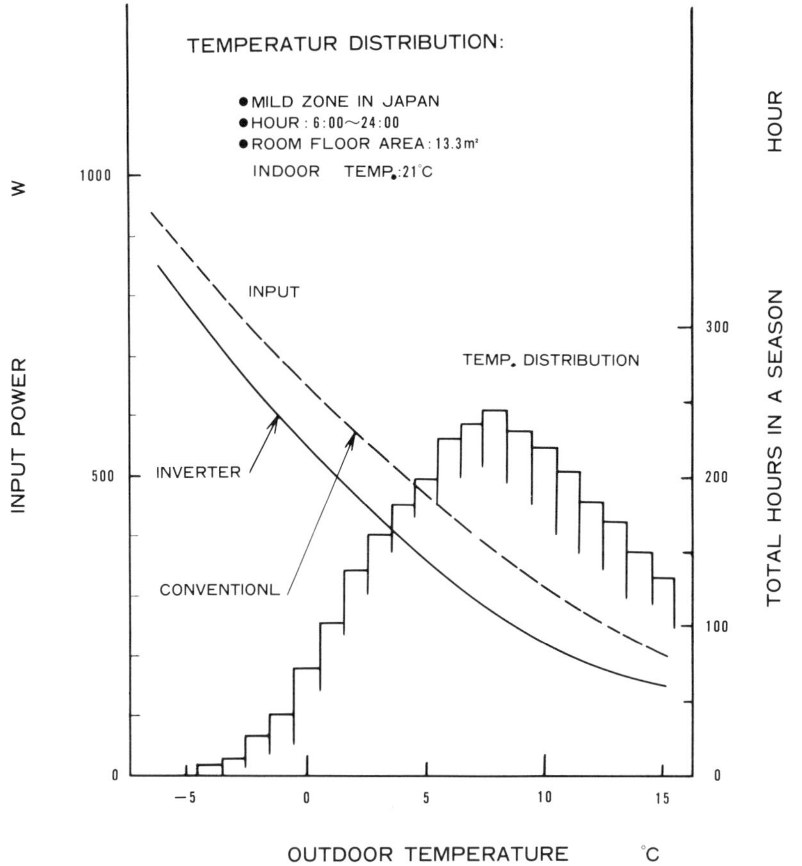

Figure 9. Comparison of inverter and conventional heat pump input and heating season temperature distribution.

improved heating capacity.

Needless to say, variable capacity control itself is not only possible by the inverter system. Although there are many variable capacity control systems, the inverter system is the most relevant considering its functions, cost and reliability at the present time. The inverter systems so far as existing in the Japanese market are applied to air-to-air heat pump air conditioners. Why is the inverter not used yet in the air-to-water heat pump? The answer is simple. This system is not popular in Japan now. The effectiveness of variable capacity control for the air-to-water heat pump is basically the same as that of the-air-to air heat pump. When heat load fluctuates widely, then the capacity control system works more effectively. In the case of the water-to-air heat pumps, they can be used for central heating systems, in which heat load fluctuations in a day are relatively small since their environment is heated almost all the time. On the contrary, the air-to-air heat pump room air conditioners are used as individual heating room-by-room or in segmented areas, and in general, they are operated not much in daytime but in the evening and nighttime, in Japan. Their heat load fluctuations are relatively large. Supposing the variable capacity control is applied to the air-to-water central heat pump system, some individual control method, in which each room temperature can be controlled separately, will be required in order to derive its merits effectively.

Anyway, the variable capacity control will be an essential part for getting energy savings, comfort and producing products with high additional values.

The basic objective in the technical area for application to heat pumps is improvement of COP and HSPF. There are still many aspects to be researched and developed, the main items of which are as follow.

Compressor COP. This is an essential factor to improve the system COP and HSPF. Although rapid progress might not be expected, they will be improved step by step with the items mentioned before.

Heat exchanger(air-to-air). The fin and tube heat exchanger has been improved recently by applying aluminum slit-fins, inner grooved tubes and hydrophilic treated fins. Especially, the hydrophilic treatment is very useful to shorten the defrosting operation time since deiced water falls instantly without staying between fin plates. The defrosting operation time effects the actual system HSPF, and non-frost or frost resistance heat exchangers, are expected.

System control. System control techniques have been improved rapidly by using micro-processors and various sensors, and more effective and lower cost system control techniques are expected.

Heat storage. Recently a new variable capacity control heat pump has been developed, which incorporates an inverter and new refrigeration cycle using heat storage. In this system, the maximum heating capacity at starting the heat pump is larger than that of the conventional inverter heat pump because the heat which is stored during nighttime with low price electricity is utilized. In a cold climate area, a large heating capacity is required to shorten the time to reach the set temperature. As the heat in the heat storage is used during defrost operation, the defrosting time is quite short so that the room temperature drops are less than normal during the defrosting operation. By using this heat storage concept and inverter, the range of heating capacity becames wider than that of the inverter only. This heat storage technique is significant and further research and development in regard to the refrigeration cycles with the heat storage techniques, as well as heat storage devices and its materials, will advance considerably major achievement in the near future.

Compound technology. Various variable capacity control technologies will be necessary for improving the COP and HSPF and heating capacity of heat pump. As an example, a compound system of an inverter, heat storage and capacity control refrigeration such as the gas injection techniques will be considered. The system combining the units of multi indoor fan coil units and one outdoor unit which is integrated with several compound technologies is to be introduced gradually into the market. This kind of system will work very effectively to meet the needs of a wider capacity control range as an alternative to the central system.

CONCLUSION

Variable capacity control is considered to be one of most important technologies for the wider application of heat pumps in the market since it has great potential merits such as energy savings, application for more comfortable environment and additional functions of various kinds.

At the present time, the inverter system can be recommended as the most useful technique for variable capacity control. Judging from the recent trends of advanced micro electronics technology, the inverter application will become more widely accepted and be utilized more easily for various heat pump systems.

The inverter system employed in the air conditioning will, however, be continually improved by futher refinements. Futher research work and subsequent developments are necessary in every direction to attain results, including reducing cost and improving the efficiency of each component and system in order to increase effectiveness. But it can definitely be said that the encouraging efforts to overcome obstacles promises future development.

REFERENCES

1. Y.Shinma et al.,"Inverter Control Systems in the Residential Heat Pump Air Conditioner",paper HI-85-31 No.2 ASHRAE Trans. pp1541-1552,June,1985

2. H.Itoh,"Improvement of a Heat Pump Room Air Conditioner by Use of Pulse-Motor-Driven Expansion Valve" No.2944 ASHRAE Trans. Jan.,1986

3. T.Itami et al.,"An Experimental Study of Frequency-Controlled Compressors," Proceeding of the 1982 Purdue Compressor Technology Conference, July 21-23,1982, pp297-304

4. A.Futakawa,"Improvements in Compressor with Special Emphasis on Interesting Developements in Japan,"Proceeding of the 1982 Purdue Compressor Technology Conference, July 21-23,1982, pp339-351

5. E.Sakurai and J.F.Hamilton,"The Prediction of Frictional Losses in Variable-Speed Rotary Compressors,"Proceeding of the 1982 Purdue Compressor Technology Conference, July 21-23,1982, pp339-351

6. K.Shimizu et al.,"Energy saving of Medium Size Package Air Conditioner," Japanese Association of Refrigeration,Trans. Nov.,1984 pp71-74

Chapter 9

Prospects for Improved Components

Richard G. Maier

ABSTRACT:

This paper reports on the situation of components for heat pumps and heating plants in Europe. Due to decreasing sales figures, interesting results of new developments have not fully been implemented. Examples such as rotary-type compressors, special condensers for hot water preparation, and low noise fans will be described. A considerable number of new developments are not very cost increasing and have led to remarkable improvements, especially in the field of diagnosis and service. Here we will report on sophisticated controls of the heat pump as the whole plant as well.

INTRODUCTION

Improvements of components must always be seen in relation to their environment, where special requirements exist which will influence special developments. Therefore first of all I am going to make some important and indeed essential remarks about the type of heating system which is most common in Western Europe. As can be seen in FIGURE 1 a hydronic central-heating system fired by an oil-furnace is used. Some elements of such a configuration seem to be of minor importance at first sight but in reality have a deciding influence on the effectiveness and reliability of the heating plant.

HEAT PUMPS: Prospects in Heat Pump Technology and Marketing (Proceedings of the 1987 IEA Heat Pump Conference), Edited by Kay H. Zimmerman © 1987 Lewis Publishers, Inc., Chelsea, Michigan 48118. Printed in U.S.A.

1st: the heat itself is produced by an oil- (or gas-) boiler, which not only provides space heating but also hot water ("Hot Water" means domestic or household hot water). Thus, considering a long list of safety factors while calculating the heat demand, the boiler as well is overdesigned at least by a factor of two.

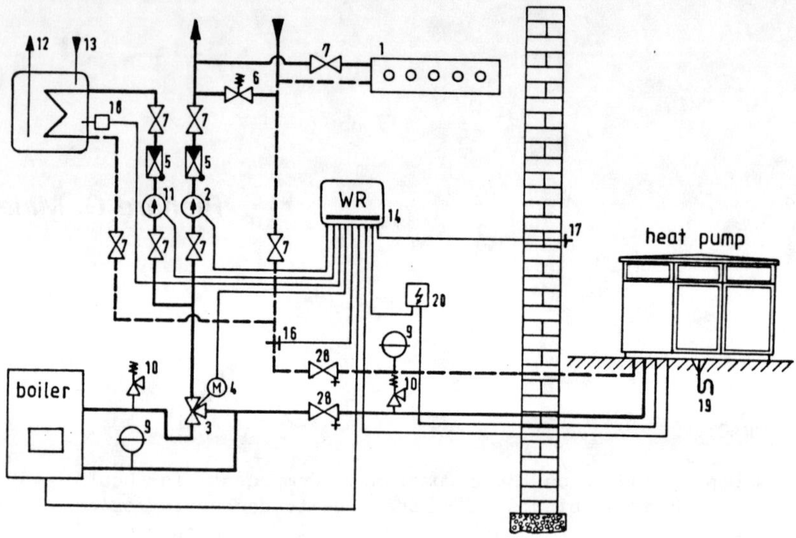

1:Floor heating (example)
2:Heating water circulating pump
3:Three-way mixing valve
7:Hand-operated valve
11:Hot water loading pump
14:Heat pump and plant control
16:Heating water temperature sensor
17:Outdoor temperature sensor
4:Motor of mixing valve
5:Non-return valve
6:Emergency valve
9:Expansion vessel
12:Cold water supply
13:Hot water outlet
18:Hot water thermostat
19:Condensed water discharge

FIGURE 1: Usual heating plant with hot water preparation by boiler and heat pump.

2nd: because the boiler is overdesigned for its job it becomes unnecessary to consider whether the circulating pump is strong enough to ensure a proper mass flow of water through the radiators. In many cases the mass flow is also reduced by thermostatic controlled valves, which are often without bypass holes.

3rd: a need for cooling in the European summer is so seldom, that almost no one is going to install a fan-coil instead of the usual radiator. As a fashion these are becoming flat and wide but with little depth. Here is the reason that in Europe heating water temperature on cold days goes up to 70 to 80 °C (some decades ago, 90 °C was the designing value).

4th: the noise disturbance of the oil-burner is the problem of the house owner, not of the neighbour.

If one tries to place a normal air to water heat pump into such a heating plant, he will fall head over heels into a never ending list of problems, because this ordinary heat pump will fail to serve the demands for the following reasons:

-the heating capacity is well below the heating demand and far below the capacity of the furnace. If there are provisions made that the hot water is also prepared by the heat pump, the user sometimes has to wait longer than he expects until the house is warmed up again - especially when defrosting is necessary.

-the mass flow of the heating water is usually too small to ensure heat transfer from the heat pump to the radiators. High pressure or some other cut-outs will consequently lead to a failure or malfunction of the heat pump

-the normal heat pump has an outlet temperature of + 55 °C if the evaporating temperature is not too low, let us say not less than - 30 °C. In case of coldness of - 10 °C or less, the limits of the temperatures on the high pressure side will be reached, especially in the valves of the compressor. Sometimes even the desired 55 °C cannot be attained, an awful event for the housewife. In any case we do not recommend a heat pump operating for long heating periods under conditions described.

-the fan outside the house forcing some thousand cubic meters air per hour through the evaporator disturbs the neighbor's peace and quiet and not the owner's.

Now certainly the impression occurs that heat pumps are not suited to fit into such heating systems. This is of course only partly true.
On the other hand one can look for improved future components, design and control to match the requirements better. But in no way can we see monovalent air-to-water heat pumps without a backup system equalling the extraordinary high heating capacity of an oil-furnace of some 25 and more kilowatts at a reasonable price.
In the bivalent mode which is superior to all others in West-Germany, the heat pump can be switched off at rather moderate ambient temperatures from - 5 to + 3 °C depending on the regulations of the utilities or personal wishes. Therefore no serious problem will result from too low evaporation temperatures. However, attention should be given to defrosting.

DEFROSTING CONTROL

The following problems have to be solved in an optional manner:
-maximum efficiency between defrosting cycles
-defrosting cycles as few as possible
-defrosting time as short as possible
-drying of evaporator fins obligatory

The first task can be solved by a defined fin distance which should be neither too small nor too wide. Our measurements and field tests have shown that with too narrow fins spacing ice and hoar-frost grows rapidly and make the heat transfer worse and hinders the air flow through the evaporator earlier than wider fin spacing of 2,5 mm and more do. A fin spacing of 5 mm would be very welcome but costs would increase astronomically.

The second task is to reduce the defrosting cycles and is of course related to the first task. The worst, but cheapest method, is to defrost regularely after a certain pre-adjusted time, for instance every hour. This method is not the best one, because often the evaporator will be defrosted while no ice at all is present. On the other hand it can happen with foggy weather that the evaporator is full of ice in a remarkably short time before defrosting is begun again.
There are quite a lot of opinions about when to start defrosting. Many producers take the signal from a low-pressure cut out. A very interesting method is to take the increase of the fan motors resistance as a trigger which is more or less a function of the growth of ice and decrease of air flow (1). We decided after a series of tests to make the actual defrosting time the measure of running time between defrost-cycles (2). The results are encouraging. Compared to conventional methods we were able to reduce the energy consumption per heating period by approximately half the original amount. Typical seasonal operation times are:
compressor running time : 2.1 hours
defrosting time : 3 to 5 minutes

DEFROSTING METHOD

At this point I would like to make some comments about the defrosting method itself. In Western Europe with almost no exception two methods are common: "hot gas" defrosting
"reverse cycle" defrosting

We have made experiments with both methods and additionally with an improved 'hot gas' method (3). Our results are measured at the following conditions: incoming air: 7 °C/ water outlet: 35 °C
evaporation: - 6 °C/ condensation: 40 °C

As a measure of the "effectiveness" we consider the fraction of the relative time of defrosting, which is defined as the defrost time divided by the sum of defrost and heating times. This value is limited by our regulation according to DIN 8900 to a maximum of 15 %.

Equipment	Absolute defrost time (min)	Effective-ness (%)
Hot gas	10.5	0,3
Improved hot gas	7.2	0,5
Reverse cycle	3.5	1,0

These results allow us to hold to the reverse-cycle. But care must be taken in order not to destroy the condenser (which is the evaporator in the defrost mode). To avoid this, we have incorporated independent elements of safety:

- temperature control of the heating water
- defrost security program (patents pending)
- mass flow control of heating water by separate safety guard
- temperature control of evaporator body.

We admit that in exceptional cases all these safety provisions were not sufficient, mainly when the installer or the user did not take pains enough to follow exact instructions.

MASS FLOW OF HEATING WATER

We have found that the mass flow of the heating water very often is small- even too small. To avoid high pressure cut outs, in modern plants by-pass-valves together with a storage tank are recommended. They insure enough water flow and enough storage capacity to keep the heat pump running a definite and necessary period of around 6 to 10 min. In addition this equipment provides the heat pump with a sufficient volume of hot water during defrosting. In case of malfunction a microprocessor control (which is the usual control for all heat pumps) signals that insufficient water flow has come through. In this case, the user should call the installer for help.

HOT WATER HEAT EXCHANGERS

Many problems still arise from the heat exchangers in the hot water boilers, because no circulating pump is working. All the heat must be transferred by thermal convection in the laminar region with its low heat transfer coefficients. Especially in summer, when the ambient temperature is high, the heating capacity of the heat pump is at a maximum but this cannot be transferred to the hot water. The heat pump control will signal high presure fault, the user will have a refreshing cold shower instead of a comfortable warm one. To overcome this problem we have developed (patents pending) a special design of heat exchanger which has a 3-tube condenser within the heat pump. FIGURE 2 shows how the condensating refrigerant first heats up the heating water which serves only as a heat transfer medium to the hot water tube, which is the inner one. Of course, provisions must be taken to avoid separation of minerals such as chalk on the walls of the inner tube. This type of condenser is widely used in small heat pumps which are run from a simple 220 V single phase plug in socket.

reaction: $Ca^{++} + 2(HCO_3) \rightarrow CaCO_3\downarrow + H_2O + CO_2\uparrow$

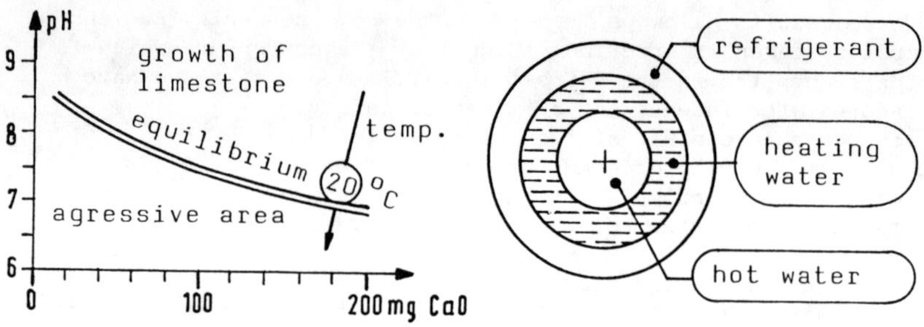

FIGURE 2: Hot water preparation without mineral layer forming

ELECTRICAL INSTALLATION

At this point some remarks about the electrical installation habits in West-Germany may be of interest. From several reasons which are more or less pertinent we have a wide variety of regulations of many utilities to meet. The strictest are:
- -three phase 380 V alternating current required
- -extra electricity meter
- -sometimes new switchboards
- -very strict limitation of starting current
- -not more than three starts per hour
- -electricity switch-off, up to three times a day, for up to two hours
- -sometimes special tariff meter
- -temperature limit of operation, for instance greater than + 3 °C

To overcome most of these regulations we developed a small 220 V single phase heat pump with a nominal power input of 1.4 kW designed as an indoor unit for preparation of hot water and some space heating as well. This heat pump is a genuine German heat

pump with full defrosting and control facilities. There can be no doubt that matching most of the above listed requirements led to extraordinary high costs and to a remarkably high price of course.
In case of heat pumps beyond 1.4 kW nominal power input there is no release from the above mentioned regulations. For these cases we developed microprocessor controls which trace defects, gives us supervising and indicating signals to control operation and to assist service. Only one example should show in TABLE 1 how we achieve a supervisory diagnosis of a heating plant with almost no extra costs. It will be most convenient to the service personal to have in addition to this control task a signalising display which indicates function or faults or malfunction without a compiling procedure. We are sure that this way will be in general the next step to a better control of the heating plant.

TABLE 1: Diagnosis of operating mode and of malfunctions of brine-to-water heat pump in a heating plant with hot water preparation

PULSE Code	Green lamp indicates	Red lamp indicates
1	Minimum stand-by time, Minimum cycle number	High pressure fault of of hot water circuit
2	Minimum interruption time, utilities I/II Main power supply breakdown	High pressure fault of heating water circuit
3	High pressure diagnosis, hot water circuit	Low pressure cut out
4	High pressure diagnosis heating water circuit, 10 minutes limit	Malfunction of brine flow
5	dito, but 2 hours	Defect of starting current device
6	Low application limit of brine	Defect of brine circulating pump
7	High pressure refrigerant cut out	
8	Freezing protection, evaporator	
9	Heating water limitation	
10	No heating demand	
11	Boiler in heating mode	

In our opinion we will have in the near future better sensors for pressure and temperature to control function much better than with a great number of discrete On/Off-switches as indicated in FIGURE 3 which shows the usual heat pump refrigeration cycle.

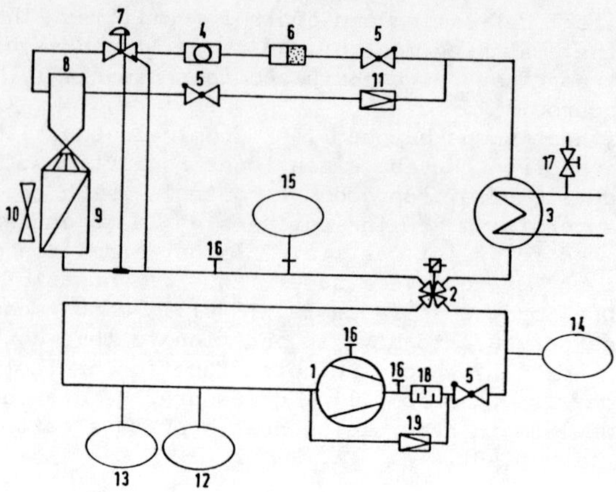

1:Compressor
2:2/4-way reversing valve
7:Expansion valve (thermostatic)
12:Low pressure cut out
13:Defrosting start switch
14:High pressure cut out
15:Defrosting end switch
17:Degassing valve (heating water)
19:Pressure relief capillary

3:Condenser
4:Sight glass
5:Non-return valve
6:Filter dryer
8:Refrigerant distributor
9:Evaporator
10:Fan
11:By-pass capillary tube
16:Service valve
18:Muffler

FIGURE 3: Typical refrigerant cycle of heat pump with control devices

NOISE OF HEAT PUMP FANS

Up until now, we have dealt mostly with factors independent of the place where the heat pump is to be installed, but some comments are necessary about the noise caused, while moving enormous volumes of air through the evaporator. More from the Japanese- than from the US-producers we have learnt, that the noise is not annoying you but your neighbour. The Japanese Air-conditioners and heat pumps offered on the European market in 1980 and 1981 were rather quiet so we had to learn how to make quieter fans. We tried to improve the noise levels by applying the evolutions theory which is nature's favourite method. FIGURE 4 shows in an astonishing manner how nature first produces a rather complicated, even funny shape of the fan blade before it comes to a fairly simple shape, which is approximately 2...3 decibels (A) less than the original fan was.

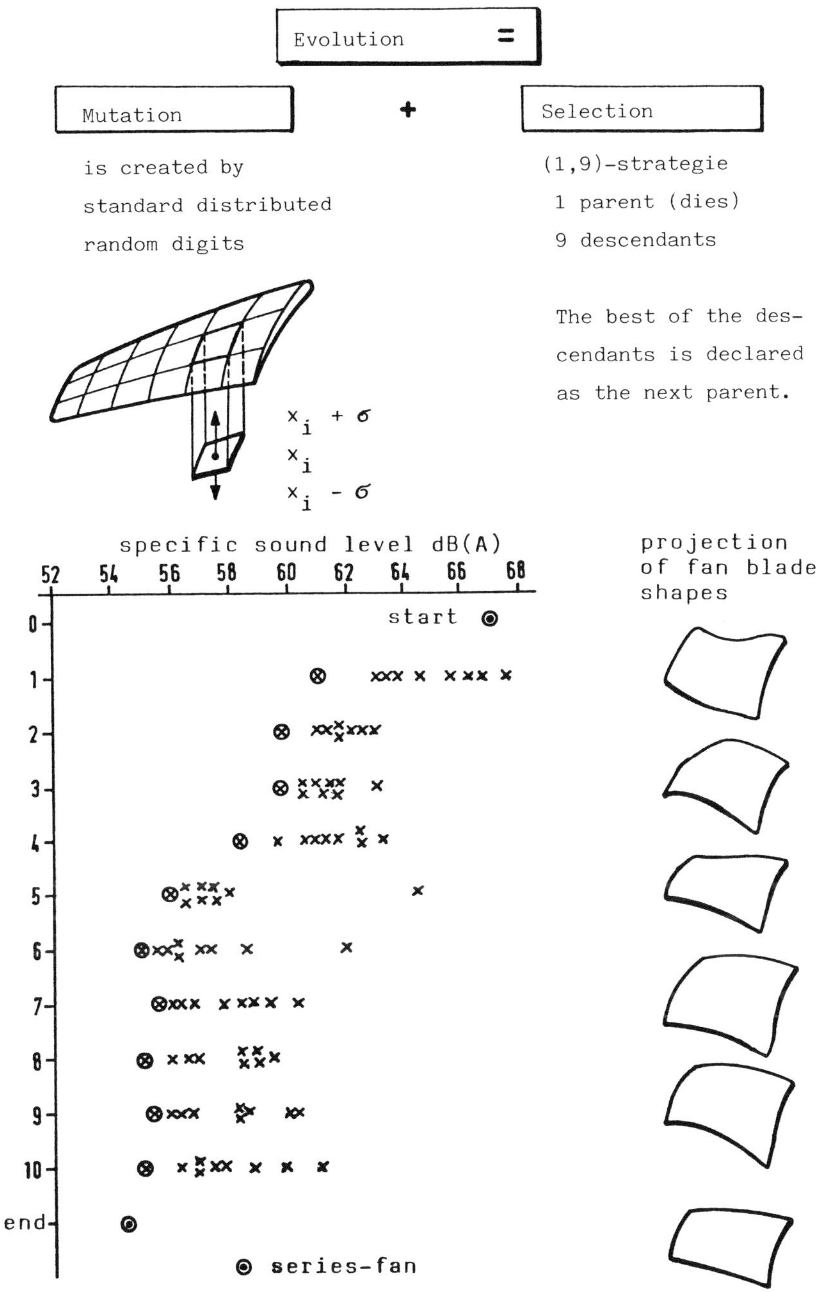

FIGURE 4: improvement of fan noise by evolution's strategy

One very important possible development of the refrigerant cycle should be mentioned: the variation of output.

VARIATION OF HEATING CAPACITY

Especially with air to water heat pumps the heating capacity decreases when the heating demand increases. Consequently it is necessary to control heating capacity according to the demand. In the first approach (which is by far the cheapest) it is a simple on/off control. Depending on the heating system and the climatic region/area one can get:
- a seasonal performance factor (SPF) of about 2.5,
- a seasonal running time of about 2000 hours
- an average running time of about 1 hour per "on"-switching

One aspect has most interest in a more continuous control:
- an increase of the SPF

which comes from less losses as in the on/off-mode and the overdesigned heat exchangers in the less than nominal operation mode. The reduction of dynamic heat transfer losses is according to (5) and to our own measurements neglibible as long as the operating time of one cycle is more than about ca. 10 min. This - on the other hand - is guaranteed by the limitation of three cycles per hour due to utility wishes. We have calculated and confirmed by extensive measurements in some heating plants that a maximum increase of 20 % of the SPF can be achieved if a continuous variable control is used compared to a simple on/off control.
The first to hand solution from a cost/profit point of view is to make use to two compressors in a heat pump unit or of a 2/4-pole motor. With such a two stage control we can approach the theoretical value quite close, maybe about 70 to 80 %. The most interesting features are hot water preparation and defrosting with half the capacity of the normal mode.

A still better SPF can be reached by a continuous control with:
-non-aceotropic mixtures of refrigerants or
-frequency control of compressor speed.

With non-aceotropic mixtures and with 2 or 3 storage tanks one can cover a capacity range between 20 and 100 % depending on the relation of vapour pressure as shown in FIGURE 5. As we also can see from this figure, the costs of heat pumps with such refrigerant mixtures will at any rate be higher than with homogenous refrigerants. There are still unresolved problems mainly with lubrication, phase-separation and control of the process.

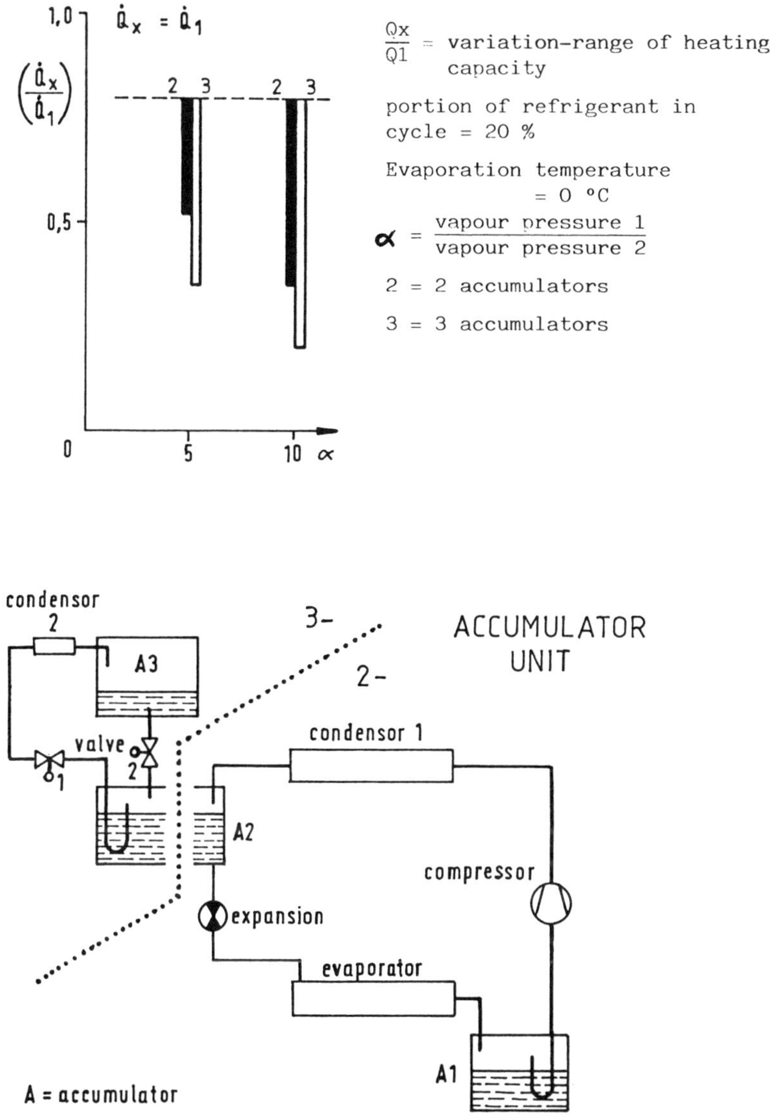

Figure 5: variation-range of heat pump output and block-diagram using non-aceotropic mixtures

Easier but within a limited range of capacity control seems to be the use of a frequency inverter. TABLE 2 indicates the most important features of such a device.

TABLE 2: Advantages and Disadvantages of frequency inverters combined with air-to- water heat pumps.

ADVANTAGES	DISADVANTAGES
Starting current not exeeding nominal current	Material portion about twice as much as for ordinary heat pumps
Starting at high pressure possible	High extra costs for good radio frequency protection
Overcurrent protection switch unnecessary	Must not run with reciprocating compressors below a limited minimum speed
Frequency range from Zero to 50 or 70 Hz	Behaviour of long time low speed operation difficult to judge
Starting with minimal speed is is a plus because of slowly degassing of refrigerant	Utilization of full range of control not possible (vibration, lubrication)
fan can be controlled in the same manner as compressor	Long running time up to 6000 hours per year
At single phase alternating current capacitors for starting and operation can be omitted	Little saving of operation costs, bad cost/profit-ratio
Cycling decisively reduced	
Increase of SPF up to 20 % against simple On/Off control	

Assuming the technical problems can be solved remains an additional amount of about 5000,-- DM to the end-user which cannot be payed back by additional savings in a reasonable period of time. It is said that Japanese makes of air conditioners already include such an inverter. As we can analyze at the moment, this has two important reasons: According to FIGURE 6, compressors with air-conditioning equipment have to operate within the very limited area "A/C" in the High Back Pressure (HBP)-field. This enables the rotary-(and scroll) compressors to be used which covers the operating range easily.

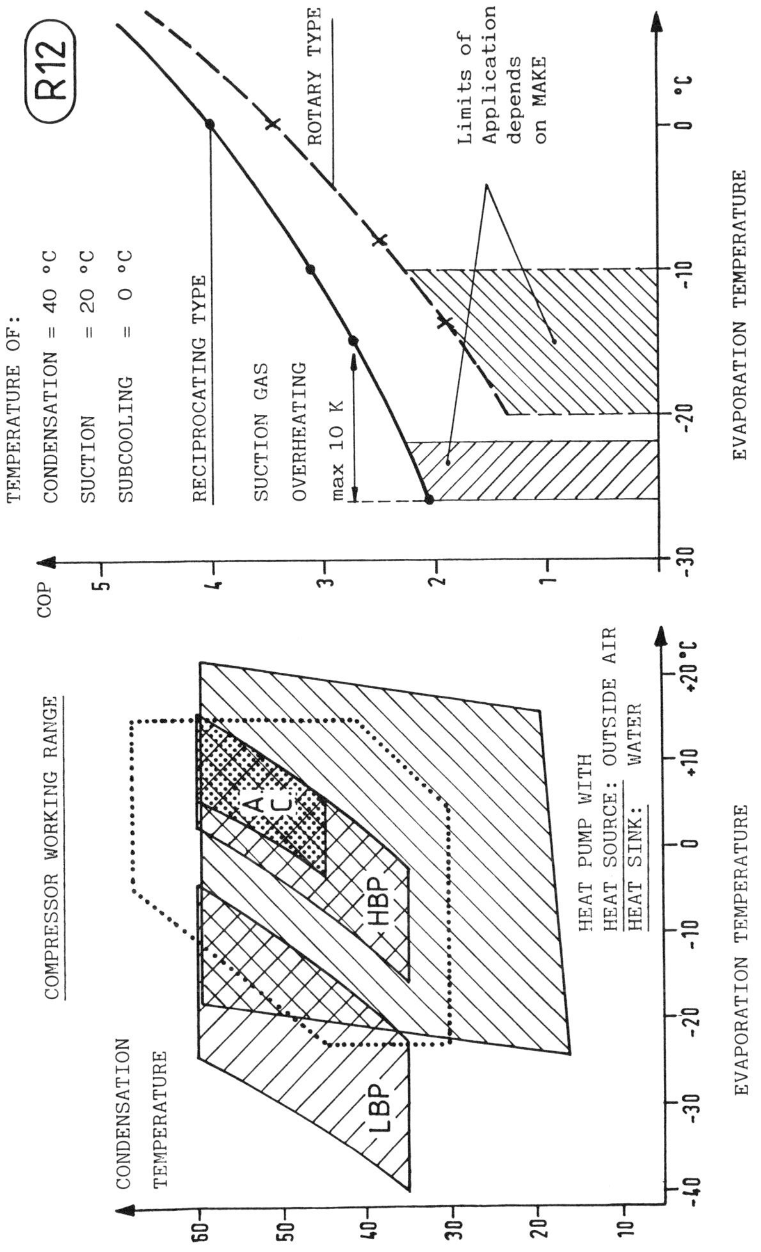

FIGURE 6: Compressor working range and comparison of reciprocating and rotary type

Because of the superior behaviour of the rotary compressors in case of speed variation it is an unavoidable step of further development to combine both techniques. We can see from FIGURE 6 that air to water heat pumps must run in a much wider field of pressure values.
In our SIEMENS R+D department we studied the behaviour of rotary and reciprocating type compressors. FIGURE 6 shows also a comparison between common types. There is not much difference regarding the COP of both makes, but as you can see, the rotary type is better at small pressure differences. A West-German heat pump runs most of the time at high pressure differences either to heat the house at low ambient temperatures or to prepare hot water. In this case (which covers the predominant portion of our heating season) the reciprocal type is superior. A better behaviour of the rotary type compressor may have been achievedin the meantime, but on the other hand the heat pump turnover in Europe is at such a low level that intensive research in this field has been reduced.

SUMMARY

After defining the requirements of the European market by describing an ordinary heating plant fired by an oil-furnace, we tried to show that some improvements have been implemented within the last years. We mentioned:

- a rather sophisticated control of the heat pump and the heating plant
- starting current limitation
- special developments as heat exchangers, evaporators, fans and condensers
- possibility of rotary and frequency controlled compressors

The great problem of cost/profit -ratio does not encourage R+D into heat pumps at the moment. The European market wants heat pumps with a reasonable efficiency giving good seasonal performance factors. It wants high reliability under running conditions which are within the allowed limits. Most of all it wants low prices.

(1) J. GROFF, Carrier Corporation
private information

(2) H. Kühnlein, Proceedings of IIF-Meeting
3. - 7. Sept. 1984, Vol 1, p. 153

(3) J. Reichelt, Proceedings of IIF-Meeting
5. - 8. Aug. 1986, West Lafayette,
Indiana, USA

(4) K.D. Müller, A. Rimkus, SIEMENS AG
unpublished results

(5) H. Kruse, B. Upmeier, see (2), p. 239

(6) H. Gutbier, SIEMENS AG
unpublished results

Chapter 10

The Latest Compressor Technologies for Heat Pumps in Japan

Katsumi Matsubara, Kazutaka Suefuji, and Hiroaki Kuno

ABSTRACT

To meet the demands of the air conditioning market, two types of rotary displacement compressors have been developed. One of them is the hermetic type scroll compressor with a new operating mechanism and the other is the semi-hermetic type screw compressor with a new rotor profile. With volumetric and adiabatic efficiencies superior to the reciprocating compressors, they can meet heat pump requirements and their simple mechanism and high reliability are applicable to speed controlled compressors such as those with inverters. They are expected to be applied more widely in the air conditioning market in the future.

HEAT PUMPS: Prospects in Heat Pump Technology and Marketing (Proceedings of the 1987 IEA Heat Pump Conference), Edited by Kay H. Zimmerman

1 INTRODUCTION

Since the oil crisis in 1973, energy and resource conservation has been one of the most important subjects in Japan, as in many other countries. Much effort has been exerted to develop new technologies for energy conservation and utilization of alternative energy sources.

As air conditioners consume a significant amount of energy, the Japanese goverment enacted a new law in 1979 that required the EER of air conditioners to be higher than a specified value.

On the other hand, consumers have demanded more efficient air conditioners and alternatives to heating oil.

As a result, heat pump air conditoners have been emphasized because of their dual cooling and heating functions. Also, their energy saving performance during heating is superior to that of oil burning equipment.

The compressor accounts for about 90 % of the air conditioner's electric power consumption and improvements in compressor efficiency have a remarkable effect on the EER of the air conditioner. Therfore, many efforts have been concentrated on developing a new compressor with high efficiency, small size and light weight.

In recent years, conventional reciprocating compressors in Japan have been replaced by many types of rotary displacement compressors, because the rotary compressors better meet the above demands. The types of compressors available in the Japanese air conditioning market are shown in Fig. 1.

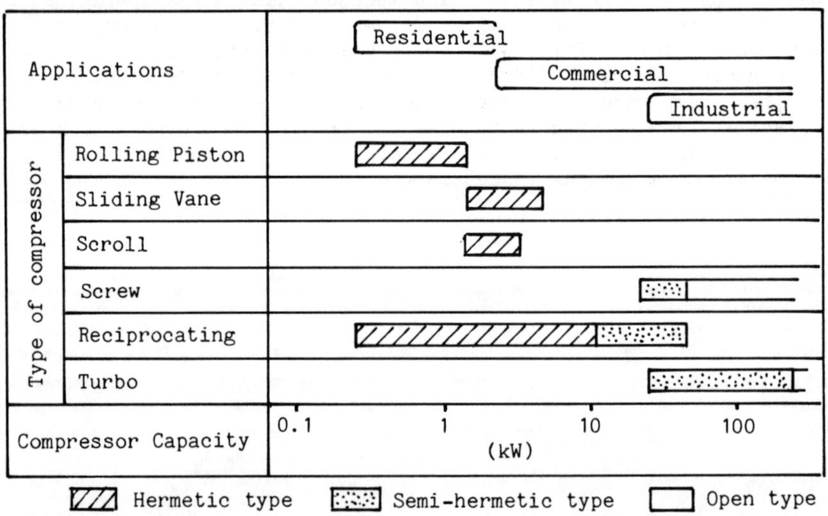

Fig. 1. Compressors that are available in the Japanese air conditioning market.

2 COMPRESSOR CHRACTERISTICS REQUIRED FOR HEAT PUMPS

2.1 Operating Pressure Range

Japanese heat pumps have been developed as cooling/heating air conditioners, because Japan is an island country which ranges widely from the sub-tropical zone at 27°N to the frigid zone at 45°N, and 85 % of the population lives in regions with less than 2,500 heating degree days. As for the heat source, air sourced air coditioners have become the main type used due to their easy installation.

As a result, the compressors mounted in the heat pumps are required to operate in wide pressure ranges, as shown in Fig. 2.

2.2 Capacity Control

The relationship between heat pump capacity, cooling/heating load and ambient temperature is shown in Fig. 3. The load varies widely and the heat pump must adjust the capacity to the load.

There are several methods of capacity control, such as a compressor on/off system, a mechanical unloader compressor, a multi-compressor system and a compressor speed control system. Among these methods the compressor speed control system with an inverter is ideal for heat pumps because it enables continuous changes in capacity. The capacity changes of a speed controlled compressor which varies from 0.5 to 1.5 times nominal speed are shown by the dark lines in Fig. 3. Capacity changes of the on/off controlled compressor are indicated by the dotted lines.

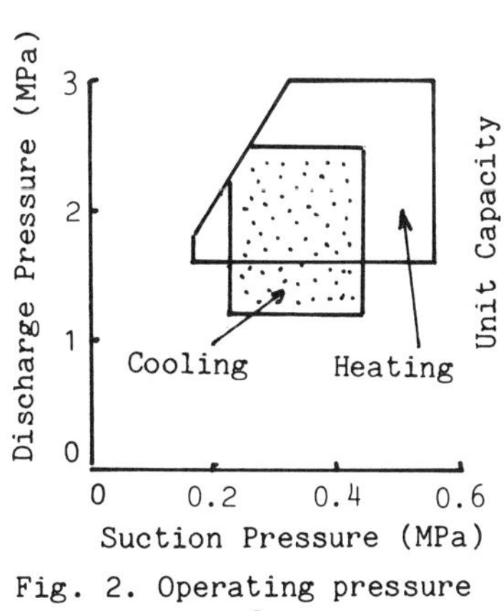

Fig. 2. Operating pressure ranges of compressors. (Refrigerant; R-22)

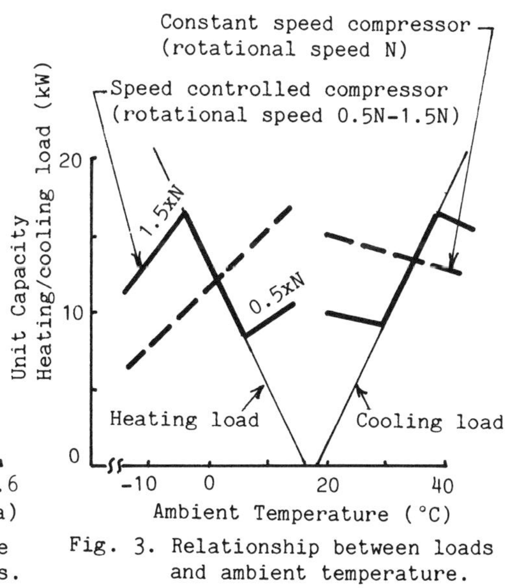

Fig. 3. Relationship between loads and ambient temperature. (3.7kW Heat Pump)

2.3 High Efficiency

The simulation results of operating modes in a speed controlled compressor and an on/off controlled compressor which are mounted in packaged type air conditioners are shown in Fig. 4. Real circles show operating hours of the speed controlled compressor and dotted circles show those of the constant speed compressor.

The results indicate that;

(1) The speed controlled compressor operates mostly in a low pressure ratio and low rotational speed ranges. Therfore, improvement of adiabatic efficiency in the low pressure ratio and low rotational speed ranges is important for energy conservation.

(2) The improvement of volumetric efficiency in the high pressure ratio and high rotational speed ranges is required to deliver high level heating capacity.

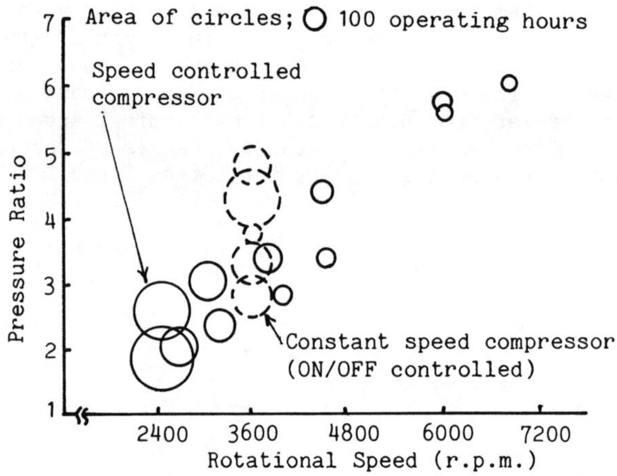

Fig. 4. Annual operating hours of compressors

3 SCROLL COMPRESSORS

3.1 Scroll Compressor Principle

The idea of the scroll compressor is not a new one. It was first patented in the United States in 1905, but not fully developed for practical application because of the lack of precise production techniques and a good mechanism to support the orbiting scroll. These problems have been solved and scroll compressors have been used in heat pump air conditioners since 1983.

The scroll compressor gas compression process is shown in Fig. 5. The main scroll compressor components are a fixed scroll and an orbiting scroll which have identical involute spiral walls. They are assembled with a 180 phase-difference and the centers are seperated by the distance of the orbit radius to form a series of working chambers. The orbiting scroll is coupled with a crankshaft and an anti-rotational coupling such as an Oldham ring. The orbiting scroll is driven by the crankshaft and orbits around the center of the fixed scroll at a fixed angular orientation. The contact points between the two spirals shift along the spirals towards the center. As the working chambers are gradually reduced in volume, a gas is compressed. The compressed gas is exhausted through a discharge port at the center of the fixed scroll.

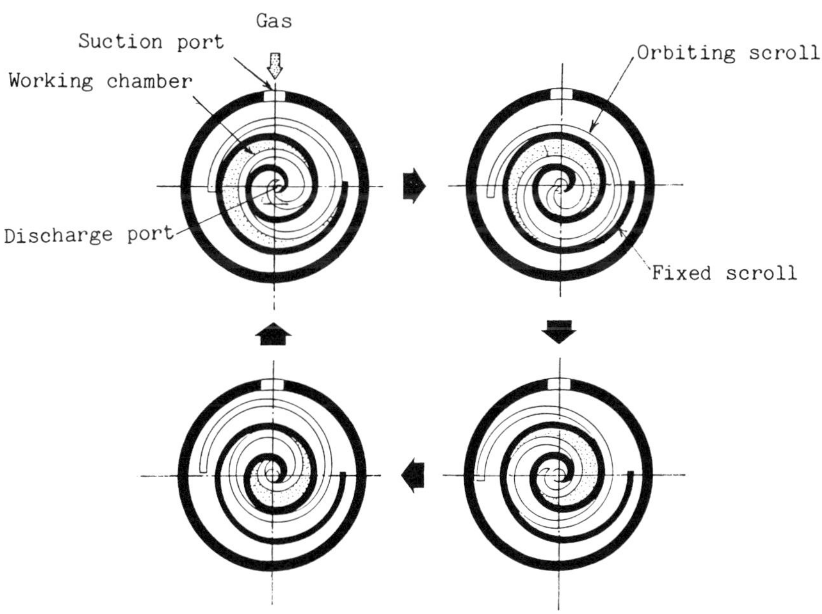

Fig. 5 Principle of scroll compressor.

3.2 Scroll Compressor Applications for Air Conditioners

3.2.1 Hermetic type scroll compressor

A cutaway view of a hermetic type scroll compressor developed for residential and commercial air conditioners is shown in Fig. 6. The scroll assembly is housed in the top of a cylindrical hermetic casing. The two-pole refrigerating compressor drive motor is located below the scroll assembly. The fixed scroll and the motor staitor are mounted on a frame member which is attached to the inner wall of the hermetic casing. The Oldham coupling fits between the back of the orbiting scroll plate and the frame member.

3.2.2 Advantages of scroll compressors

(1) High efficiency

The efficiency of a scroll compressor is compared with that of a conventional reciprocating compressor in Fig. 7. Great improvement of volumetric and adiabatic efficiency is achieved over a wide range of operating pressure ratios. The efficiency improvement in the high pressure ratio range is especially suited for heat pump air conditioners.

The higher efficiency is due to the following;

(a) Refrigerant gas flows directly into the suction chamber, which avoids an addition of heat to the suction gas. There is no process analogous to the re-expansion in the reciprocating compressors. The pressure difference between the two neighbouring working chambers is much smaller than the total pressure difference of the discharge and suction pressures. Thus, very small gas leakage loss and high efficiency are achieved.

(b) All points on the orbiting scroll move in synchronization with the small radius. This results in low friction loss due to the low rubbing speed.

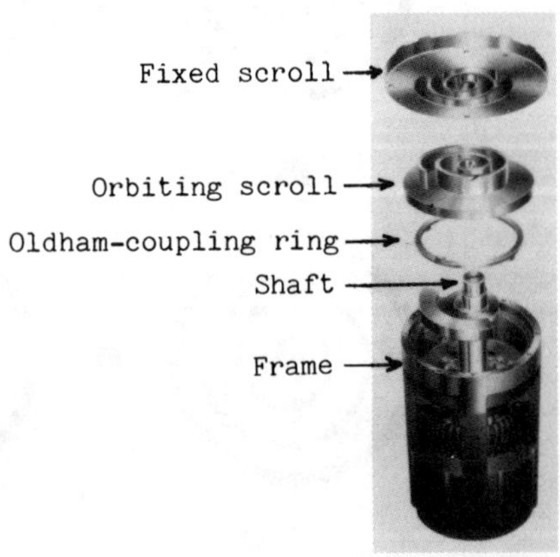

Fig. 6 Structure of hermetic type scroll compressor.

(2) Low vibration / noise level
A scroll compressor requires no valves and discharge gas flow is relatively continuous. The compression process is performed slowly and two to three compression processes go on simultaneously. This results in smooth operation and little fluctuation in shaft driving torque. Driving torque fluctuation in a scroll compressor is about a tenth of that in a reciprocating compressor. The smooth operation makes it possible to drive a scroll compressor at higher rotational speeds, and it is applicable to speed controlled compressors such as those with inverters. Scroll compressor characteristics at different rotational speeds are compared with those of the reciprocating compressor in Fig. 8..

(3) High reliability
The simple mechanism, few moving parts, low vibration and low rubbing speed, coupled with the fact that no valves are required give the scroll compressor high reliability.

(4) Small size and light weight
The outer dimensions and weight of the hermetic type scroll compressor are reduced by 40 % in volume and 15 % in weight compared with the conventional reciprocating compressor.

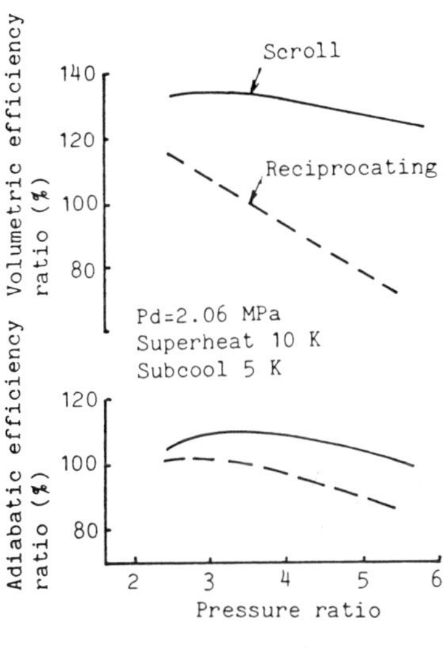

Fig. 7 Compressor characteristics for pressure ratio.

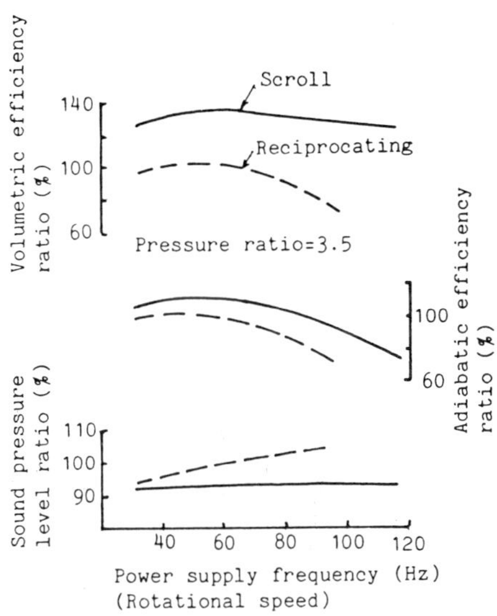

Fig. 8 Compressor characteristics at different rotational speeds.

* Data of the reciprocating compressor at pressure ratio 3.5 and frequency 60 Hz are taken as 100 %.

4 SCREW COMPRESSORS

4.1 Semi-hermetic Type Screw Compressors

Screw compressors have been used in the refrigerating and air conditioning markets for more than two decades, but their acceptance in air conditioning equipment has been limited to heavy commercial and industrial applications. They were open type compressors, and it was more difficult to make small capacity screw compressors competitive with conventional reciprocating compressors, both in efficiency and manufacturing costs.

However, more effective and smaller capacity semi-hermetic type screw compressors that are acceptable for light commercial apprications have been developed and mounted in heat pump air conditioners since 1981.

A cutaway view of a semi-hermetic type screw compressor is shown in Fig. 9. The compressor essentially consists of a low pressure portion and a high pressure portion. The former is made up of a cast iron casing contaning a pair of rotors and a drive motor, while the latter consists of a discharge chamber containing a separator element for oil separation. A slide valve mechanism is provided for capacity control and is controlled hydraulically by an unloader piston. No oil pump is required for lubrication or unloader piston actuation, because oil is distributed throughout the compressor by the pressure difference between the suction and discharge portions.

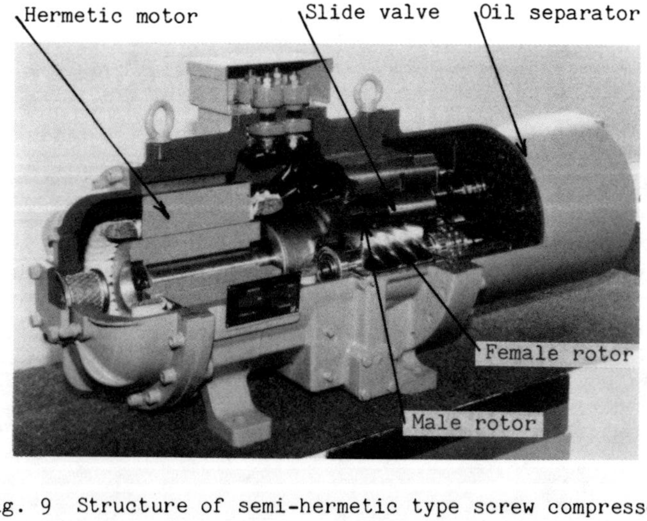

Fig. 9 Structure of semi-hermetic type screw compressor.

4.2 New Rotor Profile

The efficiency of screw compressors is determined by internal gas leakage. Compressed gas leaks through internal clearances between the rotors themselves and between the rotors and the casing, and thus, the rotor profile is the most influential factor in achieving higher efficiency. Many efforts have succeeded in developing the new rotor profile shown in Fig. 10.
Higher efficiency is achieved because of;

(1) The five male to six female lobe combination makes the pressure difference between the two neighbouring working chambers smaller and reduces gas leakage.
(2) The gas leakage passages of the new rotor profile are reduced by 19 % in the blow hole area and 22 % in seal line length compared with the conventional rotor profile.
(3) The new rotor profile is suitable for manufacturing and makes precise cutting possible.

The performance of the newly developed semi-hermetic type screw compressors with new profile rotors has been greatly improved, as shown in Figs. 11 and 12. A major part of power loss in the conventional compressor is due to internal leakage, and leakage loss is greatly reduced in the new compressor. The improvement of adiabatic efficiency is greater in the higher pressure ratio ranges.

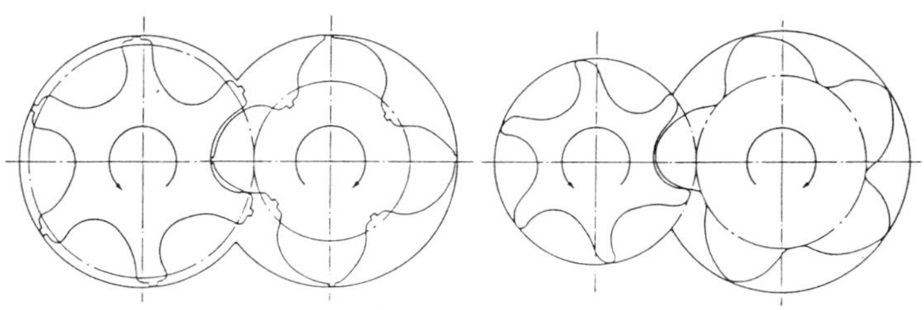

Fig. 10 Rotor profile

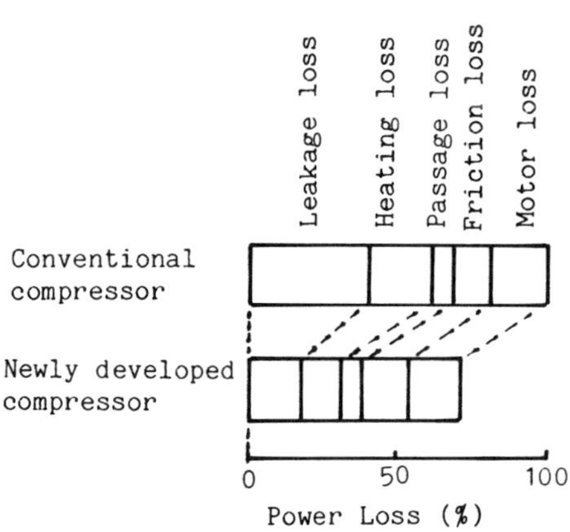

Fig. 11. Comparison of power loss.

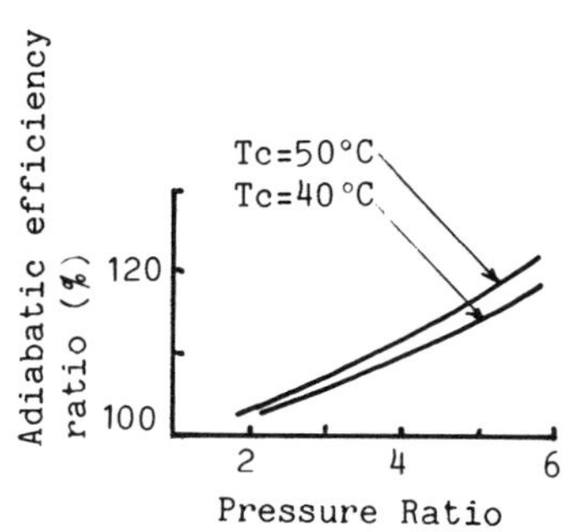

Fig. 12. Characteristic of screw compressor for pressure ratios

5 HEAT PUMPS WITH SPEED CONTROLLED COMPRESSOR

5.1 Speed Control by Inverter

Recent developments of commercial electric equipment have made it possible to use speed controlled compressors with inverters for air conditioners. The inverters have been utilized in room air conditioners since 1982 and in packaged air conditioners since 1983 in Japan. At present, inverter utilization is limited to residential and light commercial air conditioners having capacity under 7.5 kW because of power transistor cost. However, rapid improvements in electric technologies will make it possible to expand the use of speed controlled compressors up to larger capacity air conditioners in the near future. Changes in maximum power supply frequency are shown in Fig. 13.

5.2 Heat Pump Performance Improvement.

5.2.1 Rapid start up

As conventional heat pump air conditioners could not deliver the necessary heating capacity at a start up and at low ambient temperatures, they had to include electric heaters to make up for the deficient heating capacity. Heat pumps with speed controlled compressors can temporarily improve the heating capacity and will lessen the consumer's dissatisfaction. Changes in room temperature at start up are shown in Fig. 14. The heat pump with a compressor operating at 1.5 times the nominal operating speed can shorten start up time to half that of the unit with a constant speed compressor.

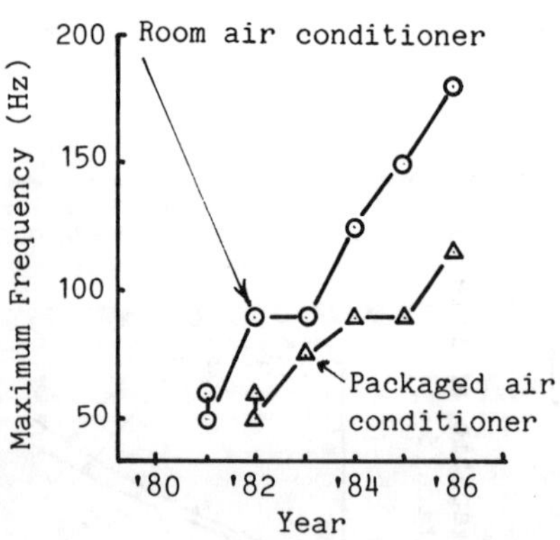

Fig. 13. Development of the maximum power supply frequency.

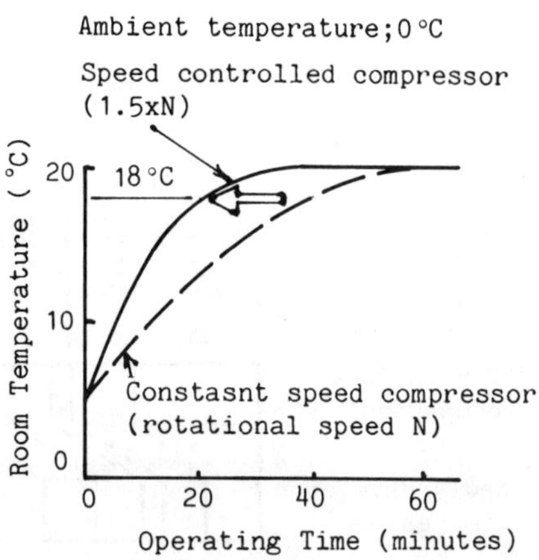

Fig. 14. Heat pump performance of start up.

5.2.2 Defrosting

Conventional heat pumps exchange a heating operation with a cooling operation in defrosting, and this results in a reduction of room temperature. Heat pumps with speed controlled compressors can solve this problem because the units are operated both in heating and defrosting at the same time. The system and the improved characteristics are shown in Fig. 15. When defrosting, a stop valve in a bypass line is opened and part of the high temperature refrigerant gas, which is delivered by the compressor operating at the highest speed, flows to an outdoor heat exchanger for defrosting. At the same time, the rest of the refrigerant gas flows to an indoor heat exchanger to continue heating. This system can prevent room temperature reduction.

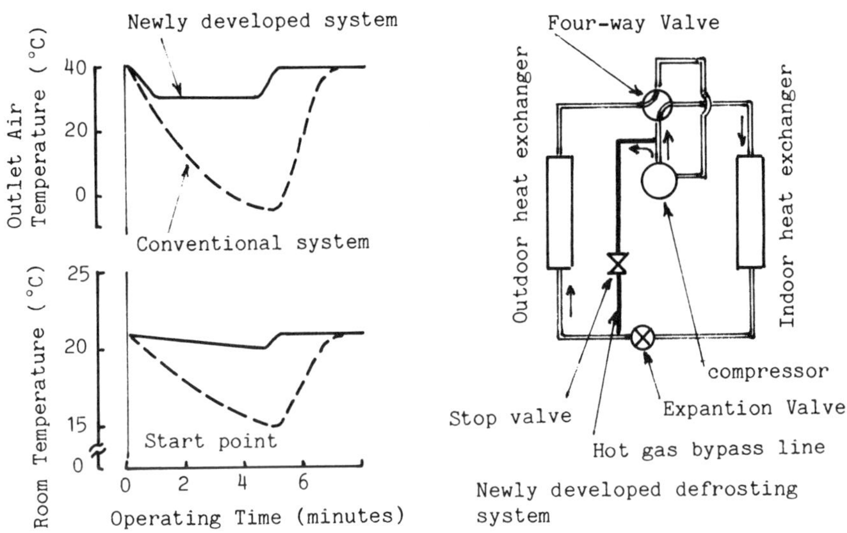

Fig. 15. Development of the heat pump with speed controlled compressor.

6 CONCLUSION

(1) With the spread of heat pump air conditioners and speed controlled compressors, operating conditions and requirements in compressors have become more stringent year by year.
(2) Japanese manufactures have made many R & D efforts and have developed new types of rotary displacement compressors to improve compressor performances.
(3) Rotary displacement compressors have many chracteristics matching present requirements and they are exepted to be applied more widely in the air conditioning market in the future.

REFERENCES

1. K Tanaka and E.Miura, "Heat Pumps in Japan 1975-1990," IEA Heat Pump Conference pp.195-206, May 1984
2. N.Arai and K.Matsubara, "Scroll and Screw Compressors; The Latest Compressor Technology for Air Conditioning and Refrigeration," Hitachi Review Vol.34,pp.141-146, June 1985

Chapter 11

Dynamic Loss Reduction with Electric Heat Pumps

Charles E. Bullock

ABSTRACT: Although much is known and has been published about the design and performance of electric heat pumps under steady-state operating conditions, much less has been documented regarding their performance under dynamic or non-steady-state conditions. The two principal effects which can degrade system performance are (1) on/off cycling and (2) outdoor heat exchanger frost build-up and removal. This paper reviews current knowledge of the magnitudes of these two effects and the success of efforts to reduce them. On/off cycling with conventional heat pumps increases the seasonal energy consumption by about 5 - 10% in heating and 10 - 15% in cooling. By employing procedures suggested in the paper, such as lighter weight heat exchangers, variable-speed compressors, or refrigerant isolation valves, these losses can be reduced by as much as 50%. The outdoor coil frost/defrost seasonal energy penalty has been measured from as low as 5% to as high as 20%. By using design principles outlined in the paper, it should be possible to reduce this factor to less than 5%. Recommendations are given for further work in both areas and the probable benefits that can be achieved.

INTRODUCTION

Over the past 30 years, the electric heat pump has become established as an efficient and dependable means for heating and cooling occupied spaces, residential as well as commercial. The level of understanding of the complex fluid flow and heat transfer processes occurring during the steady-state operation of a typical heat pump has risen to the point where manufacturers can routinely design units for maximum steady-state efficiency, minimum fabrication cost, or some optional combination of these two factors.

HEAT PUMPS: Prospects in Heat Pump Technology and Marketing (Proceedings of the 1987 IEA Heat Pump Conference), Edited by Kay H. Zimmerman Chelsea, Michigan 48118. Printed in U.S.A.

Probably as early as when the first heat pump was field- or lab-tested under varying load conditions requiring the on/off operation of the unit, it was recognized that capacity as well as efficiency can be reduced somewhat in real applications. In the case of outdoor air-source heat pumps operating in the heating mode, the gradual accumulation of frost on the outdoor evaporator coil was also found to reduce the unit's performance. Since these two effects, i.e., on/off cycling and outdoor coil frost/defrost, both involve the transient behavior of the system and involve even more complex fluid flow and heat transfer phenomena than in steady-state operation, our understanding of them has been limited and has lagged far behind that for steady-state. In the past decade, however, significant progress has been made by numerous researchers in analyzing and quantifying these phenomena and in exploring ways to reduce their effects on real system performance.

This paper will summarize the principal findings of these prior studies and will indicate areas where more work is justified.

REVIEW OF ON/OFF CYCLIC PHENOMENA

Prior to the start-up of a heat pump or air conditioner, most (as much as 80%) of the refrigerant migrates to the evaporator because of its low temperatures [1,2,3,4]. After the compressor is started and has reached steady-state, however, most of the refrigerant (again as much as 80%) will be in the condenser, or high-pressure side. The start-up of an electric heat pump is thus followed by a transient period during which the distribution of refrigerant within the system as well as its temperatures and pressures and also the temperatures of all of the system components undergo rapid and fairly drastic changes while approaching their steady-state values. The mathematical modeling of the refrigerant-side phenomena has proven to be extremely difficult.

Although several very good dynamic simulations of heat pump systems have been completed, they are very complex and thus far have been used largely for illustrative purposes [5,6,7,8]. These models as well as purely experimental studies have shown that the various temperatures and pressures do not approach their steady-state values at exactly the same rates although they are more or less interdependent. More details can be found in the references cited. In spite of the complexities of the refrigerant-side transients, it is possible to describe the external performance characteristics (heating and cooling capacities, power consumption) in simple yet accurate terms. Figure 1 shows typical capacity and compressor power consumption curves for a unit operating at a duty cycle of 50% (equal "on" and "off" time periods). Each start-up period is characterized by a fairly uniform approach to steady-state capacity which can be described fairly accurately by a simple first-order time constant. Mulroy and Didion [3] suggest that a two-time-constant representation may be slightly more accurate. The start-up period is accompanied by a loss of output capacity as

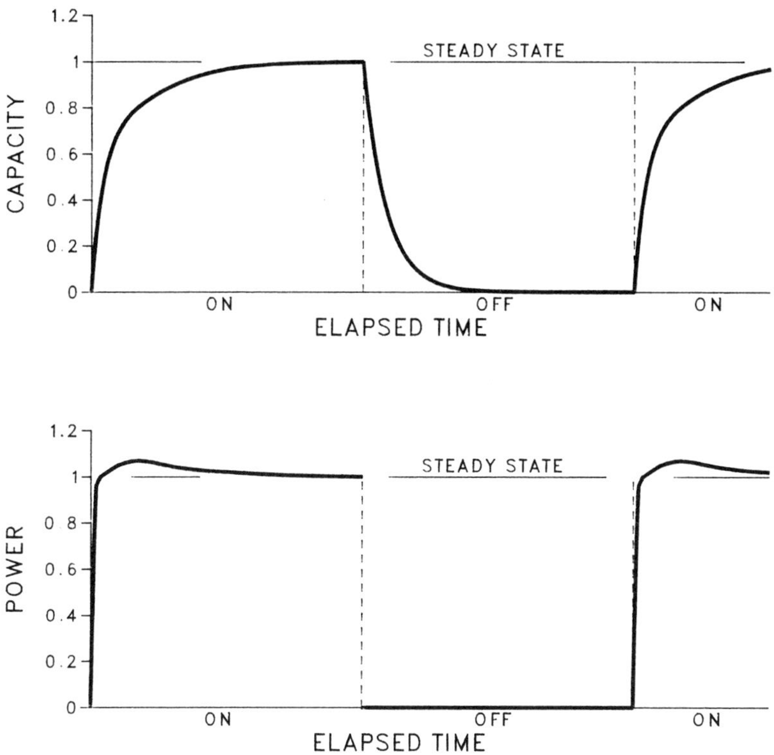

Fig. 1. Typical capacity and power transients.

compared to the capacity that could have been realized if steady-state conditions had been achieved at the instant of start-up.

Immediately following compressor shut-down, some of the capacity "lost" during start-up can be recovered if the indoor air blower (or water pump in a hydronic heating or cooling system) continues to run, as indicated in Figure 1.

Compressor power also has a transient response on start-up as indicated by the example curve in Figure 1. Power typically rises quickly to a peak (which can be 5 - 10% higher than the steady-state value) and then gradually returns to its steady-state value. As Miller and Jaster [11] point out, this peak is likely due to the higher density of refrigerant gas (or liquid) coming from the evaporator (or accumulator) during start-up. The magnitude and duration of the peak are affected by the outdoor temperature as well as the off-period length.

Air-to-water heat pumps are appropriate for hydronic heating systems, which are common in Europe. Bergman [9] reports that start-up and shutdown transients in such system are essentially negligible. This appears to be a consequence of the closed-loop type of heat distribution piping network rather than any difference in heat pump design. During an off period, the water loop temperature drops substantially, so that at the next start-up the heat pump's capacity and efficiency are enhanced compared to steady-state. This type of heat pump has the same frost/defrost problems as air-to-air units, however.

The net or average capacity, power and efficiency of a heat pump or air conditioner can be seen to depend not only on the response rate of the equipment itself but also upon the length of the on/off cycles themselves. Longer cycles will naturally diminish the effects of start-up transients. Cycle length is affected by four principal factors:

1. The heat pump or air conditioner output capacity relative to the building load.

2. The heat pump or air conditioner transient response rate.

3. The building thermal mass C, which can be expressed as:
   $C = M \times Cp$,
   where: M = building mass, kg
   Cp = specific heat, kJ/(kg.K)

4. The room thermostat cycling characteristics, principally its deadband (or differential) width ΔT.

Figures 2 - 4 illustrate the relative significances of these factors. The curves are based on an analysis similar to that used by Miller and Jaster in [11]. They are expressed in terms of (1) the part-load factor (PLF) or cyclic efficiency, which is the ratio of the average cyclic coefficient of performance (COP) to the steady-state COP, and (2) the load factor LF, which is the ratio of the average cyclic output capacity to the steady-state capacity. In all these figures, the following typical values were used for the parameters, unless otherwise noted: Heat pump time constant TC = 1 min., Building thermal mass factor $C/Q_s = 10\ K^{-1}$, $\Delta T = 1K$.

Figure 2 shows the influence of heat pump capacity relative to building load as expressed by the load factor LF. At steady-state, LF=1, and PLF=1.0. However, as the load drops, so does PLF, falling nearly linearly until LF=0.2, then dropping rapidly toward zero for lighter loads. Figure 2 also shows the influence of the heat pump transient response. As the time constant TC increases, PLF also decreases. Figure 3 shows the effects of building thermal mass C, expressed relative to the steady-state heat pump capacity Q_s. As C is increased, the cycling rate decreases, lowering the influence of start-up transients and raising the cyclic efficiency. Figure 4 shows the effects of room thermostat differential ΔT. As ΔT is increased, the cycling rate drops, also raising the cyclic efficiency. Figures 3 and 4 show that factors unrelated to the heat pump design itself can significantly affect its cyclic response. Recent field studies by Baxter and Moyers [10] and Miller and Jaster

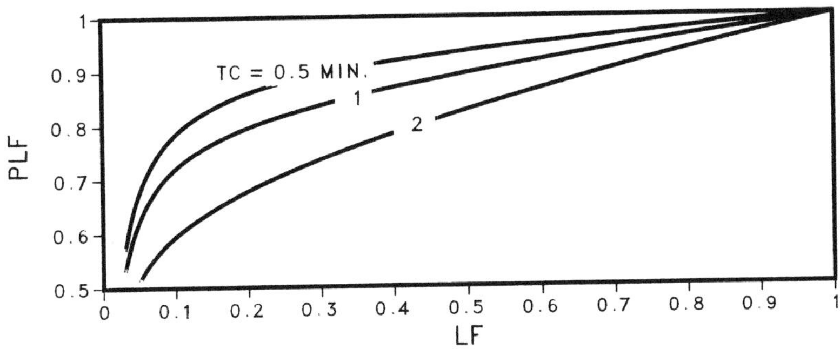

Fig. 2. Heat pump time constant effect.

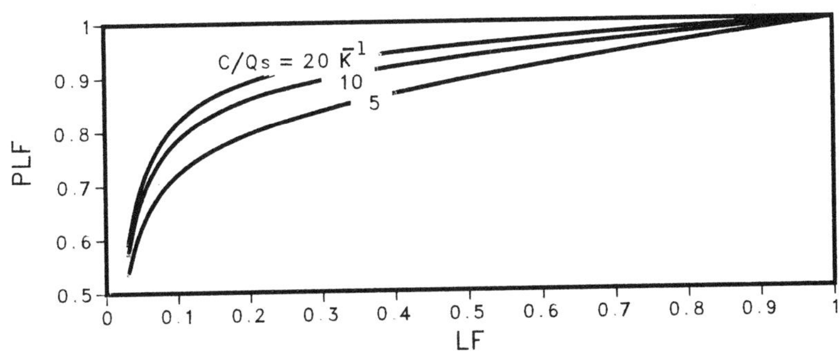

Fig. 3. Building thermal mass effect.

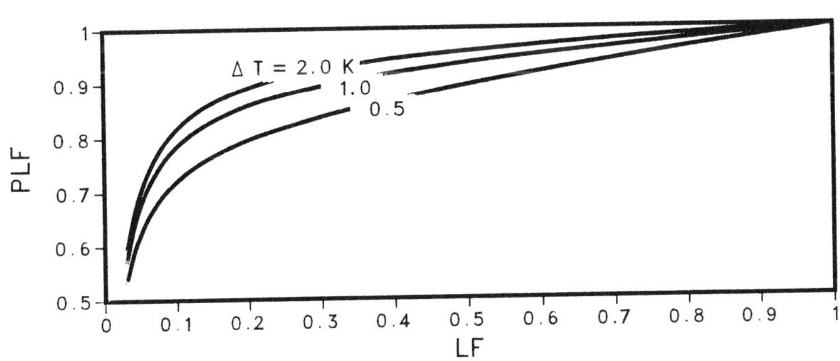

Fig. 4. Room thermostat differential effect.

[11] showed much lower cyclic rates (less than 1 cycle/hour) occurring in actual installations than would be expected from usual design practices, evidently with no adverse comfort problems. This indicated large values for C/Q_S and ΔT. In these cases, the net effects of cyclic losses were very small. The main point of Figures 2 - 4 is that all of the factors noted are very important, and each must be considered in actual applications in order to minimize cyclic losses. These figures also clearly show that, in order to improve the cyclic efficiency of a heat pump or air conditioner, the following can be done:

1. Avoid equipment oversizing; i.e, keep LF as large as possible.

2. Design or select the heat pump or air conditioner with a small start-up time constant.

3. Maximize building thermal mass if possible. The effective thermal mass can be influenced by the distribution system layout and by thermostat location.

4. Install or adjust the room thermostat to provide as large a room temperature swing as possible, without sacrificing comfort.

5. Although not addressed above, off-cycle power consumption should also be minimized. Creswick [12], for example, measured seasonal losses of 4% in one case.

Considerable attention has been given by researchers as well as manufacturers to the second item above, i.e., minimizing the heat pump and air conditioner transient response time. The following recommendations have resulted from the work to date:

1. Minimize the mass and refrigerant-side volume of all components, especially the evaporator coil [1,12,13].

2. Minimize refrigerant charge, especially in the condenser and liquid line [1].

3. For heat pumps, eliminate or reduce the size of the suction line accumulator. Preferably, design component volumes so that refrigerant can be stored on the high side during the off-period so that the need for a suction accumulator is reduced. If an accumulator is required, keep it small [1] and consider adding heat during start-up to accelerate refrigerant vaporization [12,14].

4. Keep liquid refrigerant out of the evaporator during the off period [1,12]. The most obvious way to do this is with automatic shut-off valves that close when the compressor stops. Young [14] proposed that if an electric expansion valve is used, it should be opened wide to increase

refrigerant flow to the evaporator on start-up to avoid "starving".

5. Delay the indoor blower shut-off until a minute or two past the compressor shut-off time in order to recover residual indoor coil heating or cooling. Mulroy [15] has shown that this is only effective when the indoor coil is outside the conditioned space. A continuously-operating indoor blower is not recommended because of its energy consumption.

Considerable interest has also been shown in recent years to variable-speed compressors [16,17,18,19]. Although variable-speed compressors do not reduce the start-up losses per cycle, they allow the system to adjust its output capacity to meet changing load conditions, thus drastically reducing the total number of on/off cycles per year and thereby reducing the significance of start-up transients. Variable-speed compressors also improve the system steady-state efficiency at light loads since they will operate at low speeds, resulting in lower refrigerant flow rates, more effective heat transfer, lower pressure ratios, and lower compressor powers.

A certain amount of on/off cycling is inevitable in the course of a heating or cooling season, even with variable-capacity systems. Local climate, equipment sizing, and the transient characteristics of the equipment and building will all influence the magnitude of the cycling effects. Analytical studies [20,21,22,23] as well as field tests [10,19] have generally shown that, with conventional heat pumps, on/off cycling increases seasonal energy consumption by about 5 - 10% in heating and 10 - 15% in cooling. By employing some or all of the improvements described above, these losses can be cut in half.

NEED FOR FURTHER WORK TO MINIMIZE CYCLIC LOSSES

1. Refrigerant charge control: Store refrigerant on the high side during the off-cycle. There is a need for an alternate to shut-off valves for refrigerant isolation to avoid compressor start problems. We also need a quicker way of supplying liquid refrigerant to evaporator (and suction accumulator, if there is one).

2. Heat exchangers: We need lighter, more efficient designs.

3. An alternate to a suction line accumulator is needed.

4. Equipment/building interaction research: We need more information on acceptable temperature swings, cycling rates.

REVIEW OF FROST/DEFROST PHENOMENA

When the outdoor air temperature drops below about 5°C, the heat transfer surfaces of the outdoor coil of an air-source heat pump can drop below 0°C, leading to the formation of frost on the surfaces. Frost build-up creates two problems: the gradual degradation of the heat pump's heating capacity and the need to periodically defrost the coil. Numerous investigators have studied the formation of frost on simple geometries (flat plates, cylinders, etc.); O'Neal and Tree [24] have presented an excellent review of work to date. Although useful relations for frost thermal conductivity and density (which are usually interdependent), growth rate and heat transfer rate are available, there are still many knowledge gaps and a high level of uncertainty in applying the data. These studies, however, have generally agreed that the frost layer grows faster and becomes thicker as the surface temperature drops and/or the air humidity and velocity rise. The formation of frost on a finned-tube evaporator coil is further complicated by variations in the surface temperature and geometry as air passes through the coil, often leading to non-uniform frost growth.

Stoecker [25] was one of the first to publish data from tests with multi-row refrigeration-duty evaporator coils, indicating that heat transfer is enhanced slightly with light frost but that heat transfer and airflow both drop as heavy flow builds up. Merrill [26] concluded that frost builds up on a heat pump coil in two stages, slow at first, then much faster. As O'Neal and Tree [24] point out, frost tends to grow fastest at the leading edges of the fins, gradually progressing deeper into the coil along the airflow path. Tests by Miller [27] showed that, with a typical, medium efficiency design heat pump, both capacity and COP were improved with higher air humidities at non-frosting conditions, but both were reduced under frosting conditions. The transition from a wet to a frosted surface occurred at about 5°C in this case. This agreed with the findings of Imaiida, et al. [28]. Analysis of published data as well as laboratory and field tests with conventional 2.5 ton (8.8 kW) heat pumps led Miller and Jaster [11] to conclude that heating capacity and compressor power both drop approximately linearly with time as frost accumulates. The drop-off in each item could be correlated by the simple product $\Delta W \times t$, where ΔW is the difference in the humidity ratio of air entering the evaporator coil and the humidity ratio of saturated air at the coil surface temperature and t is elapsed time. Miller and Jaster also suggested that spine-fin (also called slit-fin or pin-fin) coils may frost up faster than plate fin coils. Experience at the author's company, however, has indicated that system design, particularly as it affects the refrigerant evaporating temperature, has a far greater influence than fin type. For comparable fin densities and refrigerant- and air-side conditions, the frosting characteristics of plate fin and slit-fin coils are very similar. In all cases, frost has several important effects upon the system:

1. Frost condensation increases the overall heat transfer rate by adding latent heat flux to the existing sensible heat

flux. This can amount to 10% or more. The effect is, of course, minimized at low outdoor air temperatures.

2. The frost layer thickness adds an element to the overall thermal resistance between the refrigerant and the air.

3. The surface roughness of frost can improve heat transfer slightly for thin layers, but the net effect is a reduction in overall heat transfer as the frost layer grows, as shown by Stoecker [25]. As Barrow [29] points out, however, these latter two effects are generally small in comparison to the following.

4. The frost layer reduces the airflow passage area between fins, thus tending to increase the local air velocity and the air-side pressure drop. For typical fan-coil systems, the net effect is a drop in the total airflow rate as the system re-balances. The drop in airflow may reduce the heat transfer coefficient and will certainly lead to a drop in the overall heat transfer rate which, in turn, will cause a constant-speed compressor to rebalance with a lower refrigerant temperature in the evaporator. This is the most important effect of frost and is the primary reason why heating capacity, compressor power and COP all drop as frost builds up.

The effects of frost on system performance can be reduced by:

1. Increasing the size of the outdoor coil. This will raise both the evaporating refrigerant temperature and the coil surface temperature, which will reduce the frosting rate. It is also advisable to use one-row coils in order to spread the frost more uniformly over the total fin surface. Fortunately, this is the trend that the residential heat pump industry is following in the evolution of higher efficiency unit designs. In their tests of a mid-1970s vintage heat pump, Kelly and Bean [30] reported that the frost/defrost degradation factor was approximately 0.74 below 5°C. (The frost/defrost degradation factor is equal to the ratio of the integrated average COP over one full frost/defrost cycle to the instantaneous no-frost COP.) Since then, more efficient heat pumps have been designed with larger heat exchangers (and more efficient compressors), which have exhibited lower frosting tendencies. Figure 5 shows this trend schematically. The "low efficiency" curve represents designs typical of the 1970s with SEER (Seasonal Energy Efficiency Ratio) ratings of approximately 6-7 Btu/Wh. The "medium efficiency" curve applies to units marketed in the early 1980s with SEERs of about 8-9. The "high efficiency" curve represents today's top-of-the-line units, with 10-11 SEERs, which will be tomorrow's typical units. The clear trend toward less and less frost/defrost penalty is obvious.

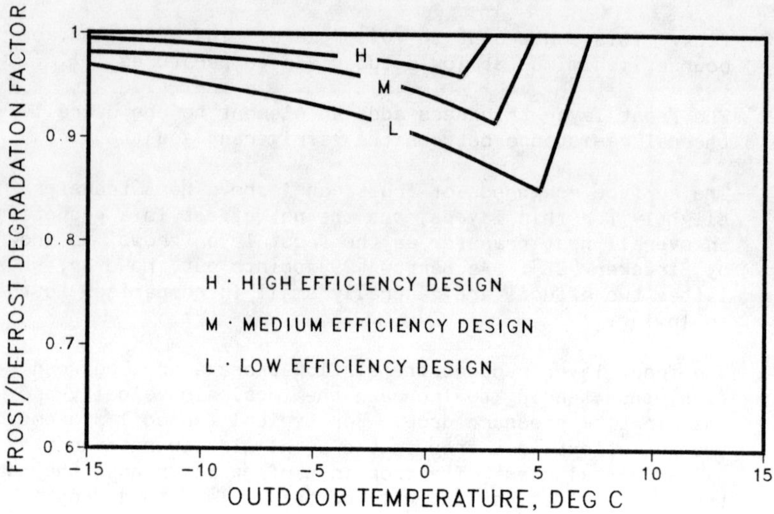

Fig. 5. Frost/defrost degradation factors.

2. Increasing the evaporator airflow rate. This will also raise the coil surface temperature.

3. Using coils with wider fin spacings [14,25,31], e.g., 12 fins per inch or less. This will allow thicker frost layers to develop before airflow is affected.

4. Sizing compressors and evaporator coils for higher evaporating temperatures during those conditions when frost is most important (-5°C to 5°C). Heating loads tend to be less, allowing the compressor to operate continuously at lower capacities and higher evaporating temperatures, again lowering frost build-up rates. This can also be accomplished with variable-speed compressors.

Defrosting of the evaporator coil is typically accomplished by reversing the heat pump cycle, i.e., by temporarily using the outdoor coil as a condenser and the indoor coil as an evapporator. To offset the indoor cooling effect, some resistance heat tempering is generally used. As indicated by Figure 6, this is not a very energy efficient method. The need for resistance heat tempering (COP=1) adds a significant energy penalty. Typically, frost/defrost accounts for about 10% of the total heating season energy with one-third or more of this due to resistance heat tempering. Passive defrost, in which the compressor is idle and the coil defrosts by natural convection, is more efficient but requires a much longer defrost time [31] and is not effective at outdoor temperatures below 0°C. The use of some form of energy storage as the source of defrost energy is the most efficient scheme, but it requires additional hardware and controls. Thus, the reverse cycle approach remains as the predominant choice in today's units.

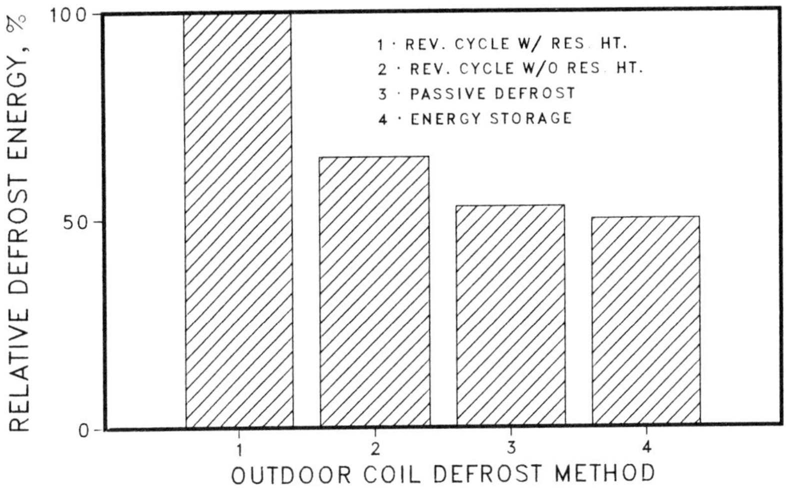

Fig. 6. Defrost energy comparisons.

Several methods of initiating the system reversal for defrost have been proposed. Most heat pumps today utilize the time-temperature approach, which initiates a defrost after a fixed interval of compressor run time during which the evaporator coil temperature is below a set value, such as 1°C. This can result in defrosts with little or no frost and also with excessive frost. "Demand" defrost methods, on the other hand, have been proposed in order to initiate defrosts only when the coil is sufficiently frosted. These methods have included measurement of the coil air-side pressure drop, the air-to-coil temperature difference, compressor or outdoor fan motor current [14] or some other indicator. Several studies, such as [32,33,34], have concluded that the defrost initiation method has very little net effect upon the overall system energy efficiency, however. A method which allows the system to run longer between defrosts will also have to run longer (and at lower COPs) because its average heating capacity is lower and will also require longer defrost times. Demand defrost methods are often chosen, however, because they can drastically reduce the number of defrost cycles, with presumed benefits to system reliability.

Regardless of the method chosen, the defrosting efficiency can be improved most readily by reducing the time required for coil defrosting. This can be accomplished by:

1. Designing or selecting the coil/compressor system for low frost loading, following the suggestions given previously.

2. Designing the outdoor coil with several parallel refrigerant circuits for even distribution of the hot gas to the coil.

3. Making sure the coil, particularly the bottom, is dry after each defrost. This can be done with a vertical (or nearly vertical) coil, a desuperheat/subcooling circuit at the bottom of the coil [35] or the use of a surface-active agent to promote water drainage [36]. Care must be taken, however, with the use of a desuperheat/subcooling circuit to avoid penalizing system performance during the normal cooling or heating modes.

4. Eliminating the suction line accumulator, if possible, in order to avoid its thermal lag effects.

5. Modifying the system design or operation to insure quick return of condensed liquid to the indoor coil. This can be done by opening the expansion valve or using a bypass line [14].

6. Using variable-speed compressors, there may be advantages to increasing the compressor speed just prior to a defrost in order to raise its temperature and thus serve as an additional heat source during defrost [36].

As mentioned previously, the use of resistance heat for indoor supply air tempering should also be avoided or at least minimized in order to improve the overall defrosting efficiency.

The net effects which outdoor coil frost have over a full heating season depend upon the local climate as well as the design and the operation of the heat pump system. Analytical studies, such as [13] and [23], have estimated that the seasonal penalty for frost/defrost losses at less than 10%. Field tests of mid-1970s heat pumps at four northern U.S. locations reported by Groff and Reedy [37] all showed seasonal frost/defrost losses less than 9%. More recent field tests reported by Baxter and Moyers [10] and by Creswick [12] showed frost/defrost losses of 8 - 14%. Recent studies reported by EPRI [38,39] showed seasonal frost/defrost losses ranging from 4% to 21%, the latter involving the use of excessive strip heat for defrost tempering. Results reported by Bouma [29] for an air-to-water heat pump were quite similar. By employing design principles described above and especially minimizing the use of strip heat, it should be possible to reduce the frost/defrost penalty to less than 5%.

NEED FOR FURTHER WORK TO MINIMIZE FROST/DEFROST LOSSES

Research is needed on:

1. Evaporator fan/coil combinations -- the effects of fin shape and density, number of rows, and the shape of the fan pressure/flow characteristic.

2. Coil circuiting -- more uniform refrigerant distribution during defrost.

3. Coil water condensate drainage -- the cost-effectiveness of surface-active agents.

4. Alternate defrost energy sources.

5. Refrigeration system design -- alternate to suction line accumulator.

REFERENCES

1. N. Tanaka, M. Ikeuchi, and G. Yamanaka, "Experimental Study on the Dynamic Characteristics of a Heat Pump", ASHRAE Trans. 88(2), 323-331, 1982.

2. W. E. Murphy and V. W. Goldschmidt, "Transient Response of Air Conditioners - A Qualitative Interpretation Through a Sample Case", ASHRAE Trans. 90(1B), 997-1008, 1984.

3. W. J. Mulroy and D. A. Didion, "Refrigerant Migration in a Split-Unit Air Conditioner", ASHRAE Trans. 91(1A), 193-206, 1985.

4. W. A. Miller, "The Laboratory Evaluation of the Heating Mode Part-Load Operation of an Air-to-Air Heat Pump", ASHRAE Trans., 91(2B), 524-554, 1985.

5. J. Chi and D. Didion, "A Simulation Model of the Transient Performance of a Heat Pump", Int. J. of Refrigeration, 5(3), 176-184, May 1982.

6. J. W. MacArthur, "Analytical Representation of the Transient Energy Interactions in Vapor Compression Heat Pumps", ASHRAE Trans., 90(1B), 982-996, 1984.

7. W. E. Murphy and V. W. Goldschmidt, "Cyclic Characteristics of a Typical Residential Air Conditioner - Modeling of Start-Up Transients", ASHRAE Trans. 91(2A), 427-444, 1985.

8. W. E. Murphy and V. W. Goldschmidt, "Cycling Characteristics of a Residential Air Conditioner - Modeling of Shutdown Transients", ASHRAE Trans., 92(1A), 186-202, 1986.

9. A. Bergman, "Influence of Intermittent Operation on Heat Pump Effectivity. Laboratory Tests", Report No. BFR-R-111-1985, Swedish Council for Building Research, 1985.

10. V. D. Baxter and J. C. Moyers, "Air-Source Heat Pump: Field Measurement of Cycling, Frosting, and Defrosting Losses, 1981-83", Report ORNL/CON-150, Oak Ridge National Laboratory, November 1984.

11. R. Miller and H. Jaster, "Performance of Air-source Heat Pumps", Report EPRI EM-4226, Electric Power Research Institute, November 1985.

12. F. Creswick, "Research on Residential Air-Source Heat Pump Dynamic Losses at ORNL", IEA Newsletter, 3(4), 3-6, December 1985.

13. U. Bonne, et al., "Electric-Driven Heat Pump Systems: Simulations and Controls", ASHRAE Trans., 86(1), 687-705, 1980.

14. D. J. Young, "Development of a Northern Climate Residential Air-Source Heat Pump", ASHRAE Trans., 86(1), 671-686, 1980.

15. W. J. Mulroy, "The Effect of Short Cycling and Fan Delay on the Efficiency of a Modified Residential Heat Pump", ASHRAE Trans., 92(1B), 813-826, 1986.

16. Y. Shimma, T. Tateuchi, and H. Sugiura, "Inverter Control Systems in the Residential Heat Pump Air Conditioner", ASHRAE Trans., 91(2B), 1541-1554, 1985.

17. T. Senshu, et al., "Annual Energy-Saving Effect of Capacity-Modulated Air Conditioner Equipped with Inverter-Driven Scroll Compressor", ASHRAE Trans., 91(2B), 1569-1584, 1985.

18. M. Hori, I. Akamine, and T. Sakai, "Seasonal Efficiencies of Residential Heat Pump Air Conditioners with Inverter-Driven Compressors", ASHRAE Trans., 91(2B), 1585-1598, 1985.

19. F. A. Creswick, "New and Future Heat Pump Technologies", paper presented at APPA/EEI/EPRI/NRECA Seminar on Meeting Customer Needs with Heat Pumps, Kansas City, Mo., April 8-9, 1986.

20. C. E. Bullock and W. R. Reedy, "Heat Pump Cyclic Performance and Its Influence on Seasonal Operation", paper presented at Third Heat Pump Technology Conference, Oklahoma State University, April 1978.

21. V. W. Goldschmidt, G. H. Hart, and R. C. Reiner, "A Note on the Transient Performance and Degradation Coefficient of a Field Tested Heat Pump - Cooling and Heating Mode", ASHRAE Trans., 86(2), 368-375, 1980.

22. V. W. Goldschmidt, "Effects of Cyclic Response of Residential Air Conditioners on Seasonal Performance", ASHRAE Trans., 87(2), 757-770, 1981.

23. B. B. Bittle and V. W. Goldschmidt, "Effect of Cycling and Frost Formation on Heat Pump Performance - Implications of Cyclic Test Data", ASHRAE Trans., 89(2B), 743-754, 1983.

24. D. L. O'Neal and D. R. Tree, "A Review of Frost Formation in Simple Geometries", ASHRAE Trans., 91(2A), 267-281, 1985.

25. W. F . Stoecker, "How Frost Formation on Coils Affects Refrigeration Systems", Refrigeration Engineering, 65, 42-46, February 1957.

26. P. Merrill, "Heat Pumps On-Off Capacity Control and Defrost Performance Tests Using Demand Defrost and Time-Temperature Defrost Controls", ASHRAE Trans., 87(1), 381-393, 1981.

27. W. A. Miller, "Frosting Experiments for a Heat Pump Having a One-Row Spine-Fin Outdoor Coil", ASHRAE Trans., 90(1B), 1009-1025, 1984.

28. T. Imaiida, et al., "Development of Demand Defrost Control System for Residential Heat Pumps", ASHRAE Trans., 91(1B), 1527-1540, 1985.

29. H. Barrow, "A Note on Frosting of Heat Pump Evaporator Surfaces", Heat Recovery Systems, 5(3), 195-201, 1985.

30. G. Kelly and J. Bean, "Dynamic Performance of a Residential Air-to-Air Heat Pump", Paper Presented at Second Heat Pump Technology Conference, Oklahoma State University, October 1976.

31. J. W. J. Bouma, "Frosting and Defrosting Behavior of Outdoor Coils of Air-Source Heat Pumps; Condensed Report", TNO Report 81-01480, Netherlands Organization for Applied Scientific Research, February 1981.

32. D. Mueller and U. Bonne, "Heat Pump Control Solutions Using Advanced Microelectronic Technology", ASHRAE Trans., 86(2), 461-474, 1980.

33. B. Zakrzewski, "Optimal Defrost Cycle for the Air Cooler", Intl. J. of Refrigeration, 7(1), 41-45, 1984.

34. C. Tantakitti and R. Howell, "Air-to-Air Heat Pumps Operating Under Frosting Conditions on the Outdoor Coil", ASHRAE Trans., 90(1B), 827-842, 1986.

35. A. Trask, "Ten Design Principles for Air-to-Air Heat Pumps", Paper Presented at Second Heat Pump Technology Conference, Oklahoma State University, October 1976.

36. E. Kuwahara, T. Kawamura, and M. Yamazaki, "Shortening the Defrost Time on a Heat-Pump Air Conditioner", ASHRAE Trans., 92(2), Paper No. 2974, 1986.

37. G. C. Groff and W. R. Reedy, "Investigation of Heat Pump Performance in the Northern Climate through Field Monitoring and Computer Simulation", ASHRAE Trans., 84(1), 767-785, 1978.

38. C. E. Bullock, R. E. Hough, and S. C. Pederson, "Monitoring of Improved Air-Source Heat Pumps", Report EPRI EM-3978, Electric Power Research Institute, June 1986.

39. C. E. Bullock, R. E. Hough, and S. C. Pederson, "Monitoring of Central and Room Heat Pumps", Report EPRI EM-4674, Electric Power Research Institute, July 1986.

Chapter 12

Dynamic Electrical Control for a Variable-Capacity Heat-Pump Air Conditioner

Fumio Matsuoka

ABSTRACT: This paper describes electronic control technologies for heat pumps in Japan.

The requirements for heat pumps are:

1) Comfortable air conditioning
   - High temperature transpiration
   - Defrosting characteristics improvement
2) Fundamental performance improvements
   - Heat capacity increase
   - High seasonal energy efficiency ratio

This paper discusses how the following control items should be realized in order to cope with the above-said requirements.

1) Frequency control of the inverter rotary compressor
2) Opening area control of the electronic expansion valve
3) Fan speed control of the indoor heat exchanger
4) 4-way valve, solenoid valve control

HEAT PUMPS: Prospects in Heat Pump Technology and Marketing (Proceedings of the 1987 IEA Heat Pump Conference), Edited by Kay H. Zimmerman Chelsea, Michigan 48118. Printed in U.S.A.

INTRODUCTION

The application of microelectronics is currently a true challenge to refrigeration and heat pump engineering. Microelectronics and, subsequently, automation and information processing are subject to revolutionizing changes. The cooperation of refrigeration technologists with electronics specialists, control engineers and data processing experts is vital if the complex processes arising from the use of microelectronics are to be mastered.

Electronics in its general connection with information is the inward determinant of technology at the highest level. Its proper potential will evolve more and more in the time to come since it will lay the foundations for overcoming the constraints of the immediate work process and, moreover, it constitutes the main resource for revealing the unbounded advantages of economic intensification.

The concept of intelligence simply implies the capacity to acquire and apply knowledge. It provides the maintenance of programmed goals within a given margin even under varying conditions. The control of unstable processes and systems is of greater importance. What we call the "self-organization" of systems is becoming technical reality at this present time. A new phase in man-machine relations has thus been inaugurated. The computer integrated sensor concept will be used in ever greater complexity and it will revolutionize the art of developing intelligent techniques and man-machine relations in its substructure. On the impulses of this evolution in decentralized feedback and feedforward structure, the dynamic interchange impacts all other hierarchical levels.

HEAT PUMP REQUIREMENTS IN JAPAN

In Japan, very few homes have central heating in all rooms throughout the day during the winter. It is normal for heating to be supplied for individual rooms during specific time zones. Under these conditions of use, it is necessary to create a comfortable temperature environment in which the indoor temperature is raised as quickly as possible as soon as the heating operation is started.

Furthermore, winters in Japan are characterized by a high level of humidity and so frost necessarily forms on the outdoor heat exchangers. As a result, defrosting operation is an indispensable requirement and even during defrosting operation, it is necessary for the user-oriented indoor temperature environment to maintain the state of comfort.

Unlike cooling operations, heating operations are accompanied by characteristic requirements and these can be divided generally into problems associated with reducing the start-up time and into improvements in the defrosting characteristics.

Comfortable Air Conditioning

In order to enhance the degree of comfort by reducing the start-up time, the compressor must be made to start up rapidly by an inverter. However, there is an upper current of the inverter, and methods of controlling the inverter frequency for rapid start-up at around the upper current limit or below this limit constitute an important key technology.

When a start-up operation is initiated, the indoor air temperature is still low, and so special measures need to be taken in order to reduce the amount of air flow volume by the indoor heat exchanger so that the user does not feel a draft.

When it comes to discussing control with comfort maintained during start-up, how to control the frequency of the above-mentioned compressor inverter and how to control the air blown by the indoor heat exchanger are problems targeted for solution through technical development.

In order to improve the degree of comfort during defrosting and in order to avoid unnecessary defrosting operations, important themes for consideration are presented by improvements in sensing technology dealing with methods of sensing the frosting and methods of sensing the end of defrosting, and also by the configuration of the refrigeration cycle, which need not employ an indoor heat exchanger as a low-temperature evaporator even during defrosting, and by control geared to returning operation to normal heating as soon as possible the end of the defrosting.

Improvements in Fundamental Performance

In order to do full justice to the functions which will enhance the degree of comfort as mentioned above, the following improvements need to be made to the basic performance in the refrigeration cycle.

Improvements in heating seasonal performance factor

The question as to what needs to be done about the control of the electronic expansion valve so as to keep the amount of refrigerant superheat at the compressor inlet in the optimum level not only during continuous operation but also during start-up and after defrosting remains to be answered.

Reduction of start-up time

The frequency is rapidly accelerated in order to reduce the start-up time. The way in which the frequency is to be accelerated must be below the upper current limit allowed by the inverter by establishing a correlation with the time constant of the refrigeration cycle response.

Reduction of defrosting time

The question as to what is to be done about the configuration of the refrigeration cycle for removing the frost formed on the outdoor heat exchanger in the shortest possible time and about the control of the 4-way valve, solenoid valve, electronic expansion valve, fan and compressor for the defrosting operation itself remains to be answered.

CONFIGURATION OF HEAT PUMP AIR CONDITIONERS

Given below is an outline of the control targets and configuration of the refrigeration cycle in order to cope with the requirements placed on heat pump air conditioners.

As shown in Figure 1, the elements which configure the refrigeration cycle link up in sequence the compressor with inverter frequency converter, check valve, 4-way valve, indoor heat exchanger, electronic expansion valve and outdoor heat exchanger. During a heating operation, the refrigerant circulates as shown by the solid lines in Figure 1. At this time, the control targets are the frequency of the compressor, the electronic expansion valve and the indoor heat exchanger fan.

During a defrosting operation, the refrigerant circulates as shown by the broken lines in Figure 2. The high-temperature, high-pressure gas refrigerant from the compressor is bypassed to the solenoid valve side, a part of the gas refrigerant finds its way to the accumulator, the remaining greater part enters the outdoor heat exchanger through the check valve, and the frost is defrosted. The refrigerant used for the defrosting is then liquefied, it flows into the accumulator, mixes with the above-mentioned gasified refrigerant and is sucked again into the compressor. At this time, the control targets are the solenoid valve, electronic expansion valve, indoor heat exchanger fan, outdoor heat exchanger fan and compressor frequency.

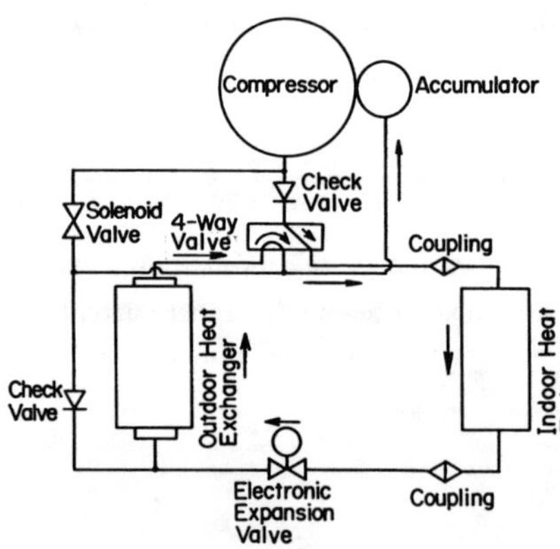

Fig. 1. Heating operation.

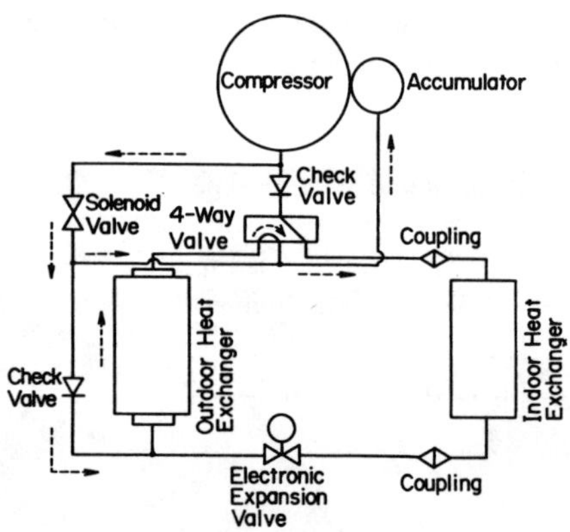

Fig. 2. Defrosting operation.

DYNAMIC CHARACTERISTICS OF REFRIGERATION CYCLE

The control which is to be exercised when a heating operation starts or when, in other words, the air conditioner has reached the regular operation mode, has been the subject of varied research and practical application. This is why this paper focuses specifically on the transient characteristics of the refrigeration cycle and mentions important points to be borne in mind for microcomputer-based control.

Compressor

Figure 3 shows the ageing characteristics of the primary current flowing to the compressor motor when the compressor frequency is raised by 12 Hz (from 60 Hz to 72 Hz) at one stroke. Figure 4 shows the ageing characteristics of the load torque on the refrigerant side applied to the compressor at this time. Figure 5 shows the ageing characteristics of the superheated refrigerant gas at the compressor inlet.

The ageing characteristics of the primary current virtually resemble the ageing characteristics of the load torque applied to the compressor.

The peaks in these ageing characteristics tend to appear between 30 and 60 seconds from the response start time, and they then dip momentarily before they gradually increase again.

The time constant applying up to these peaks is 3 sec. for the primary current and 16 sec. for the load torque. The time constant for the response after the peaks are reached is high with 21 sec. for the primary current and 60 sec. for the load torque. In comparison, the superheat has a more lenient response with a 70 sec. time constant.

The current is increased by 2 A and the superheat temperature rises by 7.5 deg.C by accelerating the compressor frequency by 12 Hz.

Electronic Expansion Valve

Figure 6 shows the ageing characteristics of the superheat when the opening of the electronic expansion valve is suddenly increased by 10% from regular operation. Conversely, Figure 7 shows the ageing characteristics of the superheat when the opening of the electronic expansion valve is suddenly reduced by 10% from regular operation. In both cases, the time constant is 90 sec. However, when the electronic expansion valve is open, the amount of refrigerant circulating tends to increase and the superheat temperature falls by 2 deg.C. Conversely, when the electronic expansion valve is closed, the amount of refrigerant circulating tends to decrease and the superheat temperature rises by 4.3 deg.C.

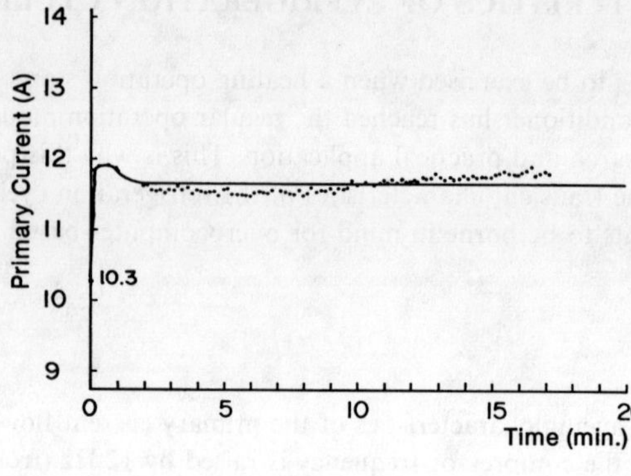

Fig. 3. Primary current.

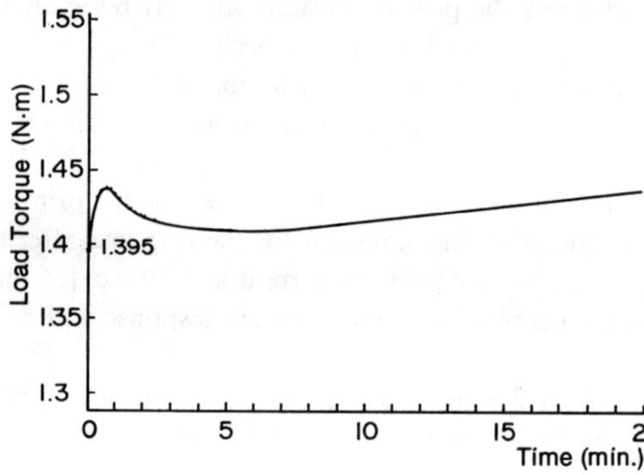

Fig. 4. Load torque.

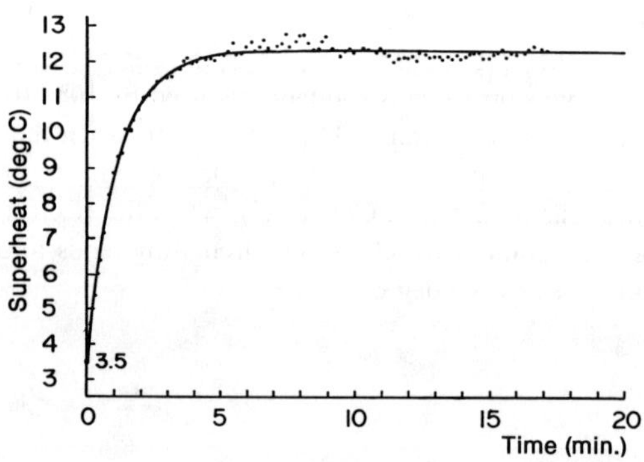

Fig. 5. Superheat.

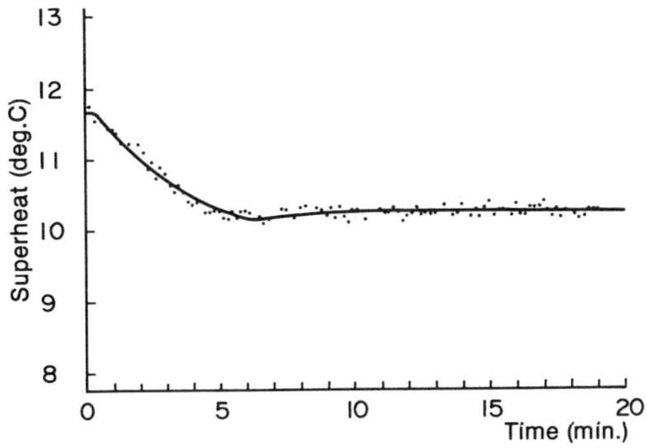

Fig. 6. Superheat (expansion valve open).

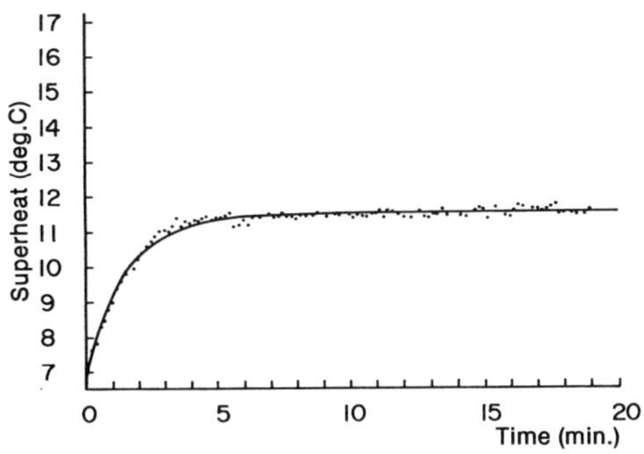

Fig. 7. Superheat (expansion valve closed).

DC Fan Motor

In order to enhance the degree of comfort indoors during the start-up of heating, the air blown into the indoor heat exchanger is kept down to the absolute minimum in order to prevent the user from feeling a draft, when the room temperature is still low. As the time passes and it becomes possible for warm air to be blown out, the fan DC fan motor used for the fan is accelerated and the amount of air is increased.

Figure 8 shows the ageing changes in the temperature of the defused indoor air when the air flow is increased by the DC fan motor from 6.13 m^3/min. to 7.84 m^3/min., and Figure 9 shows the ageing changes of the superheat.

The temperature of the indoor air blown out falls by 3.5 deg.C at a time constant of 72 sec.

On the other hand, the superheat rises by 7 deg.C at a time constant of 93 sec.

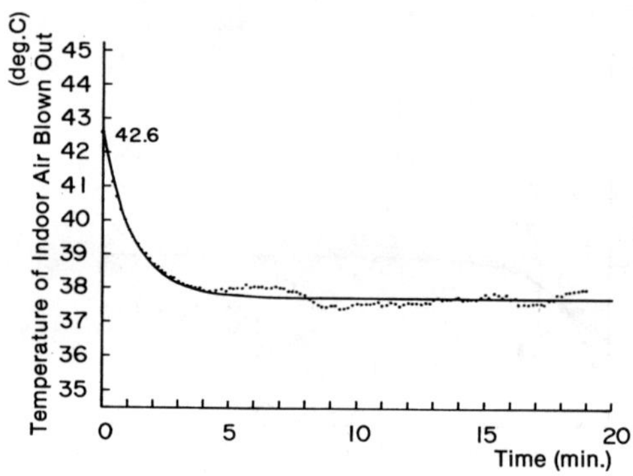

Fig. 8. Temperature of indoor air blown out.

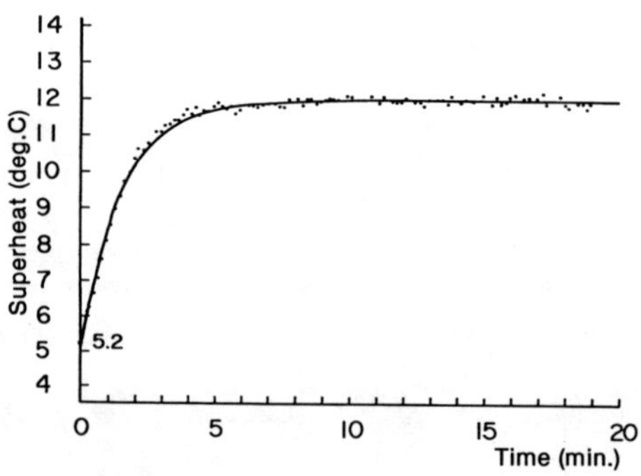

Fig. 9. Superheat.

Cyclic Time Constants and Correlation between Actuators

By changing the actuators (simply speaking, the frequency Hz, electronic expansion valve opening EV and DC fan motor speed DC of the indoor heat exchanger) which configure the refrigeration cycle, the responses (compressor primary current I, superheat SH of refrigerant gas taken in by compressor and temperature Ti of indoor defused air) come to have alternately different time constants, as shown in Figure 10. Even so, the time constants and gains come in a very wide range, depending on the environment, operating mode and operating history.

It is now beginning to be necessary to study the constantly changing time constants and gains even during operation and to conduct intelligent control.

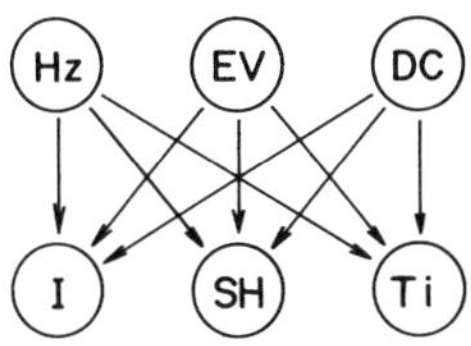

Fig. 10. Correlation between actuators.

Preview Tracking Algorithms

(1) Past data relating to operational amounts of the ever-changing actuators and past data relating to the responses should be stored.

(2) Past data relating to responses should be used to predict future trends.

(3) New target trends should be determined by means of the deviation between the predicted trends and target trends.

(4) Actuators should be operated with respect to the new target trends.

(5) The operational amounts of the actuators and response data should be updated and fed back to (2).

In (1), the past data to be stored will differ for each actuator, and each response and the time constants (in hours and minutes) should be provided.

The future trend predictions in (2) should be based on three-point problems.

The anticipated value of "$\lambda \cdot t$" should be used for the new target trends in (3).

ELECTRONIC CONTROL BY MEANS OF HOLONIC DATA PROCESSING SYSTEMS

Using a control system for an ecosystem as a model, it is possible to exert balanced and harmonious control, bearing in mind the correlation between the individual elements with their individual functions and the overall system.

Figure 11 shows a study of the ageing changes and gain using a buffer control subroutine for estimating the gain and a change buffer control subroutine.

After the new target trends have been determined with preview tracking algorithms by means of the correlation command subroutine, the ignition output of the neurons is produced by the nervous activity and principle of neurodynamics model which is being used with the latest developments in artificial intelligence (AI).

In the balance control subroutine, the above correlation commands and system commands based on future control methods are generalized and integrated and commands are issued by which the overall system harmonizes with the various actuators.

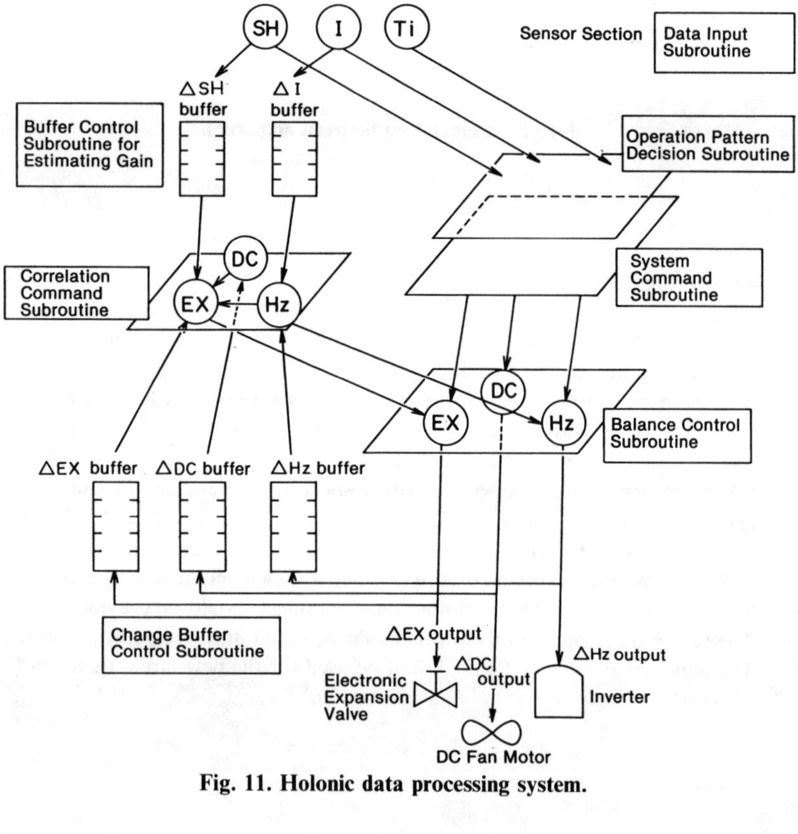

Fig. 11. Holonic data processing system.

EXPERIMENTAL EFFECTS

Figure 12 shows the ageing characteristics of the primary current with a rapid start heating operation, which is one means of enhancing the degree of comfort.

This control makes it possible to halve the amount of time it takes for warm air to be blown out, compared with the time taken in the past.

Furthermore, as shown in Figure 2, the defrosting cycle is such that the high-temperature refrigerant, which passes through the check valve at the compressor outlet and through the indoor heat exchanger and which flows to the electronic expansion valve, collects inside the indoor heat exchanger since the electronic expansion valve is closed even during the defrosting period, and for several minutes warm air can be blown out to indoors.

Even when it comes to the defrosting of the outdoor heat exchanger, the high-temperature refrigerant gas which has left the compressor passes directly through the solenoid valve and is used for defrosting. As a result, the conventional defrosting time is halved.

Even when operation returns to heating upon completion of the defrosting, the refrigeration cycle is characterized by a cooling of the indoor heat exchanger and so it is possible to blow high-temperature air into the room immediately.

To deal with rapid changes in the operating mode such as these, the electronic expansion valve serves to control the superheat of the refrigerant gas at the compressor inlet with a view to protecting the liquid flowing back to the compressor and, while sudden torque fluctuations in the compressor are being monitored by the primary current, the temperature of the air blown out from the indoor heat exchanger, whether during heating start-up or defrosting operations, serves to provide the user with comfort through the control of the DC fan motor.

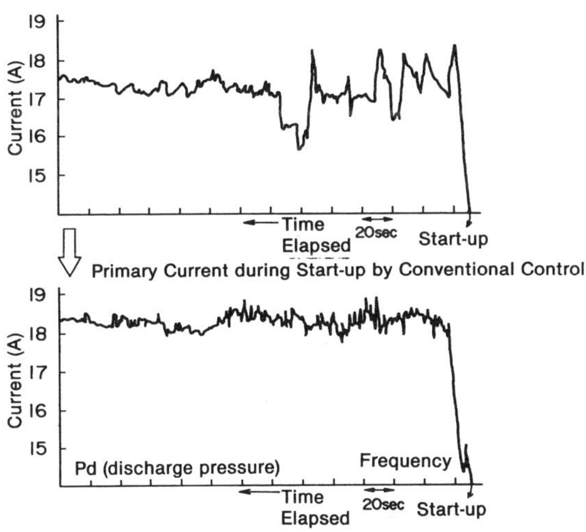

Fig. 12. Primary current during start-up by holonic control.

CONCLUSION

It has been shown that as electronic technology advances, the demands on inverter air conditioners to adopt intelligent and self-learning functions are steadily increasing and are being made possible. Using microcomputers as the brains, progress in man-machine interfacing is being realized and in the future it will surely be possible to create even more comfortable environments.

REFERENCES

1. S. Nowotny, "Possibilities and Trends in the Application of Microelectronics to Refrigeration and Heat Pump Technology," *Refrigeration Science And Technology*, I.I.R., Commission B2, 1984/2

2. E. Herlitzius, "Sociological Aspects of the Application of Microelectronics: Relations Between Man and Technique," *Refrigeration Science And Technology*, I.I.R., Commission B2, 1984/2

3. Keiji Sakaino et al, *The Study of Dual Cylinder Rotary Compressors,* 1986 Purdue Compressor Technology Conference, P292-P304

4. Felix Blaschke, *Das Prinzip der Feldorientierung, die Grundlage fuer die TRANVEKTOR-Regelung von Drehfeldmachinen,* Siemens-Z, 45 (1971) S.757-760

5. F. Matsuoka, "An Analysis of a New Type of Refrigerant Cycle," *Proceedings of the 1982 Purdue Compressor Technology Conference*

6. Kenichi Yoshimoto et al, "Application of the Preview Tracking Control Algorithm to Serving of Robot Manipulators," *Proceedings of 1st I.S.S.R (1983)*

7. R.E. Bellman and L.A. Zadar, "Decision-Making in a Fuzzy Environment, " *Management Science* Vol. 17 No.4 12, 1970

8. B. Auron & E.A. Feigenbaum, *The Handbook of Artificial Intelligence,* Vol 2, 1982

9. Takamoto Saito et al, "The Operation Refrigeration System Analysis Used of Simulation Programs," *Annual Report of Engineering Research Institute,* Faculty of Engineering, University of Tokyo, Vol 38, Sept. 1979

Chapter 13

Current Status and Future Potential of Nonazeotropic Mixed Refrigerants

Horst H. Kruse

ABSTRACT

The application of non-azeotropic refrigerant mixtures in refrigeration systems, because of energy saving aspects, already has been proposed at the beginning of the century. Especially for heat pumps this aspect and others led ten years ago to the proposal to use mixed refrigerants for this special application of refrigeration. Besides the energy saving aspects the additional advantages for this application could be a capacity control by concentration shifting, an adaption of capacities, temperatures and pressures to the desired values with given hardware to allow the use of low cost, mass produced hardware components.

In several research laboratories special investigations on mixed refrigerants concerning thermophysical data and their behaviour in heat pumps had been done.

The paper will sum up those results gained up to now and evaluate the predicted advantages concerning further heat pump developments. Special fields of favourable mixture application will be shown.

INTRODUCTION

Although the application of non-azeotropic refrigerant mixtures in refrigeration systems has been proposed since the beginning of this century and a lot of investigations concerning those systems with mixed refrigerants and regarding their thermophysical data have been done since that time first in the

HEAT PUMPS: Prospects in Heat Pump Technology and Marketing (Proceedings of the 1987 IEA Heat Pump Conference), Edited by Kay H. Zimmerman

seventieth non-azeotropic refrigerant mixtures were proposed especially for heat pump application.

Whereas in refrigeration systems up to that time there was no great incentive to use refrigerant mixtures instead of pure fluids as working media because of lacking competition to the vapour compression systems by other processes the scene changed drastically after the first energy crisis in 1973. Although already marketed before, from that time on the heat pumps gained much more interest in order to save energy especially in the field of space heating. Here the heat pumps have to compete with the conventional low first cost heating systems like firmnesses. So proposals were made (1), (2), (3) to use the possibilities of nonisothermal phase change of non-azeotropic refrigerant mixtures in heat pumps in order to decrease the irreversibilities in the heat exchangers of heat pump systems and consequently save energy as proposed earlier for refrigeration systems (4), (5). Depending on the temperature change of the working fluid in the evaporator and condenser 25% increase in energy efficiency for the refrigerant mixtures R12/R114 and about 12% for the mixture R13B1/R114 have been predicted at that time for instance for temperature gradients of 10 K in the evaporator and condenser according to Figure 1 (3).

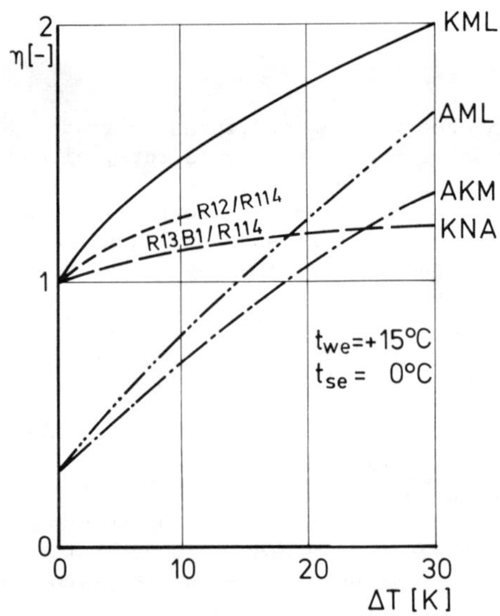

KML – Solution cycle with compressor

KNA – Non azeotropic refrigerant mixture compression cycle

AML – Resorption cycle

AKM – Absorption cycle

Fig.1. Relative energy efficiency of different possibilities to realize the Lorenz-Cycle.

At that time also, based on patents for variable capacity refrigeration sytems (6), (7), (8) a proposal (1) was made to use the properties of non-azeotropic refrigerant mixtures especially for the very important capacity control of heat pumps where the heating capacity has to be changed according to the outdoor temperature. Continuous capacity control by a composition change of the mixture can contribute to avoid transient system losses occuring in heat pumps with on/off control.

Further, based on extensive measurements of refrigerant mixtures in refrigeration systems (9) the use of non-azeotropic refrigerant mixtures for adaption of the heat pump to the required capacity and working conditions by a tailormade refrigerant had been proposed also (10). The background of this proposal was not to develope special refrigerants and hardware components for heat pump working conditions but to use tailormade refrigerant mixtures in order to apply the mass produced components of refrigeration systems under the other working conditions of heat pumps.

These possibilities of non-azeotropic refrigerant mixtures as energetic improvements by using the gliding temperatures in the evaporator and condenser, continuous capacity control by composition shifting, capacity adaption and adjustment to the application limits of the components by using a tailormade refrigerant mixture have led in the following years to a growing interest in these fluids for heat pump application as shown in Figure 2.

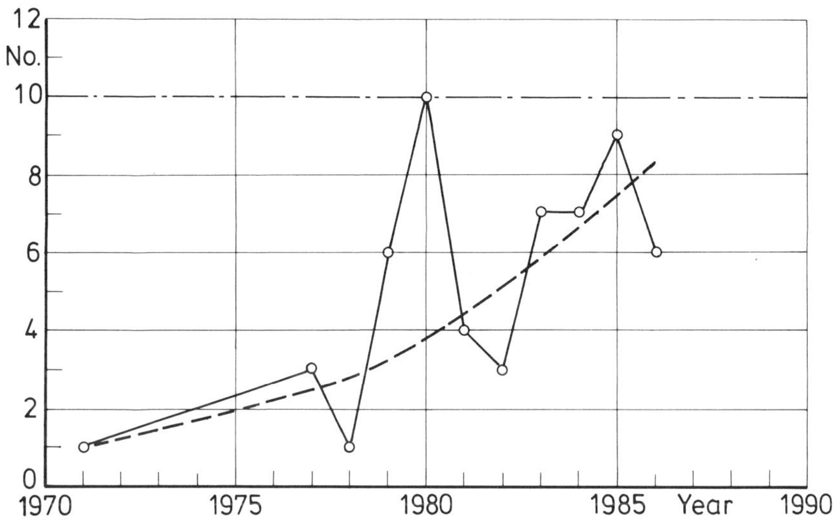

Fig.2. Number of publications about heat pump systems with non-azeotropic mixtures

Also the International Energy Agency created a special working group in its Annex VI to update the state of the art and investigate the potential of the application of refrigerant mixtures for heat pumps. Three reports have been prepared by this working group about this topic (11), (12), (13).

Further, the IEA Heat Pump Center published a bibliography about non-azeotropic refrigerant mixtures as working fluids in compression heat pumps (14). Finally the ASHRAE published a Technical Data Bulletin on Advances in Non-azeotropic Refrigerants for Heat Pumps (15).

All these publications show the great interest in non-azeotropic refrigerant mixtures for heat pumps offering by their adaptable composition a new dimension concerning the layout and design of vapour compression systems and the amount of work which has been done since the first proposal to use those fluids in heat pumps. The question is what results according to these proposals have been gained and what are the future aspects of refrigerant mixtures for heat pumps.

CURRENT STATUS OF REFRIGERANT MIXTURES FOR HEAT PUMPS

Because of the advantages of non-azeotropic refrigerant mixtures of heat pumps after a thorough experimental investigation (16) the research and development mainly has concentrated on three binary mixtures R12/R114, R22/R114 and R13B1/R152a. Mixtures with R11 show some problems with oil solubility and hence lubrication so that R114 instead of R11 is the better choice in normal heat pump temperature range for space and tap water heating whereas in the higher range for industrial heat pumps R114 is prefered against R11 and R113 as well because of stability problems of the refrigerants themselves.

Besides those three binary non-azeotropic mixtures also ternary mixtures have been proposed for heat pump application in patents like R13/R22/R11 (17) and investigated like R12/R114/R11 (18), R23/R22/R114 (19) and R13B1/R12/R152a (20). They can offer special advantages over binary mixtures which shall be discussed later.

All research and development of non-azeotropic mixture application in heat pumps can be divided into two subsections, namely into the investigation of the thermophysical data of the fluids and the evalution of the potential of mixture application in heat pumps.

Thermophysical Data of Refrigerant Mixtures for Heat Pumps

At the beginning of the heat pump mixture research very little information about the behaviour of the mixed fluids for heat pumps was available because that time those investigations had been made especially for lower temperature cooling application. Therefore an intensive research of the state and transport properties of those mixed refrigerants for heat pumps was necessary. Up to now a consibarable knowledge has been gained in this field so that a better theoretical evaluation and preliminary design of heat pumps with mixed fluids can be made.

State properties

Whereas in the first evaluation of mixture potential for heat pumps because of lacking information on fluid properties the blend was considered as an ideal mixture. After measurements of the state properties it was possible to use equations of state predicting also the non ideal behaviour of those mixtures.

Besides of a lot of experimentally investigated properties of refrigerant mixtures (21) the three important refrigerant mixtures for heat pump application have been investigated by measurements of the state properties in different laboratories and presented in publications for R12/R114 (22) (23), R22/R114 (24) and R13B1/R152A (25), so that was it possible to establish equations of state to predict the thermodynamic properties of those mixtures.

Ternary Mixtures for Heat Pump Application have been investigated experimentally for the mentioned ternary mixture R13B1/R12/R152a (20) and R22/R12/R114 (21) besides some other ternary mixtures not investigated up to now for heat pump application. Also for the measured data for ternary mixtures equations of state have been applied to predict their state properties.

In order to predict the thermodynamic properties of refrigerant mixtures and use the equations of state for performance evaluation of the heat pump system the equations must not be not too complicated since additional mixing rules have to be applied as compared to the equations for pure fluids. Therefore the very often used Martin-Hou-Equation for pure refrigerants with its 12 individuell parameters gets too complicated and very often simpler equations with reasonable accuracy are sufficient for that purpose.

The Benedict-Webb-Rubin-Equation in the modified form of Lee, Kessler and Plöcker (LKP) is used very often as well as modifications of the simpler Redlich-Kwong-Equation for instance in the modification of Soave (RKS). The advantage of this R-K-S equation is the possibility of a solution according to a cubic root. This equation therefore is used very often for the prediction of refrigerant mixtures for heat pumps as well as for binary blends (25), as also for ternary mixtures (20). Since this very simple R-K-S equation does predicts very well the vapour properties but very poor the liquid properties of the mixed fluids efforts were made by Carnahan, Starling and De Santis leading to an improved but also again more complicated equation (27), (28). Using the measured data to adapt the equations of state to the properties of the mixture and predicting in the whole application range those properties thermodynamic diagrams for the non ideal blends have been calculated and published for instance for mixtures of R12/R114, R22/R114 and R13B1/R152a (23), (29), (30), (25), (31), Figure 3 (23). Those diagrams allow a rough estimation of performance of a mixed refrigerant cycle for heat pumps when no thorough calculation using the equations of state is possible.

Transport properties

The transport properties as viscosity, thermal conductivity, surface tension and diffusion coefficient are important for the heat transfer between fluid and heat exchanger wall. Very poor information about those properties is available in the literature, only concerning the important non-azeotopic mixtures for heat pumps a few data of the viscosity of the fluid R22/R114

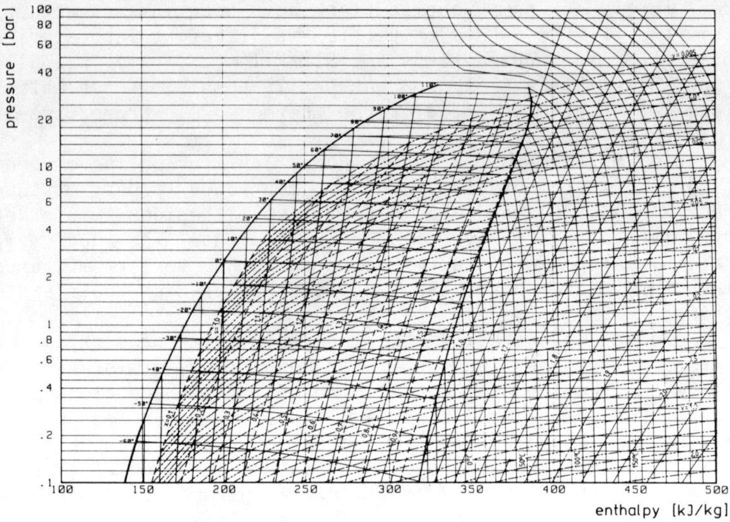

Fig.3. p-h diagram of R-12/R-114 (70% R-12)

are available (32). Besides of viscosity of those refrigerant mixtures also the viscosity of a blend of the non-azeotropic refrigerant mixture R12/R114 together with oil had been investigated in the oil rich region (33). Therefore currently very little information about the transport properties is available. Concerning the heat transfer coeffients as influenced by those transport properties in contrary more investigations have been made.

Heat transfer coefficients

The main work on heat transfer investigations concerning refrigerant mixtures has been made for the evaporation and only one for the condensation.

The heat transfer measurements during evaporation in the earlier investigations have been made namely under pool boiling conditions (34). The measurements show a great degradation of the heat transfer coeffient of a non-azeotropic refrigerant mixture as compared to its pure components.

For heat pumps the heat transfer coefficients of flow boiling during evaporation is more important because of the mainly used dry expansion evaporators in those systems. Here investigations (35), (36) show that a degradation in the range of 30% can be observed which decreases with increasing mass flow because of better remixing effects with high turbulency. On the other hand with high mass flow rates the pressure drops decrease and counteract the advantage of gliding temperatures of non-azeotropic refrigerant mixtures. So to a certain extent also for flow boiling inside tubes a degragation of the heat transfer coeffient has to be accepted.

Flow boiling in falling films show better behaviour with only 15% degradation as a very thorough investigation (37) show. So the design of a falling film evaporator shows a very good performance for non-azeotropic refrigerant mixtures concerning the heat transfer, but normally can be applied only for larger systems.

For condensation also a degradation of the heat transfer can be expected to a certain degree as far as first measurements show (38). The reason is a separation of vapour and liquid mainly in the midrange of the condenser (40). The degradation of the heat transfer coeffient is in the order of approximately 30% in maximum.

Concerning the degradation of the heat transfer coeffients for evaporation and condensation it should be stated that according to parameter studies (39) 30% lower heat transfer coeffients both on condenser and evaporator side ot the heat pump system lead to a decrease of the coeffients of performance of 2% by the degradation in the heat exchangers each. It should be taken in mind if the evaluation of the performances of heat pumps with mixtures are concerned.

Evalution of the Mixture Potential in Heat Pump Systems

As explained earlier the advantages of mixture application in heat pump systems can be the possibility of energy saving by applying gliding evaporating and condensing temperatures, by continuous capacity control and by adapting hardware components concerning capacity and application limits to the special working in heat pumps.

Energy saving

One of the main incentive to propose refrigerant mixtures for heat pumps was the aspect of possible energy saving by using the gliding evaporating and condensing temperature in order to decrease irreverasibilities of the heat exchange created by the temperature differences in the heat exchangers. This evaluation has been made as well theoretically with computer programs for those systems with mixtures as well as also by measurements made with refrigerant cycles at special test arrangements as also in real heat pump systems.

Computer programs for heat pump systems using non-azeotropic refrigerant mixtures have been written (39), (40), (41) and applied in order to predict the potential of energy saving for heat pumps of different application cases as for instance for space heating, tap water heating and industrial heat generation. As already Figure 1 showed the possible amount of energy saving depends on the temperature change of the secondary fluids in the evaporator and condenser to which the temperature change of the non-azeotropic refrigerant mixture has to be adapted in an ideal manner. This normally is difficult to obtain because the secondary fluids change their temperature according to their sensible heat along the heat exchanger in a linear manner whereas the refrigerant mixture shows a non linear behaviour during phase change. In order to obtain as parallel as possible temperature profiles of both secondary fluids and the refrigerant mixture in the heat exchangers the composition of the mixture has to be chosen that way that the temperature lift during evaporation or condensation is nearly the same like those in the secondary fluids. The maximum width Δ tmax between dew and boiling line lies in the region of a 50% composition and depends on the distance of the normal boiling points Δt_s at atmospheric pressure (42).

So if this 50% composition has the same temperature change as the secondary fluids a maximum of the coefficient of performance (COP) by a good parallelity of the temperature profiles in the heat exchangers can be expected at this composition. For mixtures with higher temperature changes between dew point and boiling point at the midths of the lense two compositions should show the optimum COP values one the left hand and the other on the right hand side of this maximum width at those compositions where the temperature change during evaporation and condensation is the same as in the secondary fluids. Indeed those computer programs as established to predict the heat pump performance with non-azeotropic mixtures show these results. Simpler computer programs (41) used ideal behaviour of the mixtures itself and did not take into account their behaviour concerning heat transfer coeffients and pressure drops whereas later developed physical models (42) include those effects. In general especially the simpler computer programs which allow extensive parameter studies indicate that for approximately 10 K temperature change in the secondary fluids by making use of the gliding temperatures of non-azeotropic refrigerant mixtures a COP improvement roughly 15% should be gained. The computer evaluation of ternary mixtures show according to figure (18) for the COP that the highest COP is obtained by a binary mixture of the three components so that other effects of ternary mixtures should be more interesting than the energy saving aspect. The question is to what extent the theoretical predictions of energy saving with non-azeotropic refrigerant mixtures in heat pumps have been confirmed by experiments.

In the last 10 years a lot of experimental evaluations have been made with special heat pump test rigs or with real heat pumps in order to confirm the predictions or not. The result of those measurements is that very different amounts of improvements of the COP have been measured and very contradictory results have been obtained. The main reason for the contradicting results is the kind of comparison of test results with pure fluids and mixed refrigerants. There are two extrem possibilities to compare those tests assuming on the one hand by applying the same evaporator and condenser inlet temperatures that the heat exchangers have infinite areas as well for pure fluids as for mixtures and on the other hand with finite areas both the same size for pure and mixed refrigerants, Figure 4. The first case is very favourable for the mixture

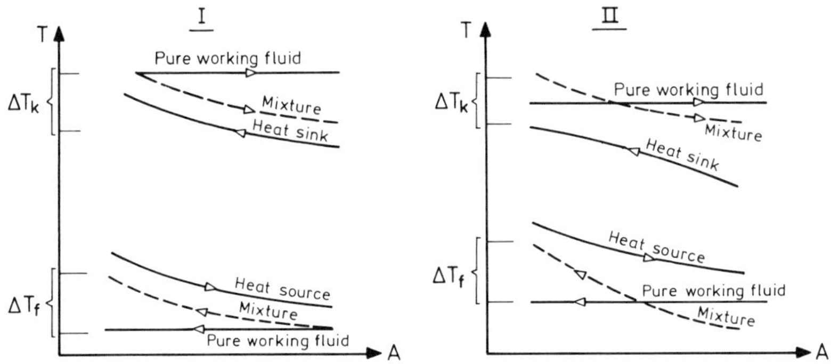

Fig.4. Two ways of comparing a mixture and apure working fluid (A = heat transfer area)

because it allows for the mixture an infinite small temperature difference during heat exchange whereas the pure fluid has still a given difference because of the pinch point. In the second case where for both fluids the heat exchanger size is the same and, assuming that the heat transfer coeffients are also the same than the mean logarithmic temperature difference must be also the same for a given heat load. This results in nearly the same COP or a slightly smaller COP if the degradation of heat transfer coefficients and certain pressure drops are taken into account. The conclusion of this is that in order to get energetic improvements with non-azeotropic refrigerant mixtures the heat transfer areas must be enlarged to a certain size as given by economical aspects where the smaller running costs as given by a better COP lead to reasonable payback time for the higher capital costs of the heat exchangers.

So the different test results ranging from for instance from 30% COP improvement for a mixture of R22/R114 (16) to 0% or a small degradation of the COP can be explained by the special test conditions. Both cases show the limits of the possible range of COP between a maximum value with infinite areas and a minimum value with the same heat exchanger size. The problem evaluating experimentally refrigerant mixtures in heat pump systems is that with the composition although the capacity changes. Changing capacity leads to changing heat loads in the heat exchangers and consequently changing temperature differences so that a comparison to the pure fluids with the same heat exchanger area cannot be made.

Taken into account all experimental results as known up to now to evaluate the energetic improvement of refrigerant mixtures in heat pumps it can be stated that in general not the predicted values could be achieved.

Besides of the difficulty to enlarge in tests the heat exchanger size it is nearly impossible to maintain experimentally the same heat loads with pure fluids and mixtures of different compositions without any other difficulties. Adapting the heat load by the mass flow via compressor speed changes pressure drops in the lines and compressor efficiency too. The right way would be to use for each mixture an optimized heat pump system. This is nearly impossible experimentally. The only possibility is to evaluate with a good and confirmed heat pump model for non-azeotropic refrigerant mixtures the performance theoretically.

Also a reason for the discrepancy between calculations and tests is the fact that the computer models do not take into account the real behaviour of the system components when using refrigerant mixtures, especially the heat exchangers with their degradation in heat transfer coefficients and their pressure drops and the compressor with its different behaviour concerning its efficiency. Especially a better knowledge about the compressor efficiency is necessary.

Measurements have been made and computer programs established which take those effects into account (43), they show that for instance for the mixture of R22/R114 according to Figure 5 only slightly changings of compressor effiency with the mixture can be stated. On the other hand according to the applied

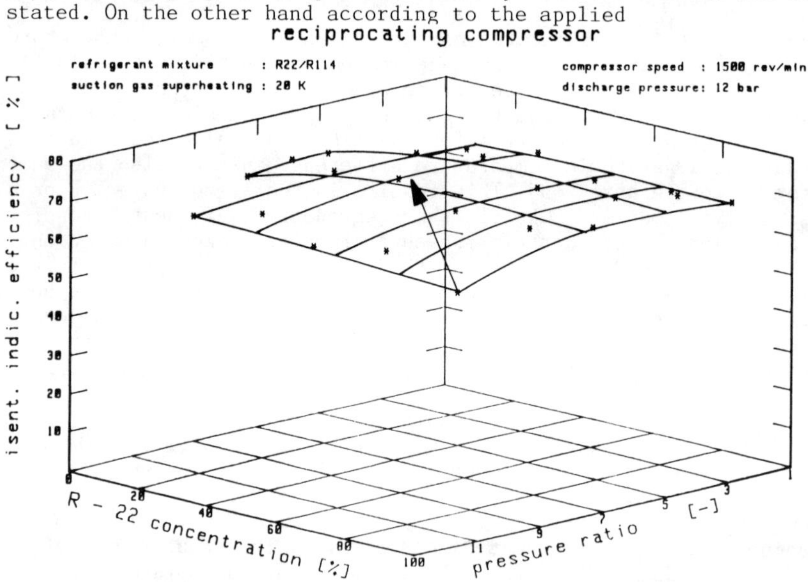

Fig.5. isentropic indicated efficiency as a function of pressure ratio and R-22 concentration

working conditions as given by the gliding temperatures the pressure ratio also is changed. Changing pressure ratio has also an effect on compressor efficiency and on the isentropic compression work. So the different behaviour of a compressor running with a pure refrigerant on the one hand and a refrigerant mixture on the other depends as well as on the influence of the fluid onto the efficiency for instance by influencing the pressure drops in the valves as also by the applied external pressure ratio leading to different efficiencies and isentropic works. Hence it is very important to include into the models the influence of pressure ratio and mixture compostion on the compressor efficiency.

The conclusion concerning the potential of energy saving when using refrigerant mixtures in heat pumps is that an overall optimized system including the heat exchangers and the compressor has to be investigated especially for the mixed fluid. This can only be done if a theoretical heat pump modell includes all real effects of the mixture onto the components.

Capacity control

Another incentive to use refrigerant mixtures was the possibility of capacity control by a continuous composition shift. This has experimentally demonstrated with a rectifying column 5 (33). A wide range capacity control with a constant system COP is possible taking not into account the additional heat supply to the sump of the columne. This system was investigated for internal combustion engine driven heat pumps where the waste heat of the engine can be used to heat this sump and consequently does not effect the coeffient of performance. For smaller heat pumps other systems have been proposed (44), (45) with the refrigerant mixture R13B1/R152a with a single distillation stage in the suction line or in the liquid line or both. Whereas with binary mixtures only small amounts of composition change in the region of 10% could be stated (45), by using a ternary mixture of R23/R22/R114 a greater capacity range of 28% to 100% could be achieved with the last mentioned system (19).

Adaption of capacity and applications limits

An interesting aspect when using non-azeotropic refrigerant mixtures is the possibilitiy to use mass produced hardware components for refrigeration systems and adapt them according to the wanted capacity or the allowed application limits when running in heat pump systems. For instance using a mixture of R22/R114 the capacity can be changed between the values of 100% R22 and of R12 at a mass friction of around 40% R22/R114 with nearly the same COP (46), also the application limits can be changed by applying a refrigerant mixture instead of a pure component for a given hardware. Running a screw compressor with a maximum casing pressure of 13 bar with a mixture of R12/R114 instead of a pure refrigerant R12 higher condensing temperatures than 56 C for the pure fluid could be obtained (33). Also the torque of such a compressor can be adapted to the driving torque of internal combustion engines if not the right compressor size is available. So in this way tailormade refrigerants for certain given hardware components can contribute to a better economy of heat pumps.

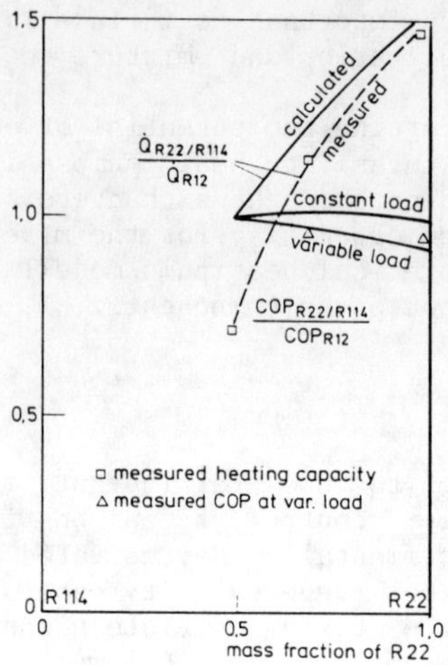

Fig.6. Comparsion of measured and calculated heating capacity and COP for heat pump application

Practical problems

When substituting pure refrigerant by non-azeotropic refrigerant mixtures problems have been predicted very often concerning leakage and oil behaviour. An investigation published recently (47) showed that simulating leakages in the two phase region at the condenser the receiver and the evaporator leading to a diminishing of the charge of about 50% in the first 2 components and 12% in the eveporator no remarkable composition change could be measured. No problems with leakage should not occur when using non-azeotropic refrigerant mixtures instead of pure fluids.

In order to design adequately heat pump systems with refrigerant mixtures a good knowledge of the behaviour of the oil together with the mixture is necessary as far as the viscosity and solubility later are concerned. It has been shown that from the data of the mixture of oil with each component the behaviour of the so called ternary oil refrigerant mixture can be predicted very well (33). So a good basis for system design from this point of view is also available.

FUTURE POTENTIAL OF REFRIGERANT MIXTURES FOR HEAT PUMPS

When trying to evaluate the future potential of non-azeotropic refrigerant mixtures in heat pumps it is favourable to judge their application concerning the three main advantages which they offer, namely energetic improvement, capacity control and adaption of hardware components regarding capacity and application limits.

Further, concerning an evaluation of the potential the different types of heat pumps have to be considered. The

International Energy Agency in its Annex VI has defined by a questionaire several heat pump profiles of heat pumps of different sizes which can roughly be divided into three cases, namely heat pumps for space heating, tap water heating and industrial heat pumps. For those three cases under the different aspects of the advantages which refrigerant mixtures do offer the evaluation concerning a future potential shall be made.

Heat Pumps for Space Heating

According to the IEA questionaire in most cases the temperature change in the secondary fluids of evaporator and condenser are less than 10 K. In this case only small improvements of COP by using the gliding temperatures of non-azeotropic mixtures in comparison to pure fluids theoretically can be gained and have been confirmed by experiments. Also an increase of the heat exchanger area to double the size, see Figure 7, does not increase remarkable the COP further. So, under the assumption of small

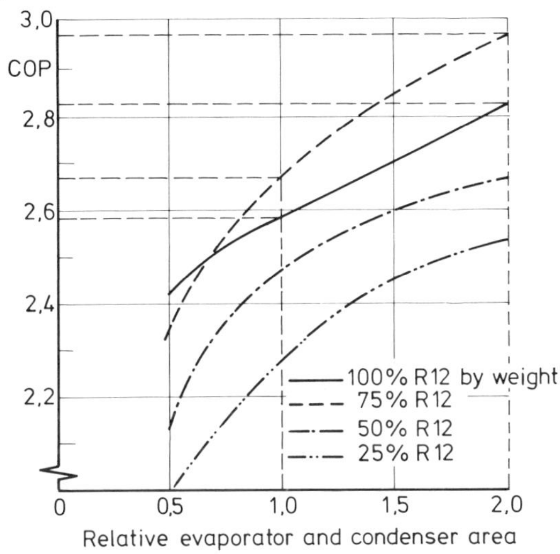

Fig.7. COP of pure refrigerant R-12 and blends R-12/R-114 versus heat exchanger size (space heating case) (39)

gliding temperatures according to the IEA questionaire there is only very small potential for improvements in the future. Higher temperature changes in the secondary fluids as for instance in older hydrolic radiator systems with 20 K difference on the condenser side could improve the situation but especially in new heating systems these temperature differences are very small so that it can be stated that for space heating heat pumps the potential in the future as far as energy saving aspects are concerned is small.

The aspect of a continuous capacity control on the other hand is for those heat pumps very important since very often the heat source is ambient air and changes its temperature in contrary to the heating demand. In those cases the heat pump has to be designed so that at a certain balance point its heating capacity can cover the heating demand. At higher outdoor temperatures a capacity control is required which mainly is done by on/off control. It has been shown that capacity control by composition shifting is a possibility of non-azeotropic refrigerant mixtures. A very wide capacity range be controlled by a rectifier but this for small residential heat pumps is normally a too expensive device. So only very simple configurations which are based on the distillation process can be used because of cost considerations. Here the future potential of refrigerant mixtures lies in the possibility of designing a better system than the mainly investigated distillation process with R13B1/R152a. One way for improvement in the future could be the use of a ternary mixture (49) which seems to be a better fluid for capacity control. Other ways are thinkable and under investigation.

The future potential of a capacity control using non-azeotropic refrigerant mixtures depends further mainly on the development of marketable speed inverter control equipment. If a non-expensive inverter control for refrigeration compressors is available then the incentive to use non-azeotropic refrigerant mixtures for capacity control will be diminished. It is questionable if an additional mixture control to an possible inverter control will be reasonable from a cost point of view.

Another future potential for heat pumps for space heating purposes is the possibility of using a refrigerant mixture for special adaptation concerning the capacity and application limits of the compressor. Whereas for smaller heat pumps a lot of components of different sizes are available for larger compressors it could be advantageous to adapt the capacity by using a suitable refrigerant mixture instead of pure fluid. So if for instance R114 gives a too low capacity and R12 a too high one a mixture of both fluids could use a given hard ware component for the required capacity as for instance for large turbo compressor. Also the application limits for all sizes of compressors can be extended by using a refrigerant mixture for instance in the case if the condensing temperature is too high for R12 application and when using R114 the volumetric capacity of this refrigerants lead to a too large compressor. So also a mixture could be a good choice to adapt the application limits of a compressor to the required temperatures.

In total it can be stated that the future potentials of heat pumps for space heating purposes are capacity control and adaption of hard ware components by using non-azeotropic refrigerant mixtures.

Heat Pumps for Tap Water Heating

According to the mentioned questionaire tap water heat pumps normally have higher temperature gradients then heat pumps for space water heating purposes. So here in future there could be a better potential for energy saving purposes. On the other hand very often the design of tap water heat pumps are so that at least the condenser because of being inside the storage has only a very small temperature gradient because of the stratification of the water. In these cases tap water heat pumps show nearly no energetic potential for improvements by using non-azeotropic heat pumps. For tap water heat pumps the capacity control because of the small size of those heat pumps should be in future normally an on/off control or maybe later a inverter control. It is felt that a capacity control using control devices by shifting composition of a non-azeotropic mixture is too expensive for those small heat pumps. Here no future potential for non-azeotropic mixture can be seen. Because of the limited temperature of the tap water and the heat source the possible extension of application range of heat pump compressors cannot offer a special advantage in this field. Also, because for tap water heat pumps hermetic compressors of small sizes are used which are offered with a lot of various swept volumes no advantage can be seen to adapt the capacity by using an non-azeotropic mixture instead of a pure refrigerant in tap water heat pumps with those small hermetic compressors.

Industrial Heat Pumps

According to the IEA questionaire the industial heat pumps offer very often temperature changes in the secondary fluids of 15 - 20 K. As can be seen from figure 47. Those temperature changes lead to energetic improvements as predicted by computer programs in the range of about 12%. So here, if special cases with high temperature changes in the secondary fluids can be detected the application of non-azeotropic refrigerant mixtures can offer a favourable potential in the future. Enlarging the heat exchanger areas it can be seen from Figure 8 for industrial heat pumps the mixture increases still in COP when the pure refrigerant levels out already.

Here the investment cost for greater heat exchanger sizes can be contribute to a better economy by lower running cost. Computer calculations have shown that the payback period when using a adequate mixture as compared to R114 for an industrial heat pump a decrease of up to 10% can be gained. This amount of economic improvement seems too small to influenced decision making in industry for the use of a heat pump with non-azeotropic refrigerant mixtures. Although there is a remarkable energy saving potential when using those fluids the question is if the incentive in the industry is great enough from the economy point of view. Additional advantages have to be created by the use of non-azeotropic refrigerant mixtures. The composition shifting when using non-azeotropic refrigerant mixtures can be an important aspect of future application of those fluids in industrial heat pumps because additional energy saving from the system efficiency can be gained if a continuous control can be achieved. For larger industrial heat pumps also more sophisticated capaciy control devices like rectifying columns can be used which are able to change the capacity of the heat pump continuously in a wide range. If those heat pumps are driven by electrical motors this advantage of non-azeotropic refrigerant mixtures can be an important future potential for industrial heat pumps. If they are driven by internal combustion engines then because of the possible speed control in a wide range this is a hard competition to the method of composition shifting.

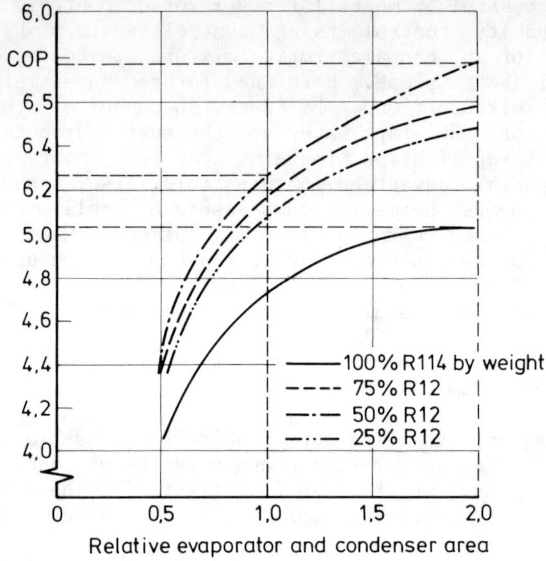

Fig.8. COP of pure refrigerant R-114 and blends R-12/R-114 versus heat exchanger size (industrial heat pump) (39)

Another advantage of non-azeotropic refrigerant mixtures, the possible hardware adaption concerning capacity and application limits as discussed earlier can be a important future potential especially for industrial heat pumps. In this case not big series of hardware components like compressors are available on the market and very often the steps of possible swept volumes are very large. So a tailormade refrigerant in this case can adapt the same compressor to special required capacities. Industrial heat pumps normally have to cover a wide range of capacities according to the given situations in industry and the hardware components have to be adapted in a wide range. Also it has been shown that the temperatures of heat source and sink vary in a wide temperature range. This also can influence as well the capacity of the components as also the application limits for given compressors with certain pure refrigerants. Also here a continuous extension of the application limits can be made when using non-azeotropic refrigerant mixtures.

SUMMARY

Summing up all future potentials of non-azeotropic refrigerant mixtures it can be stated that heat pumps for space heating mainly can apply refrigerant mixtures for capacity control in an advantageous, if a low cost device for a wide manner range capacity control by composition shifting can be designed.

Tap water heat pumps offer only some potential if they show larger gliding temperatures in the secondary fluids which not in all cases is valid. So in total the potential of applying non-azeotropic refrigerant mixtures in tap water heat pumps seems to be limited.

Industrial Heat Pumps normally have high temperature differences in the heat source as well as in the heat sink if they transport sensible heat. A remarkable amount of energy saving by using non-azeotropic refrigerant mixtures is possible but at the recent energy prices the economic advantage is not so great that a decision in industry for those heat pumps can be expected alone by this possibility. Additional the possible capacity control which allows for larger systems more sophisticated devices with wide range capacities and special tailormade refrigerants which can adapt the hardware of an industrial heat pump to the very different situation in each case in industry can be a great advantage to use non-azeotropic refrigerant mixtures in those systems. So the conclusion is that the best potential for using non-azeotropic refrigerant mixtures in future should be in industrial heat pumps.

REFERENCES

/1/ Kruse, H. — Energy Saving by Using Suitable Refrigerants for Heat Pumps in Europe ASHRAE Sminar International Day Seminannual Meeting Chicago 1977

/2/ Kruse,H. Jakobs, R. — Die Bedeutung der nichtazeotropen Zweistoff-Kältemittel beim Einsatz in Wärmepumpen und Kälteanlagen KI Klima + Kälte-Ingenieur 7-8/1977

/3/ Lotz, H. — Behaviour of Single- and Manifold-Medium Heat Pumps Vortrag auf der Wärmepumpentagung Essen (Germany, F.R.) 1977

/4/ Haselden, G.G. — Refrigeration Cycles Providing Cooling over a Temperature Range Proc. Inst. of Refrigeration, London,49 (1952/53) p. 136/141

/5/	Haselden, G.G. Klimek, L.	An Experimental Study of the Use of Mixed Refrigerants for Non-Isothermal Refrigeration J. Refr. 1 (1958)4, pp. 87-89
/6/	Etherington, T.L.	Variable Capacity Refrigeration System US Patent 2 951 350 Cl. 62-149 (1960)
/7/	Schwind, H.	Multiple Fluid Refrigeration System US Patent 2 938 362 CI. 62-149 (1960)
/8/	Herrick, C.S.	Variable Temperature Refrigeration US Patent 2794 329 Cl. 62-115 (1957)
/9/	Mc Harness, R.C. Chapman, D.D.	Refrigeration Capacity and Performance Data for Various Refrigerants, Azeotropes and Mixtures ASHRAE Trans. 67 (1961), pp. 441-464
/10/	Jakobs, R. Kruse, H.	The Use of Non-Azeotropic Refrigerant Mixtures in Heat Pumps for Energy Saving IIF - IIR Commission B2 Delft (The Netherlands) -1978-3
/11/	Berntsson, T. Ljungvoist, K. Moser, F. Schnitzer, H.	High Temperature Working Fluids and Non-Azeotropic Mixtures in Compressor Driven Heat Pumps IEA Annex VI, State of the Art Report
/12/	Berntsson, T. Schnitzer, H.	Study of Working Fluid Mixtures and High Temperature Working Fluids for Compressor Driven Systems IEA Annex VI, Final Report Part A 1985
/13/	Berntsson, T. Schnitzer, H.	Study of Working Fluid Mixtures and High Temperature Working Fluids for Compressor Driven Systems IEA Annex VI, Final Report Part B 1986
/14/	Herrmann, J.	Non-Azeotropic Refrigerant Mixtures as Working Fluids in Compression Heat Pumps IEA Heat Pump Center, Edition 2, December 1985, Report No: HPC-B1
/15/	-	Advances in Non-Azeotropic Mixture Refrigerants for Heat Pumps ASHRAE Technical Data Bulletin ASHRAE Annual Meeting at Honolulu, Hawaii, June 1985
/16/	Jakobs, R.	Die Verwendung von nichtazeotropen

Zweistoff-Kältemitteln in Wärmepumpen
Deutscher Kälte- und Klimatechnischer Verein
Forschungsbericht Nr. 3

/17/ Rojey, A. — Heat Production
UK Pat. Appl. GB 2 068 996 A. (1981), 6 pp.

/18/ Moser, F., Schnitzer, H. — Mixtures of Two or More Refrigerants as Working Fluids for Compression Heat Pumps
VDI Berichte 539, ORC-HP-Technology, Proc. Int. VDI-Seminar, Zürich (1984), pp. 117-131

/19/ Vakil, H.B. — Means and Method for Modulating and Controlling the Capacity of a Vapor Compression Cycle Device
US Patent 4,384,460 (1983)

/20/ Wright, E.G. — Prediction of Refrigerant Ternary Mixture Properties Using the Redlich-Kwong-Soave Equation of State
ASHRAE Technical Data Bulletin
ASHRAE Annual Meeting at Honolulu, Hawaii, June 1985

/21/ Schulz, U.W. — Thermodynamic Properties of Multicomponent Refrigerants
IIF - IIR Commission B1, B2, E1, E2
Purdue (USA) 1986

/22/ du Pont — Vapor Pressure and Liquid density of "Freon" Propellants
E.I. du Pont de Nemours 1957

/23/ Küver, M., Schroeder, M., Kruse, H. — Measurement and Calculation of p-v-T-x Data of the Binary Mixture R12/R114
IIR-XVIth International Congress of Refrigeration (Paris 1983)
Proceedings Tome II

/24/ Hackstein, G. — Ein Beitrag zur experimentellen Bestimmung thermodynamischer Stoffwerte an halogenierten Kohlenwasserstoffen, deren Gemischen sowie deren Gemische mit Kältemaschinenöl
Dissertation, Universität Dresden, 1976

/25/ Connon, H. — Estimation and Application of Thermo-

	Drew, D.	dynamic Properties for a Non-Azeotropic Refrigerant Mixture IIF -IIB Commission B1, B2, E1, E2 Essen (FRG) -1981-5
/26/	Küver, M. Kruse, H.	The Application of Non-Azeotropic Refrigerant Mixtures in Two Temperature Refrigerators IIF - IIR Commission B1, B2, E1, E2 Purdue (USA) 1986
/27/	Morrison, G.	The Importance of Including the Liquid Phase in Equations of State of Non-Azeotropic Refrigerant Mixtures ASHRAE, Transactions, 1985
/28/	Morrison, G. Mc Linden, M.	Two Refrigerant Mixtures and the Hard Sphere Fluid ASHRAE Technical Data Bulletin ASHRAE Annual Meeting at Honolulu, Hawaii, 1985
/29/	Pedersen, P.H.	Stofegenskaber for R12/R114 kolemiddelblandinger Refrigeration Laboratory, Technical University of Denmark, Lyngby, March 1982, Report F67
/30/	Fushimi, K. Daikin Ind. Ltd.	Heat Pump with Non-Azeotropic Refrigerant Mixture IEA Annex VI, Final Report, Part B
/31/	Domanski, P.A. Didion, D.A.	Equation of State-Based Thermodynamic Charts for Non-Azeotropic Refrigerant Mixtures ASHRAE Transactions, 1985
/32/	Heide, R.	An Instrument for Measuring the Viscosity of Refrigerants and Refrigerant Mixtures Sci.Tech.Froid (1980) 1, pp. 99-104 10 ref.
/33/	Schroeder, M.	Entwicklung und Erprobung eines preisgünstigen Schraubenverdichters für die Kältemittel R12 und R114 sowie deren Mischungen BMFT-FB-T 86-022, Februar 1986
/34/	Jungnickel, H. Wassilew, P. Kraus, W.E.	Investigations into Heat Transfer in Boiling Binary Refrigerant Mixtures Proc. XVth Int. Congress of Refr. Venice, Sept. 1979, Paper No. B1 -47

/35/ Pedersen, H. — Heat Transfer to Evaporating Refrigerant Mixtures
EFP81, Projekt 7.5, Teknologisk Instituts Forlag 1983

/36/ Radermacher, R.
Ross, H.
Didion, D.A. — Experimental Determination of Forced Convection Evaporative Heat Transfer Coefficients for Non-Azeotropic Refrigerant Mixtures
ASME-Paper 83-WA/HT-54, 1983, 16 ref.

/37/ Berntsson, T.
Berntssion, K.M.
Panholzer, H. — Heat Transfer of Non-Azeotropic Mixtures in a Falling Fim Evaporator
ASHRAE Technical Data Bulletin
ASHRAE Annual Meeting at Honolulu, Hawaii, 1985

/38/ Stoecker, W.F.
Kornota, E. — Condensing Coefficients when Using Refrigerant Mixtures
ASHRAE Technical Data Bulletin
ASHRAE Annual Meeting at Honolulu , Hawaii, 1985

/39/ Berntsson, T.
Schnitzer, H. — Study of Working Fluid Mixtures and High Temperature Working Fluids for Compressor Driven Systems
IEA Annex VI, Final Report Part A
Sept. 1985
Chapter 5, Munch Bertsson, K.
Berntsson, T.
"Technical Sensitivity Analysis"

/40/ Connon, H.A. — A Generalized Computer Program for Analysis of Mixture Refrigeration Cycles
ASHRAE Transaction 1983

/41/ Domanski, P.A.
Didion, D.A. — Simulation of a Heat Pump Operating with a Non-Azeotropic Mixture
ASHRAE Technical Data Bulletin
ASHRAE Annual Meeting at Honolulu, Hawaii, 1985

/42/ Lorenz, A. — Application of Binary Refrigerant Mixtures in Compression Refrigeration Systems
Luft-Kältetech. 9 (1973) 6, pp. 296-301

/43/ Quast, U.
Kruse, H. — Experimental Performance Analysis of Reciprocating Compressor Working with Non-Azeotropic Refrigerant Mixtures
1986 Int. Compressor Engineering Conference -at Purdue

/44/ Vakil, H.B. — New Concepts in Capacity Modulation Using Non-Azeotropic Mixtures
IIR XVIth International Congress of Refrigeration (Paris 1983) Proceedings Tome II

/45/ Cooper, W.D. — The Use of Mixed Refrigerants in Air-to-Air Heat Pumps
ASHRAE Transactions 1982

/46/ Kruse, H.
Küver, M.
Quast, U.
Schroeder, M.
Upmeier, B. — Theoretical and Experimental Investigations of Advantageous Refrigerant Mixture Applications
ASHRAE Technical Data Bulletin
ASHRAE Annual Meeting at Honolulu, Hawaii, 1985

/47/ Blaise, I.C.
Dutto, T. — Some Practical Results Obtained With Non-Azeotropic Mixtures of Refrigerants in High Temperature Heat Pumps
IIR Commissions E2
Meeting August 5-8
Purdue (USA), 1986

Chapter 14

What Needs To Be Known About Working Fluids to Calculate Coefficients of Performance

Georg Alefeld

ABSTRACT: The efficiencies of compressor heatpumps are usually calculated by enthalpy balances. Complete and accurate enthalpy diagrams are needed.

Starting from First and Second Law an equation for the COP has been derived into which fluid properties enter only as non-dimensional parameters. For freons the knowledge of only one parameter is sufficient to predict COPs quantitatively.

INTRODUCTION

The common procedure for the calculation of COPs of compressor heatpumps or refrigerators is as follows: First, complete enthalpy-pressure or enthalpy-entropy or temperature entropy diagrams must be established. By applying First Law and using the enthalpy charts, the COP can be determined.

Everybody, trained in basic thermodynamics, knows that the COP of a heatpump must and will obey the Second Law. But where has the Second Law come in? The answer is very simple: By using the experimentally determined data charts of real fluids, the requirements of Second Law have been taken into account, although in a rather indirect way. Nevertheless the term "first law efficiency" for such a COP is at least misleading.

The statements of Second Law are independent of any fluid properties. In this presentation /1/ a different approach is made: Second Law is taken into account from the very beginning, independently of any special properties of the fluid to be

HEAT PUMPS: Prospects in Heat Pump Technology and Marketing (Proceedings of the 1987 IEA Heat Pump Conference), Edited by Kay H. Zimmerman

considered. By having introduced the requirements of Second Law in a direct way, the still needed knowledge about fluid properties is reduced to a few nondimensional parameters. These paramaters can be used to characterize and evaluate fluids. In text books the Second Law is usually presented in verbal formulations. We will use Second Law in an algebraic form by making use of the fact, that entropy is a state function i.e. by entropy balances. This "Second Law Analysis" which will be presented should not be confused with an availability or exergy analysis, for which the term "second law analysis" is commonly used. On the contrary: We will show that our analysis does not have the shortcomings of the exergy analysis, namely to be dependent on an undefined ambiguous environmental temperature.

THE COP OF COMPRESSOR HEATPUMPS USING FIRST AND SECOND LAW

Figure 1 shows schematically the energy flow in a compressor heatpump. For the external temperatures i.e. those of the heat carrier fluids, small letter symbols and for the internal temperatures capital letters are used. In case of working fluid-mixtures the internal temperatures have a certain spread as well. In Figure 2 and Figure 3 the compression process is displayed for fluids with differing properties at the dew line, namely with $ds/dT < 0$ (e.g. R 22) or with $ds/dT > 0$ (e.g. R 114).

The First and the Second Law can be written as:

I. Law: $$w + q_1 = q_2 \tag{1}$$

II. Law: $$\frac{q_2}{t_2} - \frac{q_1}{t_1} = \sum_{i=1}^{n} \delta s_i \tag{2}$$

where δs_i = irreversible entropy increase, q_1, q_2, $w > 0$.

All quantities are actually time derivatives (fluxes) and refer to one mass unit of circulating working fluid. The right-hand side of Equation (2) is the sum over all entropy creating steps of the cycle, including the heat transfer to the carrier fluids. All contributions δs_i are therefore positive. The terms q_2/t_2 and q_1/t_1 are the entropy changes per unit time for the heat carrier fluids and have to be determined as follows:

$$q_i = \int_{t_i^1}^{t_i^2} c_i dt \qquad \frac{1}{t_i} = \frac{1}{q_i} \int_{t_i^1}^{t_i^2} \frac{c_i dt}{t} \tag{3}$$

where c_i = heat capacity of the carrier fluid.

Figure 1: Internal (T), external (t) temperatures and heat flow for a compressor heatpump.

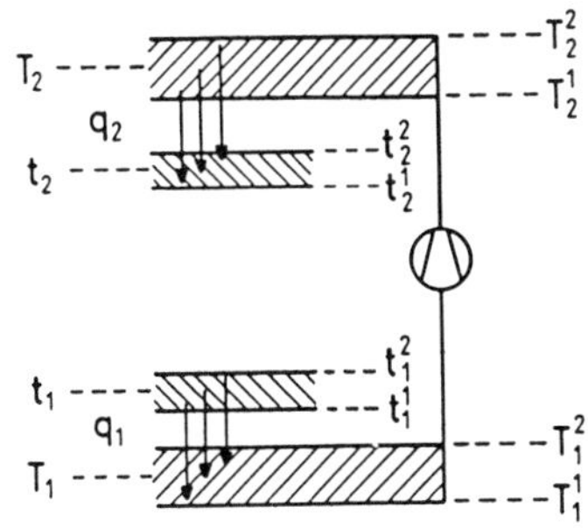

Figure 2: Temperature-entropy diagram for a fluid with ds/dt < 0 at the dew line.

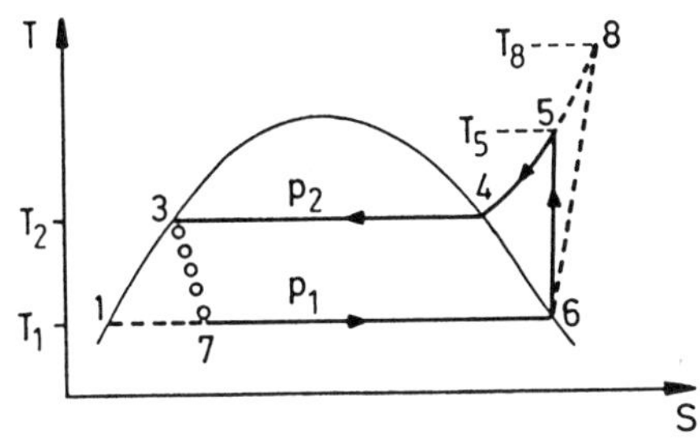

Figure 3: Temperature-entropy diagram for a fluid with ds/dt > 0 at the dew line.

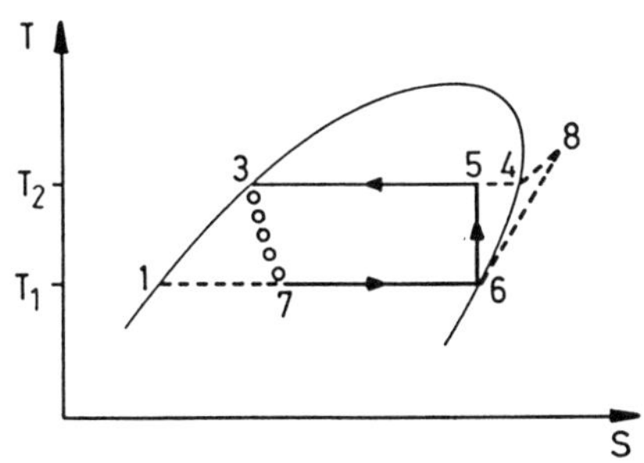

The temperatures t_1 and t_2 are thus precisely determined averages, which will be called "entropic averages". For most cases the middle temperature is a good approximation:

$$t_i \approx (t_i^1 + t_i^2) / 2 \tag{4}$$

The following irreversible process steps can be identified:

δs_1 = throttling 3 - 7
δs_2 = desuperheating between 5 and 4, i.e. flow of heat to the temperature T_2 after isentropic compression
δs_3 = ccmpression and desuperheating between 8 and 5, i.e. flow of heat created by irreversible compression to the temperature level T_2
δs_4 = heat transfer at the condenser, i.e. between T_2 and t_2
δs_5 = heat transfer at the evaporator, i.e. between t_1 and T_1
δs_6 = pressure drops by flow in pipes etc.
δs_7 = subcooling of liquid fluid and superheating of gaseous fluid
δs_8 = open-ended listing

It is important to note that the individual contribution δs_i are not independent from each other. For example, the irreversibilities of the compressor are increasing q_2 and thus δs_4 (but not δs_5!). A change in δs_1 (e.g. by using an expansion turbine or a different fluid) influences the amount of δs_5.

We will now determine the COP and quality factors for refrigeration. A conversion to COPs for heatpumping is straight forward.

Eliminating q_2 from Equation (1) and (2) yields the following equations for the work and the COP η for refrigeration:

$$w = q_1 \frac{t_2 - t_1}{t_1} + t_2 \sum_1^n \delta s_i \tag{5}$$

$$\eta = \frac{\eta_r (t_1, t_2)}{1 + \frac{t_1 \cdot t_2}{t_2 - t_1} \frac{\Sigma \delta s_i}{q_1}} \tag{6}$$

where $\eta_r (t_1, t_2) = \dfrac{t_1}{t_2 - t_1}$ = reversible or Carnot efficiency for refrigeration.

Completely reversible Rankine process ($\Sigma \delta s_i = 0$) operate according to the Carnot efficiency

$$\eta(\text{rev}) = \eta_r \tag{7}$$

with the temperatures t_2 and t_1 defined in the Equations (3).

A quality factor $g(t_1, t_2)$ for the process, relatively to a process with infinite large heat exchanger surfaces, an isentropic compressor, working with a hypothetical fluid without superheating and negligible specific heat capacity of the liquid ($ds/dT = 0$ at the bubble and dew line) can be defined as:

$$g(t_1, t_2) = \frac{\eta}{\eta(rev)} = \frac{1}{1 + \frac{t_1 \cdot t_2}{t_1 - t_2} \frac{\Sigma \delta s_i}{q_1}} \tag{8}$$

The irreversibilities $\Sigma\delta s_i$ are weighted by process temperatures t_1 and t_2, in contrast to similar quality factors, derived from the exergy concept into which an ambient temperature t_u enters. If Equation (5) is written as

$$w = w_0 (t_1, t_2) + \Sigma w_i \tag{9}$$

one finds for the quality factors $g(t_1, t_2)$:

$$g(t_1, t_2) = \frac{w_0(t_1, t_2)}{w} \tag{10}$$

According to Equation (5) and (9) the work required to produce a certain refrigeration power q_1 is determined by the minimum value $w_0 = q_1(t_2 - t_1)/t_1$ and additional work caused by irreversibilities. It should be noted that this extrawork is determined by $t_2 \, \Sigma\delta s_i$ which is not the exergy loss. t_2 can widely differ from the ambient temperature t_u.

The irreversibilities δs_i in Equation (5) and (6) are partly caused by fluid properties as δs_1 and δs_2, and partly by design, as e.g. size of the heat exchanger surfaces or properties of the compressor. First, the irreversibilities dependent on the design, will be considered. For simplicity the entropy increases δs_6 to δs_8 are ignored at present.

For the heat transfer in the condenser and evaporator the following equations hold:

$$\delta s_4 = \frac{q_2}{t_2} - \frac{q_2}{T_2} \tag{11}$$

$$\delta s_5 = \frac{q_1}{T_1} - \frac{q_1}{t_1} \tag{12}$$

For fluid mixtures the internal temperatures T_1 and T_2 are exactly defined averages similar to Equation (3). Inserting Equation (11) and (12) into Equation (1) and (2) and eliminating q_2, yields:

$$w = q_1 \frac{T_2 - T_1}{T_1} + T_2 \sum_1^3 \delta s_i \tag{13}$$

The Carnot factor is now determined by the internal temperatures T_1 and T_2, and only 3 entropy sources remain. The irreversibilities δs_3 caused by the compressor can be calculated in two steps:

$$\delta s_3 \text{ (compression)} = s_8 - s_6 \tag{14}$$

$$\begin{aligned} \delta s_3 \text{ (desuperheating 8-5)} &= (h_8 - h_5)/T_2 + s_5 - s_8 \\ &= (1 - g_C)w/T_2 + s_6 - s_8 \end{aligned} \tag{15}$$

where g_C is the isentropic efficiency of the compressor, defined as:

$$g_C = \frac{w(rev)}{w} \tag{16}$$

The sum of Equation (14) and (15) yields:

$$\delta s_3 = (1 - g_C)\ w/T_2 \tag{17}$$

This result is immediately plausible: The additional work $(1 - g_C)w$ caused by nonisentropic compression finally ends as heat at the temperature level T_2. To calculate the entropy change the irreversible compression process is replaced by a reversible process and the addition of heat at the temperature level T_2.

With Equation (17) one finds for the work w and the COP η:

$$\begin{aligned} w &= \frac{1}{g_C}\left[q_1 \frac{T_2 - T_1}{T_1} + T_2\ (\delta s_1 + \delta s_2)\right] \\ &= \frac{1}{g_C}\left[w_0\ (T_1, T_2) + w_1 + w_2\right] \end{aligned} \tag{18}$$

$$\eta = g_C \cdot \frac{\eta_r(T_1, T_2)}{1 + \dfrac{T_1 T_2}{T_2 - T_1}\ \dfrac{\delta s_1 + \delta s_2}{q_1}} \tag{19}$$

where $\eta_r\ (T_1, T_2) = T_1/(T_2 - T_1)$.

The value of η calculated with Equation (19) is identical to that of Equation (6). Again a quality factor $g(T_1, T_2)$ for the process, in comparison to a process with identical internal condenser and evaporator temperatures, with a reversible compressor and working with a hypothetical fluid without superheating and throttling losses can be defined:

$$g\ (T_1, T_2) = \frac{\eta}{\eta(rev)} = g_C \cdot \frac{1}{1 + \dfrac{T_1\ T_2}{T_2 - T_1}\ \dfrac{\delta s_1 + \delta s_2}{q_1}} \tag{20}$$

This quality factor is the product of the quality factor of the compressor and that of the fluid g_f with

$$g_f = \frac{1}{1 + \frac{T_1 T_2}{T_2 - T_1} \frac{\delta s_1 + \delta s_2}{q_1}} = \frac{w_0(T_1, T_2)}{w_0 + w_1 + w_2} \tag{21}$$

The quantities $\delta s_1/q_1$ and $\delta s_2/q_1$ are characteristic for the individual fluids: They can either be calculated analytically or taken from data charts according to the following equations:

$$\delta s_2 = \frac{h_5 - h_4}{T_2} - (s_5 - s_4) \tag{22}$$

This term is exactly zero for fluids of the type shown in Fig. 3.

$$\delta s_1 = \frac{h_3 - h_1}{T_1} - (s_3 - s_1) \tag{23}$$

The refrigeration power q_1 is given by:

$$q_1 = h_6 - h_7 = r - (h_3 - h_1) \tag{24}$$

where r = heat of evaporation/kg at T_1).

Using Equations (22) to (24) in Equations (18) and (19), the work w and the COP η can be calculated.

These formulae must yield exactly the same results as the commonly used equations:

$$w = \frac{1}{g_K} (h_5 - h_6) \tag{25}$$

$$\eta = g_K \frac{h_6 - h_7}{h_5 - h_6} \tag{26}$$

since no approximation has been made in the derivation. This can be proven rigorously /1/. Although Equations (18) and (19) look more complicated at the first view, they have several important advantages which will be discussed immediately.

Using Equation (19) the COP for heatpumping η_{HP} can be written as:

$$\eta_{HP} = 1 + g_C \frac{T_1}{T_2 - T_1} \frac{1}{1 + \frac{T_1 T_2}{T_2 - T_1} \frac{\delta s_1 + \delta s_2}{q_1}} \tag{27}$$

With

$$\eta_{HP}(rev) = \frac{T_2}{T_2 - T_1} \tag{28}$$

the quality factor g_{HP} (T_1, T_2), corresponding to Equation (20) is now given by:

$$g_{HP}(T_1,T_2) = \frac{\eta_{HP}}{\eta_{HP}(rev)} = \frac{1 + \eta}{1 + \eta_r(T_1, T_2)}$$

$$= 1 - \frac{T_1}{T_2} (1 - g\ (T_1, T_2)) \tag{29}$$

A similar relation can be derived from Equation (6) for the quality factor g_{HP} (t_1, t_2), which includes the heat transfer:

$$g_{HP}\ (t_1,t_2) = \frac{1 + \eta}{1 + \eta_r(t_1,t_2)} = 1 - \frac{t_1}{t_2} (1 - g(t_1, t_2)) \tag{30}$$

The quality factors for the heatpump g_{HP} are somewhat larger than those for refrigeration g, since the additional heat output caused by irreversibilities is now part of the product q_2.

In Table 1, results for four typical fluids are shown: NH_3 belongs to the group of fluids according to Figure 2, R 114 to the group of Figure 3.

Line No. 1 shows the reversible efficiency (Carnot efficiency), calculated with internal temperatures, whereas in line 2 reversible efficiencies are calculated with external temperatures, taking an "entropic" temperature difference $T_2 - t_2 = 7$ K and $t_1 - T_1 = 7$ K. Line No. 3: the refrigeration power per kg circulating fluid; Line 4: work per kg fluid for a reversible process between T_1 and T_2; Line 5 and 6: additional work due to the irreversibilities in throttling and desuperheating, in brackets: ratio between the additional work and the reversible work of line 4. For NH_3 the throttling and the desuperheating cause about the same amount of additional work; for R 22 the

Table 1: Coefficient of performance η, reversible efficiencies η_r, irreversibilities δs_i and quality factors g (effectiveness) for four working fluids.
Input parameters: $T_2 = 30°C$; $T_1 = -20°C$;
$T_2 - t_2 = 7$ K; $t_1 - T_1 = 7$ K; $g_C = 0,7$;
$g_f = \eta/g_C \cdot \eta_r(T_1,T_2)$; $g(T_1,T_2) = \eta/\eta_r(T_1,T_2)$;
$g_{HP}(T_1,T_2) = (1+\eta)/(1+\eta_r(T_1,T_2))$;
$g_{HP}(t_1,t_2) = (1+\eta)/(1+\eta_r(t_1,t_2))$;

		R 114	R 12	R 22	NH_3	Units	Equ.
1	η_r (T_1, T_2)	5,06	5,06	5,06	5,06		(19)
2	η_r (t_1, t_2)	7,22	7,22	7,22	7,22		(6)
3	q_1	94,17	114,86	160,38	1098,1	kJ/kg	(24)
4	w_0 (T_1, T_2)	18,6 (100)	22,7 (100)	31,7 (100)	217,0 (100)	kJ/kg %	(18)
5	$w_1 = T_2\ \delta s_1$ w_1/w_0	5,45 (29,3)	5,4 (23,8)	7,1 (22,4)	25,5 (11,7)	kJ/kg %	(18)
6	$w_2 = T_2\ \delta s_2$ w_2/w_0	--- ---	0,1 (0,4)	1,0 (3,1)	23,4 (10,8)	kJ/kg %	(18)
7	g_C	70,0	70,0	70,0	70,0	%	(16)
8	η (= COP)	2,75	2,85	2,82	2,89		(19,26)
9	g_f	77,3	80,4	79,6	81,6	%	(21)
10	g (T_1, T_2)	54,0	56,3	55,7	57,2	%	(20)
11	g (t_1, t_2)	38,1	39,4	39,0	40,0	%	(8,10)
12	g_{HP} (T_1, T_2)	61,9	63,5	63,0	64,1	%	(29)
13	g_{HP} (t_1, t_2)	45,6	46,8	46,5	47,4	%	(30)

desuperheating causes an effect of only 3.1% (or 2,5% of the total work), for R 12 of 0,4% and for R 114 it has disappeared completely. On the other hand the reduction in desuperheating is compensated by an increase in the throttling-contribution (line 5), so that the COP (line 8) does not differ strongly for the four fluids. Line 9 shows, that the fluids by themselves would reach about 80% of Carnot efficiency. It is the compressor efficiency g_C = 0,7, which finally reduces the effectiveness to about 55% (line 10). In line 11, the effectiveness in respect to external temperatures are shown, and the last two lines give the analogous values as calculated for heatpump operation. The sequence from line 9 to 11 demonstrates that the main reduction in effectiveness is not fluid-caused but caused by the compressor and the heat transfer. From line 5 it can be deduced that an expansion turbine instead of a throttle causes increasing improvements from NH_3 to R 114.

Equations (18) and (19) together with Equations (22) to (24) have the following advantage compared to the standard Equations (25) and (26): With the latter equations the COP is determined by the differences of large experimental numbers which, therefore, must be known with higher precision than required for the COP. With Equations (18) and (19) the COP is predominantly determined by the Carnot factor $\eta_r(T_1, T_2)$, which for single component fluids is precisely known once the internal temperatures are given. The fluid dependent denominator in Equation (19) is only a correction (for $T_2 - T_1 = 50$ K of about 25%, see Table 1), and therefore the accuracy requirements are reduced. The absolute value of the heat of evaporation is of minor relevance for the COP. Only ratios of fluid properties, like $\delta s_1/r$, with the heat of evaporation in the denominator, enter into the Equation (19). This will become even more evident, when analytic equations for the COP are derived. Equation (19) shows furthermore that the COP is only weakly dependent on the absolute values of the pressures. These need to be known only with such a precision, that the type of compressor can be chosen. The isentropic efficiency for a fluid must be anyhow measured separately.

ANALYTIC EQUATIONS FOR THE COP

Analytic equations would have the advantage that the dependence and sensitivity on fluid parameters become more transparent. Furthermore, fast calculations or estimates are possible. The enthalpy and entropy differences in Equations (22) to (24) can be calculated using the mean specific heat capacities c and c_p. The following relations hold:

$$q_1 = r - c\,(T_2 - T_1) \qquad (31)$$

$$T_2\ \delta s_1 = c\ T_2 \left(\frac{T_2 - T_1}{T_1} - \ln \frac{T_2}{T_1} \right) \approx \frac{c}{2} \frac{(T_2 - T_1)^2}{T_1} \qquad (32)$$

$$T_2\ \delta s_1 = c_p\ T_2 \left(\frac{T_5 - T_2}{T_2} - \ln \frac{T_5}{T_2} \right) \approx \frac{c_p}{2} \frac{(T_5 - T_2)^2}{T_5} \qquad (33)$$

We thus find for the COP η:

$$\eta = g_c\ \eta_r \frac{1}{1 + \frac{1}{2} \dfrac{\dfrac{c\ T_1}{r} \dfrac{1}{\eta_r} + \dfrac{c_p\ T_1}{r} \dfrac{1}{\eta_r} \left(\dfrac{T_5 - T_2}{T_2 - T_1} \right)^2 \dfrac{T_1}{T_5}}{1 - \dfrac{c\ T_1}{r} \dfrac{1}{\eta_r}}} \qquad (34)$$

The efficiency and the quality factors are, in general, determined by three nondimensional fluid parameters.

$$\frac{c\ T_1}{r}, \quad \frac{c_p\ T_1}{r}, \quad \frac{T_5 - T_2}{T_2 - T_1} \left(= \frac{\text{superheating}}{\text{temperature lift}} \right) \qquad (35)$$

For not too large temperature lifts $T_2 - T_1$, the superheating is proportional to $T_2 - T_1$:

$$T_5 - T_2 = (T_2 - T_1) \left(\frac{r\beta}{c_p} - 1 \right) \qquad (36)$$

where β = thermal expansion coefficient, with $\beta = 1/T$ for ideal gases) /2/.

Except for H_2O and NH_3 the superheating term can be neglected, compared to the throttling term. For fluids with the properties of Figure 3, it is anyhow exactly zero. Thus the efficiency for practically all freons is determined by one parameter, namely $c\ T_1/r$.

With $\delta s_2 \approx 0$, Equation (34) can be written as

$$\eta = g_C \, \eta_r \frac{1 - \dfrac{c\,T_1}{r} \dfrac{1}{\eta_r}}{1 - \dfrac{1}{2} \dfrac{c\,T_1}{r} \dfrac{1}{\eta_r}} \qquad (37)$$

For $\eta_r \gg cT_1/2r$ one finds a very simple equation for the COP of a compressor refrigerator:

$$\eta \approx g_C \left(\eta_r - \frac{1}{2} \frac{c\,T_1}{r}\right) \qquad (38)$$

The Carnot efficiency η_r is lowered by a certain value, which is fluid-dependent, and then multiplied with the compressor efficiency.

In Figure 4 this dependence is shown, using the numbers of Table 2.

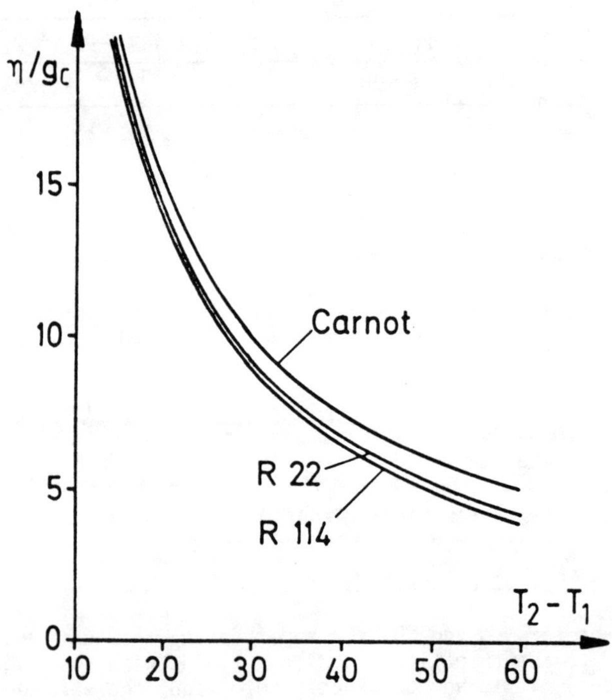

Fig. 4: COP for two fluids according to Equation (38) with cT_1/r from Table 2

The more precise Equation (37) would lower the values for $T_2 - T_1 \approx 50$ K by only 5%, which is hardly noticable on this scale.

Table 2: Fluid parameter cT_1/r for different refrigerants (T_1 = 273 K); η from Equation (37) with T_2= 30°C, T_1= -20°C, g_C = 0,7.

	R 114	R 12	R 22	NH_3	H_2O
cT_1/r	1,99	1,75	1,60	1,09	0,46
η	2,77	2,87	2,92	---	---

The efficiencies in Table 2 are calculated with Equation (37), using T_1 = 253 and ignoring the temperature dependence of c and r. They agree very well with those of Table 1 except for R 22, which is 3% larger since superheating has been neglected. For a temperature rise of 50 K the fluid-dependent term amounts to only 20% of the Carnot factor. The needed accuracy is thus a factor 5 lower than desired for η.

CONCLUSION

In this paper rather simple equations for the COP of compressor heatpumps have been derived. The fluid properties are characterized by a small set of nondimensional parameters. For new fluids of the freon type only cT_1/r needs to be determined.

Long chain halocarbons which are highly fluorinated, are considered as potential high temperature working fluids. Due to the large numbers of internal vibration modes of such molecules, the ratio cT/r will increase. On the basis of the above equations one must expect that the efficiency of such fluids will be reduced.

For the COP of mixtures of fluids only the entropic averaged condensation and evaporation temperature (Equation 3) and the effective ratio cT_1/r needs to be known.

Using the analysis, as in this paper has been done, equations for COPs and criteria for COP improvements for the following problems can be found /1/:

1. liquid fluid expansion turbine
2. liquid fluid injection for H_2O or NH_3
3. heat exchange between gaseous and liquid fluid
4. fluid mixtures
5. compressor heatpump with an absorbing fluid

For the latter two cases details about throttling must be specified. The analysis has further yielded an equation by which the investment costs for heat exchanger surfaces at the condenser and evaporator can be minimized. This is an interesting example of the impact of entropy on economy.

Exergy is the potential work which could be gained if heat of a certain temperature were converted into work in a reversible power station working against the lower temperature t_u. Exergy losses are considered to be indicative for the effectiveness of a process step. The exergy loss is proportional to the entropy increase δs_i multiplied by t_u.

The choice of t_u as ambient temperature is rather ambiguous and may be dependent on the season or on other parameters. Instead of asking for the loss in potential work due to irreversibilities we consider the following questions to be of more relevance: How much additional work must be supplied to make the process run in spite of irreversibilities. The answer is given in Equations (5), (13) or (18). This work is also proportional to the entropy produced but the proportionality factor T_2 differs, in general, widely from the ambient temperature t_u. With T_2 = 150°C for a high temperature heatpump this work is, at least, 50% higher than what is called the exergy loss.

According to Equation (18) the entropy increases δs_1 and δs_2 must furthermore be multiplied by the reciprocal compressor efficiency. Neither the exergy loss nor the absolute values for the entropy production is sufficient to characterize and compare the effectiveness of a process step, since the individual entropy sources are in general not independent from each other. The proportionality factor, by which the entropy production has to be multiplied, depends on the process considered and on the product which the process delivers.

REFERENCES

/1/ This presentation is an extract of a more detailed paper of the author on: "The COP of Compressor Heatpumps from First and Second Law", submitted for publication

/2/ M.P. Bertinat, "Fluids for High Temperatures Heatpumps", Int. Journal Refrig. 1986, Vol. 9, pp. 43-50

Section IV

Heat Pumps for Industrial Applications

Chapter 15

Vapor Recompression Systems: Different Compressor Technologies and Various Examples of Industrial Applications

Philippe P. Letout

ABSTRACT : Different types of compressors can be involved in mechanical vapor recompression system (M.V.R). The most common are centrifugal compressors, which are widely developed in this paper.

Axial compressors are used with the highest vapor flows when volumetric compressors are installed on the smallest units ; the different types of volumetric compressors are :

- lobe compressors
- twin-screw compressors
- liquid ring compressors
- piston or reciprocating compressors,

At present, the last three types are not much in use.

The main characteristics of all these types of compressors are given, and their performances are compared.

The second part of the conference presents six (or so) M.V.R. plants chosen to cover the largest range of :

- industrial activity branches
- European countries
- industrial processes and components
- compressors powers and types.

Each example sets out the main features of the original plant and of the new plant with M.V.R. A schematic diagram and an economic comparison are given.

HEAT PUMPS: Prospects in Heat Pump Technology and Marketing (Proceedings of the 1987 IEA Heat Pump Conference), Edited by Kay H. Zimmerman © 1987 Lewis Publishers, Inc., Chelsea, Michigan 48118. Printed in U.S.A.

VAPOUR COMPRESSEUR :

In the M.V.C. evaporation, distillation and drying plants, the compressor is used to such in vapour, compress it and reintroduce it in the process as heating steam.

The type of installation defines the paramaters used for the selection of the compressor.

These different paramaters will be analyzed in this paper.

a) General considerations.

M.V.C. is used in many different industrial processes ; it is therefore not easy to clearly define the requirements.

In some cases, compressor design is tailored to the specific requirements of the user ; in sectors where operating conditions are well known standard design compressors are used.

The choice of a compressor depends on specific process conditions such as :

. Nature of the vapour.
Generally, the vapour is steam : but it can be chemical solvents, and it is important to known clearly its physical, chemical and thermodynamic properties.

. Erosion.
As the vapour to be compressed is saturated, fine water droplets may appear and erode various parts of the machine. Design considerations include materials, limitation of high local velocities, installation of water traps,...

. Corrosion.
Appropriate - at times, expensive - materials need to be used to withstand corrosive vapour.

. Fouling.
The vapour can be impregnated with particles or solutes which, as result of superheating during the compression process, deposit in the machine adversely affecting the performance. Provision for cleaning devices may sometimes be necessary.

. Pressure level.
whenever sub-atmospheric pressures are used, it is necessary to prevent air ingress to avoid the presence of non-condensible gases in the circuit, which reduce heat exchange in the evaporators. When higher than atmospheric pressure is used, leaks may occur; these are only significant if the discharge vapours are toxic or pollutant.

Depending on the flow rate and pressure rise to be achieved, the following compressors will be used :

. Turbocompressors.
Raise the fluid velocity. The resulting energy surplus is transformed into pressure at the impeller outlet. These machines are used for high and medium flow rates, while the compression ratio per stage is relatively low.

. Volumetric compressors.
The pressure rise is achieved by compression of a gas volume trapped in a chamber. The pressure ratio depends on the chamber geometry change and, therefore, can be high. On the other hand, volumetric flow rates are low as a result of the discontinuous fluid flow.

Table 1 gives the mean characteristics of commonly used compressors.

Category	Type	Volumetric flow rate	Pressure rise rate per stage
Turbo-compressors	— centrifugal — axial	7,000 to 500,000 m^3/h	1.8 to 2 1.2 to 1.8
Volumetric compressors	— reciprocating (piston)	< 2,000 m^3/h	4 to 6
	— rotary • lobe • screw • liquid ring	 100 — 15,000 m^3/h 1,000 — 25,000 m^3/h < 8,000 m^3/h	 $\Delta p \cong 1$ bar 3 to 6 1.4 to 1.5

Table 1 — Compressor characteristics

The following paragraphs give more details on each type of compressor.

b) Turbocompressors

1) Centrifugal compressors

These are the most widely used for high and medium capacity MVC-installations. The operating principle is shown in figure 2.

Gas enters axially in the impeller and is expelled by force. It passes through a diffuser to the volute and enters the high pressure circuit.

In most cases, impellers are of open helico-radial types.

With this type of compressor, it is possible to achieve pressure ratios ranging from 1.8 to 2 per stage and isentropic efficiencies ranging from 0.7 to 0.8.

By placing one, two, three or even four impellers in series on the same casing, the pressure ratio increases by a high proportion.

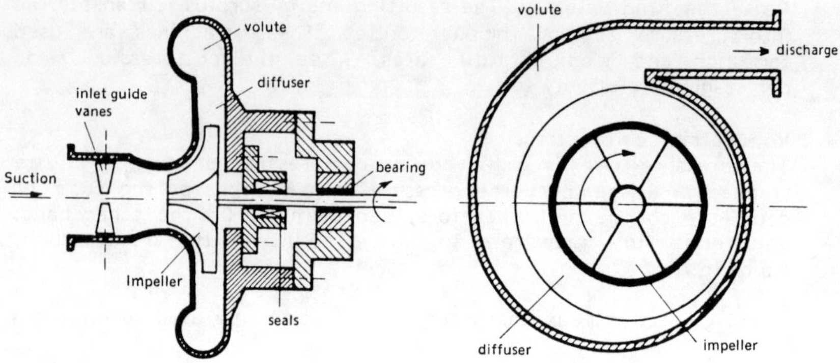

Fig. 2 Centrifugal compressor

The volume flow rate is proportionnal to the impeller size and speed, and the pressure difference increases as the square of peripheral speed.

This shows the advantage of having a high peripheral speed, though within the limits of the material resistance to the strain induced by the centrifugal force.
With a high quality material, it is possible to build more efficient unites and to minimize the number of compression stages and consequently the capital investment.

In practice, peripheral speed does not exceed 420 to 450 m/s.

Implellers are made of high-alloy steel, aluminium of even titanium, using moulding, welding or milling techniques. Impellers in composite material with carbon fibre and casings in glass fibre have appeared recently on the market.

Rotors are generally mounted on tilting pads hydrodynamic bearings and shaft sealings are provided to limit leakage along the shaft. Volutes are cast or welded, depending on the manufacturers.

To accomodate any change in the operating conditions, the compressor is equipped with either inlet guides vanes (mobile blades) (fig. 2) or variable pitch blades in the diffuser ; operating condition can also be modified by adjustable speed drive. These control systems enable satisfactory efficiency to be achieved under varying service conditions, and avoid operation of the compressor under transient or surge conditions.

Centrifugal compressor suction pipes are generally equipped with water traps, to prevent erosion.

2) Axial compressors

These compressors are composed of a succession of fixed and mobile blades mounted on concentric stator and rotor (fig. 3).

After being directed by the first fixed blades row, the gas flow is sucked by the first mobile blades impeller :

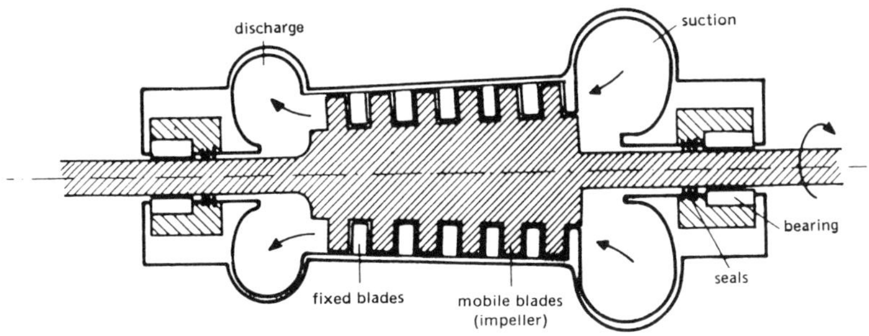

Fig. 3 Axial compressor

Then it is straightened by the second row of fixed blades for further compression by the second impeller and so on.

Multi-stage arrangements are easy to realize : but the pressure ratio per stage is relatively low, usually ranging from 1.2 to 1.8.

These machines are particularly suited to very high flow rates (for example, more than 200.000 m3/h) and attain high efficiencies. They are difficult to control because departure from rated conditions leads to large variations in performance. Special care is needed to eliminate water droplets from the vapour, as this compressor type is susceptible to erosion, especially at the leading edge of the blades.

In spite of their compactness and good efficiency, axial compressors are not frequently used in MVC because of their high cost.

c) Volumetric compressors

The principle of this type of compressor is derived from air compressor technology. The big difference lies in that, for MVC applications, it is imperative to remove any oil traces in the vapour circuit ; the moving parts must therefore be contactless or lubricated by the fluid.

1) Lobe-type compressors

At present, the most commonly used type of compressor for MVC duty.

They consist of two conjugated profile rotors driven by external synchronization pinions so as to move without contact.

To minimize internal leakage, the pressure difference between suction and discharge is limited to 1 bar. So, the isentropic efficiency, though not high, is at an acceptable level (0.4 to 0.65). The flow rate range varies from about 100 to 15.000 m3/h of inlet vapour.

Maximum admissible discharge temperature is about 150°C ; therefore, provision for desuperheating by water injection into the pressure chamber is often necessary.

This injection results in the formation of a water film which has the favourable effect of cleaning the rotor and reducing internal leakage.

The merits of this type of compressor are : stability and simplicity, good resistance to high temperature and vacuum ; attractive price due to standardisation. In addition, the inlet and outlet pressure levels tend to adapt to the degree of evaporator fouling.

The main disadvantage is that their sound level is high.

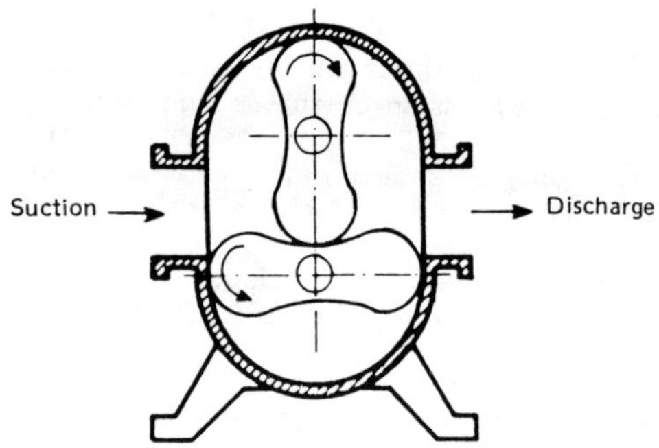

Fig. 4 Lobe-type compressor

2) Dual-screw dry-running compressors.

Are made of two non lubricated intermeshing helical screws driven by external synchronizing pinions (figure 5). Clearance between screws decreases continually from inlet to outlet port, so that the sucked fluid is compressed as it transfers to discharge.

Rotor speed is always high to minimize internal leakage but, as diameters are small, peripheral speeds are low, usually ranging from 80 to 120 m/s.

The pressure ratio is about 4 and isentropic efficiency levels range from 0.5 to 0.75. A pressure ratio of 6 is possible but with reduced performance. Flow rate ranges about from 1,000 to 25,000 m3/h.

The use of this type of equipement is restricted to temperatures of about 180 ° C is steady state condition. Desuperheating by injection of water in the casing can be envisaged ; its effect is the same as for the lobe compressor.

This equipement is not yet commonly used for M.V.C. because it is expensive. Extension of the use of this type of compressor should reduce its cost.

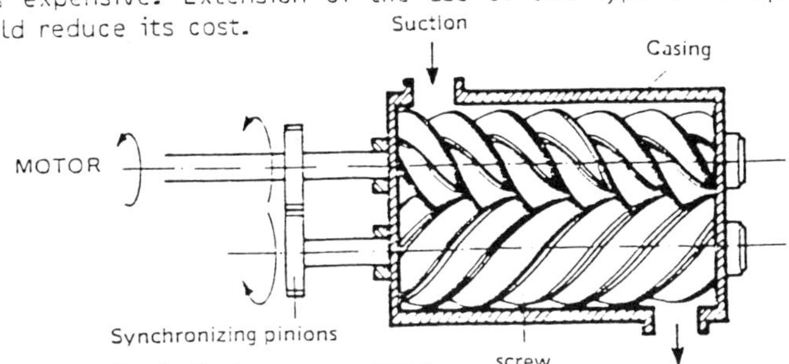

Fig. 5 Dual-screw compressor

3) Liquid ring compressors.

consist of a blade rotor rotating in an eccentric stator ; a continually fed liquid ring is sprayed into the stator, which ensures compressor tightness and alternative volume variation of compression chambers (figure 6).

The chambers are connected to the suction duct during the increasing volume phase and communicate with the discharge duct when the confined gas has been compressed.

These compressors can withstand liquid surges and can operate with highly polluted vapour. They have the advantage that the compression process is carried out with little superheating as the ring is supplied with liquid at a temperature slightly lower than the boiling temperature under discharge pressure.

The flow rate does not exceed 8,000 m3/h ; their isentropic efficiency is rather low (0.2 to 0.5). They are not often used for M.V.C.

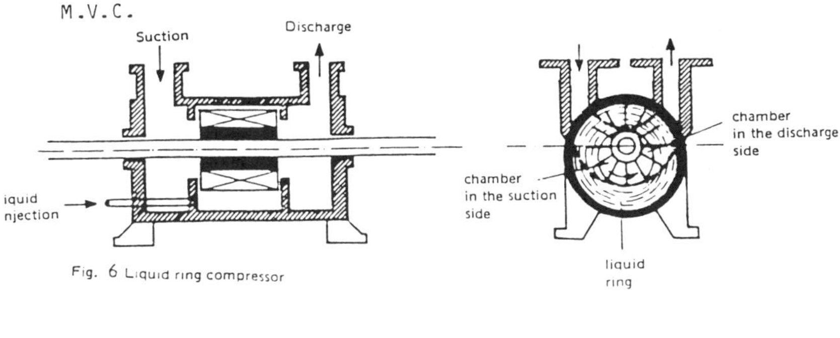

Fig. 6 Liquid ring compressor

4) Reciprocating compressors.

Figure 25 gives the principle of this compressor, which is well known in various industrial applications.

Machines with several pistons operating in parallel or in series are in common use. The maximum stage pressure ratio is high 4 to 6, and the isentropic efficiency ranges from 0.5 to 0.67.

The compressor is not often used for M.V.C. because oil free pistons must be used, which are expensive ; in addition, the valves are sensitive to liquid surges.

They are used for low volumetric flow rates (for example, less than 2,000 m3/h).

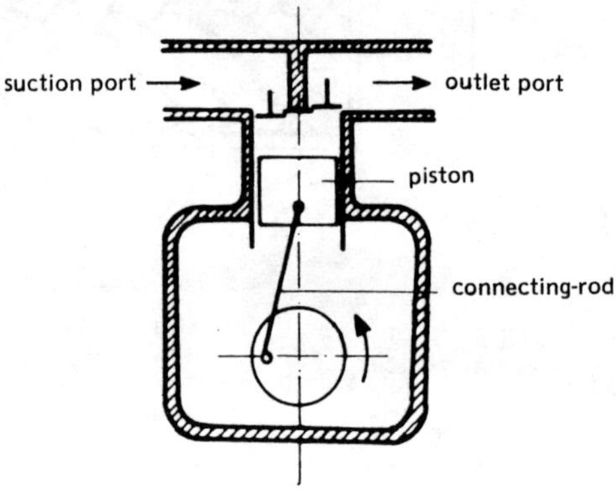

Fig. 25 Reciprocating compressor

3.3. CURENT EXPERIENCE

Some initial reservations about introducing new technology could have been expressed by industries not used to compressor pratice (food industry), but experience todate shows that these reservations were not juistified, and the following conclusions can be drawn:

- Provided careful attention is paid at the desing stage to the dimensioning of components and the optimization of the process, the integration of a compressor into an evaporation process line raises no major technical problem. Very satisfactory technical and economic balances haves been obtained. It is essential to involve an evaporator manufacturer to achieve such good results.

- Machines have proved to be reliable, and little downtime is due to unavailability of compressors. Automatic control systems well suited to the requirements are being rapidly developed, improving operational security.

- Manufacturers and user are very concerned by the risk of erosion caused by droplets and particle entrainment. Experience shows that there are only a few cases of deterioration due to erosion and that fouling problems arising in some processes can be resolved by water injection and washing cycles, which have proved very efficient.

EXAMPLE 1

HEAT RECOVERY FROM A LIQUID EFFLUENT

SOLVAY - Couillet Plant (Belgium)

PROCESS REQUIREMENTS

To valorize a large quantity of thermal energy generated by a chemical process and previously dissipated in ambient air.

ORIGINAL PLANT

The hot liquid effluents generated by chemical process were cooled in air-cooler.

NEW PLANT

It comprises two cooling water loops with a unit of 1,200 m3/h The water is cooled from 92 ° C to 89 ° C by expansion into two flash tanks at 0.7 bar (absolute pressure).

A centrifugal compressor raises the pressure of the steam generated to 1.1 bar (absolute pressure) ; the compressed steam is passed into the low pressure steam main for use around the factory. Cooled water leaving the flashing tanks is recycled in the cooling circuit. A supply of fresh water compensates for the steam (13.5 tonnes per hour) conveyed to the steam main.

M.V.C. was started up in June 1985.

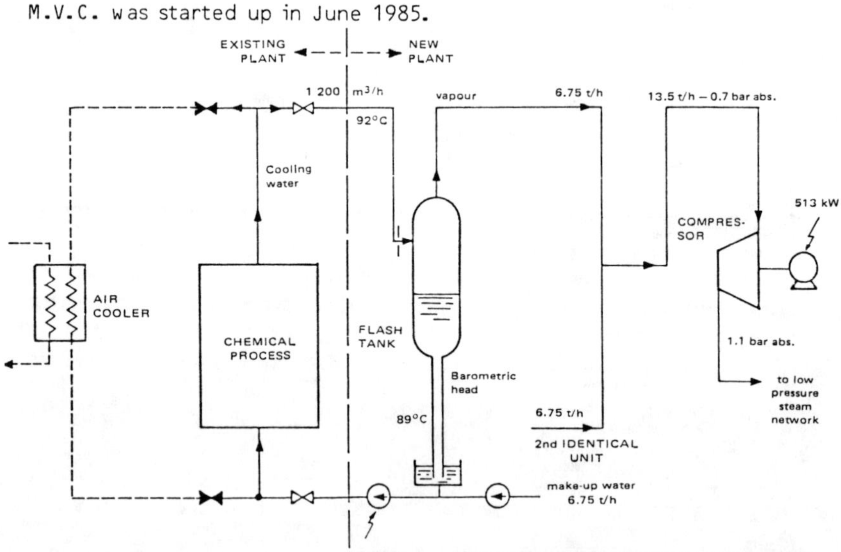

SCHEMATIC DIAGRAM OF THE UNIT WITH MVC

ECONOMIC COMPARAISON

Based on the mean prices when the investment was decided :

Foreseen operating time	h/year	8,000
Steam production by heat recovery	t/h	13.5
Primary energy saving	GJ/h	36.8
Additional electricity consumption	kWh/h	750
Primary energy cost	FB/GJ*	250
Payback time	years	1.6

* Net Calorific Value

Engineering : Internal to the Compagny

Compressor manufacturer :: ALSTHOM RATEAU (France)

EXAMPLE 2

ETHANOL DISTILLATION

SUNTORY Ltd - Osaka (Japon)

PROCESS REQUIREMENTS

The "wine" containing 6.5 % weight of ethanol is distilled in a column, to produce an alcohol containing 94 % weight of ethanol. This alcohol is intended for whisky production.

CONVENTIONAL PLANT

The 94 % ethanol vapours leave the column head at 78° C under atmospheric pressure. They are condensed in a tubular condenser fed with cold water. The waste liquid leaving the column bottom contains water, essentially 3 700 kg/h low pressure steam are supplied into the column bottom.

MODIFIED PLANT

The vapour from the column head is condensed in a tubular condenser/evaporator. Make-up water is expansed, the vaporized at 73° C and 0.37 bar (absolute pressure) in the evaporator. The water vapour is compressed in a screw compressor up to 1.13 bar (absolute pressure), desuperheated to 105° C, the admitted in the column bottom, as heating fluid.

SCHEMATIC OF THE MVC PLANT

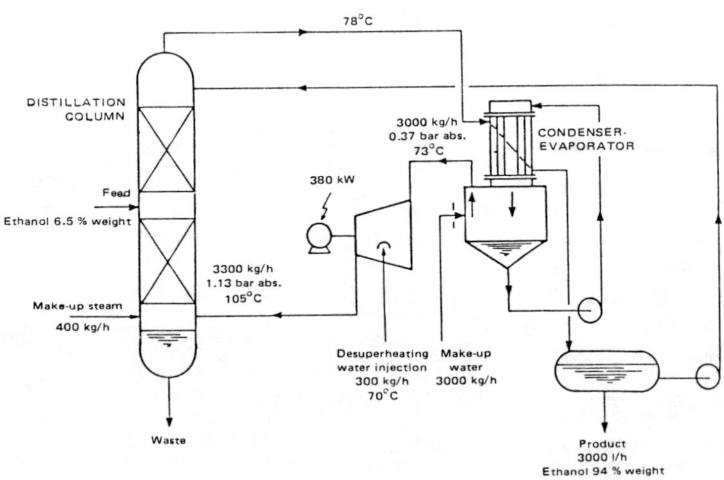

ESTIMATED ECONOMIC COMPARISON

(Starting up in 1986)

		CONVENTIONAL PLANT	NEW PLANT with MVC
Live steam consumption	t/h	3.7	0.4
Electricity consumption (excluding pumps)	kWh/h	0	380
Operating hours	h/year	8 000	8 000
Steam cost	Yens/t	5 900	5 900
Electricity cost	Yens/kWh	—	16
Annual energy cost	Millions Yens/year	174 640	67 520
Energy saving	Millions Yens/year	—	107 120
Additional investment	Millions Yens	—	170 000
Pay-back time on additional investment	years	—	1.6

Engineering : MYCOM and SUNTORY (Japan)

Compressor manufacturer : MYCOM (Japan)

EXAMPLE 3

WORT BOILING

Brasserie du Pêcheur (Brewery) - Schiltigheim (France)

PROCESS REQUIREMENTS

To maintain boiling for one and a half hours under atmospheric pressure, to bring about interaction between the wort constituents and to evaporate 1.5 tonnes of water per hour for a 75 tonnes brew. This process is discontinuous.

ORIGINAL PLANT

This comprises a double-bottom copper with a 40 m2 heat exchanger area, the heat requirements being met by condensing steam at 2.5 bar (absolute pressure) in the double-bottom.

MODIFIED PLANT

Using the existing copper, with the chimney and all opening sealed to eliminate air ingress, the vapour is now collected in a lobe-type blower. The vapour is compressed to 1.85 bar (absolute pressure) then condensed in the double-bottom copper. The additional heat exchanger used for wort preheating is adequate to compensate for the pressure reduction. Condensates are used to produce hot water before discharge at 20° C to the sewer.

The plant was commissioned in 1984.

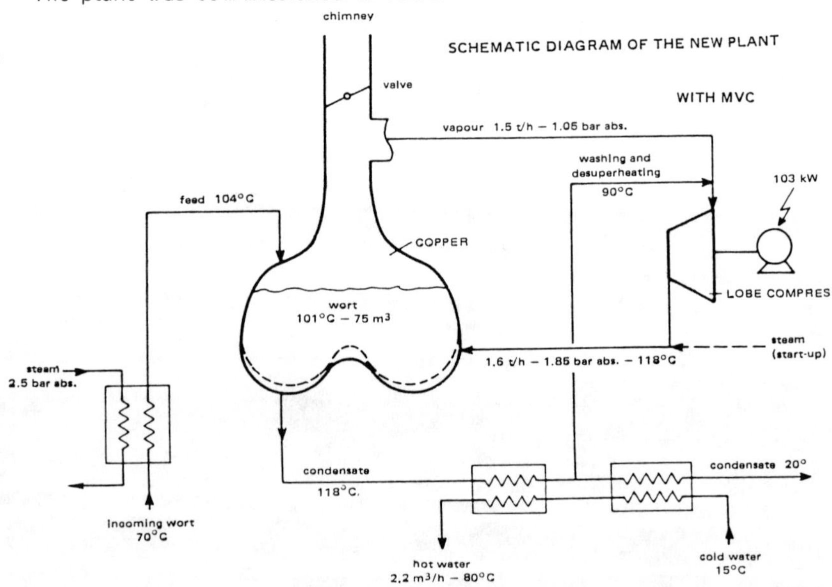

ECONOMIC COMPARISON

Based on 1984 prices :

Oil cost : 1.85 FF/litre
Electricity cost : 0.54 FF/kWh (1)
Brews per year : 875
Operating time : 1,312 hours per annum

		ORIGINAL PLANT	NEW PLANT with MVC
Fuel-oil consumption for boiling (2)	t/year	185	13
Hot water production (80°C) by heat recovery	m^3/year	0	2,890
Fuel-oil saving on hot water production	t/year	—	30
Total fuel-oil saving	t/year	—	202
Additional electricity consumption	kWh/year	—	136,000
Investment cost of MVC	kFF	—	1,200
Payback time on investment (3)	years	—	4

Comment

(1) The electricity cost is a mean annual rate including peak consumption.
(2) Steam consumption to raise wort to the boiling point is the same in both plants and is ignored in the comparison.
(3) Payback time is relatively high because of low annual operating hours.

Engineering : NORDON (France)

Compressor manufacturer : HIBON (France)

EXAMPLE 4

PRECONCENTRATION OF SUGAR BEET EXTRACT

GUIGNICOURT Sugar Works (France)

PROCESS REQUIREMENTS

To pre-concentrate the sweet extract from sugar beet before this is sent for crystallization. The works produce about 70,000 tonnes of white sugar per year.

ORIGINAL PLANT

A five evaporator was used to concentrate the sugar beet extract and also to provide the various steam flows required for the other processes in the works, different pressures being achieved by tapping off the multiple effect evaporator at different points. Energy conservation reduced the steam requirements through-out the works to such an extent that there was surplus of steam production in the evaporator, which had to becondensed with considerable energy loss.

NEW PLANT

To enable the water evaporated in the multiple effect to constantly match the amount of vapour taken away, it was necessary to pre-concentrate the sweet juices with a wide margin on the evaporating capacity, varyling between 30 and 80 tonnes per hour.

A pre-concentrator using a centrifugal compressor was thus installed.

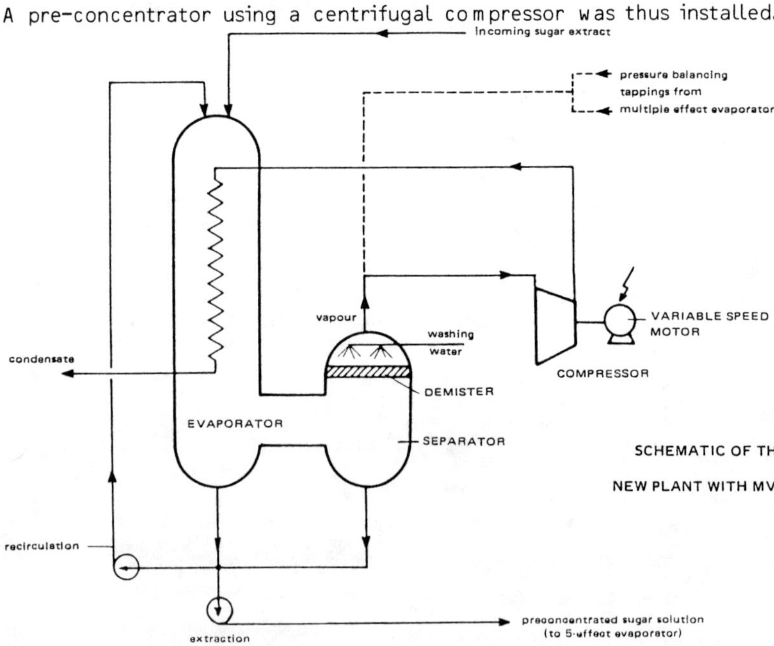

SCHEMATIC OF THE NEW PLANT WITH MVC

ECONOMIC COMPARISON

The total investment amounted to FF 6 millions (in 1983). The installation of M.V.C. enables the total boiler steam production to be reduced by 10 %, which meant that the investment was paid off in 4 sugar campaigns (about 8,000 hours in total), even though the sugar works use low cost coal for steam production.

Engineering : KESTNER (France)

Compressor manufacturer : ALSTHOM RATEAU (France)

Chapter 16

Vapor Recompression Systems: State of the Art in Europe — Economic Analysis and Optimization

Jean-François Reynaud

ABSTRACT : The wide variety of applications are described under four main headings : concentration and crystallization by evaporation, distillation, drying and thermal effluent valorization. Practical and technological aspects of mechanical vapor recompression are emphasized.

Engineering considerations are given, i.e. : technical aspects (characteristics of the product to be treated, choice of the compressor, ...), optimization of the heat exchange surface area and of the compressor saturation temperature rise (to reduce annual overall expenditure), plant maintenance.

Finally, the results of a recent investigation on the existing M.V.R. plants in Europe are given and the trends of development in the near future are shown.

1 - WHAT IS MECHANICAL VAPOR RECOMPRESSION ?

Mechanical Vapor Recompression (M.V.R.) proceeds from the same principle as the COMPRESSION HEAT PUMP. Vapor generated by an industrial process can, by this means, be used as heating-steam in the same - or an adjoining - process. This vapor is compressed, and thus suited to ensure heat exchange, by increasing its condensation temperature (Figure 1).

Mechanical power absorbed by the compressor is only a small part of the thermal energy supplied when condensing compressed steam. Normally, boiler steam is no longer required, except for start-up.

HEAT PUMPS: Prospects in Heat Pump Technology and Marketing (Proceedings of the 1987 IEA Heat Pump Conference), Edited by Kay H. Zimmerman

Figure 1 : Mechanical vapor recompression compared to compression heat pump

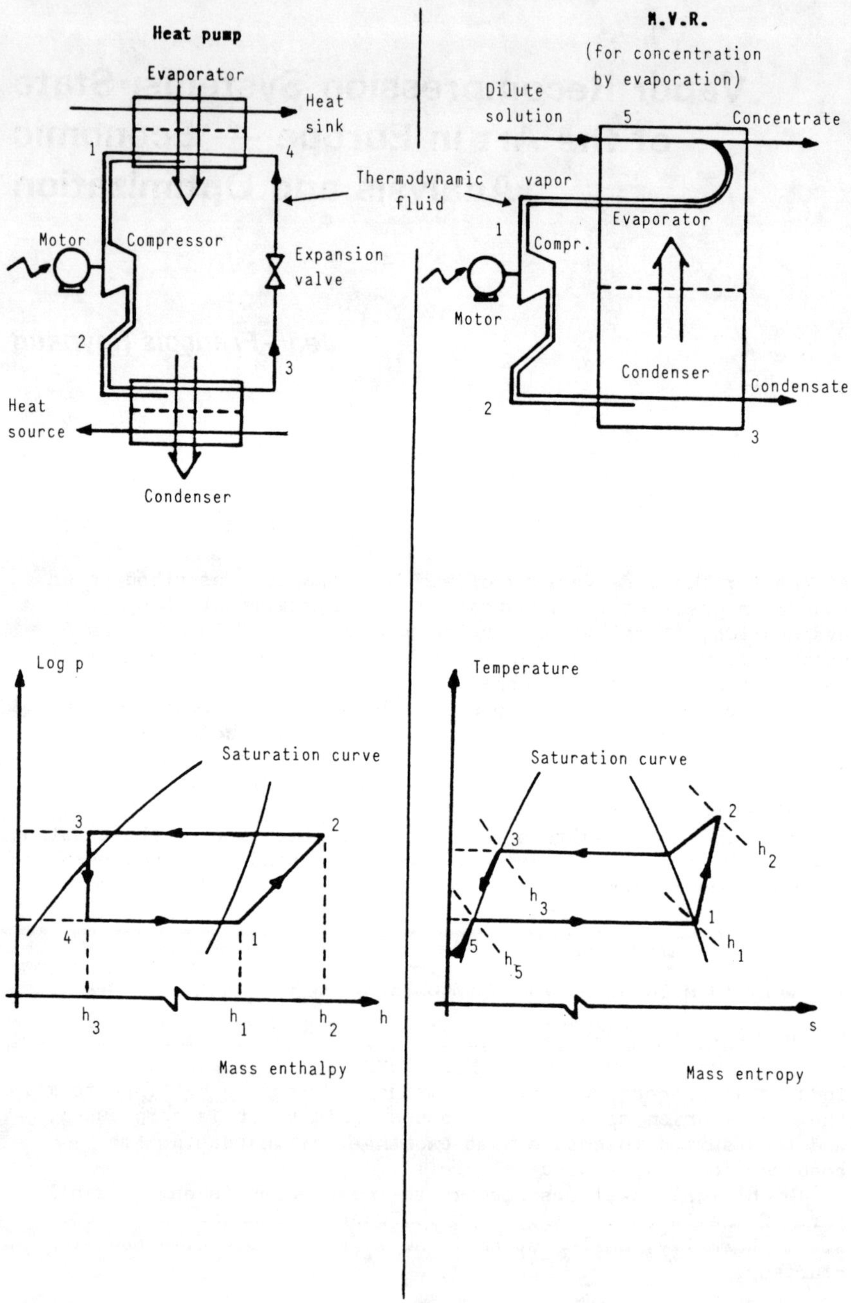

Though the first M.V.R. plants were constructed during the last century, a large increase in the number of plants only began at the end of the 70s ; in spite of a very low power consumption, this expansion was only made possible thanks to an extensive availability and competitive cost of electricity, and to the development of efficient and reliable steam compressors.

Steam Compressors

EJECTORS (or JET COMPRESSORS) have been in use for many years to perform static compression of low pressure vapor by expanding high pressure steam generated by a fuel boiler (Figure 2).

Although ejector performance has been improved, this reliable and hardy device still consumes a significant amount of expensive energy ; furthermore, it is very noisy and the usable range for induced vapor is rather limited.

In units equipped with M.V.R., the ejector is used :

- as a stand-by device. A condenser must then be installed ;
- to reduce eventual make-up steam consumption. The ejector operates in parallel with the compressor, or on another evaporator body. Make-up steam is used as motive steam.

MECHANICAL COMPRESSORS are used to suck in all (or nearly all) the vapor leaving a process, to compress it and reintroduce it as heating steam. Depending on the desired flow rate and pressure rise, the following compressors are used :

. Turbocompressors : raise the fluid velocity. The resulting energy surplus is transformed into pressure at the impeller outlet. These machines are used for high and medium flow rates, while the compression ratio per stage is relatively low. CENTRIFUGAL compressors are most commonly used, while the largest plants are equipped with axial compressors.

. Volumetric compressors : the pressure rise is achieved by compression of a gas volume trapped in a chamber. The pressure ratio depends on the chamber geometry change and, therefore, can be high. On the other hand, volumetric flow rates are low as a result of the discontinuous fluid flow. Lobe-type compressors are the most common volumetric compressors. Some plants are equipped with twin-screw or piston non-lubricated compressors, or liquid ring compressors.

The following chapter, by Mr. LETOUT, will develop compressor characteristics and their performance.

2 - APPLICATIONS OF MECHANICAL VAPOR RECOMPRESSION

It is estimated that there are 1500 to 2000 MVR plants throughout the world, a large part of them involving very low mechanical power to produce small flows of very pure water for pharmaceutical and other industries.

The wide variety of existing or potential applications of MVR can be grouped in four main headings :

Figure 2 : Concentration by evaporation

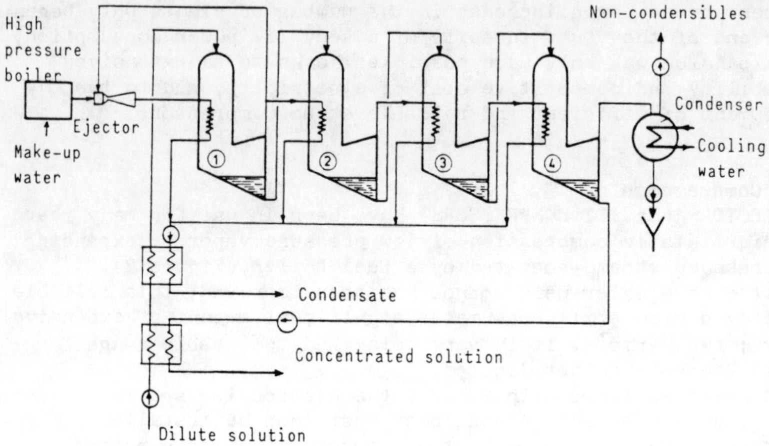

a) Four-effect thermo-compression evaporator

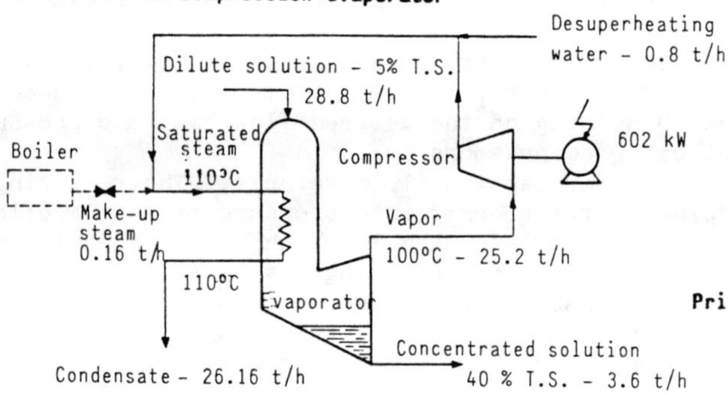

Principle

b) Single-effect mechanical-compression evaporator

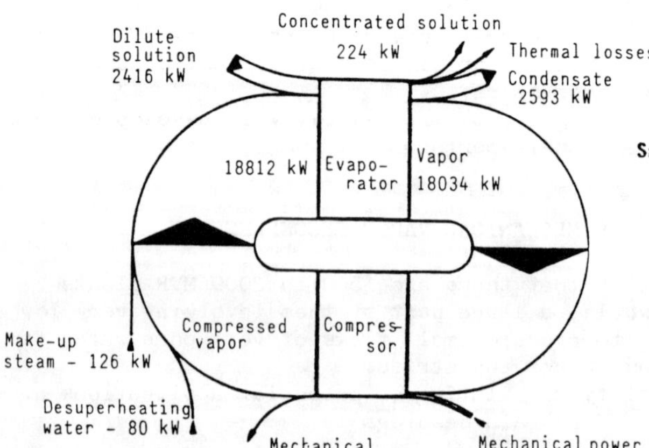

Sankey diagram

(figures given as an example)

2.1. Concentration and Crystallization by Evaporation

These are certainly the most widespread techniques and the most common in terms of energy consumption to remove a solvent from a solution. This explains that they include the largest number of MVR plants.

In view of the wide range of MVR installations, only the most common ones will be quoted :

- concentration of milk and by-products,
- concentration of black liquor (pulp and paper industry),
- salt crystallization,
- desalination of sea water to produce "distilled" water,
- effluent concentration to recover fresh water, to valorize the concentrate (sold as animal feed, for example), and/or for nature conservation (distillation residues, wool washing water, etc)...

Conventional plants are MULTIPLE EFFECT evaporators, each effect being fed by the vapor produced by the previous effect, operated at a higher pressure and temperature.

THERMO-COMPRESSION consists in drawing out a part of the vapor leaving the first effect by an ejector, to heat the evaporator, as shown in Figure 2. The reduction in heating vapor consumption approaches that resulting from the addition of one effect to the evaporator.

In any case, the amount of vapor evaporated by the last effect (about the same amount as that of live steam feeding the first effect) must be condensed.

MECHANICAL VAPOR RECOMPRESSION aims to recover all (or nearly all) the energy contained in the vapor for recirculation in the evaporator as heating steam (Figure 2). Single or multiple effect evaporators can be equipped with MVR, the higher the number of effects, the larger the compression ratio to be performed by the compressor and the smaller the steam flow to be compressed. There is usually no need for a condenser.

Table 1 shows, in descending order of energy consumption, various typical evaporator arrangements (the information in this table is derived from experimental data, studies carried out by manufacturers and bibliographical literature).

Evaporators and crystallizers : most of the evaporators used for MVR duty are of the falling film tubular type. Some are plate evaporators, in which the plates are welded, thus reducing leakage risks. Forced circulation evaporators are used when scaling or fouling can occur ; they are also used for continuous crystallization.

2.2. Distillation

Distillation columns are used in many chemical plants. Distillation aims to separate two or more miscible solvents, according to their volatility under a given pressure. Distillation requires heating the column bottom liquid and condensing the column head vapor (which is at a lower temperature than the bottom liquid). MVR aims to recover energy contained in the head vapor to

Table 1	SPECIFIC CONSUMPTIONS for concentration by evaporation taking into account the boiler, condenser and auxiliaries		
Type of evaporator	(a) kg live steam per tonne of water evaporated (lb live steam per lb of water evaporated)	(b) kWh electricity per tonne of water evaporated (kWh electricity per lb of water evaporated)	(a) + (b) Total energy required in MJ per tonne of water evaporated (in B.T.U. per lb of water evaporated)
1 effect	1,100 - 1,300 (1.1 - 1.3)	10 - 20 (4.5 - 9)	2,790 - 3,320 (1,200 - 1,430)
2 effects	630 - 670 (0.63 - 0.67)	5 - 10 (2.3 - 4.5)	1,590 - 1,710 (680 - 740)
1 effect with ejector	450 - 550 (0.45 - 0.45)	3 - 7 (1.4 - 3.2)	1,130 - 1,400 (490 - 600)
3 effects	400 - 450 (0.4 - 0.45)	3 - 7 (1.4 - 3.2)	1,010 - 1,150 (430 - 490)
2 effects with ejector	320 - 400 (0.32 - 0.4)	3 - 7 (1.4 - 3.2)	810 - 1,020 (350 - 440)
4 effects	300 - 340 (0.3 - 0.34)	3 - 7 (1.4 - 3.2)	760 - 870 (330 - 370)
3 effets with ejector	200 - 300 (0.2 - 0.3)	3 - 7 (1.4 - 3.2)	510 - 760 (220 - 330)
5 effets	240 - 260 (0.24 - 0.26)	3 - 7 (1.4 - 3.2)	610 - 670 (260 - 290)
4 effects with ejector	150 - 230 (0.15 - 0.23)	3 - 7 (1.4 - 3.2)	380 - 600 (160 - 260)
5 effects with ejector	130 - 170 (0.13 - 0.17)	3 - 7 (1.4 - 3.2)	340 - 450 (150 - 190)
6 effects with ejector	115 - 140 (0.115 - 0.14)	3 - 7 (1.4 - 3.2)	300 - 370 (130 - 160)
7 effects with ejector	105 - 125 (0.105 - 0.125)	3 - 7 (1.4 - 3.2)	270 - 340 (120 - 150)
1 effect with M.V.R. (electric motor)	0 - 20 (0 - 0.02)	13 - 25 (6 - 11)	60 - 140 (30 - 60)
2 effects with M.V.R. (electric motor)	0 - 20 (0 - 0.02)	11 - 25 (5 - 11)	50 - 120 (20 - 50)
3 effects with M.V.R. (electric motor)	0 - 20 (0 - 0.02)	11 - 25 (5 - 11)	50 - 120 (20 - 50)
Equivalent 1 kg steam = 2.5 MJ; 1 kWh electricity = 3.6 MJ; 1.B.T.U./lb = 2,326 MJ/tonne			

heat the column bottom. This is only possible when the temperature difference is limited to, say, 50 to 80°C (90 to 140°F), so that efficiency is acceptable and conventional steam compressors can be used. Figure 3 shows schematic diagrams available for MVR duty. At present, only diagrams a, c and d are used in practice :

- diagram a : column head vapor compression and condensation in reboiler,
- diagram c : column head vapor condensation to evaporate demineralized water. Generated steam is compressed and directly injected in the column bottom. This arrangement is applied in some of the ethanol/water distillation plants (to produce alcoholic beverages),

Figure 3 : Schematic diagrams of distillation with mechanical vapor recompression

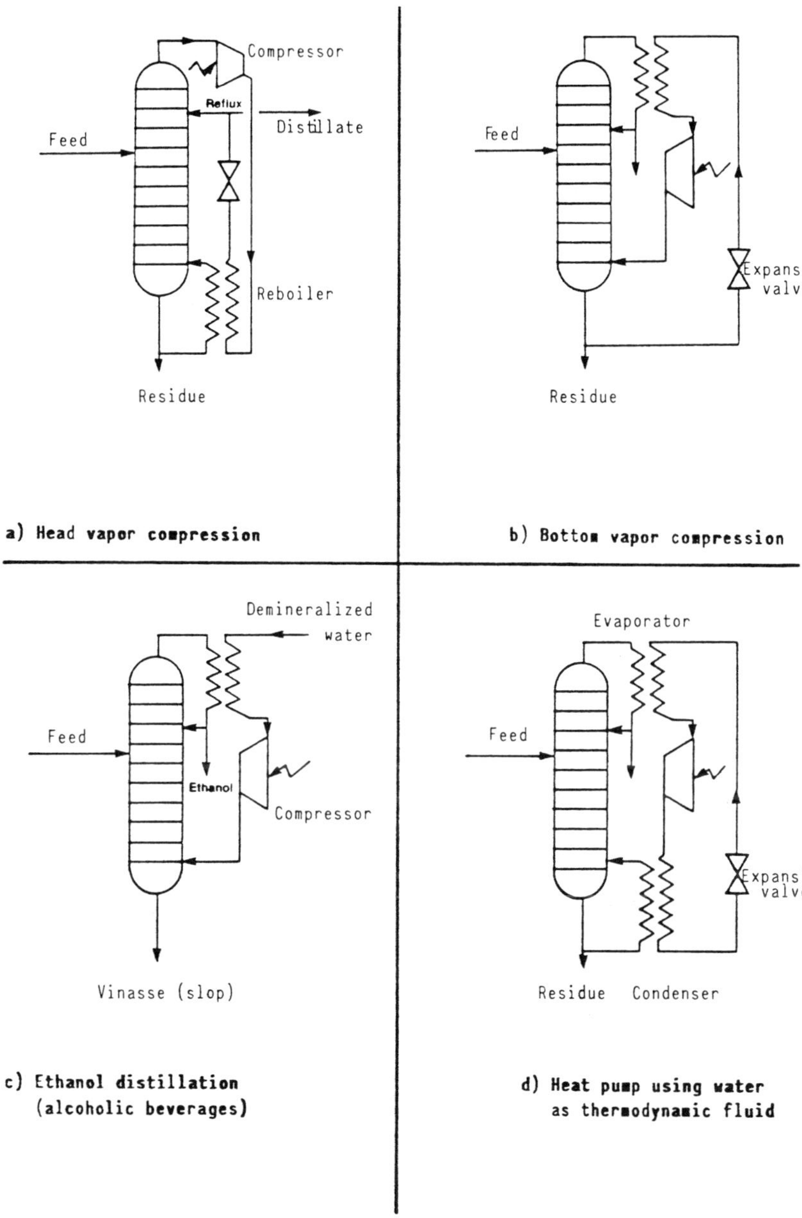

- diagram d : heat pump using water as thermodynamic fluid when physical, chemical or thermodynamic reasons prevent head or bottom vapor from being compressed. This design is less efficient than the previous ones.

Average energy savings figures cannot be given because of the very different distilled liquids and operating conditions. A detailed example will be stated in the following chapter. In addition to ethanol/water distillation already quoted, we can refer to propylene purification (separation from propane), cyclohexane distillation (for Nylon production), C4/C5 mixture distillation (for synthetic rubber production)...

Distillation columns : tray and packing distillation columns can be used for MVR. Packings appear to be more attractive thanks to their low pressure drop, but trays allow for faster restart in case of compressor breakdown.

As for evaporators, distillation columns can be fitted in multiple effect configuration, energy cost being reduced according to the number of effects. However, multiple effect columns carry severe constraints.

2.3. Drying

Drying aims to eliminate a solvent (most often, water) contained in a non volatile compound, by evaporation. Drying is very expensive in terms of energy.

Therefore, it is necessary to eliminate as much water as possible by centrifuging, pressing, evaporating, ..., before introducing the product or the solution into the drier. Drying generally needs high temperature difference, for physical and economical reasons. MVR implementation is therefore difficult because a high compression ratio is required.

At present, the first two MVR plants (as far as we know) using drying by ebullition are being commissionned in Europe :

. animal scrap drying and sterilization in Belgium : the animal scraps are sterilized and dried in a tube evaporator equipped with a twin-screw steam compressor. Electricity consumption is expected to be $26.\ 10^{-3}$ kWh/lb of water evaporated, plus a little steam make-up corresponding to $6.5.\ 10^{-3}$ lb fuel oil per lb of water evaporated ;

. sugar beet pulp and alfalfa drying in France : an air-tight drier with a sieve belt, heated by superheated steam and using a centrifugal compressor. Electricity consumption is expected to be $82.\ 10^{-3}$ kWh per lb of water evaporated without any steam make-up during continuous operation.

2.4. Thermal effluents valorization (heat recovery)

This heading concerns very different types of processes, whose common point is to generate liquid or gazeous hot (generally 80 to 130°C, 180 to 270°F) effluents or products, which have to be cooled before being directed to a downward process or disposed as waste. The purpose of MVR is to produce heating steam (often passed into the low pressure steam main) by compressing vapor generated by effluent or product cooling.

- Liquid effluents : unless they are very corrosive, aqueous effluents are flashed in 1, 2 or 3 stages to generate low pressure vapor or even vapor under atmospheric pressure (e.g. : amine cooling after regeneration, thermal recovery on effluents leaving wood defibrators). Non aqueous or very corrosive effluents are cooled in a heat exchanger generating vapor from demineralized water (e.g. : sulphuric acid cooling between absorption towers).
- Gazeous effluents : in certain processes, in particular chemical reactions, hot gases have to be cooled. As previously, the effluents are introduced in a heat exchanger to generate vapor (e.g. : gazeous acetaldehyde cooling after catalytic reaction ; heat recovery from exothermic reaction : hydrating quick lime). Other processes generate steam near atmospheric pressure, which is pure enough to use no heat exchanger before compression : wort boiling,...).

3 - ENGINEERING CONSIDERATIONS

3.1. Technical aspects

Generally speaking, all types of evaporators, reboilers, driers, cookers, flash tanks, etc, can be operated in conjunction with MVR. In practice, the choice depends very much on :

- characteristics of the product to be treated (flow, concentration, chemical composition, boiling point rise, temperature, viscosity, tendency to corrosion, erosion, fouling, foaming,...),
- process requirements (flow variations, automation, operating hours...),
- energy scheme in the factory (hot water needs, in-house cogeneration, process location...),
- economic aspects (investment capacity, new or partially refurbished plant, fuel costs, electricity costs, amortization rules...).

If process requirements allow for a cheaper exchange surface to be used, a large surface area with a lower temperature difference is justified ; MVR applications are then economically very attractive. In this respect, a new type of evaporator is being developed in France, with an exchange surface made of thin plastic tubes to minimize investment and power consumption.

The choice of compressor depends on :

- Nature of the vapor : generally, the vapor is steam ; but it can be a chemical solvent, and it is important to know clearly its physical, chemical and thermodynamic properties.
- Corrosion : appropriate (and often, expensive) materials need to be used to withstand corrosive vapor.

- Erosion : as the vapor to be compressed is saturated, fine water droplets may appear and erode various parts of the machine. Design considerations include materials, limitation of high local velocities, installation of water traps, etc...
- Fouling : the vapor can be impregnated with particles or solutes which deposit in the machine as result of superheating during the compression process and adversely affect the performance. Provision for cleaning devices may sometimes be necessary.
- Pressure level : whenever sub-atmospheric pressure is used, it is necessary to prevent air admission to avoid the presence of non-condensible gases in the circuit.
- Boiling point rise : an increase in the concentration of the solution means an increase in its boiling temperature (in comparison to the pure solvent). This elevation corresponds to a higher pressure rise of the compressor.

3.2. Optimization

Plants must be optimized to reduce overall annual expenditure (amortization plus energy costs). The technical and economic optimization lies essentially in the choice of the saturation temperature rise and hence of the compressor pressure rise in the case of MVR processes.

This will be illustrated with a simplified example of an existing MVR plant in France, whose analysis and engineering were carried out by the KESTNER Company (Lille, France) in 1979-1980.

The purpose was to concentrate and crystallize 100 t/h (61.2 lb/sec) of dilute effluents from chemical treatment of ores. The operation was subdivided into :
- evaporation of 88 t/h (53.9 lb/sec) water until saturation of solution,
- evaporation of 8 t/h (5.4 lb/sec) water for salt crystallization.

The economic bases were :
- fuel-oil cost 450 F/t (1 tonne = 2,205 lb)
- mean electricity cost 0.135 F/kWh
- amortization rate : 10 %/year, interest rate : 12 %/year.

Figure 4 gives a part of the results of this analysis. MVR was immediately installed on the concentration unit. The crystallizer has been operated with a steam ejector for two years (because of the uncertainty of a totally new chemical process) ; a centrifugal compressor replaced it in 1982. Both compressors are still in very satisfactory use according to the operating staff. The overall specific energy consumption is 19 kWh per tonne of water evaporated ($8.6 \cdot 10^{-3}$ kWh/lb).

3.3. Maintenance

With MVR, the maintenance of the compressor and of the motor involves :
. a weekly servicing (1 h/week), checking and lubrication,

Figure 4 : Optimization of an evaporation plant (actual case history in France) :
- Top : Mutiple-effect evaporator
- Bottom : Mechanical vapor recompression evaporator

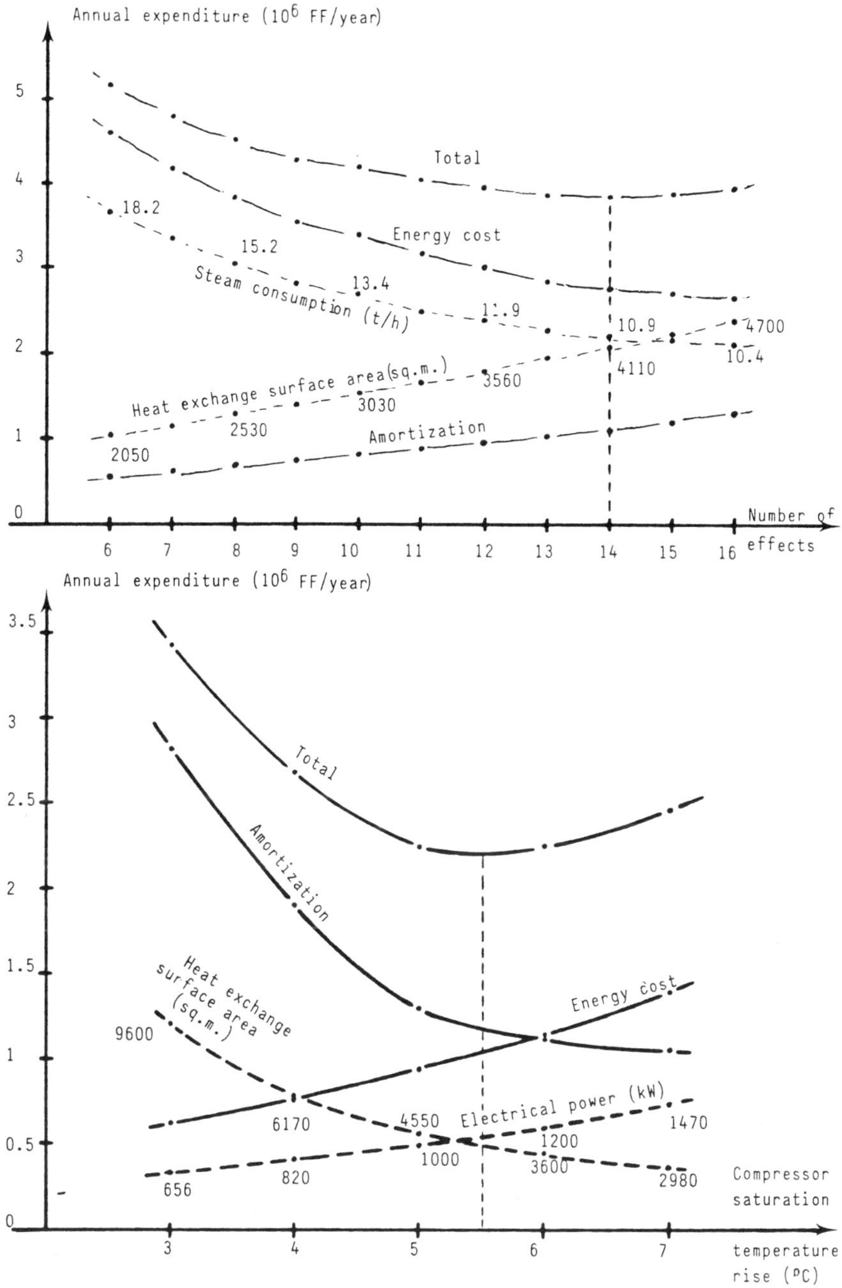

. annual servicing of the compressor by a specialist (approx. 50 hours),
. annual servicing and electric motor cleaning (approx. 40 hours).

The maintenance costs for the compressor and its electric motor normally account for 2.5 to 3 % of the investment cost of this unit, i.e. not more than 1 % of the total investment.

4 - EXISTING MVR PLANTS IN EUROPE AND TRENDS OF DEVELOPMENT

4.1. Existing plants

A recent investigation was undertaken by the International Union for Electroheat on the existing MVR plants in Europe. Though the answers received are limited, some lessons have been drawn, which are to be considered as tendency guides. They are presented in schematic form in Figure 5. The diagrams represent 433 installations with a total absorbed power of 267.4 MW (199,400 HP). These figures do not include the smallest plants to produce pure water for pharmaceutical and other industries.

Concerning distillation, these figures include 10 MVR plants in France, for a total absorbed power of 8.9 MW (6,670 HP), and 10 plants in Great Britain, for a total absorbed power of 6.7 MW (4,970 HP), 18 of them being driven by electric motors.

The general conclusions we can draw for MVR in Europe are (rough figures) :

- two thirds of the plants were installed after 1979,
- the food industry and chemical industry each account for 40 % of the total absorbed power,
- the number of plants can be classed as follows : 80 % for concentration by evaporation, less than 2 % for crystallization by evaporation, 5 % for distillation, less than 1 % for drying and 8 to 10 % for thermal effluents valorization,
- unit powers range from several dozen HP to 5000 HP, the average power being 830 HP,
- the percentage of driving machines is distributed as follows : 89 % electric motors, 8 % steam turbines (all of them commissioned before 1979) and 3 % combustion engines,
- centrifugal compressors account for 80 to 85 % of the number of plants,
- nearly all the plants were engineered and commissioned by manufacturers of evaporators, distillation columns...

Various significant examples will be outlined in detail by Mr. LETOUT in the following paper.

4.2. Trends of Development

At present, there appear to be a small number of favorable decisions, if compared to the strong increase during the years 1981 to 1985. This seems to result from two causes : companies' low investment capacity and low fuel costs which decrease the

Figure 5 : Existing M.V.R. plants in Europe

Analysis of the number of installations surveyed by industrial activities

(433 plants)

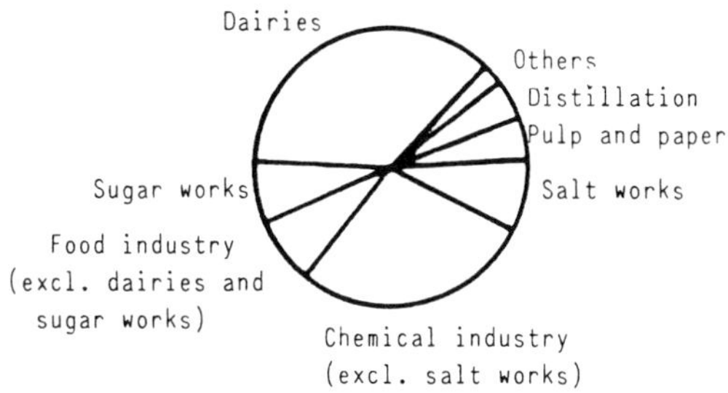

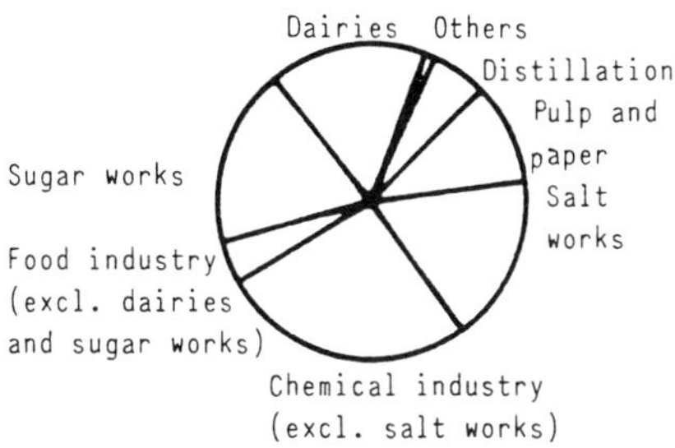

Analysis of the power of installations surveyed by industrial activities

(267.4 MW = 199,400 HP)

Analysis of the power of installations surveyed by countries :

CHEMICAL INDUSTRY
excluding salt works
(69.9 MW = 52,100 HP)

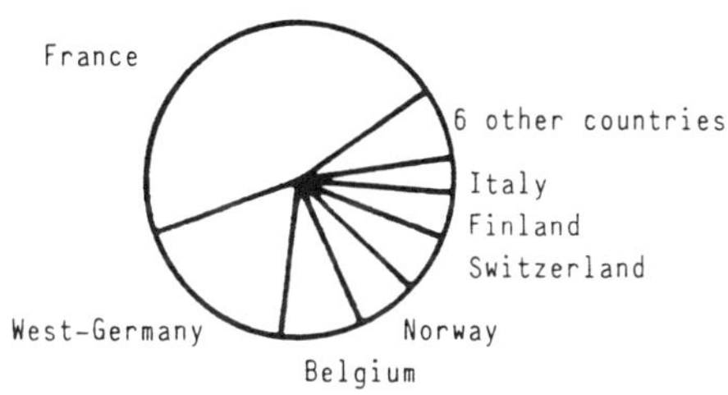

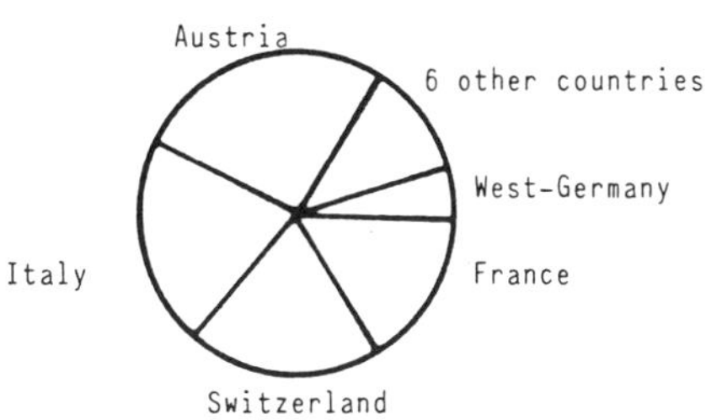

SALT WORKS
(45.9 MW = 34,200 HP)

Figure 5 (continuation) :

SUGAR WORKS
(48.6 MW = 36,200 HP)

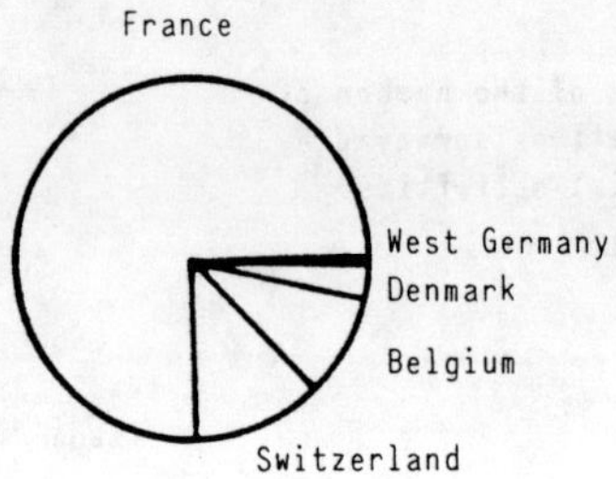

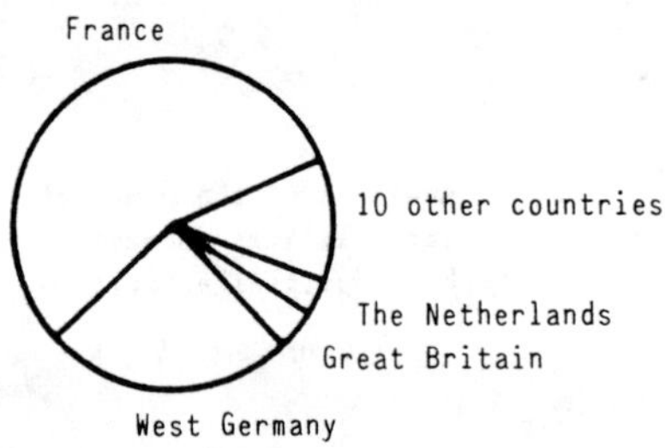

DAIRIES
(45.4 MW = 33,800 HP)

FOOD INDUSTRY
Excluding sugar works
and dairies
(11.6 MW = 8,600 HP)

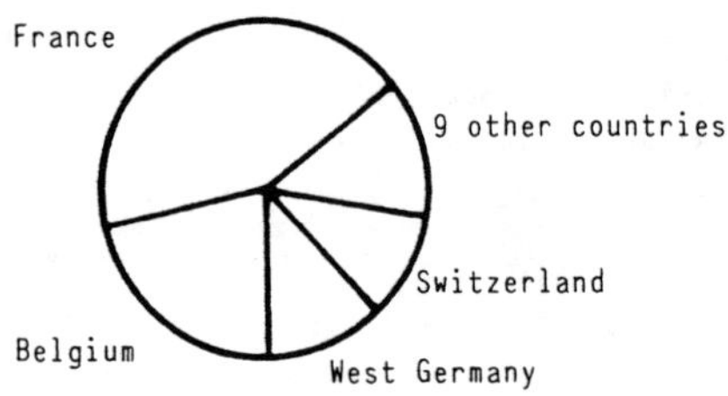

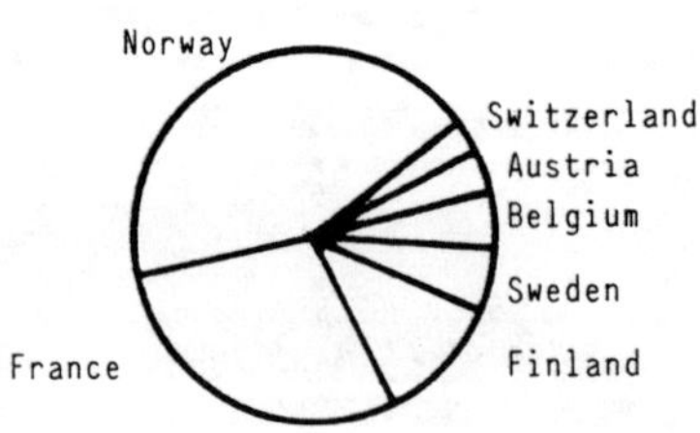

PULP AND PAPER INDUSTRY
(27.7 MW = 20,600 HP)

competitiveness of MVR plants equipped with electric motors.

The main activity branches where MVR continues to develop are the milk, chemical, and to a minor extent, pulp and paper, and sugar industries.

Waste treatment plants are also full of promise, but their development is linked to the "pressure" exerted for environmental protection. Drying applications should also be numerous if the industrial prototypes being commissionned keep their promise.

Under the present economic conditions, we can forecast a number in the range of 40 to 60 new plants per annum in Europe (power exceeding about 100 HP). The number of new MVR plants will remain closely linked with fuel-oil prices. This number could be greatly increased if investment costs are reduced, because much remains to be done in plants involving little flows.

5 - CONCLUSIONS

Experience to-date shows that initial reservations about introducing new technology in industries not used to compressor practice were not justified. The following conclusions can be drawn from current experience :

. MVR projects must be studied without any prejudice : plants have been installed thanks to very economical operating prospects, even though product flow or compressor power or operating hours were very small.
. Provided careful attention is paid at the design stage to the dimensioning of components and the optimization of the process, the integration of a compressor into a process line raises no major technical problem. Very satisfactory economic balances have been obtained. It is essential to involve a specialized company to achieve such good results.
. Machines have proved to be reliable, and little down-time is due to unavailability of compressors. Automatic control systems are being rapidly developed, improving operational security.
. Experience shows that there are only a few cases of deterioration due to erosion or corrosion and that fouling problems arising in some processes can be resolved by water injection and washing cycles.

Partial Bibliography

Mechanical vapor compression for concentration by evaporation, International Union for Electroheat working group "Process heat recovery", Paris, 1986

5 CONCLUSIONS

Chapter 17

Heat Pumps for Waste Heat Recovery in the Pulp and Paper Industry

M. K. Azarniouch and J. Romagnino

ABSTRACT: The application of the high temperature heat pump in the pulp and paper industry would have an important potential to accomplish the following objectives:

- recovery of substantial portion of prime energy used in the mill, and
- allowing a higher degree of white-water recycle, with reduction in fresh water usage, fibre loss and pollutants discharged.

At present our work centers around process integration studies to assess the technical and economic feasibility of various steam-generating heat pump applications for waste heat recovery in major sectors of the pulp and paper industry.

INTRODUCTION

The production of newsprint or bleached kraft pulp requires gross energy inputs of 20-28 GJ/tonne. Since the work actually required to produce the newsprint or pulp from wood represents less than 10 to 20% of the gross energy provided, 80 to 90% of the energy ends up in liquid and vapour phase waste streams discharged by the mills. Most of this is theoretically recoverable with, e.g., a heat pump in the form of steam. The actual technically and economically feasible coefficient of performance (COP) depends on the temperature rise (product vs. source) to be achieved. In most cases, to achieve satisfactory economic return, it is desirable to select only streams which have sufficiently high temperatures. It is also generally desirable to keep the COP above 3 for optimum energy and financial economy. In any case, the application of heat pumps offers great opportunities for recycling low level waste enthalpy in usable form, thus reducing

HEAT PUMPS: Prospects in Heat Pump Technology and Marketing (Proceedings of the 1987 IEA Heat Pump Conference), Edited by Kay H. Zimmerman

both the net energy input, the discharge of waste streams by the mills, and the cost of production.

In most cases the energy used for driving the heat pumps is electricity. Gas or diesel engines might also be used with even greater efficiency in situations where availability and cost make this advantageous.

BACKGROUND

The major incentives for reducing or possibly eliminating the discharge of process waters in the pulp and paper mill operations are:

(i) potential energy savings,
(ii) potential fibre savings,
(iii) reduction of pollutants discharged into the receiving waters,
(iv) reduction of fresh water requirements.

Furthermore, gaseous discharges represent another important waste heat source which could potentially provide substantial energy savings, if the heat can be up-graded with a steam-generating heat pump.

Newsprint Mill

A comprehensive survey [1] of more than half of the North American and Scandinavian groundwood/newsprint mills has shown that effluent discharges average 58 m^3 per ton of newsprint. Fresh water addition to the groundwood mill for stone cooling varies from 0 to 20 m^3 per ton of newsprint, the average being 6 m^3.

However, newsprint mills are expected to "tighten up" more and more their operations by increasing recycle of white waters. One of the consequences of achieving a high degree of process water recycle is an increase in the overall system temperature. In order to compensate for this upset, cooling of recycled white waters is required. This offers an excellent possibility for a waste heat recovery system such as the heat pump which could be used to extract "free" heat from recycled white waters at elevated temperature in order to produce process steam.

A case in point is the large amount of heat generated in the grinders of the stone groundwood mill. Over 90% of the power input into the grinders of the stone groundwood pulping in newsprint mills is converted into heat, most of which ends up in the white water, which in turn has to be cooled in order to maintain an acceptable temperature level in the white-water system. A recent study [2] has shown that close to 1/3 of the 5 GJ/t used in producing stone groundwood pulp can be recovered as process steam if a high temperature heat pump is used. With an annual production of over 5 million tons of stone groundwood pulp in Canada, the following benefits can potentially be derived [1,2] if the proposed technology could be successfully developed:

(i) Energy savings - 1.25 million barrels of oil equivalent per year by the newsprint industry.

(ii) Net operating gains - \$5-6/t newsprint.
(iii) Reduction in fresh water usage - 6 m^3/t newsprint (10% reduction).
(iv) Fibre savings - 36,000 tonnes per year by the newsprint industry.
(v) Reduction in suspended solids discharge - 8.4 kg/t newsprint (35% reduction).
(vi) Reduction in BOD_5 discharge - 6.4 kg/t newsprint (50% reduction).

The dryer section of the paper machine uses most of the thermal energy essential to the paper making process; the steam requirement amounts to approximately 5 GJ/t newsprint. Most of this heat appears in vapour-laden air which is exhausted through some form of heat recovery system to atmosphere. Usually, air-to-air heat exchangers are used in an open-air cycle to pre-heat incoming air with the hot vapour-laden exhaust air. These conventional heat recuperators recycle only some of the sensible heat back, and nearly all of the latent heat is rejected. However, a heat pump, by providing both a surface that is sufficiently cool for condensation of the outgoing vapours to occur and a means to transfer this latent heat to a higher temperature, is also able to provide a latent heat recycling mechanism free of the heat and mass balance restraint of vented systems. This potential has been recognized [3-5], and it has been suggested that using a de-humidification heat pump to cool the dryer hood exhaust air at 58°C from conventional economizers down to 50°C and heating the incoming fresh air to 100°C would reduce the steam consumption of drying by 6%. Further improvement in the energy recovery could be obtained with the use of a largely closed air cycle using heat pump dehumidification drying of recycled air.

It has been suggested that this technology is today technically and economically feasible provided electricity is available at low cost (i.e. hydro-power) or the heat pump is driven by gas or diesel engine.

Although the above studies have shown reasonable economic potential for the heat pump in major newsprint mill operations, full-scale applications have not yet been widely accepted in the industry.

Kraft Mill

Potential major sources for heat recovery in kraft mills are:
a) bleach-plant effluent,
b) digester blow heat,
c) recovery- and hog fuel boiler flue gases.

a) Bleach plant effluent

The combined bleach-plant effluent has a temperature of approximately 52°C and its enthalpy is in the order of 11 GJ/t pulp [6]. On the other hand, the steam demand of the bleach plant is approximately 3.5 GJ/t [7].

At present, the fresh water usage of the bleach plant averages 87 m^3/t with an internal water reuse of only 29% [8]. Trends toward partial closure (e.g. 75 to 90%) by increased effluent recirculation combined with high consistency operation [7,9] could provide an important source for heat recovery with a heat pump (e.g. 3 GJ/t).

Based on an annual Canadian production of 5 million tonnes of bleached kraft pulp the following benefits [8,10] could potentially be derived if the proposed technology could be successfully developed:

(i) Energy savings - 2.25 million barrels of oil equivalent per year by the kraft industry.

(ii) Reduction in fresh water usage - 53 m^3/t pulp (34% reduction).

(iii) Fibre savings - 145,000 tonnes per year by the kraft industry.

(iv) Reduction in suspended solids discharge - 0.35 kg/t pulp (7% reduction).

(v) Reduction in BOD_5 - 1.4 kg/t pulp (28% reduction).

b) Digester blow heat

The blow heat system of batch digesters produces flashed steam in the order of 2 GJ/tonne of pulp. Condensers and accumulators are typically already available. In a typical jet condenser system the steam from the blow tank is condensed in a direct contact condenser and hot contaminated water at about 95°C is produced. In some mills, this hot contaminated water is used directly for washing purposes, however, it is invariably contaminated with black liquor and malodorous sulphur compounds. Either this contaminated hot water or the flash steam itself provides a potential source of heat recovery with a heat pump to produce clean process steam at elevated pressure and temperature.

c) Flue gases

Scrubbers are becoming common on kraft recovery boilers because (i) they enable the mill to increase recovery furnace load to 10 to 30% above the design capacity, while still maintaining both odour and particulate emissions at the required regulatory levels, and, (ii) heat recovery in the form of 70°C water can partly cover the operation of the process. The recoverable heat can be up to 5.0 GJ/t pulp out of the 13.2 GJ/t [7] available in organic solids. However, it is unlikely that the total amount of heat recoverable as hot water can be used with credit for fuel savings. Incorporating a heat pump into the flue gas heat recovery system could provide (i) process steam (e.g. 1.5 GJ/t), and, (ii) hot water, thus allowing a higher energy recovery from the flue gas.

Scrubbers can also be used on hog fuel boiler flue gases to produce heat pump hot water source at about 70°C. Where hog fuel dryers are used the vapour in the flue gas from the dryer can be a heat pump source via a direct contact condenser or scrubber.

In addition to requiring waste heat sources, the application of industrial heat pumps is constrained by steam end uses. Standard pulp and paper operations utilize process steam generally at three levels. These are:

- 240 kPa - building heating
- 510 kPa - evaporators, dryers, air heaters for paper machines, steam coils, stock injection heating, bleach plant processes.
- 1.22 MPa - final paper machine dryers, digester cooking/ injection, recovery boiler air preheating.

Obviously, the 240 kPa application is limited in its use. The 1.22 MPa needs are significant but the 510 kPa level is easier to achieve, widely needed, and represents the best target value.

The brief outline of the energy availability and energy end-use discussions in the previous sections indicate that a number of applications for heat pumps in the pulp and paper industry exist which have broad interest.

Industrial Heat Pump Systems

The principal elements of a heat pump are a condenser, an evaporator, a compressor and an expansion device. A suitable working fluid is fed to the evaporator, where it absorbs heat at low temperature and, in doing so, is completely vaporized. The low temperature vapour is compressed which requires the input of external work. The work done on the vapour raises its pressure and temperature to a level where its energy becomes available for use at a higher level of enthalpy than that of the original source of heat. The vapour flows through a condenser where it gives up its energy as it condenses to a liquid. The liquid is then flashed and fed back to the evaporator where the cycle repeats.

The performance of heat pump is ordinarily described in terms of the coefficient of performance (COP) which is defined as:

$$\text{COP} = \frac{\text{Heat delivered}}{\text{Work input}}$$

To determine whether it would be worthwhile, in a given instance, to replace a direct source of heat (such as an oil-fired boiler) with a heat pump, it is necessary to calculate the minimum COP at which the heat pump would have to operate to achieve a net saving of energy. The following is an example:

Assume: Efficiency of conventional fossil fuel boiler - 85%
Overall efficiency of power plant, including transmission losses - 30%
Efficiency of electric motor driving the heat pump compressor - 0.95%

Then: $$\text{COP} = \frac{0.85}{0.30 \times 0.95} = 3.0$$

Thus the COP of the heat pump would have to be greater than 3.0 in order to achieve a net energy saving as compared to steam generated by a boiler operating at the same efficiency of 85%.

The above assumptions cannot be used if hydro-electric power is used to drive the heat-pump compressor.

One major problem with vapour-cycle heat pump systems for industrial applications is the selection of the working fluid for high temperature cycles. Halogenated hydrocarbons such as R114 (dichlorotetrafluoroethane) used in a number of industrial heat pumps and in refrigeration cycles are suitable for operating temperatures of up to about 120°C and could allow to produce low pressure steam.

At present, the only fully proven commercially available industrial heat pumps operate with freon R114 as working fluid and generate hot water up to about 98°C. It is likely that steam generating freon heat pumps will be developed in the near future. However, it is certain that steam generated with these heat pumps will not have sufficiently high pressure to be widely usable in the pulp and paper industry as process steam. It has to be combined with a steam compressor.

Sweden has tabled a proposal by ASEA-STAL to the International Energy Agency (IEA) to test a steam generating R114 heat pump to recover heat from humid air of a paper machine. The dryer part of the machine uses steam at about 350 kPa. The existent conventional heat recovery system on the machine has a scrubber unit according to Flakt's standard for paper machine heat recovery systems. It delivers a water flow of 300 kg/s at a temperature of 42.5°C, recuperating approximately 8 MW from the humid air.

The heat pump is based upon the standard concepts used by ASEA-STAL in a large number of district heating plants. It is a R114 process using a two stage turbo-compressor. The refrigerant is evaporated at a temerature of 31°C and condensed at 110°C. The refrigerant condenser is cooled by feed water from the mill system and thereby steam is boiled in the condenser at a temperature of 105°C or about 120 kPa. 11.5 MW is produced in form of this low pressure steam.

The steam pressure is increased to 350 kPa in a two stage turbo compressor with guide vane control and a by-pass for part load. Between the compressor stages and after the last stage water is injected to cool the steam to a suitable temperature level. The 350 kPa steam is delivered to the mill's low pressure steam grid.

Such a heat recovery system would give an output of 12.8 MW at a coefficient of performance of 2.4.

An alternate option to the above approach is to use a thermocompressor to increase the low pressure of the steam generated with the R114 heat pump process. This option would have (i) higher COP, (ii) for a given output it would require less prime energy input, and (iii) would require less capital investment. The major drawback of a freon heat pump system with associated thermo-compression is its inflexibility in terms of mill steam balance, since it requires a relatively large amount of motive high pressure steam.

Whether or not steam generating freon heat pumps combined with steam compressors will provide the best solutions for different mill applications remains to be demonstrated. The technical capabilities exist and will be enhanced when freon heat pumps generating steam directly become available.

The specific volume of steam at atmospheric conditions is substantially higher than at conditions required for process steam and consequently electric compressors for steam at near atmospheric pressure are much more expensive than those for higher pressure. The need for a heat pump working fluid which (i) has a lower specific volume than steam at atmopspheric conditions, and (ii) is more stable at higher temperatures than freon R114, becomes apparent.

Methanol vapour has a specific volume much lower than that of steam at atmospheric conditions and it is stable up to a temperature of 200°C. Methanol boils at about 65.5°C. The source leaving temperature must be well above that to avoid a vacuum during operation. In most cases the source temperature fluctuates so there must be a substantial margin of safety. If a vacuum were to develop, and it were to leak into the system, an explosive mixture of air and methanol is a risk.

Westinghouse [11] has presently a development program funded by the U.S. Department of Energy, to install and test a 30 GJ/h high temperature heat pump with methanol as working fluid to produce slightly superheated steam at 535 kPa from steam at 170 kPa coming from a TMP refiner. Their unit was tested successfully in mid 1983. For testing, it has been installed in a laboratory facility in Pittsburgh, using recirculated steam at about 180 kPa as a source. When fully developed, the plan is to move the unit to an industrial location. Another high-temperature heat pump is near the completion of its development stage [12]. In this heat pump a secret working fluid with properties similar to that of methanol is used.

Based on the above discussions, the following options of steam generating heat pump systems appear to have potential to recover substantial portions of the waste heat from pulp and paper making operations:

a) A freon R114 heat pump with subsequent thermal recompression of low pressure steam.
b) A freon R114 heat pump with subsequent mechanical recompression of low pressure steam.
c) A high temperature heat pump (e.g. methanol as working fluid) to produce process steam directly.
   In this case - a heat pump with methanol as working fluid - the source leaving temperature must be at least 70°C in order to avoid vacuum during operation. Lower temperature sources could still be handled by appropriate modifications in the mill operating practices, e.g. in white-water handling and/or stock preparation. This may require an investigation of the capability of the heat pump evaporator to extract heat from streams with very high fibre content, e.g. to recover heat directly from the grinder pit of the stone groundwood mill. Alternately, certain modifications in the operation of the methanol heat pump could allow the application to lower source temperature cases, without the risk to develop a vacuum.
d) A freon R114/methanol binary system to produce process steam. Lower temperature sources can also be handled by combining a freon cycle with a methanol cycle heat pump. Heat can be removed from a source at less than 65°C by a freon evaporator. The freon could then be compressed and condensed in a heat

exchanger which also evaporates methanol. The methanol vapour could be compressed and piped to a condenser/steam generator. This freon/methanol binary system for generating steam with lower temperature sources would require 3 heat exchangers.

CASE STUDY

Mill Closure

A stone groundwood (SGW) mill was chosen for the first demonstration of the potential benefits of installing a high temperature heat pump (HTHP). A newsprint mill in northern Ontario was selected for a case study.

The PAPMOD simulation model for the newsprint mill, developed by PAPRICAN, was found to be a useful tool to reproduce mass and energy flows in the mill. PAPMOD is a steady-state, modular simulation package and is an adaptation of the GEMCS (General Engineering and Management Computation System). It was designed to be run with a variable number of flow components and with process modules of varying degrees of complexity; this allows the user to start with a simple process representation and add complexity as the process knowledge improves. Modular simulation refers to the construction of a model flow sheet out of pre-written building blocks, or "modules". The advantage of this approach is that a substantial process description library can be kept available for putting together a variety of process models with relatively little effort.

The case mill is fully integrated with the paper machines, as are TMP and magnefite mills, and has a high degree of white-water recycle. The white-water system is such that it evens out temperatures throughout the mill, thereby reducing the need to cool or heat streams. Presently, heat would be recoverable at about 60°C. Desirable target heat source temperatures are about 70-75°C. This would involve concentrating the heat to a particular stream or set of streams.

Figure 1 is a PAPMOD simulation flowsheet of the mill under study. A simplified model of the paper machine wet end was also included in the overall model in order to evaluate the effect recycled white water has on the SGW mill. Modules representing heat loss in the wet end adjust recycled white-water temperatures to their current levels.

Figure 2 represents a conceptually modified version of the same mill. The fresh water cleaning showers at the disc filters (stream #19) would be replaced with white-water showers. Stream 51 would use previously sewered white water instead of heated fresh water. Both of these changes would reduce fresh water consumption. Streams 29 and 40 would be removed to help segregate the SGW mill white water from the paper machine white-water system. Replacement of cold fresh water with warmer white water and white-water segregation between the SGW mill and the paper machines serve to intensify heat levels in the SGW mill. Stream 33 would be removed from the clean sump and most of the sewering done at the wet end, with a minimal amount at the cloudy white-water chest. Besides intensifying heat levels at the SGW

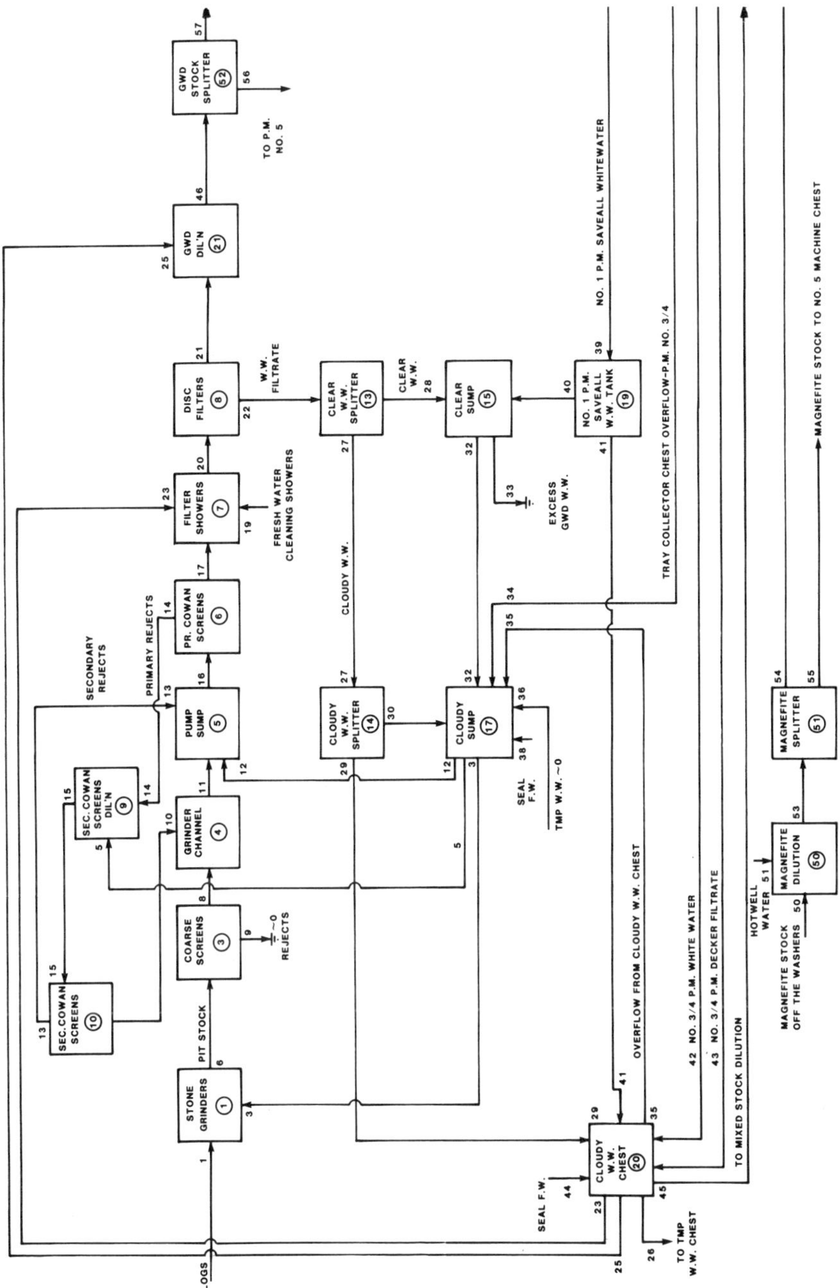

Fig. 1. PAPMOD simulation flow sheet of the case SGW mill.

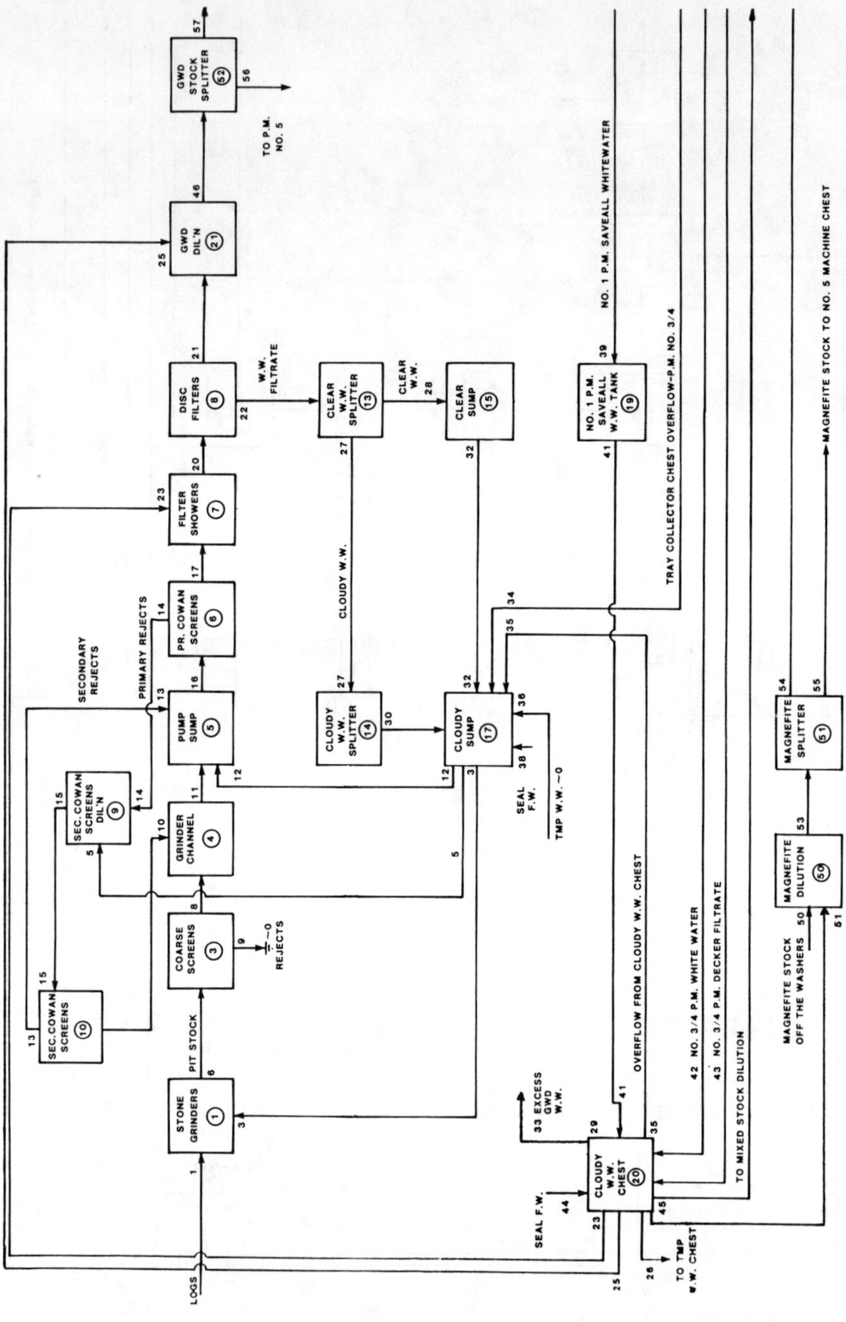

Fig. 2. PAPMOD simulation flow sheet of the modified case SGW mill.

mill, sewering at the wet end also conserves fibres, since white water consistencies would be higher with closure at the SGW mill.

Heat Pump Integration

A simplified high temperature heat pump (HTHP) computer program was also developed to assess the energy and economic benefits of the heat pump integration into the mill's white-water system. The heat pump system considered was a freon R114 heat pump with subsequent mechanical recompression of low pressure steam.

Two cases were examined. In all cases, the total capital cost for the installed heat pump system with 1 stage compressors is estimated at 5.4 x 10^6 \$CDN. Changes in pumping costs are not considered important. Payback calculations are based on producing steam at 140°C. Steam compression from 110°C to 140°C is roughly estimated to consume 54.3 kWh/tonne, using a 1 stage compressor. The cost of piping modifications are assumed to have very little impact on capital investment. An after-tax DCF-ROI (discounted cash flow return on investment) was also performed. The assumptions made for this analysis are:

1. tax rate: 50%;
2. project life: 20 yrs;
3. depreciation: 20% (declining balance); and
4. salvage value: 0.

Other data:

Operation:	350 days/yr
Electricity costs:	2.5¢/kWh
Steam cost:	\$5.00/GJ (approx. \$20 U.S./barrel oil)

Case 1 gives the results of installing a HTHP on Stream 22 in the modified SGW mill (see Figure 2). The results of Case 1 are presented in Table I. The estimation was based on a steady-state source temperature of 61.3°C and a source temperature drop of about 4°C. The heat transfer rate was about 30 GJ/h. Payback period was 6.7 years with an after-tax DCF-ROI of 9.5%.

In an attempt to improve the payback, a simulation was run, i.e. Case 2, with a HTHP on Stream 8 in the modified SGW mill (see Figure 2). The results of the Case 2 estimates are presented in Table II. The source inlet temperature was 78.9°C and the source temperature cooled down to 67°C. In this case the heat transfer rate was over 31 GJ/h. Payback period improved to 5.8 years and the after-tax DCF-ROI to 11.1%. However, this last case has a high technical risk, since the HTHP would be installed on a pulp suspension line instead of a white-water stream as in Case 1. It may be difficult to find a heat exchanger which could run satisfactorily with a pulp slurry at 3% consistency.

CONCLUSIONS

While, in general, there are vast potentials for the application of a HTHP in a variety of pulp and paper making operations the actual application is mill specific. In an integrated mill with a high degree of recycle it requires simulation of the whole mill. It often requires modifying the

Table 1. Modified SGW mill: HTHP on Stream 22 (Case 1)

Evaporator:	
Process water flowrate:	42328 MTPD @ .013% C
Heat transfer rate:	726.5 GJ/day
Inlet temperature:	61.3°C
Outlet temperature:	57.2°C
Steam produced:	452.4 MTPD
Power consumption:	
Heat pump:	3810.0 kW
Circulating pumps:	140 kW
1 stage steam compressor:	927.9 kW
Total:	4877.9 kW
Yearly power consumption:	40,974,360 kWh/yr
Cost:	1,024,360 $/yr
Steam generated:	366,080 GJ/yr
Value:	1,832,650 $/yr
Net annual benefit:	808,290 $/yr
Payback period:	6.7 yrs
After-tax DCF-ROI:	9.5%
C.O.P.:	3.1

Table II. Modified SGW mill: HTHP on Stream 8 (Case 2)

Evaporator:	
Process water flowrte:	15038.5 MTPD @ 3% C
Heat transfer rate:	749.1 GJ/day
Inlet temperature:	78.9°C
Outlet temperature:	67.0°C
Steam produced:	429.984 MTPD
Power consumption:	
Heat pump:	2799.5 kW
Circulating pumps:	140 kW
1 stage steam compressor:	907.7 kW
Total:	3847.2 kW
Yearly power consumption:	32,316,500 kWh/yr
Cost:	807,910 $/yr
Steam generated:	347,000 GJ/yr
Value:	1,737,740 $/yr
Net annual benefit:	929,830 $/yr
Payback period:	5.8 yrs
After-tax DCF-ROI:	11.1%
C.O.P.:	4.0

mill in order to get higher temperatures. Steam produced by a HTHP, besides requiring to be an economical operation, must be justified with respect to end-use and other potential heat recovery schemes. Further, as suggested by the economic analyses, running a thermodynamically efficient HTHP will not necessarily mean that we have an economical operation. However, two iportant factors would undoubtedly considerably improve the economics of the HTHP and make its implemention in the pulp and paper industry possible:

a. A substantial escalation of the oil prices from the present level of under US $20/barrel.
b. Further research and development into finding acceptable HTHP working fluids (e.g. methanol, freon R113) which could (i) reduce the cost of a heat pump system and (ii) improve its efficiency of energy recovery.

REFERENCES

1. Envirocon Limited, Effluent Recycle Practices in Groundwood/ Newsprint Operations, CPAR Project Report 593-1, March 1977.

2. Azarniouch, M.K. and Lemay, Y., "Heat Exchanger for White Water Heat Recovery Application", Pulp Paper Can., 86 (11), T347 (1985).

3. Dyer, D.F., Dryer with Low Temperature Heat Recovery System, U.S. Patent 4026035, May 31, 1977.

4. Chiniara, C. and Kruska, D., "Energy and Water Recycle in the Paper Drying Process by Heat Pump Application", Wochenblatt fur Papierfabrikation, 105 (11/12), 421 (1977).

5. Holik, H. and Brugger, H., "How Useful is the Application of a Heat Pump in a Paper Mill", Wochenblatt fur Papierfabrikation, 108 (13), 489 (1980).

6. McCrosson, F.J. and Murray, K.A., South Carolina Application for Industrial Waste Recovery Using Heat Pumps and Rankine Cycle Power Generation Systems, Project Report prepared for U.S. Dept. of Energy, Office of Coal Utilization Contract No. DE-AC09- 77ET12866, 1980.

7. Reside, D., Roche, A.A., Bouchard, D.C. and Muratore, E., "Kraft Mill Energy Management", Pulp Paper Can., 82 (11), T382 (1981).

8. Ekono, "Study of Pulp and Paper Industry's Effluent Treatment", FAO Advisory Committee on Pulp and Paper 13th Session, Rome, May 1972.

9. Sittig, M., Pulp and Paper Manufacture, Energy Conservation and Pollution Prevention, Noyes Data Corp., Park Ridge, New Jersey, U.S.A., 1977.

10. Bruley, J., The Basic Technology of the Pulp and Paper Industry and its Waste Reduction Practices, Report No. EPS 6-WP-74-3, April 1975.

11. Wolf, R.W., "The Industrial Heat Pump - An Overview", paper presented at Electro-technologies in Industry Conference, Montreal, May 1982.

12. Roth, R.C., "Generation of High Pressure Steam from Low Temperature Waste Heat Sources", Pulp Paper Can., 83 (12), T336 (1982).

Chapter 18

Operating Experience with Industrial Heat Pump Systems: Economic Advantages, Obtained by Several Actual Installations

Masahiro Wakabayashi and Shigeru Sakashita

ABSTRACT: This report outlines the distinguishing characteristics and economic efficiency of the Screw Steam Compression Heat Pump System, giving as an example an application for an Alcohol distillation plant, a representative energy-intensive industry. Additionally, successful applications over the past several years in Japan and abroad are presented to clarify facilities which can benefit from heat pump use in various production processes.

INTRODUCTION

In Japan, industry has been endeavoring to reduce primary and secondary energy consumption since the first oil crisis in 1973. When we look at energy saving in the past, however, we see that most efforts were aimed only at recovery of a portion of waste heat or just a part of "waste heat being produced": the key processes that produce waste heat were ignored. By applying a complete thermal recovery heat pump system to such processes, a remarkable effect (reduction of primary and secondary energy consumption) is obtainable. Industry has, however, adopted the heat pump system at a slow pace despite it's proven advantages.

In order for installation of heat pumps to be considered by Japanese industries, primary energy ratio (PER) of more than 1.0 times will be the level needed for acceptance. If PER is more than 1.5, an enterprise will find the heat pump of major economic significance. Industrial heat pumps are, however, sometimes installed for other reasons and even if a system promises sufficient economic advantage and achievement of short-term investment cost redemption less than two years, it is not always be adopted. Heat pump makers may be required to offer the industry

HEAT PUMPS: Prospects in Heat Pump Technology and Marketing (Proceedings of the 1987 IEA Heat Pump Conference), Edited by Kay H. Zimmerman

advantages other than purely economic merit. This is an investigation of such merits and characteristics based on a study of industrial heat pumps installed over the past several years.

- Screw Steam Compression Heat Pump System (S.S.H.P.)
  - Vapor recompression (VRC) for alcohol distillation
  - VRC for malt boiling
- Freon Heat Pump:
  - Dehumidifying/Drying heat pump
  - Paint temperature controlling heat pump
  - Malt drying heat pump

1. ACTUAL INSTALLATION RESULTS

1.1. Steam Heat Pump for Alcohol Distillation

This is an adaptive instance of the Screw Steam Compression Heat Pump (S.S.H.P.). The flow sheet (Fig. 1) and specifications (Table 1) are shown below.

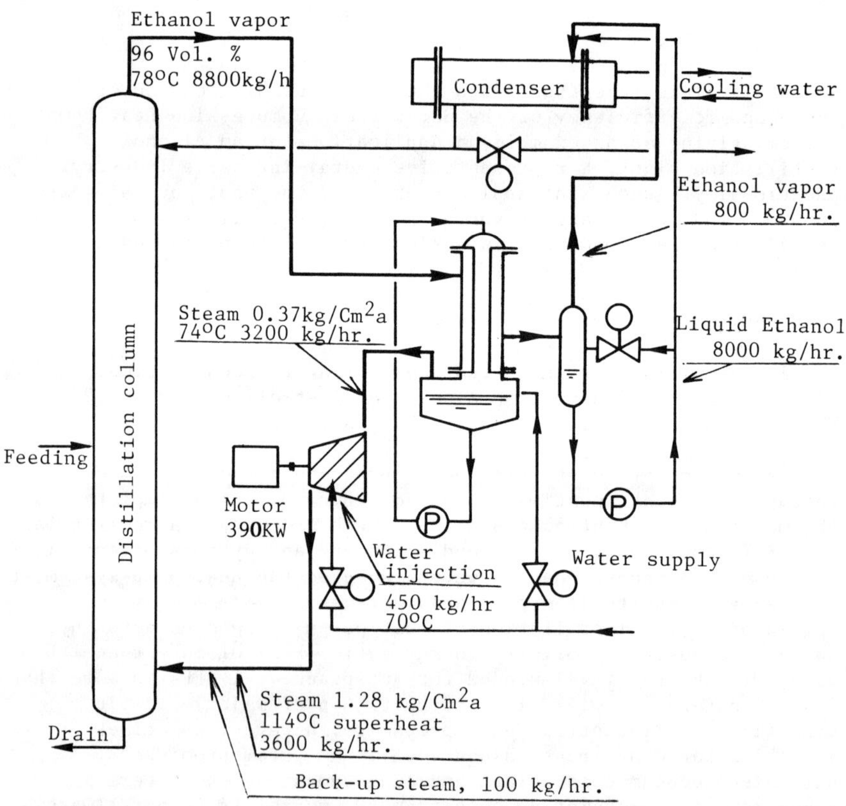

Fig. 1. VRC Flow Sheet for Alcohol Distillation

This system leads alcohol vapor (96% vol. 78°C) from the top of the distillation column to the shell side of a falling film type heat-exchanger. Water circulating inside of the tube is evaporated and the vapor is directly blown to the bottom of the column after temperature and pressure are raised by the compressor. The distinguishing characteristic of this system is that only one heat-exchange unit is needed in the heat pump cycle, resulting in simplified system configuration. Further, the employment of a high efficiency falling film type heat-exchanger also successfully reduces the range of temperature by 5 through 10°C as compared with a system (e.g., R114 High Temperature Heat Pump Cycle) that uses two heat-exchangers for evaporation and condensation.

Consequently, the C.O.P. is improved by about 1.0 and industrial water consumed is reduced by 20 m^3/h. Moreover, a dramatic reduction in the term of redemption is obtained. Regarding other characteristics, this system controls revolution of the compressor motor by means of a converter to match with the amount of alcohol production. System control (control of heating steam temperature, control of condensing alcohol, control of back-up system and four other controls), and measurement and analysis necessary for the system are all by means of an industrial micro-computor. This system showed a two year term of redemption though it involved enterprising trials.

Table 1.

Compressor Model No.	Pd/Ps	REV	G	Q	I.P.	C.O.P.
STM400L-M	1.30/0.37	6000	3600	1920	390	5.73

A	U	Tc	Te	Q
427	900	78	73	1920

Where,

Working medium: Steam
Falling film type heat exchanger
Shell side: Alcohol vapor
Tube side: Water

Pd = Discharge pressure (kg/cm^2a)
Ps = Suction pressure (kg/cm^2a)
REV = Revolution (r.p.m.)
G = Discharge steam (kg/h)
I.P. = Inverter input power (KW), including auxiliary driving power
C.O.P. = Coefficiency of performance
A = Heat transfer surface area (m^2)
U = Overall coefficient of heat transfer ($Kcal/m^2h^oC$)
Tc = Condensing temperature (^oC)

Te = Evaporative temperature (°C)

Conditions for Merit Calculation:

Unit price of electricity:	18 yen/Kw
Unit price of kerosene:	60 yen/Litre
Kerosene calorific value:	8,200 Kcal/Litre
Boiler efficiency:	0.87
Annual operating time:	8,000 hours
Date of delivery:	May, 1986

1.2. Malt Boiling Steam Heat Pump

Only a general outline is presented here as this system was introduced during a previous meeting. This simplified system, like the distillation plant, uses a heat pump providing various merits. The beer boiling process is carried out using a batch system (2,000 batches/year x 1.5 hr/batch = 3,000 hrs/year) and engine driving is advantageous taking operating time and the change in power rate for increasing facility into consideration. Industries that consume large amounts of hot water such as beer plants gain increased benefit by supplying jacket hot water to other processes.

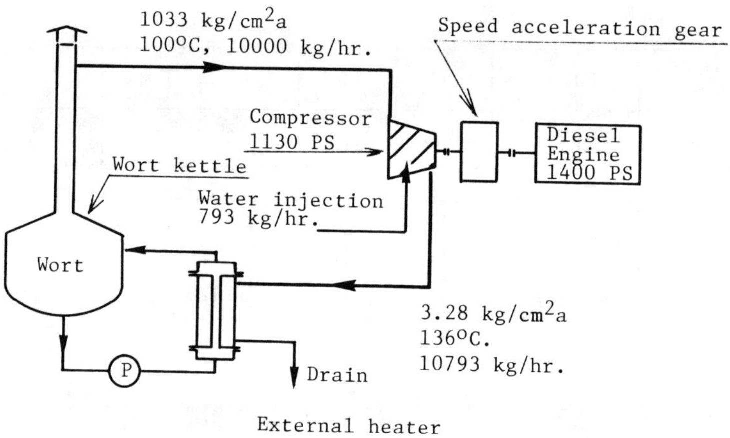

Fig. 2. VRC Flow Sheet for Malt Boiling

Table 2.

Compressor Model No.	Pd/Ps	REV	G	Q	E.P.	COP
STM510-L	2.38/1.03	3,000	13,300	5659	830	10.1

Where, Working medium: Steam
E.P.: Engine output (KW)

1.3. Dehumidifying/Drying Heat Pump

The flow sheet and specifications of the heat pump system for gelatine dehumidifying/drying are shown in Fig. 3 and Table 3, respectively. The heat pump, together with a refrigeration compressor used in this process, reduce the moisture in circulating air to 6.5g/kg at two temperature levels, 30°C and 40°C. The circulating air is cooled down by the evaporators of the heat pump and the refrigeration compressor and humidity is reduced by condensing moisture in the air. The dehumidified air is led to a drying chamber after being heated by a re-heater and the high temperature side of the heat pump. Usually the process of dehumidification is done by chemisorption (regeneration by steam) and steam and exhaust heat are all discharged into the air.

By replacing such a system with the above heat pump cycle, steam consumption is reduced from 3.2 Ton/h (1.2 Ton/h for regeneration) to 0.5 Ton/h.

The following merit calculation is based on unit prices as of May, 1983 (time of delivery), i.e., heavy petroleum 75 yen/Litre and electricity 25 yen/Kwh: 38 million yen/year (annual operating time 10 hr. x 23 days/month x 12 months/year = 2,760 hours). Investment is returned within one year. The reason for the remarkable merit is concurrent utilization of both low and high temperature sides, employment of the pre-cooler and re-heater, and selection of a reasonable temperature range (interlock with existing steam) and circulation system.

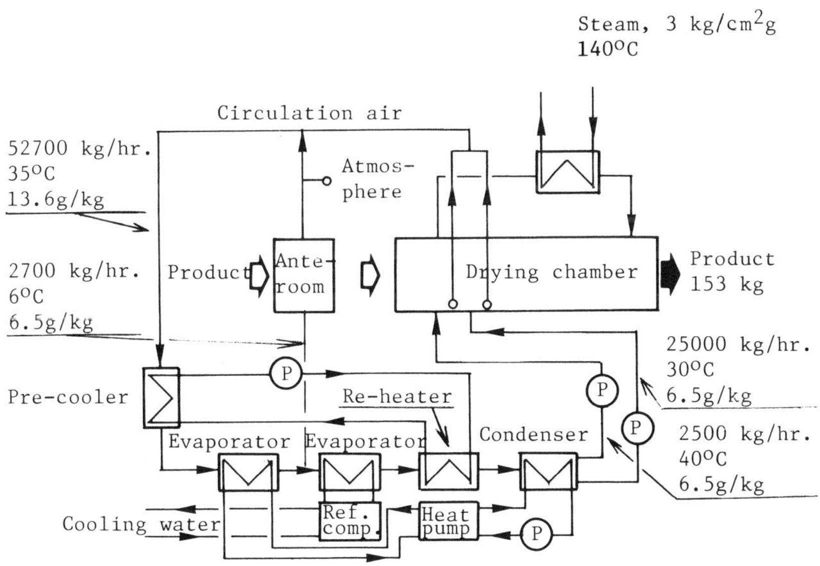

Fig. 3. Flow Sheet for Dehumidifying/Drying Heat Pump

Table 3. Specifications (Delivered in May, 1983)

Drying substance	Gelatine 153 kg/h (dry weight)
Dehumidified moisture	375 kg
Heat pump	45 Kw
Refrigeration compressor	75 Kw

1.4. Paint Temperature Control Heat Pump

This is an example of heat pump technology being adopted in an automobile painting process (Ref. Fig. 4 and Table 4). The system controls the temperature of a degreasing bath (for removing oil from the vehicle body), a chemical film coating bath (chemical film coating) and an electro-magnetic coating bath (electro-magnetic priming) on the painting line of an automobile body assembly line.

This primer coating process requires considerable precision control of degreasing, formation and electromagnetic coating at a constant temperature to secure a fine finish. The heat source is electrolytic heat in the electromagnetic coating bath and the heat sink is the degreasing bath and chemical film coating bath.

The main characteristics of this system are successful achievement of a high COP (6.58) and simultaneous utilization of a heat source and heat sink. With installation of this system, the factory reduced running costs by half and recouped their investment in about two years. Stable quality of products was, of course, also achieved. Three additional systems have since been supplied, displaying superlative performance.

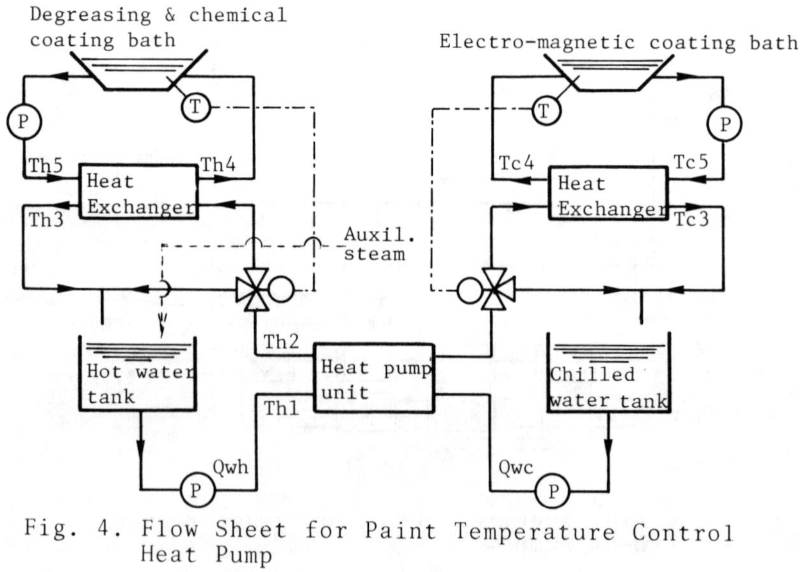

Fig. 4. Flow Sheet for Paint Temperature Control Heat Pump

Table 4. Specifications of Heat Pump
(Refrigerant: R12)

Qc/Qh	MP	R.C.	M	COP	S.C.
644/856	280	1544	1510	6.58	100

Where,

Qc = Coating capacity (Mcal/h)
Qh = Heating capacity (Mcal/h)
MP = Motor power (KW)
R.C. = Running cost (Ten thousand yen/year)
M = Merit (Ten thousand yen/year)
COP = Coefficiency of performance
S.C. = Auxiliary steam consumption (tons/year)

Conditions for Merit Calculation:

Unit price of electricity:	15 yen/Kw
Unit price of heavy petroleium:	60 yen/Litre
Calorific value:	9,000 Kcal/Litre
Boiler efficiency:	0.85
Operating time:	3,840 hrs/year

1.5. Malt Drying Heat Pump

This is an installation in France. The system and specifications are shown in Fig. 5 and Table 5. The heat pump is installed for the drying process in beer manufacturing (a high temperature drying required). The drying is carried out by batch processing (total of 20 hours/batch) which requires a set temperature pattern and seasonal adjustment for load, as shown in Table 5. Efficient operation of high Δt air at a Maximum 65°C is essential for this process.

The system is comprised of 3 compressor units. In case 85°C hot air is required, cascade operation of the No. 2 and No. 3 compressors must be carried out. As stated above, the system of three compressors has shown excellent follow-up properties to load change and high quality production has resulted. A COP of 3.94 was obtained even in the winter season when load is largest.

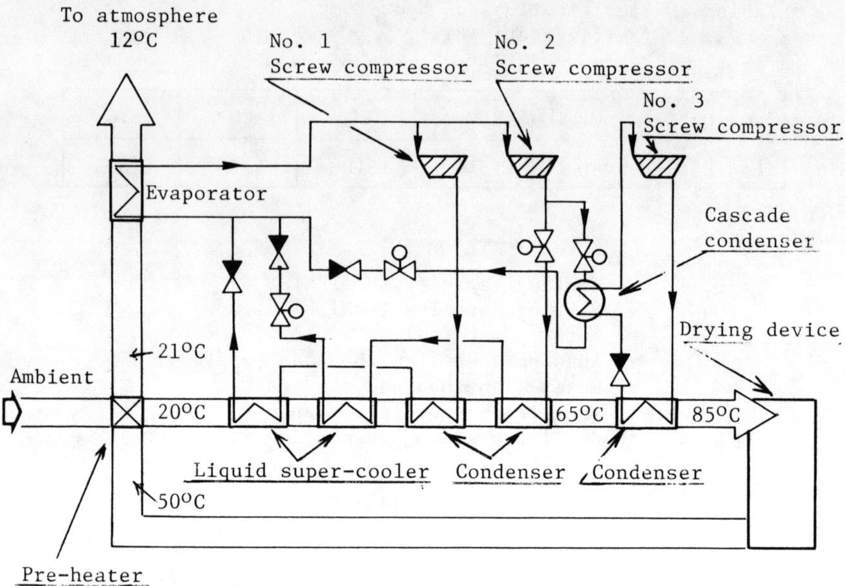

Fig. 5. Flow Sheet for Malt Drying Heat Pump

Table 5. Specifications

Compressor	Qc/Qh	Tc	Te	M.P.	REV	W.M.	C.O.P.
No. 1	1811/2242	52	6	502	2950	R12	3.84
No. 2	1140/1720	62	6	675	3300	R12	2.96
No. 3	1720/2080	90	50	429	2950	R114	5.66

Where,

Qc = Drying capacity (Mcal/h)
Qh = Heating capacity (Mcal/h)
Tc = Condensing temperature (°C)
Te = Evaporative temperature (°C)
M.P. = Motor power (KW)
REV = Revolution (r.p.m.)
W.M. = Working medium
C.O.P. = Coefficiency of performance

Table 6. Load Changes

Time (h)		Hot air temp. (°C)	Load (Mcal)		
			Summer	Spring/Autumn	Winter
0	6	60	3459	3847	4088
6	9	65	3983	4382	4612
9	12	70	3858	4288	4395
12	15	75	3300	3755	4110
15	17	80	3043	3533	3853
17	20	85	2889	3401	3646

1.6. Summary of Application Examples

The characteristics of the heat pump are presented giving as examples five applications installed over the past several years as stated above.

In brief, industrial heat pumps prove particularly attractive for the following reasons:

* Superior product quality

  With the use of proper detecting and final control elements, the high degree of control possible with a heat pump allows for more discreet, higher level product quality control.

* Improved profitability

  The heat pump provides:
  High C.O.P.,
  Quick return on investment,
  Energy, resource and labor savings of a high order.

* Harmony with production processes

  Simultaneous heat source and heat sink,
  Simplified processes,
  Selection of a driving apparatus most suitable to the production process and heat pump based on careful consideration of fuel expenses, power rates, the need for extension of power lines, noise and gas pollution.

* Reliability

  High reliability and durability of components,
  High reliability of the system as a whole (indispensable for heat pump installations).

2. Present Conditions and Future Prospects for Industrial Heat Pumps

2.1. Present Conditions

Falling oil prices in 1986 have had a significant impact on the industrial heat pump market. Oil prices have declined from the $25 per barrel level of late 1985 to below $10 per barrel in mid-1986. In Japan, fuel prices have also been dramatically lowered by the appreciation of the yen (¥), which rose in value from ¥250 to the dollar level at the end of 1985 to around ¥150 to the dollar.

Naturally, these changes have greatly lowered the cost of steam. Today, the cost of generating steam is less than half that of a year ago. It goes without saying that the cost advantages of introducing heating supply systems based on the heat pump have been considerably reduced. In fact, numerous heat pump projects have been suspended in recent months.

The following is a calculation of the economic marginal C.O.P. in the case of a motor driven heat pump, assuming steam costs to be ¥5,000 per ton in 1985 and ¥2,500 per ton in 1986 and electricity costs to be ¥20 per KWH in 1985 and ¥18 per KWH in 1986.

Motor-driven Heat Pump	1985	1986
A) Steam unit (¥/Mcal)	¥5/kg ÷ 0.55 Mcal/kg = ¥9.09/Mcal	2.5 ÷ 0.55 = ¥4.54/Mcal
B) Electricity unit (¥/Mcal)	¥20/KWH ÷ 860 ÷ 0.9 x 1,000 = ¥25.8/Mcal	18 ÷ 860 ÷ 0.9 x 1,000= ¥23.2/Mcal
C) Economic marginal C.O.P. B/A	2.83	5.12

The above calculations show that a heat pump must have a C.O.P. of at least 5.12 to be considered for introduction.

Reports have been received that the economic efficiency of heat pumps installed at several factories in this country has not been sufficiently high and they have been shut down.

The economic marginal C.O.P. is simply an indicator showing if an operation will lead to economic benefit or not. Usually, the pay back period or "P.B." is used as a rule of thumb for deciding whether a heat pump should be installed or not. In this country, 2 ~ 3.5 years is the common P.B. for investment projects. If this period is considered, the C.O.P. of a newly installed heat pump is much higher than the economic marginal C.O.P.

The economic marginal C.O.P. of a heat pump driven by an internal combustion engine (gasoline, diesel or gas turbine) is expressed as follows:

$$\text{C.O.P.} = 1/\eta E - \eta R$$

Where, ηE = Power conversion efficiency
ηR = Waste heat recovery

For instance, assuming power conversion efficiency $\eta E = 0.35$ and waste heat recovery $\eta R = 0.45$, the C.O.P. is 2.4. This value is not influenced by the oil energy unit representing a steam unit. The annual maintenance cost component increases as energy unit cost decrease, however, so the economic marginal C.O.P. is increased somewhat.

In calculating P.B., the absolute value of the energy saving merit decreases and the initial expense for the engine and other components becomes the major obstacle to installation.

A sudden increase in energy costs to the 1985 level is not expected but if we assume an annual increase of 5%, it will take at least ten years to regain the high C.O.P. enjoyed by heat pumps in 1985. We must therefore take steps to disseminate the heat pump under the currently prevailing energy situation.

Apart from national energy policy measures, it is obvious that the following two points are indispensable for general acceptance of the heat pump.

1. Raise the C.O.P. (to at least 5.5).
2. Reduce initial costs.

2.1.1. Raising the C.O.P.

There is a possibility that efficiency can be raised around 10% compared to the normal type heat pump.

* The range of temperature increase of the heat pump should be kept as small as possible. Usually, the heating/cooling systems at factories are designed based on a boiler system for the heating side and a cooling tower for the cooling side. The ΔT required for this process is large. When using a heat pump system, the ΔT should be kept as small as possible. This is because;

  a) The heating surface of the heat exchanger can be enlarged but this has the drawback of increasing initial cost.

  b) The ΔT required by the process itself should be reduced. For example, in the case of a distillation tower, modification to a low pressure loss tower such as a packed tower or double column system can be considered.

  c) An open cycle heat pump such as VRC is effective. In this case the ΔT equivalent to one heat exchanger unit can be reduced and the initial equipment cost likewise reduced by the one unit. An expensive heating medium thus becomes unnecessary.

Chemical engineering and process engineering expertise is essential to facilitate a), b) and c) above. Equally important is a clear understanding of the heat pump by the end user.

2.1.2. Reducing Initial Costs

* Efforts must be made to reduce the production costs of components such as the compressor and heat exchanger.

* The number of components, e.g., heat exchanger, should be reduced by employing systems such as the open cycle type.

2.2. Future Prospects

1986 has had a serious impact on the prospects of the heat pump. With the sharp drop in energy costs, the C.O.P. of motor-driven heat pumps jumped to 5 or more. Numerous system plans have been scrapped or postponed.

Despite the current cheap energy environment, the heat pump systems of Mycom and end user plants for next year are still economically attractive. The particulars of these systems will be reported on next time but the characteristic feature of these systems is that they are all open cycle.

The open cycle heat pump is superior to the closed cycle type in terms of both C.O.P. and initial cost.

The steam input of the open cycle heat pump is a process gas so chemical and mechanical studies become particularly important in design. The difficulty of heat pump installation under such conditions increases but teamwork between the end user's process engineer and the heat pump supplier can surmount this.

In the coming year, various types of open cycle heat pump should come into general use as end users become more knowledge about the heat pump. Meanwhile, regarding the closed cycle heat pump, increased demand for use in low temperature side foodstuffs and other industries can be expected because of the importance of low temperature in product quality.

There are a number of favorable factors at work which in the years ahead will contribute to the appeal of heat pump systems.

1. There is a consensus of opinion that oil costs will increase over the long term, thus making the heat pump alternative more attractive.

2. More and more, restrictions on the use of cooling water and increased water costs will come into play.

3. Factory managers are becoming more involved in environmental issues, particularly when facilities are located adjacent to residential areas.

If we conceptualize the factory of the future as an environmentally clean facility with a closed thermal cycle and an energy efficient system, the heat pump is extremely attractive. As a manufacturer of heat pumps we always keep in mind this image of the ideal factory in our efforts to promote general use of the heat pump.

Chapter 19

Overview of Heat Transformers in Japan

Katsuyuki Mashimo

ABSTRACT: The absorption process is considered to be one promising method to recover a low potential thermal energy.
The absorption heat transformer can upgrade the low-level waste heat and get higher temperature hot water or steam, if waste heat, cooling water and a little electric power are available.
Since 1981, around 10 heat transformers with working pair H_2O/LiBr have been installed in Japan. In this report, typical installation examples of heat transformer at industrial process are described as operating experience.
The available heat output of heat transformer is more than 45% of the waste heat input.
Also, from the economical point of view, payback period of heat transformer is within 2 years on an average.

INTRODUCTION

Basic ideas on the use of the absorption process for heat pumps were proposed as reversible absorptionsmachinen by Altenkirch during 1913-1920. (1)(2)(3)(4)(5)(6)(7)
His ideas were picked up by Nesselmann, who for some years was a co-worker of Altenkirch. In 1933 he published a comprehensive paper on the theory of heat transformation with a very systematic review of the different applications of the absorption process, according to the very clear and simple sketch. (8)(9)

HEAT PUMPS: Prospects in Heat Pump Technology and Marketing (Proceedings of the 1987 IEA Heat Pump Conference). Edited by Kay H. Zimmerman

Fig.1 shows principle of heat transformation according to Nesselmann. T0, T1 and T2 are temperature levels.

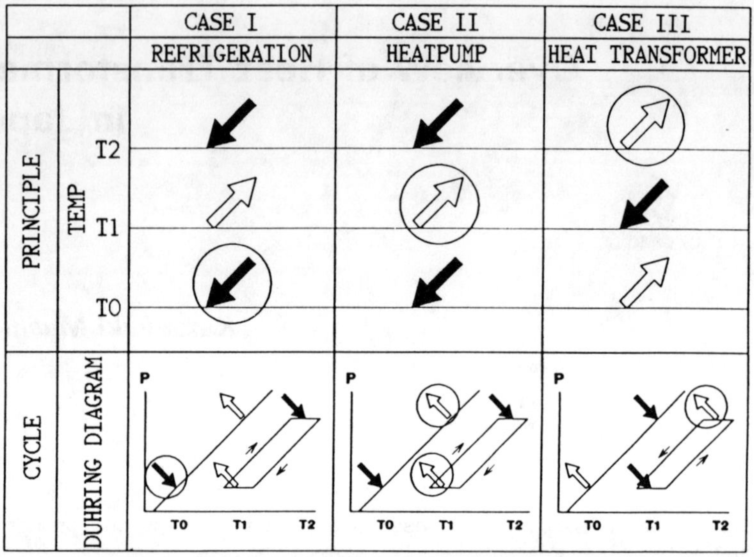

Fig.1 Principle of heat transformation

In case I, the purpose of the absorption machine is cooling at T0. -absorption chiller or absorption refrigerating machine.

In case II, the purpose of the absorption machine is heat supply at a medium temperature T1. -absorption heat pump or type I absorption heat pump.

In case III, the purpose of the absorption machine is heat supply at a high temperature T2. The generator has to be heated at a medium temperature T1 of part of this heat is rejected at the temperature T0 of the environment, part of it is transformed into heat at the high temperature T2. - absorption heat transformer or type II absorption heat pump.

The research and development of the absorption heat transformer was made in one of the "Moonlight Project" during 1976-1981, under the sponsorship of Industrial Technology Institute in Ministry of International Trade and Industry, Japan.

For the absorption heat transformer, LiBr/H_2O is used as the working fluid. Because, it is non-flammable, the working pressure is under atmospheric pressure and it has been widely used as working fluid for conventional absorption chiller.

THE HEAT TRANSFORMER CONCEPT

Principle

The heat transformer employs an absorber, evaporator, generator, condenser, solution heat-exchanger absorbent pump, refrigerant pumps and the piping to connect these. Fig.2 shows the flow diagram of the heat transformer. The principle of absorption heat transformer is to make use of the physical phenomenon where, at the same pressure, the equilibrium temperature of absorbent is higher than the saturated temperature of refrigerant, and the heat is transferred from the evaporator to the absorber.

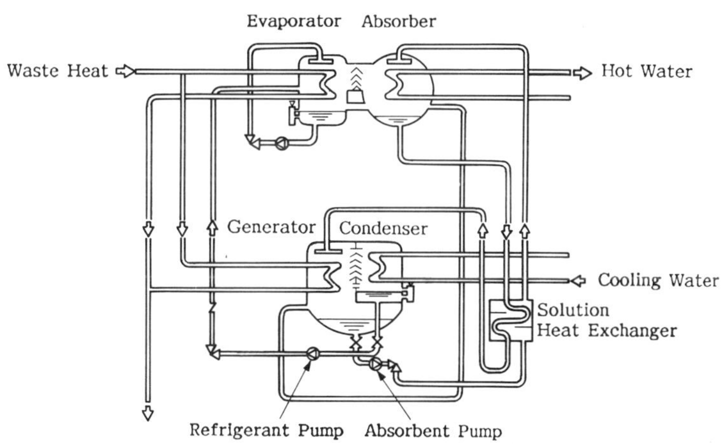

Fig.2 The flow diagram of absorption heat transformer

The waste heat is first supplied to the evaporator, When the refrigerant in the evaporator is heated with the waste heat, it is evaporated, and then in the absorber, it is absorbed in the absorbent. Heat of absorption is generated, and this is used for heating of hot water or generating of steam. The absorbent that has been diluted by absorbing the refrigerant vapor is sent to the generator through the solution heat-exchanger. The diluted absorbent is heated by the same waste heat and the refrigerant is vaporized in the generator. This process concentrates the absorbent. Now the concentrated absorbent is sent back to the absorber through the solution heat-exchanger. While the vaporized refrigerant is cooled by the cooling water, and condensed. This is sent back to the evaporator. This whole process comprises the cycle of the heat transformer.

Characteristics of Temperature rise

The higher the hot water temperature of the heat transformer, the better utilization is possible in the absorption heat transformer.

The Characteristics of temperature rise in the heat transformer is influenced by the condition of hot water, waste heat and cooling water as well as the heat transfer area, heat transfer coefficient and flow rate of absorbent in the heat transformer. Therefore when studying an application of the heat transformer, it is necessary to obtain a temperature and flow rate of hot water meeting with these conditions of temperature of hot water, waste heat and cooling water. The Fig.3 shows the temperature relationship of cooling water, hot water and waste heat. This will enable people who are interested in installation of heat transformer to estimate the characteristics easily.(10)

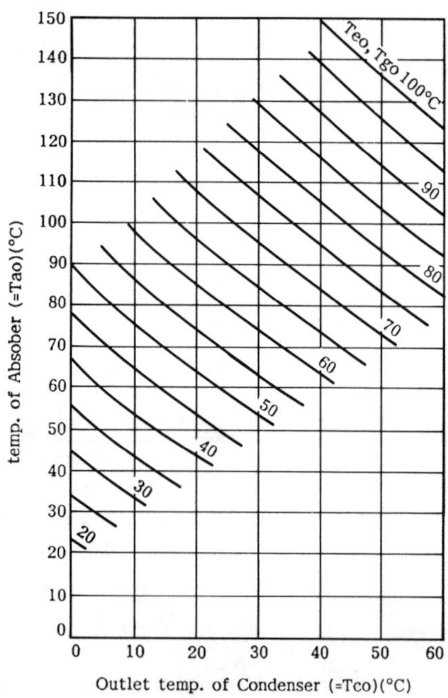

Fig.3 The characteristics of temperature rise(Relationship between Tao, Teo, and Tco)

OPERATING EXPERIENCES

Since 1981, around 10 heat transformers have been installed in Japan. Chemical factories are perhaps the best suited for introducing the heat transformer for recovery of the waste heat. Where a different level of heat, very high temperature to low temperature are used.

Fig.4 shows Installation map of absorption heat transformers.

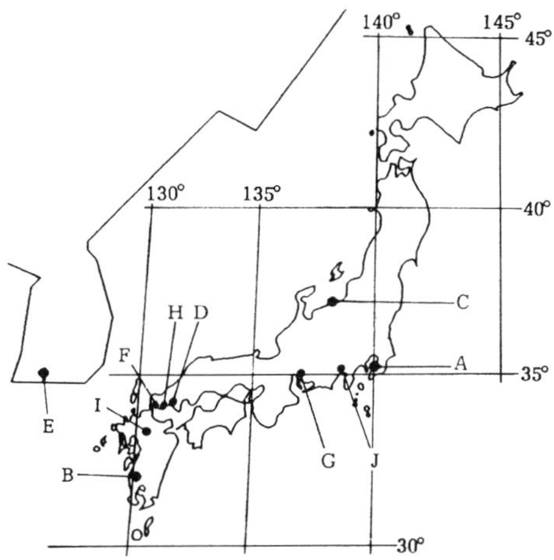

Fig.4. Installation map

Case A

The system is designed to recover the heat from an organic vapor that is discharged out of the distillation tower at a synthetic rubber factory.

The system, which has been in operation since November 1981, is composed of distillation tower that produces organic vapor, a heat transformer to produce hot water, flash tank to produce steam out of the hot water. Other equipments to supplement the system are a condensate tank that temporary keeps the discharged organic liquid condensed in the heat transformer and also a pump that sends organic liquid.

98°C organic vapor discharged at the distillation tower is taken into the evaporator and generator in the absorption heat transformer, and is condensed at 88°C inside heat transfer tubes. Then, the condensed liquid is sent to the condensate tank.
While 26°C cooling water is taken into the condenser and boosted 32°C. Then hot water(133°C) is produced by the heat transformer and steam(127°C) in the flash tank with a capacity of 3.8 ton per

hour. This steam is used as the heat source for reboiler of the distillation tower. Table 1 is the typical operating condition of the heat transformer.

Table 1 typical operating condition

General			
Location		Chiba	
User		Butadiene Rubber Plant	
Manufacturer		Tokyo Sanyo Electric Co., Ltd.	
Purpose		Steam generation for process heating by heat recovery of waste steam	
Site Condition		Pressure and gas proof area	
Start of Operation		December, 1981	
Load Factor		8000 hours/year	
Operating Conditions			
Heat Source(Waste Heat)			
Used Medium Condition		Waste Steam	
Input(Qe+Qg)	KW	5000	
Evaporator			
Inlet Temp.	°C	98	
Outlet Temp.	°C	88	(Drain)
Generator			
Inlet Temp.	°C	98	
Outlet Temp.	°C	88	(Drain)
Available Heat to Process			
Used medium Condition		Low Pressure Steam	
Output (Qa)	KW	2350	
Absorber			
Inlet Temp.	°C	127	
Outlet Temp.	°C	133	
Cooling Water			
Discharged heat (Qc)	KW	2650	
Condenser			
Inlet Temp.	°C	26	
Outlet Temp.	°C	32	
C.O.P.(Qa/Qe+Qg)		0.47	

Fig.5 is the picture of the system and Fig.6 is the system diagram.

Fig.5 picture of the system

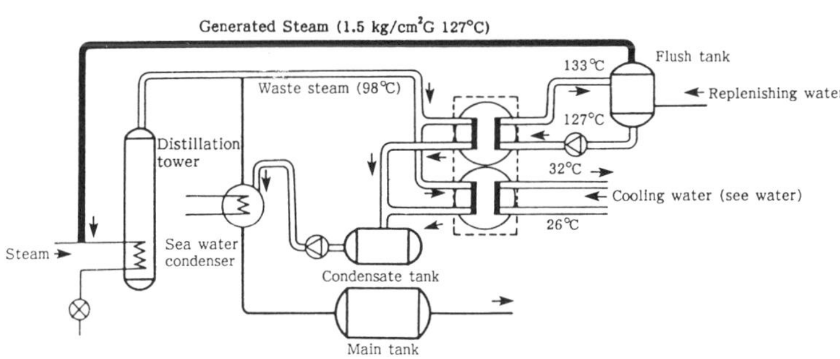

Fig.6 system diagram

Following is the energy that can be saved in a year. The unit price of the steam at that time(1981) was 5,000 yen/ton.

Energy Saving

Recovered Energy (Qam) (Expression in Monetary Unit)	¥/year	152,000,000
Running Cost (Rm)	¥/Year	3,840,000
Net Saved Energy (Qam-Rm)	¥/year	148,160,000
Pay-back Period (Approx,)	year	1

Case B

The system is designed to recover the heat from an alcohol vapor that has been wasted at an ethylalcohol distillation plant.

The system has been in operation since January 1984, composed of distillation tower that produces alcohol vapor, the heat transformer to produce hot water, flash tank to produce steam from the hot water

80°C alcohol vapor discharged at the distillation tower is taken into the evaporator and generator in the absorption heat transformer, and is condensed at 80°C inside heat transfer tubes. Then, the condensed liquid is sent to the condensate tank.
While 20°C cooling water is taken into the condenser and boosted to 30°C. Then hot water(124°C) is produced out of the heat transformer and steam(119°C) in the flash tank with a capacity of 1.4 ton per hour. This steam is used as the heat source for reboiler of the distillation tower.
Table 2 is the typical operating condition of the heat transformer.

Table 2 typical operating condition

General		
Location		Kagoshima
User		Ethylalcohol Distillation plant
Manufacturer		Tokyo Sanyo Electric Co., Ltd.
Purpose		Steam generation for process heating by heat recovery of alcohol vapor
Site Condition		Pressure and gas proof area
Start of Operation		January, 1984
Load Factor		7200 hours/year
Operating Conditions		
Heat Source(Waste Heat)		
Used Medium Condition		Alcohol Vapor
Input(Qe+Qg)	KW	1950
Evaporator		
Inlet Temp.	°C	80
Outlet Temp.	°C	80 (Drain)
Generator		
Inlet Temp.	°C	80
Outlet Temp.	°C	80 (Drain)
Available Heat to Process		
Used medium Condition		Low Pressure Steam
Output (Qa)	KW	930
Absorber		
Inlet Temp.	°C	119
Outlet Temp.	°C	124
Cooling Water		
Discharged heat (Qc)	KW	970
Condenser		
Inlet Temp.	°C	20
Outlet Temp.	°C	30
C.O.P.(Qa/Qe+Qg)		0.48

Fig.7 is the picture of the system and Fig.8 is the system diagram.

Fig.7 picture of the system

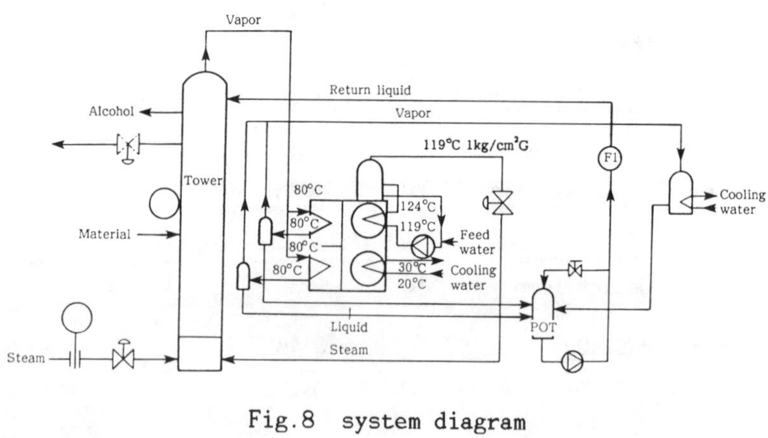

Fig.8 system diagram

Following is the energy that can be saved in a year.

Energy Saving

Recovered Energy (Qam) (Expression in Monetary Unit)	¥/year	50,040,000
Running Cost (Rm)	¥/Year	1,728,000
Net Saved Energy (Qam-Rm)	¥/year	48,312,000

For reference, the pay-back period of this system is 1.5 years.

Case C - K

The heat transformers operate mainly in the distillation plants. The other operating experiences are shown in table 3, 4 and 5. Operating case E and K are shown for reference, since case E is in Republic of Korea and case K is in Kingdom of the Netherlands.

Fig.9 is the picture of case G. The heat transformers in case F-K are vertical type which are based on the counter current heat exchanger technology.

Fig.9 picture of case G

Table 3 Operating experiences

Operating Case	C	D	E
Location User	Niigata Distillation Plant	Tokuyama Distillation Plant	Korea Distillation Plant
Manufacturer	Tokyo Sanyo Electric Co., Ltd.		
Purpose	Steam Generation	Steam Generation	Steam Generation
Start of Operation	Feb.'84	Jun.'85	May '83
Heat Source (Waste Heat)	Distillation Tower Top Vapor	Distillation Tower Top Vapor	Distillation Tower Top Vapor
Input KW	3450	2220	4890
Inlet Temp. °C	80	125 122	98
Outlet Temp. °C (Condensation)	80	91 79	88
Available Heat to Process	Low Pressure Steam	Low Pressure Steam	Low Pressure Steam
Output KW	1670	1090	2350
Inlet Temp. °C	111(Water)	134 (Water)	127 (Water)
Outlet Temp. °C	116(Water)	139 (Water)	132 (Water)
Cooling Water	Ground Water	Sea Water	Cooling Tower Water
Inlet Temp. °C	15	27	31
Outlet Temp. °C	20	30	-
C.O.P.	0.48	0.49	0.48
Pay-back Period	1 - 2 Years		

Table 4 Operating experiences

Operating Case	F	G	H
Location User	Shimonoseki Distillation Plant	Tokai Distillation Plant	Sin-Nanyo Distillation Plant
Manufacturer	Hitachi Zosen Corp.		
Purpose	Steam Generation	Steam Generation	Steam Generation
Start of Operation	Jul.'82	May '83	Sep.'84
Heat Source (Waste Heat)	Distillation Tower Top Vapor	Distillation Tower Top Vapor	Distillation Tower Top Vapor
Input KW	3450	3950	6980
Inlet Temp. °C	——	——	——
Outlet Temp. °C (Condensation)	83	80.5	100
Available Heat to Process	Low Pressure Steam	Low Pressure Steam	Low Pressure Steam
Output KW	1660	1880	3330
Inlet Temp. °C	100 (Water)	100 (Water)	90 (Water)
Outlet Temp. °C	111 (Steam)	112 (steam)	143 (Steam)
Cooling Water	Cooling Tower Water	Cooling Tower Water	Cooling Tower Water
Inlet Temp. °C	28	32	31
Outlet Temp. °C	32	36	37
C.O.P.	0.48	0.48	0.48
Pay-back Period	1 - 2 Years		

Table 5 Operating experiences

Operating Case	I	J	K
Location	Amaki	Fuji	The Netherlands
User	Hot Waste Water Recovery Plant	Distillation Plant	Distillation Plant
Manufacturer	Hitachi Zosen Corp.		
Purpose	Hot Water Supply	Steam Generation	Steam Generation
Start of Operation	Dec. '84	May '85	Sep. '84
Heat Source (Waste Heat)	Hot waste Water	Distillation Tower Top Vapor	Distillation Tower Top Vapor
Input KW	720	580	13780
Inlet Temp. °C	95	——	——
Outlet Temp. °C (Condensation)	85	95	100
Available Heat to Process	Hot Water	Low Pressure Steam	Low Pressure Steam
Output KW	350	270	6420
Inlet Temp. °C	95 (Water)	25 (Water)	127 (Water)
Outlet Temp. °C	120 (Water)	131 (Steam)	150 (Steam)
Cooling Water	Cooling Tower Water	Cooling Tower Water	Cooling Tower Water
Inlet Temp. °C	32	32	24
Outlet Temp. °C	37	37	40
C.O.P.	0.49	0.47	0.47
Pay-back Period	1 - 2 Years		

CONCLUSION

The heat transformer as described in the above examples could contribute greatly to improving energy efficiency at plants. However much efforts are necessary to develop more efficient, higher temperature rise absorption heat transformer.

Before ending this paper, great thanks should be acknowledged to Hitachi Zosen Corp.

REFERENCES

1. E.Altenkirch, "Reversible Absorptionsmachinen", Zeitschrift fur die gesamte Kalte-Industrie, 20(1), 1-9 January 1913.
2. Ibid , 20(6), 114-119 June 1983.
3. Ibid , 20(8), 150-161 August 1913.
4. Ibid , 21(2), 21-24 February 1914
5. Ibid , 25(7), 49-53 July 1918.
6. Ibid , 27(4), 77-81 July 1920.
7. E.Altenkirch and W.Niebergall, "Verfahren und Vorrichtung zum Betrieb einen Warmepumpe", FRG Patent 953378, November 1956.
8. W.Raldow, New Working Pairs for Absorption Processes, Swedish. Council for Building Research, 2, 1982.
9. K.Mashimo, "New Organic Absorption Heat Pump" New Technology symposium Part 3, JMA'86 (1), 23-34, March 1986.
10. M.Sano, H.Kusakabe and Y.Ikumi, "Absorption Heat Pump for Upgranding of Industrial Waste Heat" I.I.R., Comission E2, Paris(France), September 1983.

Chapter 20

Industrial Heat Recovery with Heat Transformers — Practical Applications and Development of Advanced Systems

Horst Bokelmann

ABSTRACT: In many industrial thermal processes high-grade energy is emitted as waste heat to the atmosphere without being utilized so far. A heat transformer is a device which enables the recovery of these waste heat flows elegantly and profitably. As it raises the temperature level of the recovered heat without thereby consuming considerable quantities of mechanical, electrical or thermal energy, it allows rejection of the recovered heat to the same process. Heat transformers nowadays "on sale" are capable of achieving a maximum output temperature of 150°C, a temperature lift of 50 K and a COP of 0.48. Examples of practical applications and operational experience are given. However, many inquiries from potential customers have shown that temperature lifts to high operating temperatures and better overall COPs are often desirable or demanded. Consequently. a development program ("Advanced Heat Transformer") was started. Advanced heat transformers utilize new working fluids and/or multistage sorption plants including combinations of sorption heat pumps and heat transformers. This paper presents some of these advanced systems and analyzes their possible application.

INTRODUCTION

Although we are living an energy price down it is easy predicting that energy saving will be of booming interest shortly as, on the one hand, energy prices will rise again on medium-term basis and, on the other hand, it seems convenient to lower both the primary energy consumption and the el. energy demand for reasons of environmental protection.

HEAT PUMPS: Prospects in Heat Pump Technology and Marketing (Proceedings of the 1987 IEA Heat Pump Conference). Edited by Kay H. Zimmerman

Industrial-scale thermal transformation processes are best suited for high energy savings. Strong efforts were made to cut the energy consumption in industrial processes during the past decades. However, as before, major energy flows are being ceased to the atmosphere without being used/optimally exploited.

The waste heat utilization of a works is said optimal if the primary energy consumption is minimized. Yet it proves difficult to fulfill this as such trivial requirement. Generally, the primary energy consumption cannot be reached by heat insulation measures and simple heat exchangers for preheating and utility water heating alone. The requirement can be met almost 100 % if there is a possibility to directly refeed part of the waste heat flow to the heat emitting process. This necessitates lifting of the temperature level of the waste heat flow.

There are different ways to rise the temperature level, namely steam jet compressors, vapour compressors, absorption heat pumps and heat transformers. Whereas the first three of the above-named systems all consume high-grade energy themselves (which needs be supplied in form of primary energy possibly), the heat transformer can do without third energy. On the other hand, the heat transformer utilizes half of the supplied waste heat flow only. The temperature lifts which can be realized with the single-stage heat transformer are superior to those of the other single-stage processes, generally.

The present paper deals with the heat transformer mainly. For better understanding a comparison is made between heat transformer and absorption heat pump and applications for a combination of heat transformer/absorption heat pump are referred to. The next paragraph is meant to give an introduction of the theoretical grounds of heat transformation.

THEORY OF HEAT TRANSFORMATION

Figure 1 shows the well-known absorption-type heat pump process in form of a steam diagram with the pertinent energy flows. Figure 2 illustrates the heat transformer process. In both processes, the solution comprising refrigerant and absorbent circulates between generator and absorber. In the generator part the refrigerant is evaporated from the solution by heat admission. The refrigerant steam enters the condenser to be liquefied while emitting heat. In the evaporator it is then reconverted to vaporized refrigerant by heat admission. The vaporized refrigerant flows into the absorber and is united with the solution coming from the generator (and of low refrigerant content). The latter is now able to absorb the vaporized refrigerant and liquefy it simultaneously. Again, heat is emitted during this liquefying process. In both processes heat is emitted in condenser and absorber while heat must be supplied to generator and evaporator. However, both pressure and temperature levels of the described part processes vary. While evaporation and absorption take place at the lower pressure level in heat pump and absorption refrigeration plant, they take place at the higher pressure level

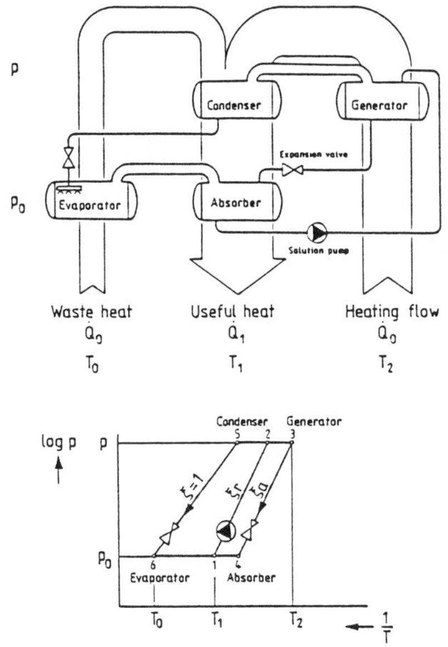

Fig. 1 Absorption heat pump

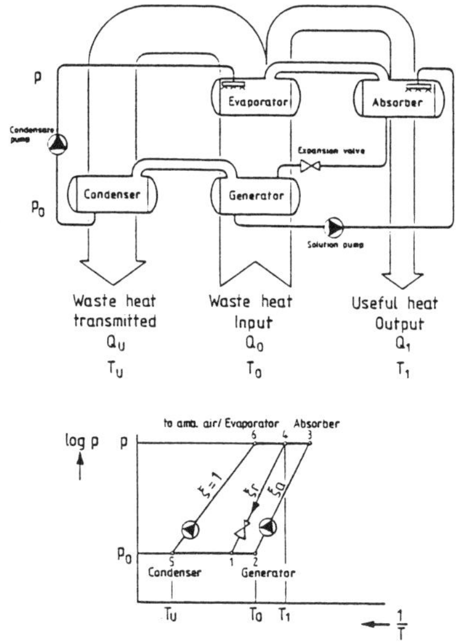

Fig.2 Process Diagram of the Heat Transformer

in the absorption-type heat transformer. I.e., heat is supplied at the highest as well as at the lowest temperature level in the heat pump and useful heat is emitted at mean temperature level, while the heat transformer is fed with heat at mean temperature level. The heat transformer is capable of emitting heat at the highest temperature level provided there is a heat sink with a lower temperature level than that of the supplied heat.

Both above described plant alternatives are suitable for industrial heat recovery, generally. In case of a heat pump waste heat would be supplied at the lowest temperature level (to the evaporator) necessitating the additional supply of high-grade heat energy to the generator. 30 to 70 % more heat (than supplied to the generator) can thus be taken from absorber and condenser. The waste heat can be supplied to the generator, too, of course. At mean temperature level heat would be emitted to the atmosphere in this case while the heat to be fed to the evaporator would serve for refrigeration. The heat transformer, on the contrary, is fed with mean temperature waste heat. Almost 50 % of the waste heat can be recovered consuming considerable quantities of high-grade energy.

HISTORICAL OVERVIEW

The basic idea of heat transformation was developped by E. Altenkirch as early as 1914 /1/. As with other ideas, he was trendsetting with this invention, too. In the twenties, the theoretical idea was translated to practical use in form of the so-called Koenemann-transformer, steam converter or pressure transformer /2/. An alternative solution, the "use of the Koenemann-transformer for splitting medium-pressure steam to high-pressure and low-pressure steam" /3/ already was what we call an open heat transformer today. (Application: heat recovery from evaporating plants and multistage evaporators) "The realization of these proposals failed for lack of suitable working fluids mainly" /4/.

Nesselmann /5/ revived the idea in 1933, and Niedergall /5/ in 1955. Whereas Altenkirch proposed to use the heat transformer for heating purposes by exploiting the temperature difference between underground water and outside air /6/ (possibly combined with a compression heat pump /7/), and whereas Nesselmann favoured the idea of power-heat coupling /8/, Niedergall /4/ proposed to use the heat transformer for steam generation in the way we do it today.

In the late sixties, investigations were made in Siberia using absorption refrigeration plants for heating and cooling. Under the test program an absorption refrigeration plant was operated as heat transformer /9/. In Western Europe, the heat transformer fell almost completely into oblivion until, in the late seventies, the topic of energy saving was up again and several European, American and Japanese scientific researchers took up the development work again (independently from one another).

During the last few years industrial-scale sorption systems for heat recovery became increasingly interesting in Europe. In Sweden /10/, The Netherlands and Germany /11/ plants are being built or under construction. All these plants were built by Japanese companies except for the plant in Dörnten (Germany) which was built by GEA in cooperation with Kawasaki Heavy Industries. This heat transformer for an animal carcass plant has been operated for abt. 2 years now to transform nearly 40 % of the waste heat of 100 ^{o}C to useful heat of 145 ^{o}C. The heat output is 1 MW. Like the other European plants the GEA heat transformer is a single-stage unit and utilizes the well-known working fluid water-LiBr. This pilot plant provided evidence that a heat transformer is capable of recovering waste heat on industrial scale /12/.

Lately, several research teams are busy pushing heat transformer development. Special efforts are being made in Japan, Sweden, France and the FRG /13/ to suit the heat transformer to the requirements of industrial application. The present state-of-the-art and promising new applications shall be presented in the following paragraphs.

STATE-OF-THE-ART OF HEAT TRANSFORMER

Figure 3 shows a typical heat transformer as built by GEA three times already. The plant shown is a single-stage heat transformer (MW-capacity range). Figure 4 shows the structural details of the heat transformer plant. At present, another plant of the above type is being installed in a chemical works. The

Fig. 3: The GEA Heat Transformer

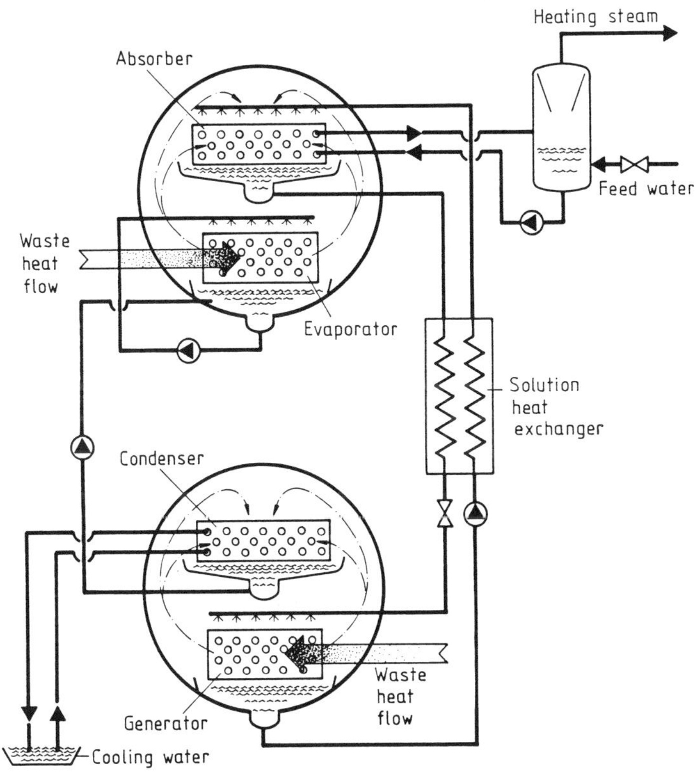

Fig. 4: Mech. design of heat transformer

installation of the heat transformer (which will have a useful capacity of 2.2 MW) is sponsored by the Commission of the European Communities.

The customary heat transformer is run with the working couple water-LiBr what involves the following advantages:
- high enthalpy of evaporation of the refrigerant water
- a rectification of the generated vaporized refrigerant is not necessary
- low plant pressure level
- high heat and mass transfer coefficients.

The following features of the working couple water-LiBr are considered a disadvantage:
- the crystallization max. of the system limits the possible waste heat temperature lift
- the solution is corrosive at high temperatures and necessitates the use of an inhibitor and limits the max. useful temperature

The left column of Table 1 shows the actual heat transformer applications restricted as explained above. However, as we know from many inquiries, more universal applications are often required. The Advanced Heat Transformer is a way to solve this problem as show the right columns of Table 1. The development of these Advanced Systems is described in the following paragraphs.

THE HEAT TRANSFORMER

ON SALE	RESEARCH AND DEVELOPMENT STAGE		
Working Fluid Water - LiBr	Working Fluid Water - LiBr	Working Fluid Organic Substances	Working Fluid Organic Substances
Single-Stage Cycle	Multi-Stage Cycle	Single-Stage Cycle	Combination of a Heat Transformer and an Absorption Heat Pump
Heat Input Temperature 70....120°C	Heat Input Temperature 50....120°C	Heat Input Temperature 50....150°C	Heat Input Temperature e.g. 100°C
Temperature Raise 30....50K	Temperature Raise > 50K	Temperature Raise > 50K	Temperature Raise e.g. 50K
Max. Heat Output Temperature 150°C	Max. Heat Output Temperature 150°C	Max. Heat Output Temperature > 200K	Heat Output Temperature 1. 90°C 2. 150°C
EEC - Demonstration Project "Heat Transformer"	Eureka Project "PACA"	EEC - Research and Development Project "Advanced Heat Transformer"	

Table 1: State-of-the-art of heat transformer

ADVANCED HEAT TRANSFORMERS

The above disadvantages of the single-stage water-Li-Br heat transformer shall be overcome by the development of up-to-date absorption systems. Developments concern both new working fluids and multistage designs.

New Working Fluids

A research team of Essen University undertook the task of finding new working fluids for absorption plant under a comprehensive research program sponsored by the EEC /14, 15/. Although the research program concerned new working pairs for absorption heat pumps mainly, some of the working fluids are suitable for use in heat transformers, too.

Under the project some 150 different working fluids were analyzed at least in some respects. Measurements made on the most promising working fluids concerned steam pressure, density, viscosity, solubility, thermal stability, phase coincidence, enthalpy of mixture and specific thermal capacity (for the temperature range of interest).

Table 2 is a comparison of the new working fluids TFE-DTG,TFE-NMP, TFE-Pyr, HFIP-NMP, PFPA-DTG with the wellknown working pair ammonia-water and water-LiBr.

The following characteristics are subjected to comparison:

- enthalpy of evaporation at $0^{o}C$: r_0
- steam pressure of refrigerant at $20^{o}C$: p_{20}
- steam pressure of refrigerant at $100^{o}C$: p_{100}
- price per kg solution
- toxicity
- boiling point difference
- data available.

Comparison properly spoken is made at a generator temperature of $100^{o}C$ and a concentration difference between the strong and the poor solution of 5 %. Table 2 specifies the three different working conditions. Without the second and third working condition a comparison between the new and the common use working fluids would not be possible as, under the first working condition, the use of ammonia as refrigerant resulted in undue high working pressures and the use of the working pair water-LiBr caused crystallization problems.

The following requirements were to be fulfilled:

- The concentration of the weak solution should be as high as possible to limit the circulating solution flow
- The reachable absorption temperature tA should be as high as possible to enable a high useful temperature
- The fictitious temperature difference in the absorber should be as low as possible to come to a high absorption temperature
- The specific mass flow rate f should be as low as possible to limit the necessary pump energy and the size of heat and mass

		General									
	Working Fluid	r_o kJ/kg	t_{kr} °C	p_{20} bar	p_{100} bar	Price DM/kg sol	Tax	Korr.	Δt_m	Rect.	Data
1.	TFE-DTG	449	250	0,07	2,9	28	yes	no	201	no	+-
2.	TFE-NMP	449	250	0,07	2,8	27	yes	no	129	(yes)	+
3.	TFE-Pyr	449	250	0,07	2,9	25	yes	no	171	(no)	+-
4.	HFIP-NMP	428	195	0,16	5,5	(2500)	less(asspt.)	(no)	145	(yes)	-
5.	PFPA-DTG	(255)	?	0,03	ca.1	295	less(as spt.)	yes Inhibitor	179	(no)	--
6.	NH_3-H_2O	1260	132	8,6	63	2	yes	partly	133	yes	+
7.	H_2O-LiBr	2500	374	0,02	1	25	no	yes Inhibitor	-	no	+

	Comparison at different Working Conditions											
	$t_E = 100$°C $t_k = 20$°C				$t_E = 100$°C $t_k = 50$°C				$t_E = 50$°C $t_k = 20$°C			
	ξ_a	t_a	Δt_a	f	ξ_a	t_a	Δt_a	f	ξ_a	t_a	Δt_a	f
1.	0,05	185	20	19	0,18	155	8	16	0,05	128	15	19
2.	0,20	ca. 210	ca. 12	16	0,52	160	10	10	0,20	1,38	5	16
3.	0,15	ca. 225	ca. 20	17	0,45	165	15	11	0,15	140	10	17
4.	(0,65)	(220)	(10)	(7)	0,82	160	8	4	(0,65)	130	12	7
5.	(0,57)	(185)	(10)	(8,5)	(0,72)	(150)	(10)	(6)	(0,57)	(125)	(10)	(8,5)
6.	-	-	-	-	-	-	-	-	0,28	125	10	14
7.	-	-	-	-	0,38	146	14	12	-	-	-	-

Table 2 : Working Fluids for Heat Transformers

transfer surfaces.

The thermodynamic properties of the working pairs PFPA-DTG and HFIP-NMP make them seem especially suitable for being used in heat transformers. However, the price of these refrigerants is that high at the moment, that their profitable utilization is not realistic. Of course, the sales prices of chemicals are rapidly and strongly fluctuating and therefore the market trend should be followed closely. As concerns the refrigerant HFIP, the market trend makes us hope we will be in a position to employ it at reasonable prices shortly.

In view of their promising characteristics, the working fluids TFE-NMP and TFE-Pyr seem best suited to substitute the wellknown working pairs. However, there will always be some applications for which water-LiBr is optimal. An advantage of the working pair TFE-DTG is that its utilization will guarantee that a rectifying system is not needed.

Multi-Stage Systems

Multi-stage systems have been of booming interest these years /16/. We refrain from giving a comprehensive survey on the individual design alternatives and prefer to present and explain some single plant solutions on the basis of specific requirements:

Realizing higher useful temperature levels: As explained above, it is difficult to come to very high useful temperatures with the working pair water-LiBr. At the same time it was evidenced that the utilization of the new working fluids enables an increase of the useful temperature level. Figure 5 shows the design data of a TFE-DTG heat transformer. A GEA test heat transformer has been operated with this working pair and the results obtained were most promising.

Using other than the above absorbent for the refrigerant TFE still increases the useful temperature level. As shows Figure 5 for the working pair TFE-Pyr, useful temperatures of 200°C and temperature lifts of 100 K seem to be realistic. However, for lack of experience with this working pair we are not in a position to guarantee for max. useful temperatures or temperature differences. Still it rests to be clarified if and under which conditions a rectification of the refrigerant vapours is needed.

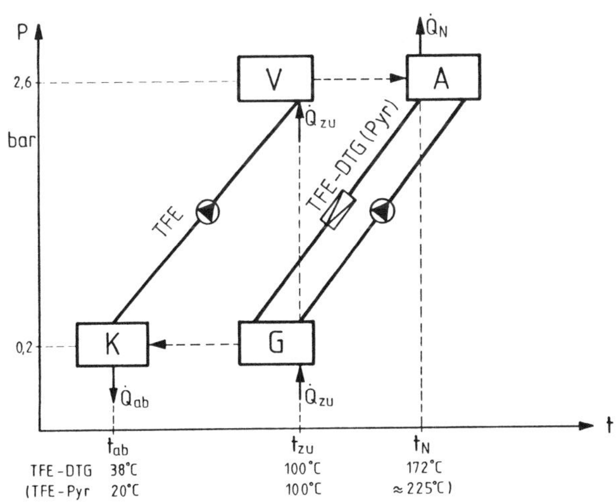

Fig 5: Design Characteristics of a TFE-DTG(Pyr) Heat Transformer

Realizing higher temperature lifts: As an example, reference is made to a temperature lift from 70 to 110°C (food industry application). A such temperature lift cannot be reached with a single-stage water-LiBr plant, but it is possible with a single-stage TFE-Pyr plant.

Should the utilization of the above new working fluids not be suitable for toxicological reasons, the desired temperature difference can be obtained with multi-stage water-LiBr plants. There are many different ways to solve the problem. Figure 6 shows a high-temperature sorption heat pump in combination with a heat transformer to come to the desired temperature difference and to a relatively high overall efficiency (abt. 0.6). Yet, this alternative solution requires the supply of heat at a relatively high temperature level.

Realizing a second useful temperature level: For many industrial processes process heat is required at a relatively high temperature level while useful heat is required at a low temperature level. This requirement can best be met by combining a heat transformer with a high-temperature heat pump (similar to the solution presented with figure 6, however, using new working fluids). The waste heat temperature is lifted from 100 to 150°C while low-temperature useful heat of 90°C is emitted in the condenser. However, high-grade heat must be supplied at a temperature level of abt. 200°C in this case. When exploiting the low-temperature useful heat the theoretical efficiency is as high as 100 %.

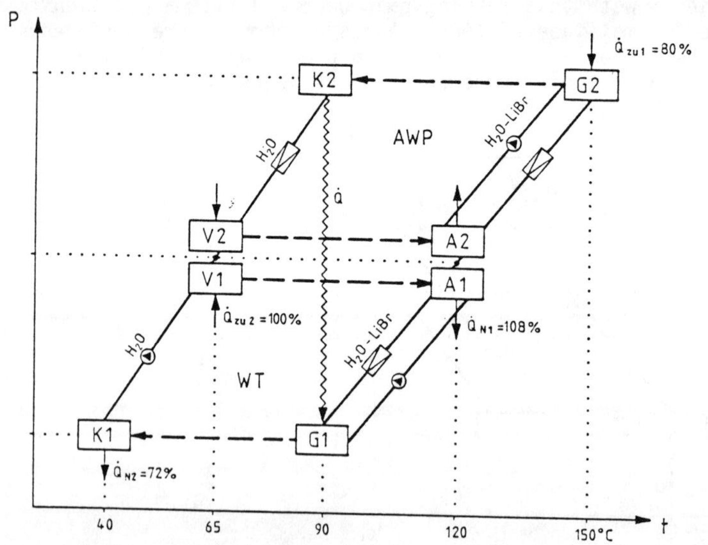

Fig. 6: Combination of an absorption heat pump with a heat transformer

Satisfying heat and cold demand: Figure 7 shows a two-stage water-LiBr sorption plant. Heat is taken up at 3°C and emitted in form of two heat flows at different temperature level. One of the two heat flows can be exploited for utility water heating at 55°C, the other for heating purposes at 75°C. The second heat flow can be switched-off during summer time when there is no need for heating. A single-stage TFE-NMP plant would suffice for this requirement; heat would be taken up at 0°C and emitted at 70°C while heating energy would be required at a temperature level of 170°C.

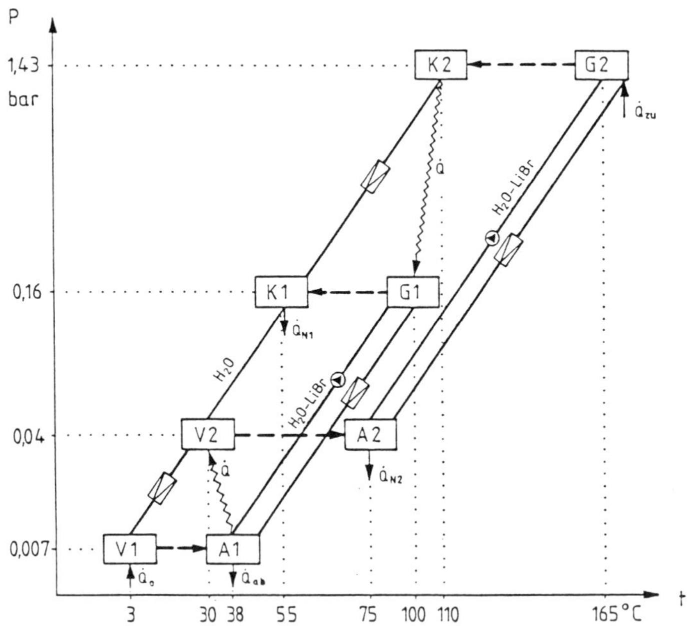

Fig. 7: Double-stage water-LiBr absorption heat pump

EXAMPLES FOR APPLICATION

For clarification, a fictitious application shall be referred to. It is assumed that two thermal processes are taking place simultaneously, one between 150 and 100°C, the other between 110 and 70°C. Each process consumes 35 % of the total energy demand. Another 20 % are needed to warm up the heating water, another 10 % are needed for utility water heating. Figure 8a shows this specific application without heat recovery measures. 100 % are needed for covering the overall energy demand.

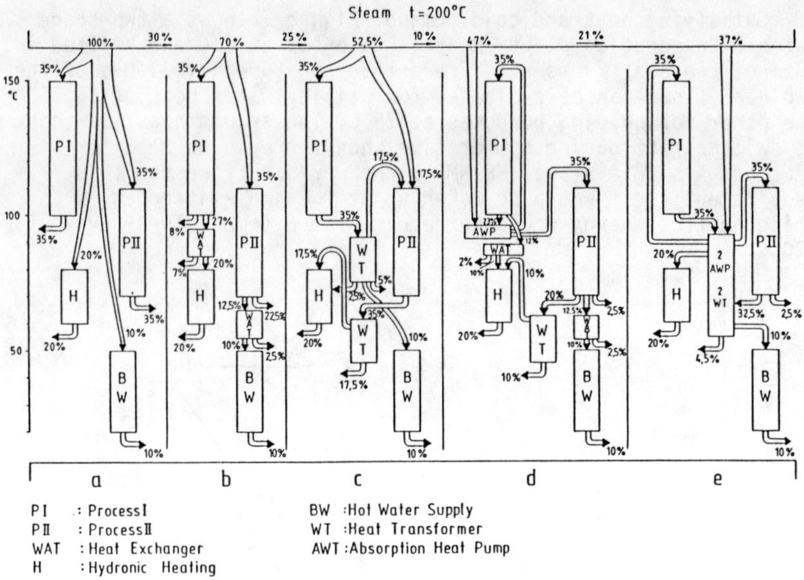

Fig 8: Examples for application

Figure 8b shows the simplest and less expensive means of heat recovery. Two heat exchangers reduce the demand for high-grade energy to 70 %. The waste heat from both processes is exploited for warming-up of heating and utility water. As there is much more waste heat available than low-temperature useful heat required, large quantities of high-grade energy are ceased to the atmosphere without being used.

Figure 8c shows a more effective, though more expensive heat recovery system. Two heat transformers serve to cut the demand for high-grade energy to 52.5 %. The transformers were assumed to have an efficiency of 50 % what is realistic in this case as the temperature differences are low.

Using a high-temperature sorption heat pump, a heat transformer and two heat excchangers enables to cut the primary energy demand to 47 % (Fig. 8d). The absorption heat pump serves to exploit part of the waste heat (first process) for heating (both processes). As the two processes take place at different temperature level it seems appropriate to have condensation and absorption (the two heat emitting part processes of the absorption heat pump) take place at different temperature levels, too, what is not too difficult from technical point of view.

Figure 8e shows a very expensive alternative solution cutting the primary energy demand to 37 %. This efficiency is reached by a combination of two absorption heat pumps with two heat transformers.

CONCLUSION

The GEA heat transformer for heat recovery on industrial-scale has become a commercial product. The single-stage unit enables the reutilization of abt. 40 to 50 % of the waste heat flows between 70 and 120°C for the same or another process at considerably higher temperature level (30 to 50 K). The single-stage heat transformer operates safely and profitably - the pay back periods for MW-plant are 2.5 years when operated full-time.

However, and as mentioned before, the reachable useful temperature and the temperature lift are restricted by the solubility behaviour of the working couple water-LiBr. Conseqently, GEA has started a second development project featuring the development of advanced heat transformers utilizing new working fluids and multi-stage cycles. Investigations will be made under the EEC R&D program and in cooperation with French companies under the EUREKA-Project.

REFERENCES

1. E. Altenkirch, "Reversible Absorptionsmaschinen," Zeitschr. ges. Kälteind. **20** (1913), 1, 114 and 150 and **21** (1914) 7 and 21,

2. L. S. Marks, "A Steam-Pressure Transformer," Mechanical Engineering **49** (1927) 6, 600-602

3. E. Praetorius, "Dampfdruckumformung durch Koenemann-Transformatoren," Arch. Wärmewirtschaft **11** (1930) 6, 191-195

4. W. Niebergall, "Absorptionsanlagen als Wärmepumpen - Wärmetransformationsanlagen," Gesundheits.-Ing. **76** (1955) 7/8, 129-137

5. K. Nesselmann, "Zur Theorie der Wärmetransformation," Wiss. Veröffentl. Siemens-Konzern **12** (1933) 2, 89-109

6. E. Altenkirch, "Verfahren zur Erzeugung von Heizwärme, nach welchem die bei der Absorption eines Gases durch eine Flüssigkeit entstehende Wärme für Heizzwecke verwendet wird," German Patent 508121, 10.01.24

7. E. Altenkirch and W. Niebergall, "Verfahren und Vorrichtung zum Betrieb einer Wärmepumpe," German Patent 953378, 29.08.50

8. K. Nesselmann, "Die Verwendungsmöglichkeiten von Absorptionsmaschinen zur Wärmekrafterzeugung," Zeitschr. ges. Kälteind. **42** (1935) 1, 8-11

9. L. M. Rosenfeld, "Investigation of lithium bromide absorption units for complex cooling and heating purposes," Proc. of the 12th Int. Congr. of Refr., 1967

10. T. Berntson, "Absorption heat pumps and heat transformers - applications and research activities in Sweden," Heat Supply in the Temperature Range up to 200 deg.C, Swedish-German Seminar, München 10./11. March 1986

11. L. Suhr, "Der Wärmetransformator - ein neuer Weg zur Energieeinsparung," Chemie Technik **13** (1984) 12

12. P. Paikert and L. Suhr, "Betriebserfahrungen mit dem ersten großtechnischen Wärmetransformator in Deutschland," KI, Klima, Kälte, Heizung **13** (1985) 4, 177-180

13. International Workshop Heat Transformation and Storage, EEC-Joint Research Center Ispra, 9.-10. Oct. 1985

14. H. Bokelmann and H. J. Ehmke, "Stoffsysteme für Absorptionswärmepumpen - derzeitiger Stand und neue Entwicklungstendenzen," gwf-gas/erdgas **124** (1983) 12, 608-611

15. H. Bokelmann and H. J. Ehmke, Working Fluids for a Sorption Heat Pump, final report of the EEC-sponsered research programm EE-031-D(B)

16. G. Alefeld and F. Ziegler, "Advanced Cycles for the Working Pair Water/LiBr," paper presented at the ASHRAE meeting, Honolulu, June 1980

Chapter 21

Future Prospects for Industrial Heat Pumps in Europe

Thore Berntsson

ABSTRACT: The future prospects for industrial heat pumps in Europe are discussed in terms of possible applications, types of heat pumps to use, present situation, economic aspects, and other aspects, such as environmental ones.

It is shown that there is a large technical potential especially in the higher range of condensing temperatures, 120-160 °C. The technico-economic comparisons have shown that there is also an economic one, which, however, is much dependent on the future electric and fuel prices. These will also influence the types of heat pumps to use. Environmental aspects will also play an important role.

The need for future R,D&D work includes development of better high-temperature heat pumps and system aspects, such as process integration studies to identify interesting applications.

INTRODUCTION

The technical conditions for heat pumps for heating of houses and communities differ very much between the various European countries, mainly due to differences in heat distribution systems and the infrastructure. In industries, however, the energy systems are often constructed principially in the same way for a given branch of industry, regardless of the country. Therefore the technical conditions for industrial heat pumps do not vary so much and a technical development in one country can more easily be transferred to other ones. Due to this fact, Europe will be considered as one market, more or less uniform, in this paper and no country-by-country discussions will be carried out. The most important differences between the countries will be the future

HEAT PUMPS: Prospects in Heat Pump Technology and Marketing (Proceedings of the 1987 IEA Heat Pump Conference). Edited by Kay H. Zimmerman

levels of fuel and electricity prices, which parameters will be discussed in this paper.

Compared with heat pumps for space heating, industrial heat pumps often have the following advantages:

- High COP:s due to small temperature lifts and/or high temperature levels.
- Long annual operation times.
- Relatively small investment cost due to large units and small distances between heat source and heat sink.
- Waste heat production and heat demand occur at the same time.

In spite of this the number of heat pump installations in industry in Europe is almost negligible compared with space heating ones. It is often argued that the main reason for this is the requirement in industry for very low pay-back periods. However, also in situations, where such requirements can be shown to be met, heat pumps have so far normally not been installed. In addition to the pay-back period requirements, other important reasons for the present situation are:

- Lack of experimental and demonstration plants in different types of industries.
- Uncertainty from the users about the reliability.
- Lack of good hardware in some types of applications.
- Lack of the combined knowledge chemical engineering/heat pump technology in industries/consultant companies, etc.

Most of these obstacles can be overcome by a further R,R&D work.

In this paper the future prospects for industrial heat pumps are discussed in terms of possible applications, types of heat pumps, present situation, economic aspects, and other aspects, such as environmental ones.

It can be discussed what an "industrial heat pump" is. In this paper it is defined as a unit, which uses only industrial waste heat as heat source and delivers all the produced heat to the industry, to the processes and/or for space heating in the industry.

APPLICATIONS FOR INDUSTRIAL HEAT PUMPS

The consumption of primary energy in industry is a big share of the total consumption in Europe, in many countries around 40 %. It would be to go too far to illustrate the conditions in each country but as an example the present situation in Sweden is shown in Figure 1. In this country, as in most other European ones, the reduction in the oil consumption in the industry has been very high, in Sweden at least 30 %, in the last 10-year period. Still, as can be seen in the figure, the consumption of oil products (Sweden has only recently started to use natural gas), is high. This indicates a big potential for further reduction with the aid of e.g. heat pumps. In Sweden the consumption in the process industry (i.e. the food, pulp and paper, chemical, cement, iron and steel industries) is 83 % of the total industrial one. Other countries may have much smaller shares of pulp and paper and/or iron and steel industries and higher shares of other process industries, but the general picture would probably be the same, i.e. the process industry contributes with a high share and the manufacturing one with a smaller share. With regard to industrial energy use, it is always dangerous to generalize too much, as this energy sector shows such a great variety. However, with this in mind, it can be stated that much of the heat needed in process industries, except in special types of industries, is in general at temperature levels between 120 °C and 160 °C, often distributed via a low- or medium-pressure steam network. Waste heat is often available at temperature levels between 50 °C and 110 °C. It is sometimes discussed to develop heat pumps for condensing temperatures up to 300 °C. The potential for heat sinks in this temperature range is probably not very high, although there are of course some types of interesting industries for this temperature level. Besides, this would require heat sources of up to or above 200 °C, which heat could probably often be used in a more intelligent way in the industry (see below the discussion on "Process Integration"). It is therefore my belief that the main interest for heat pumps in the process industry should be focused on heat sink levels below 200 °C, say up to 160 °C.

The need of heat in the manufacturing industry is generally at a lower temperature level, say between 60 °C and 100 °C. A large part of this heat, in the Swedish manufacturing industry more than 50 %, is used for space heating in the industry. Waste heat temperatures in this type of industry would vary between typically 20 °C and 60 °C.

Most industrial applications of heat pumps, installed or discussed so far, have been in such industrial unit operations where there is a simultaneous excess of heat at a relatively low temperature and a demand at a higher temperature. Unit operations of this kind are e.g. distillation, evaporation, and drying. There are obvious advantages for heat pumps in such applications, for example agreement in time between waste heat production and heat demand, often a reasonable temperature lift and small surrounding costs for the installation. This unit-operation oriented approach is recommendable in industries where the unit operation in question is the only big process unit or the main energy consumer. However, in many industries the energy system is complex and consists of many heat consumers and waste heat producers. In such plants the introduction of a heat pump in a unit operation is often directly wrong from a thermodynamical point of view, and hence also from a technicoeconomic one. Several industrial heat pumps have probably been installed in an incorrect way. The reason is the conflict between heat exchange and heat pumping in industry, which is discussed in more detail below.

Many heat pump studies in industrial plants have ended with a suggestion for an improved heat exchanger network instead of a heat pump. This shows that it is important to study possible heat pump applications in a system perspective direct from the start of the project and perform a process integration analysis. During the last 5-10 year period methods for such analyses have been developed and perhaps the most powerful one is the Pinch Technology, developed by Linnhoff et al. (1) (2) (3). This method has created a partly completely new view on the design of industrial energy systems and should be used more in the context of industrial heat pump applications in order to avoid the conflict between heat exchange and heat pumping. With the method it can be shown that most industrial systems have a "pinch" temperature, which divides the system into two parts and for which the following important statements can be made:

- No external cooling shall be done above the pinch
- No external heating shall be done below the pinch
- No heat exchanging across the pinch (Each heat unit transferred across the pinch increases the need for both external heating and external cooling by the same amount).

Three ways of introducing a heat pump into a system are shown in Figure 2. These are above, below, and across the pinch. If it is placed above the pinch the hot utility decreases only with the drive energy to the heat pump and if it is placed below the pinch the only effect is that the cold utility increases. Only if the heat pump is placed across the pinch the effect is the intended one. Many unit operations work either totally above or totally below the pinch. Several heat pumps for such operations have therefore most probably been installed in a wrong way, i.e. the energy saving could instead have been made much cheaper with heat exchanger installations.

From the above discussion it can be concluded that it is essential to make a process integration analysis of an industrial energy system before a heat pump is discussed, so that the economically feasible amount of heat exchanging is first determined.

On the other hand, the fact that there exists a pinch in most industries, normally between 80 °C and 120 °C, means that a heat pump should normally be technically feasible and the economic feasibility should be investigated in each case.

With proper process integration studies the potential will most probably show to be much higher than what has been indicated by the unit-operation oriented surveys made so far. With process-integrated heat pumps, aspects of back-up systems and flexibility during various types of failures are of course crucial. However, the experience from process integration projects in general indicates that such problems often are easier to solve than has been anticipated.

Another competitor to heat pumps, in addition to heat exchanging, is cogeneration. In large systems where back-pressure plants can be built in an economic way heat pumps will normally not be able to compete in the future. Therefore the size range of industrial heat pumps will be limited upwards to, say, around 10 MW heat output, except in situations with extremely low electricity prices. The competition between heat pumps and cogeneration based on gas- or diesel-driven engines or gas turbines in the small- and medium-size range is discussed later. Downwards there is practically no limitation for industrial heat pumps. Systems down to at least 100 kW should be of interest.

Branch of industry	Fuel consumption (TWh/year)	Oil products share of total fuel (%)
Food	5.1	89
Textile	1.0	86
Wood products	1.8	72
Pulp and paper	12.8	65
Chemical	5.4	75
Cement	6.3	43
Iron and steel	16.6	25
Manufacturing	6.6	75
	55.5	

Fig. 1. Fuel consumption in the Swedish industry 1984 (internally produced fuels in e.g. pulp and paper industry not included).

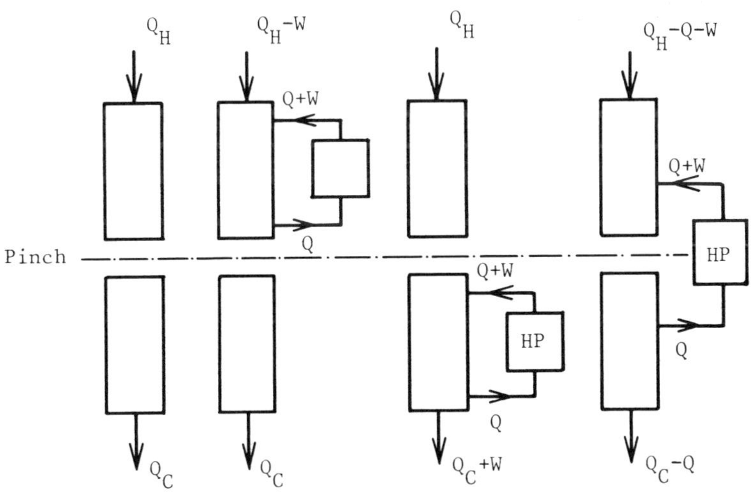

Fig. 2. Placing of a heat pump in an energy system.

Conclusions

From the above discussion the following conclusions can be drawn:

The potential for industrial heat pumps in Europe is high. Most industries will have a considerable amount of waste heat, which can be recovered, also in a situation with an optimized heat exchanger network. The size of this potential is today impossible to estimate. First, a large number of process integration studies, in which heat pumping possibilities are taken into account, must be performed.

Although possible heat sink and heat source temperatures in industry vary over a wide range, two "typical" cases can be identified:

Low-temperature case (primarily manufacturing industry):
Heat sink: 60 °C - 100 °C
Heat source: 20 °C - 60 °C

High-temperature case (primarily process industry):
Heat sink: 120 °C - 160 °C
Heat source: 50 °C - 110 °C

Of these, the high-temperature alternative has probably the highest potential.

The size range for industrial heat pumps will be large, from at least 100 kW to at least 10 MW heat output.

TYPES OF HEAT PUMPS

The types of heat pumps normally discussed in industrial applications are:

- Vapour recompression
- Open-cycle heat pumps
- Compression cycles (electric and engine driven)
- Absorption cycles
- Compression/resorption cycles

These various types are discussed briefly below.

Vapour Recompression and Open-cycle Heat Pumps

As many industrial heat distribution systems consist of a low-pressure (2-6 bar) steam grid, the direct compression of steam should always be considered as the investment cost is low and the coefficient of performance (COP) high. However, due to problems with air leakage and high specific vapour volumes, the practically possible evaporating temperature has a lower limit of about 80-90 °C (if the condensation temperature is above 120 °C) for turbo compressors and 90-100 °C for screw compressors. As many industrial heat sources have a lower temperature, this fact decreases the potential for this type of heat pump.

Open-cycle heat pumps have the same advantages as steam compressors and can be used in special cases in e.g. distillation.

Compression Cycles

Several studies, see e.g. (4), have shown that diesel or natural gas combustion engines as drives for the heat pump are economically advantageous to electric drive at an electricity/fuel price ratio of somewhere between 2 and 3 over a wide range of sizes and temperature levels. This should make combustion engine drive attractive also in industrial applications. However, according to my opinion this use will be limited at least in the near future due to the following reasons:

- The economy of combustion-engine drive is heavily reduced in high-temperature applications, as the engine cooling water for practical reasons cannot exceed approximately 110 °C in today´s units.
- The experience from diesel- and gas-driven heat pumps is that practical problems are bigger and the availability is lower than for electrical heat pumps.
- The electricity price for large consumers in industry is not extremely high even in countries with a generally high electricity price level.

For electrically driven high-temperature closed-cycle heat pumps the biggest and most well-known problem is the fact that so far no ideal working fluid for condensing temperatures above 120-130 °C has been found. An extensive work has been carried out internationally in the search for new fluids.

In this search also nonazeotropic mixtures have been studied. However, possible improvements of the COP are sometimes exaggerated. According to (5) the COP can be increased by 20-25 % at the best, normally around 10-15 % or less. Other possibilities with mixtures, such as capacity and flexibility improvements, will probably play an equally important role. So far no real good mixture, capable of working above condensing temperatures of 120 °C, has been suggested in the literature.

Due to the lack of good high-temperature working fluids, some of the larger manufacturers have suggested a two-stage solution, see Figure 3, where the lower stage produces low-pressure steam between 100 °C and 120 °C and the upper stage is a steam compressor. The main disadvantages with this solution seems to be the complexity and the relatively high investment cost.

From the discussion above it can be concluded that the success of the ongoing research for high-temperature fluids and cycles (120-160 °C) is crucial for the future introduction of industrial heat pumps.

Absorption Systems

The types of absorption systems discussed for industrial applications are absorption heat pumps ("Type I") and heat transformers ("Type II"). The principles for these are shown in Figure 4. Today´s absorption heat pumps (ammonia-water and water-LiBr) cannot work with generator temperatures above approximately 150 °C for practical reasons. This means that the heat sink temperature is maximised to slightly above 100 °C.

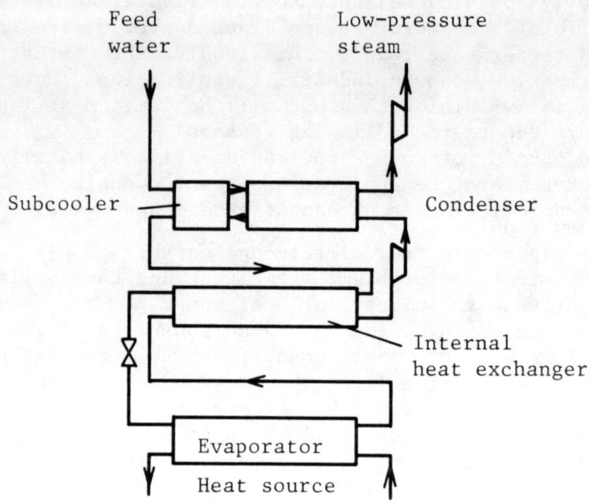

Fig. 3. Two-stage electric compression heat pump for steam production.

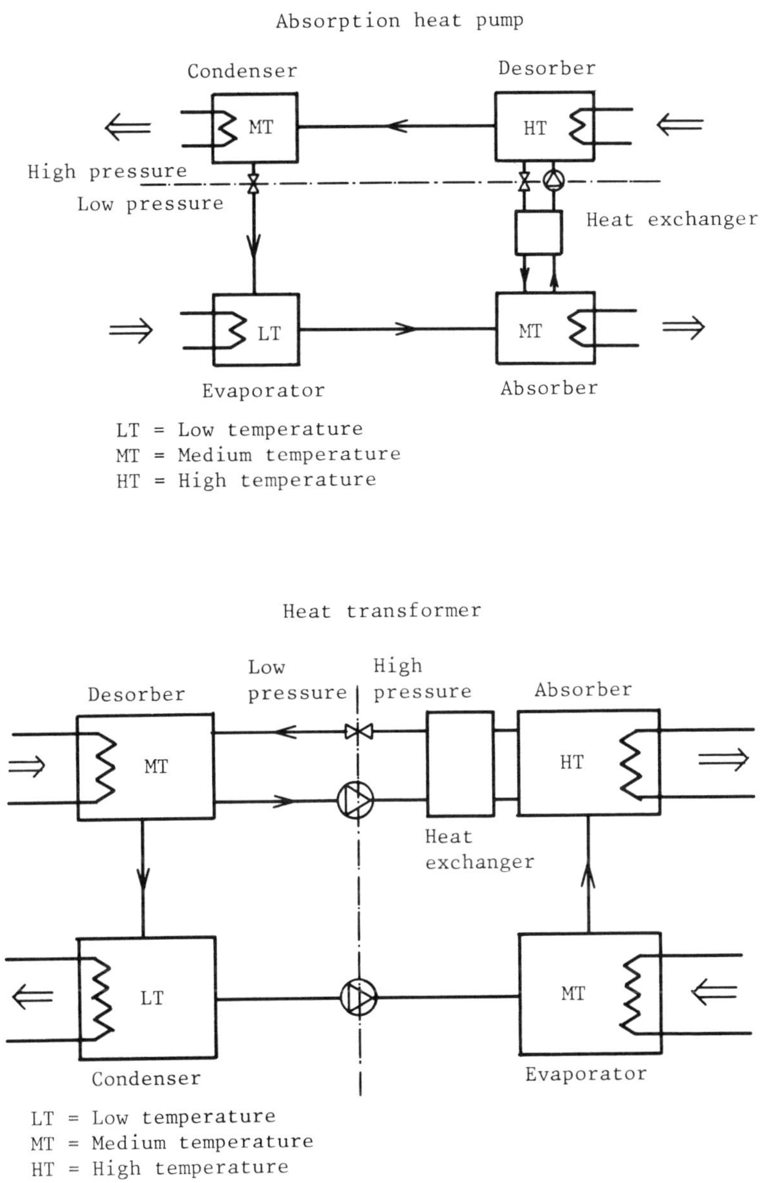

Fig. 4. Principles for the absorption heat pump (Type I) and the heat transformer (Type II) cycles.

Hence, such heat pumps cannot operate at high-temperature conditions and their future potential in industry is therefore limited. Another limitation is the possible temperature lift, which for e.g. H_2O-LiBr systems is limited to 40-45 °C.

The limiting temperature in the absorber of today´s types of heat transformers (normally water-LiBr) is approximately 150 °C and this type of heat pump can therefore be used in high-temperature applications. The most important technical aspects of this type is the restriction in temperature lift (according to approximately: $T_{heat\ sink} - T_{heat\ source} \leq 0.8 \cdot (T_{heat\ source} - T_{cond})$) and the fact that the heat amount of the heat source must be much larger, ideally double as much, than the heat needed at the heat sink side.

Some feasibility studies performed on heat transformers have indicated that their potential in industry is very limited. This is probably not quite true. When proper process integration studies are made in industry, most likely many interesting applications will be found. In process integration studies performed so far by Chalmers University in five Swedish process industries, suitable conditions for a heat transformer have been found in two places. This at least indicates a certain potential for this type of heat pump.

Compression/Resorption Cycles

The basic cycle for this type of process is shown in Figure 5. The knowledge of this and more advanced cycles of this type is today too poor for enabling any conclusions about their future possibilities. However, this type of cycles has been discussed much in recent years as they could be a good solution for a high-temperature heat pump, working up to at least 150 °C.

PRESENT SITUATION

Vapour Recompression

Vapour recompression systems are used today in most European countries. The number of units is in most countries still fairly small, except in France, where at least 150 units have been installed so far. This development has been made possible by the actions of the Electricité de France, who have initiated and supported it.

Electrically Driven Heat Pumps

The database of the IEA, Heat Pump Center, contains at present approximately 70 European industrial heat pump R,D&D projects, the majority of which are electrically driven systems. Of these, 35 are in West Germany. Sweden, so far not included in the database, has probably the largest number of industrial heat pumps installed in Europe, due to its low electricity prices. At least

130 units are in operation there at present.

Gas- and Diesel Driven Heat Pumps

A small number of units have been installed, primarily in West Germany and the U.K.

Absorption Systems

So far large absorption heat pumps (Type I) have been installed only in a few cases, in Sweden and in West Germany. The largest ones are in Sweden, two in the 6-7 MW range and two in the 27 MW range. However, none of these are purely industrial heat pumps. A few heat transformers (Type II) have been installed in West Germany and the Netherlands. The largest one, in Delfzijl in the Netherlands, has an output of 6-7 MW.

ECONOMIC ASPECTS

The most crucial parameters for the future of industrial heat pumps in Europe are of cource the investment cost and the future fuel and electricity prices. In this section the impact on the economy of these and other major parameters is discussed. Various heat pump alternatives are compared with each other and with cogeneration.

Practically all industries today make their investment decisions based on pay-back period (PBP) calculations. Therefore the PBP method is used here for the comparisons of various investment alternatives.

One parameter of major importance is the total investment cost including heat source and heat sink equipment, control system, electric transformer, etc. This cost varies from plant to plant. However, in the following comparisons only one, reasonable, total cost for each type of heat pump has been used in order to keep the number of parameters on a reasonable level. Changes in the surrounding costs will influence the various heat pump systems in approximately the same way. Anyone wanting to study the influence of this cost in the following diagrams can easily do so, as the PBP is directly proportional to the total investment cost. The investment costs used are values found to be reasonable after collecting information from existing plants in the large-scale range, approx. 5 MW heat output. For smaller units the specific cost would be higher. Due to the small number of installed units for some types the figures used are sometimes uncertain. However, the main aim with the following comparisons is to show the influence of the fuel and electricity prices and not to perform a detailed comparison between the various heat pump types. The investment costs used are shown in Figure 6. The corresponding values for cogeneration are also shown there.

As has been discussed above, industrial heat pump applications can roughly be divided into two parts, low-temperature systems

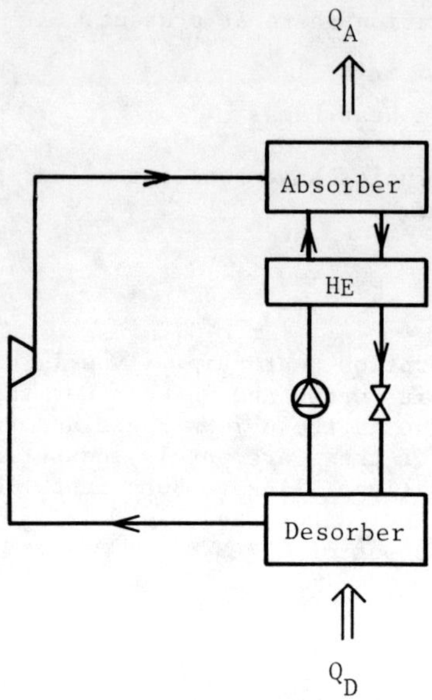

Fig. 5. The basic compression/resorption cycle. Working pair e.g. ammonia-water.

System	Total investment cost/kW heat output SEK	US$
Vapour recompression	1000	143
Electric compression cycle (R12, R500)	1500	214
Electric compression cycle (R114)	1900	271
Two-stage electric compression cycle	2500	357
Dieseldriven heat pump (using heavy-fuel oil)	1750	250
Absorption heat pump	1500	214
Heat transformer	2500	357
Gas turbine (α = 0.5)	2500	357

Fig. 6. Total investment costs for various systems.

(condensing temperatures 60-100 ^{o}C) and high-temperature systems (condensing temperatures 120-160 ^{o}C). In the following these two types of applications are discussed separately.

Low-Temperature Applications

According to the discussion above on various types of heat pumps, the ones included in the following can be used in this temperature range.

Electrically driven heat pumps

With e.g. R12 or R500 possible condensing temperatures are 75-80 ^{o}C (screw compressors) and 90-95 ^{o}C (turbo compressors). With reasonable industrial heat source temperatures, COP:s between 3 and 5 would be typical for this type. At higher condensing temperatures e.g. R114 must be used meaning a slightly higher investment cost (see Figure 6) and a slightly lower COP (not shown in the diagrams below).

Gas- or diesel-driven heat pumps

The same COP:s as above and reasonable heat recovery from waste gases, cooling water, etc. have been used in the calculations. In this size range medium-speed (700-900 rpm) engines would be used.

Absorption heat pumps

Systems with H_2O-LiBr have a COP of approx. 1.6-1.7 in a wide range of evaporating and condensing temperatures. To this must be added the boiler efficiency, here set to 0.85.

Cogeneration systems

In this size and temperature range gas- or diesel-engine based or gas-turbine based systems would be used. These systems have different investment costs and electricity-to-heat ratios. However, calculations have shown that the end result, the PBP, is approximately the same. Therefore only the results for the gas turbine, with an electricity-to-heat ratio of 0.5, are shown in the diagrams.

The PBP:s have been calculated for all the systems described above at various fuel (heavy oil or gas) and electricity prices. The two fuel price levels 100 SEK/MWh (14.3 US$/MWh) and 200 SEK/ MWh (28.6 US$/MWh) have been used, thus covering a wide range of heavy fuel oil price. (This price will probably rize to 20-28 US$/ MWh in a few years perspective.) The boiler efficiency has been set to 0.85.

In Figures 7 and 8 the PBP is shown versus the electricity-to-fuel price ratio (γ) for all systems for the two annual operating times 8000 h (A) and 4000 h (B) respectively. From these figures the following main conclusions can be drawn:

Heat pumping is in most situations more economic than cogeneration. Only at very high γ:s and low COP:s cogeneration can compete. (Note that the same electricity price has been used for the heat pumping and the cogeneration. In situations when the produced electricity is to be sold, a considerably lower price would be valid and the possible annual operating time would decrease.) At high COP:s, i.e. temperature lifts of approximately 40 ^{o}C or smaller, electrically driven heat pumps are the most interesting ones over the whole range of γ-values considered. At low COP:s, the internal-combustion driven systems would compete at γ-values above approx. 2. At these high temperature lifts absorption systems cannot be used due to technical restrictions. The most interesting area for these is probably in situations where very cheap fuel, e.g. internally produced steam, bark, peat, can be used.

For all the heat pump systems except the electrically driven ones, the most important parameter is not the γ-value but the absolute level of the oil price. The PBP versus this parameter for all systems is shown in Figure 9 for one example of electricity price, 300 SEK/MWh (43 US$/MWh). From this it can be seen that the PBP for non-electrically driven systems can be below 3 years if the oil price and the annual operating time are high . In Figures 7 and 8 it can be seen that the PBP for electrically driven systems can be 3 years or well below that if the γ-value is low or the annual operating time high.

The conclusions can be summarized thus:

- Cogeneration can compete with heat pumping in small- and medium-scale systems only at extremely high γ-values.
- There should be many situations where the PBP can become 3 years or well below for heat pumps.
- Electrically driven heat pumps are often of interest even at high γ-values, especially when the COP is high.

High-Temperature Applications

At condensing temperatures between 120 ^{o}C and 160 ^{o}C the following types of heat pumps can be used:

Vapour recompression

These systems have typical COP:s between 5 and 10 or beyond that. As discussed earlier, it is necessary with a high heat source temperature if vapour recompression shall be used. The investment cost for vapour recompression may vary much with the conditions, for example if the steam to be compressed is clean or dirty. The value used is for a situation favourable for this technique.

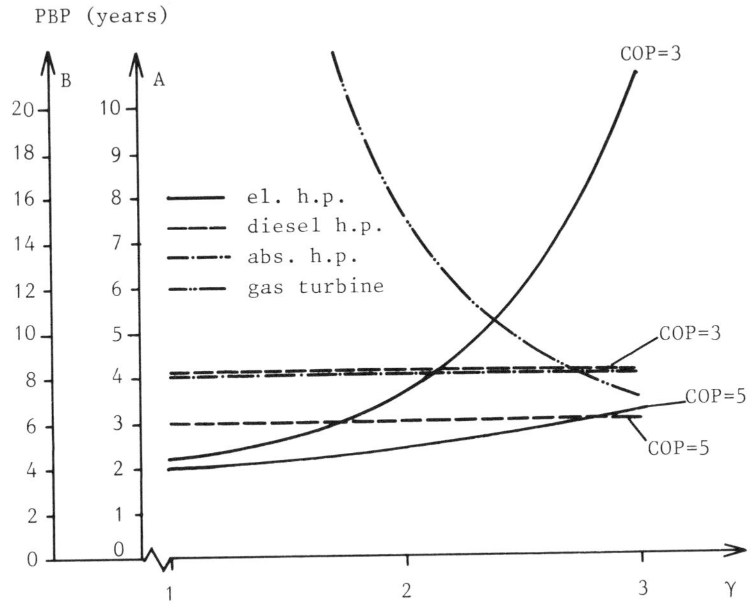

Fig. 7. PBP vs. electricity/oil price ratio (γ) for low-temperature applications. Oil price 100 SEK/MWh (14.3 US$/MWh).

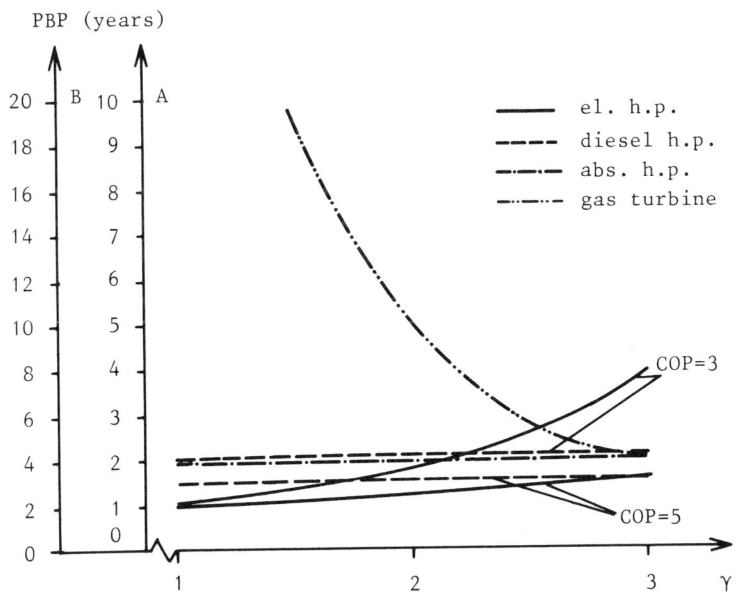

Fig. 8. PBP vs. electricity/oil price ratio (γ) for low-temperature applications. Oil price 200 SEK/MWh (28.6 US$/MWh).

Heat transformers

These have typical COP:s of 0.45-0.50 over a wide range of temperature situations. The temperature lifts are technically restricted due to the earlier discussion. It is also assumed here that the heat source amount is always double or more than the heat amount required at the heat sink. This means that a very advantageous situation technically for the heat transformer has been assumed.

Two-stage systems

This type of system was shown in Figure 5. The COP:s range between 2 and 4.

One-stage systems

At the lower end of the condensing temperature range discussed here the closed-cycle stage only of the system shown in Figure 5 can be used with e.g. R114. Typical COP:s for this cycle would also be between 2 and 4. The investment cost and hence the PBP is approximately only 0.75 of that for the two-stage system. (The curves for the R114 cycle have not been included in the diagrams.)

The PBP:s for these systems and for cogeneration are shown versus the γ-value for the two same oil prices as in the low-temperature case in Figures 10 and 11. As for the low-temperature case it can also here be concluded that cogeneration can normally not reach acceptable PBP:s. The various heat pump systems are normally not directly comparable, as they would be used in different situations concerning the evaporation temperature.

High evaporating temperatures (above approx. 90 °C)

Here all the systems can be used technically. From the figures it is clear that vapour recompression is advantageous to all the other systems at all levels of γ-values. Also the heat transformer has a low PBP and could therefore be of interest.

Low evaporating temperatures (approx. 50-90 °C)

In this range the vapour recompression cannot be used. If the temperature lift is small, so that a heat transformer can be used, this cycle is the most economic one, regardless of the γ-value. The one-stage and two-stage heat pumps can reach acceptable PBP:s if the γ-value is very low.

The influence of the absolute oil price level on these systems is shown in Figure 12. From this and from Figures 10 and 11 it can be concluded that very low PBP:s can be reached, regardless of the γ-value and the oil price, if vapour recompression or a heat transformer can be used for technical reasons. For applications

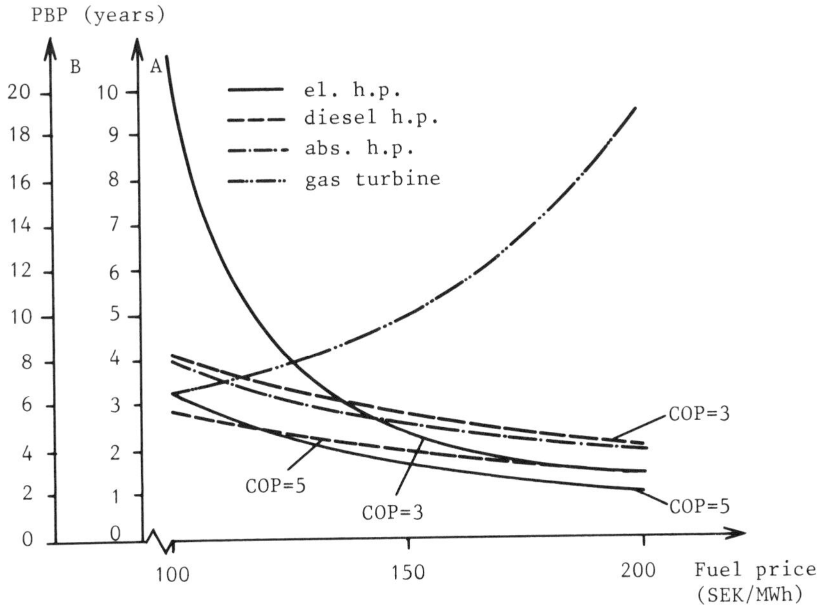

Fig. 9. PBP vs. oil price for low-temperature applications. Electricity price 300 SEK/MWh (42.9 US$/MWh).

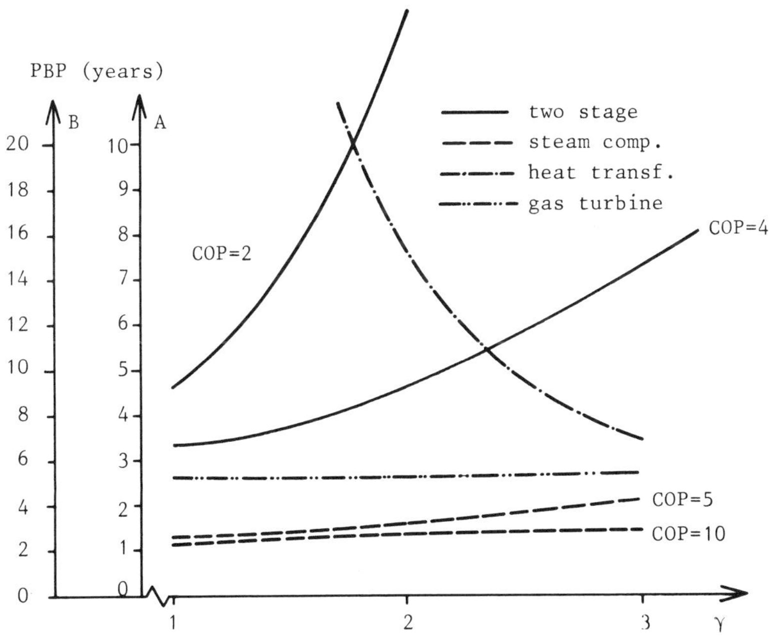

Fig. 10. PBP vs. electricity/oil price ratio (γ) for high-temperature applications. Oil price 100 SEK/MWh (14.3 US$/MWh).

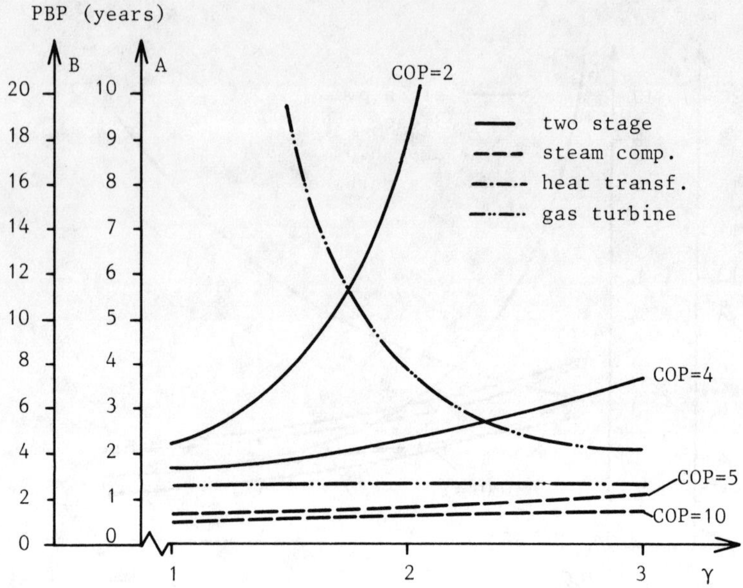

Fig. 11. PBP vs. electricity/oil price ratio (γ) for high-temperature applications. Oil price 200 SEK/MWh (28.6 US$/MWh).

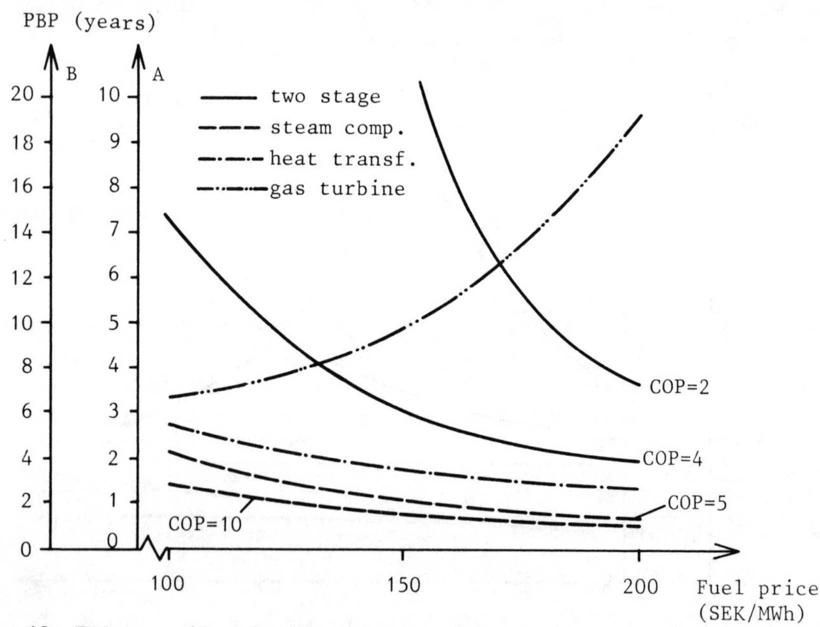

Fig. 12. PBP vs. oil price for high-temperature applications. Electricity price 300 SEK/MWh (42.9 US$/MWh).

with the probably highest technical potential in industry, evaporation at 50-90 °C and condensation at 120-160 °C, the systems suggested so far can reach acceptable PBP:s only in especially advantageous situations, high oil prices, low γ-values and/or long annual operating times.

From the various figures in this section it is clear that it is important to investigate possibilities to decrease the first cost for a heat pump installation. It is sometimes argued that the heat pump components costs will decrease when the number of installations increases. Although this is probably true, one must bear in mind that only a small part of the total cost, practically always one third or less, is related to the heat pump itself. The rest is installation, heat source and heat sink equipment, piping, control devices, engineering hours, etc. This means that somewhat lower component costs will only influence the economy marginally. It is more important to decrease the surrounding costs by e.g. developing a better combined knowledge of heat pumps, heat exchangers, installations, etc., in consultant companies and process industries.

The fact that the heat pump cost is only a small part of the total one, indicates that today´s heat pump systems are probably not optimized. A more advanced heat pump cycle, costing, say, 50 % more, would mean a total cost increase of only 10-15 %. This extra investment could enable a considerably higher COP and hence a better total economy.

OTHER ASPECTS

Influence on the Industrial System

The introduction of a heat pump into an industrial energy system influences this system in several ways. One is that the total cooling load decreases, which in many cases can mean an important annual cost reduction in terms of smaller investment and operational costs for e.g. cooling towers and an increased income due to e.g. possibilities for full production also in the summer when the cooling facilities decrease. On the other hand, the introduction of a heat pump, especially at evaporating and condensing temperatures near the pinch temperature, may sometimes necessitate an increased heat transfer area in the heat exchanger network for the units working below the evaporating temperature and above the condensing one.

Another aspect is the influence on the availability and the flexibility of the plant. A heat pump should always be introduced in such a way that there is a back-up system if the heat pump fails. If it produces steam to the existing network such a back-up system should normally be available in the form of e.g. an existing oil boiler. The temperature level of a steam network has of course normally been designed without regard of any heat pump installation. Possibilities of decreasing this temperature in at least a part of a network when a heat pump is introduced should therefore be investigated.

Financing and the Government Role

In the economic calculations shown above the payback-period method has been used, as this is the normally used one in industry today. However, this method often leads to a nonoptimal solution seen in a long-term perspective. Other ways of financing heat pump installations have been discussed recently. One promising such way would be for the industry to buy only the heat, and that the heat pump installation, service, etc. is done by someone, e.g. a private company or a governmental body, who can take higher risks and afford a longer PBP. This and other alternative ways of financing should be investigated in much more detail.

It was mentioned already in the introduction that there is a great need for industrial demonstration and experimental plants. Only with such plants in various industrial branches, decision makers in industry will start to rely on this new technique. Possibilities for governmental contributions for the financing of such plants will therefore be of high importance for the implementation of heat pumps in industry.

Environmental Aspects

The environmental aspects become increasingly important in the European countries. Problems with pollutions will in the near future without doubt be one of the crucial questions which must be solved if the industry, and the society, shall survive. This means that most probably a price will be set on emissions in the future, which will take the form of governmental grants for non-pollutant techniques, penalty costs for emissions above a certain level, etc. It is therefore important to state, that practically all the types of industrial heat pumps discussed above contribute to a high degree to a decrease of emissions from fuel-fired plants. This is true also for electrical heat pumps in industry, which have normally relatively high COP:s, even if the electricity used is produced by e.g. a fuel-fired condensing plant. The environmental impact must of course be taken into account when the introduction of heat pumps into industry is discussed.

There are two important negative environmental impacts that must be mentioned. The first one is the high production of NO_x in diesel- and gas-engine driven heat pumps. The second one is the damage of the ozone layer and the contribution to the "CO_2 effect" caused by some freons, which may make it necessary to find alternative working fluids in future heat pumps.

DISCUSSION AND CONCLUSIONS

According to the discussions above, there is a large technical potential for heat pumps in industry in Europe. The technico-economic comparisons have shown that there is also an economic one. In many cases pay-back periods acceptable to industry can be achieved. In fact, industrial applications normally compete with other energy technologies considerably better than house-heating ones. This means that in the near future the industrial field will probably be the most interesting one for the heat pump technology in Europe. The actual potential is very much depending on future electricity, oil and gas prices. Also the general future economic growth, influencing the retrofit and new-construction situation of the European industry, and environmental aspects will of course affect the future prospects for this technology.

In the introduction several obstacles for the introduction of heat pumps in industry were presented. Most of these have been discussed in this paper. According to these discussions the following R,D&D activities should take place:

Technical Aspects

- The search for alternative fluids and cycles for compressor driven systems in applications with condensing temperatures 120-160 °C and evaporating temperatures well below 100 °C must continue.
- Possibilities to find new working pairs and advanced cycles in absorption systems in order to increase the possible temperature lift and the absorber temperature must be investigated.

System Aspects

- The knowledge about possible applications in industry must be increased by performing process integration studies in various types of industries.
- Experimental and demonstration plants must be built in different industrial branches.
- Optimization studies on heat pump contra surrounding equipment costs must be performed.
- Experts in the combined field chemical engineering/heat pump technology must be educated.

ACKNOWLEDGEMENTS

The author wishes to express his gratitude to:

Prof. D. Hodgett, Electricity Council Research Centre, Capenhurst, U.K., and Dept. of Heat and Power Technology, Chalmers University of Technology, Sweden, for valuable discussions.

The Swedish National Board for Technical Development for financial support in the technico-economic part.

Dr. K. Munch Berntsson, K-Konsult, Gothenburg, Sweden, for carrying out the calculations for the technico-economic comparisons.

REFERENCES

1. B. Linnhoff and J.A. Turner, "Heat-recovery networks: new insights yield big savings", Chem. Engineering 56-70, November 2 1981.

2. T.N. Tjoe and B. Linnhoff, "Using Pinch technology for process retrofit", Chem. Engineering 47-60, April 28 1986.

3. B. Linnhoff et al., A user guide on process integration for the efficient use of energy, The Institution of Chemical Engineers, 1985.

4. K. Munch Berntsson, T. Berntsson, and D. Hodgett, Large diesel-driven heat pumps for heating purposes, Report No. 129 to the Swedish Council for Building Research, 1984. (Partly in Swedish).

5. T. Berntsson and H. Schnitzer, IEA Advanced Heat Pumps, Annex VI, State-of-the-art report and Final report Part A and Part B, 1984, 1985, and 1986.

Chapter 22

Future Prospects of Industrial Heat Pumps in Japan

Takeshi Yoshii

ABSTRACT: This paper briefly reviews the study on the application of heat pumps to industrial processes carried out by JRAIA (The Japan Refrigeration and Air-Conditioning Industry Association) supplemented with other relevant comments and conclusions.

Of the total energy consumption in the industrial sector, it is estimated that substantial part is used as heat of relatively low temperature below 300°C. Therefore it could be said that heat pumps, as an energy saving measure, have an enormous potential. Some of the most promising industrial sectors are identified and possibilities are discussed.

To open a larger market for industrial heat pumps, advanced heat pumps are expected along with the improvement in the economic situation. Some of the current research and development approaches are also described in this paper.

INTRODUCTION

The total energy consumption in Japan is roughly 1.5 x 10^7TJ. The industrial sector accounts for 64 percent of this figure, and rejects a large amount of this energy as waste heat.

The progress in industrial energy savings during the past thirteen years is remarkable. Various devices for recovering high temperature waste heat have been implemented so far. However, low temperature waste heat has been rejected, for the most part as hot water, remaining unused.

Heat pumps, as a mean to utilize low temperature waste heat efficiently, have recently obtained attention from the industrial

HEAT PUMPS: Prospects in Heat Pump Technology and Marketing (Proceedings of the 1987 IEA Heat Pump Conference), Edited by Kay H. Zimmerman Chelsea, Michigan 48118. Printed in U.S.A.

sectors [1],[2],[3],[4]. In the following paper, the present and future of industrial heat pumps are discussed. Some of the most promising industrial sectors are identified, and current research and development approaches for the advanced heat pumps are reviewed.

THE ROLE OF INDUSTRIAL HEAT PUMPS

With scarce energy resources, we depend heavily on imported energy. Furthermore the industrial sector consumes more than 60 percent of the total energy. Fig. 1 compares the energy consumption structure for several countries. It is evident that the energy savings in industry is particularly important in Japan.

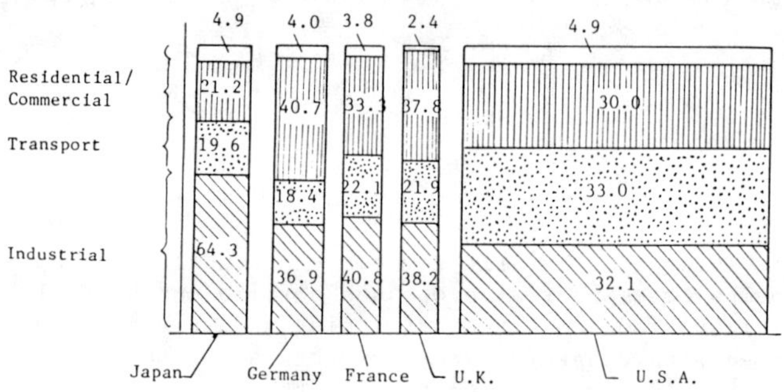

Fig. 1. **Energy consumption structure in major contries. (OECD Energy Balances 1982)**

Of the total energy consumption more than 60 percent is estimated to be used as relatively low temperature heat of lower than 300°C, as shown in Table 1. The industrial sector shares large part of this low temperature heat consumption, and at the same time rejects a large amount of low temperature waste heat.

Table 1. Temperature level of energy consumption

- 100°C	30%
100 - 315°C	37%
315 - 600°C	9%
600°C -	22%

The concept of heat pump is to pump-up low temperature heat to a higher temperature efficiently. Thus heat pumps have the capability to recover low temperature waste heat and upgrade it to a higher effective heat.

The heat pump technology available today has been fostered for many years as the technology to produce cold for refrigeration and air-conditioning. During the past thirteen years, the expensive energy prices we experienced, triggered the use of heat pumps in the residential and commercial sectors, and recently in the industrial sector.

In the industrial sector, cost-effectiveness is of prime importance for the decision of investment. Therefore the role of heat pumps in industry is to recover low temperature waste heat economically.

THE STATE OF THE ARTS OF INDUSTRIAL HEAT PUMP TECHNOLOGY

Two types of heat pumps, vapor-compression heat pump and absorption heat pumps are available today. They can be classified by application into three categories as shown in Table 2.

Table 2. Heat pump types for industrial applications

	Type		Application
1)	Vapor Re-compression Heat Pump (VRC)	....	Water Evaporation/ Concentration
2)	Heat Reclaim Heat Pump - Vapor Compression Heat Pump (CHP) - Absorption Heat Pump (AHP)	....	Simultaneous Heating and Cooling/Dehumidification Drying
3)	Temperature Boosting Heat Pump - Vapor Compression Heat Pump (CHP) - Absorption Heat Pump (AHP) - Heat Transformer (HTF)	....	Waste Heat Recovery

Vapor re-compression heat pumps are open cycle heat pumps of unit-operation used for water evaporation and concentration of solutions. Heat reclaim heat pumps are used for simultaneous heating and cooling of the processes. Temperature boosting heat pumps are used for upgrading waste heat and recycling the waste heat to process heating.

The operating temperature range of heat pumps available today are mapped in Fig. 2. The maximum output temperature of commercially available compression heat pumps, using Freon 114 as working fluid, is about 110°C. For absorption heat pumps it is about 90°C, and for heat transformers it is about 140°C.

The performance factor of heat pumps are usually expressed by coefficient of performance (COP). The COP of heat pumps varies

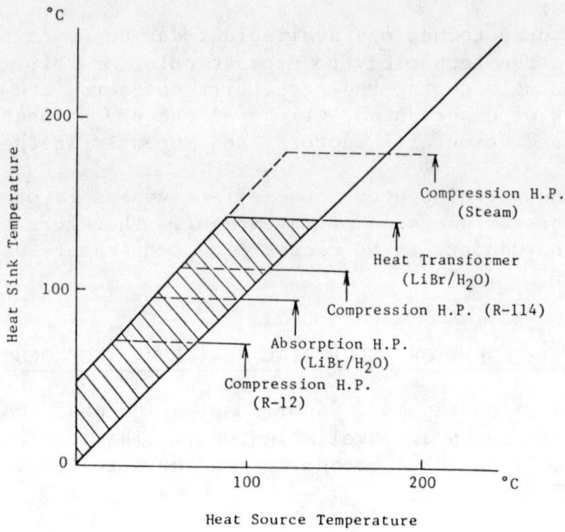

Fig. 2. Working temperature range of heat pumps.

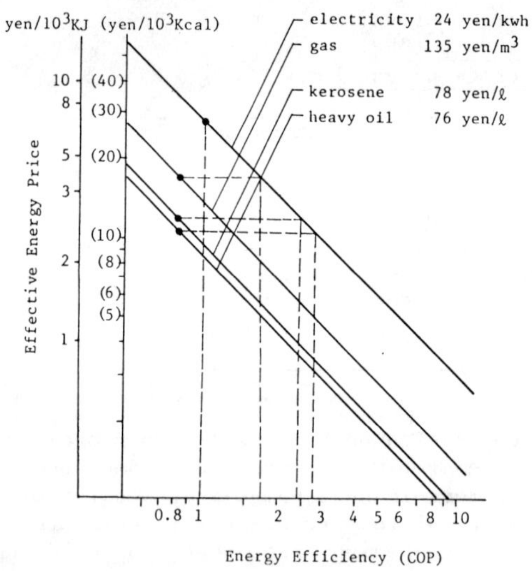

Fig. 3. Effective energy price

with the temperature lift; the smaller the temperature lift the larger the COP becomes. The COP for vapor re-compression heat pumps and heat reclaim heat pumps are usually large enough to be economically meaningful. However in the case of electrically driven simple temperature boosting heat pumps, as shown in Fig. 3, when energy prices are assumed as indicated, an economical COP is roughly 3.0. This means, in this case, that the price ratio of electricity to oil is more than 3.0, due to the fact that a considerable amount of electricity is produced in steam power plants. This follows that for the heat pumps to be energy economical, the temperature lift has to be less than 40 to 50°C with the present heat pump technology.

PRESENT MARKET SITUATION

The market for industrial heat pumps, as a whole, can be said to be in its early development. The typical applications of heat pumps found in the industrial sector at present are as follows:

- Hot water supply	:	Textile dyeing, Boiler preheating, Hotspring water heating etc.
- Space conditioning	:	Green house heating/dehumidification
- Process heat reclaim	:	Food, Chemical
- Drying	:	Food, Wood
- Concentration /Evaporation	:	Food, Chemical
- Distillation	:	Oil, Chemical

Heat pump drying of food or wood by low temperature dehumidification has the advantage of keeping the quality of the product. Evaporation or condensation with vapor re-compression heat pumps and heat reclaim heat pumps for simultaneous heating and cooling have sufficient energy savings to pay back the installed cost within 2 to 3 years.

On the other hand in the case of simple temperature boosting heat pumps, there are rather difficult situations. To make simple temperature boosting heat pump cost-effective, there are some factors to be considered:

- Efficiency (COP)
- Operating hours (hour/year)
- Installed cost (yen/kW)
- Energy price (yen/KJ)

Under current energy price situation, heat pump COPs of 3 to 4 is not enough to be justified as cost-effective and require improvement in heat pump efficiency. Operating hours have to be long enough to get sufficient energy savings per year. Installation costs have to be lowered by mass production of heat pumps and by standardization of installations.

MARKET POTENTIAL OF INDUSTRIAL HEAT PUMPS

According to the survey carried out by JRAIA, the high potential in the industrial sector for heat pumps are, as shown in Fig. 4, those which use hot water or steam produced by boiler:

- Pulp and paper
- Chemical
- Petroleum
- Textile
- Food

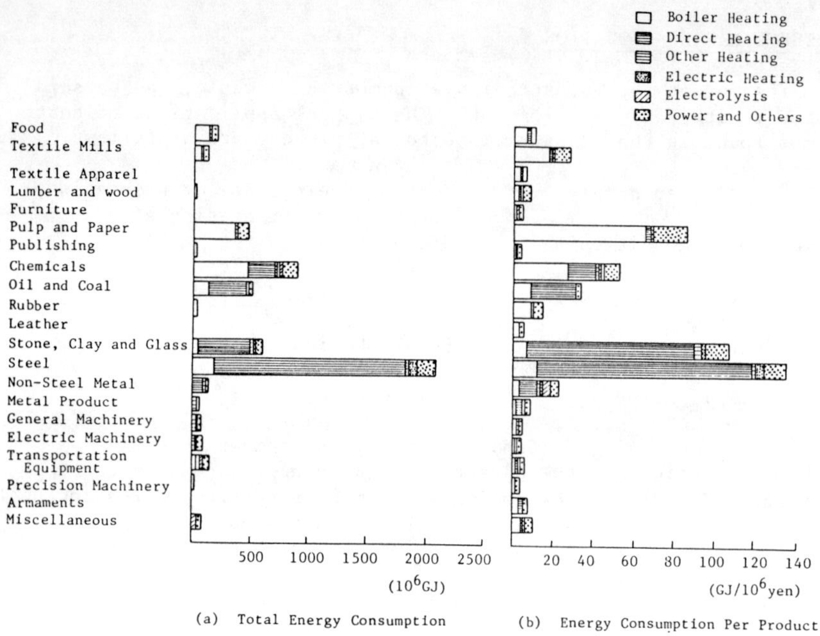

Fig. 4. Energy consumption by industrial sectors.

The temperature profile of thermal energy consumption for those energy intensive industries are shown in Table 3, heat demands are mostly in the range of 100 to 200°C. On the other hand waste heat temperatures for those industries are in the range of 40 to 100°C.

Fig. 5 shows conceptual temperature profiles of waste heat and heat demand for certain processes. The temperature lift required varies by process, ranging from 30°C to 150°C. However with the present heat pump technology available, usually it is not economical to boost waste heat temperature by more than 50°C. The process heat demand quite often exceeds the temperature level that conventional heat pumps can produce as shown in Fig. 5.

Therefore the future market of industrial heat pumps depends, to a larger extent, on technical developments attained. Another important factor is the energy situation, especially the price ratio of electricity to oil.

Table 3. Temperature levels of heat demand for industrial sectors

(%)

Industry \ Temperature	- 100°C	- 150°C	- 183°C	183°C -
Food and Tobacco	2.5	62.3	16.6	18.6
Textile	0.4	50.3	49.3	0
Lumber and Wood	1.1	9.3	6.6	83.0
Pulp and Paper	0	85.9	4.1	0
Chemicals	4.8	26.9	50.0	18.8
Rubber	0	26.3	53.4	20.4
Leather	0	100	0	0
Stone, Clay and Glass	0	85.6	14.4	0

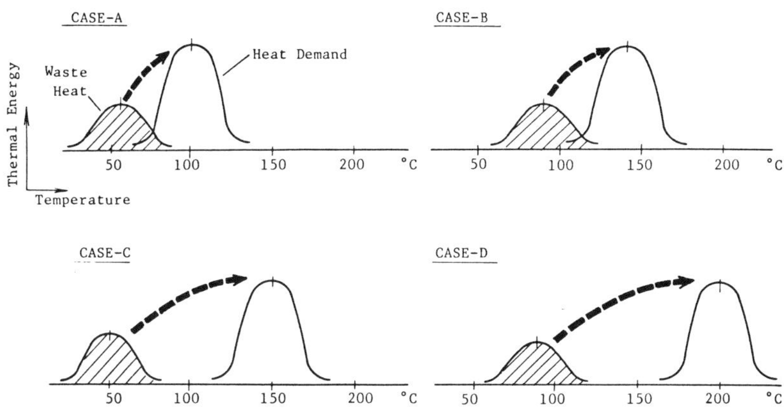

Fig. 5. **Conceptual profiles of process waste heat and heat demand.**

The long range projection of oil price is to go up in the future, on the other hand electricity price is estimated to be comparatively stable due to the increase in the share of nuclear power plants in power generation. Therefore it is expected that future energy situation is in favor of industrial heat pumps.

Although it is not possible from the available data to accurately estimate the potential market of industrial heat pumps, when conservatively it is assumed 15% of the five high potential industries' boiler heat demand is supplied with heat pumps, it amounts to 1.8×10^5 TJ.

NEW DEVELOPMENTS IN INDUSTRIAL HEAT PUMP

In order for heat pumps to penetrate the market more extensively, there are three areas of technical developments to be worked out:

1) Expand the operating temperature range, (high output temperature and high temperature lift)
2) Improve the performance (energy efficiency).
3) Matching source and sink for total system optimization.

To expand the span of operating temperature for heat pumps, it is required to get higher output temperatures up to around 200°C, and to get a larger temperature lift with higher efficiency. At the same time to make it economically competitive, higher performance and low cost is a prerequisite. It is also important to integrate the total system to best match the process. In some cases thermal storage is required in the process to offset the time lag between supply and demand.

In 1984, a national project named "Super Heat Pump Energy Accumulation" was started by the Agency of Industrial Science and Technology (AIST)[5]. This project is aimed at developing advanced heat pumps for large commercial and industrial applications:

1) Advanced vapor compression heat pumps; high performance and high output temperature.
2) Chemical heat storage and heat pump systems.
3) Total system study.

In the development of advanced vapor compression heat pumps, it is intended to double the COP and to attain higher output

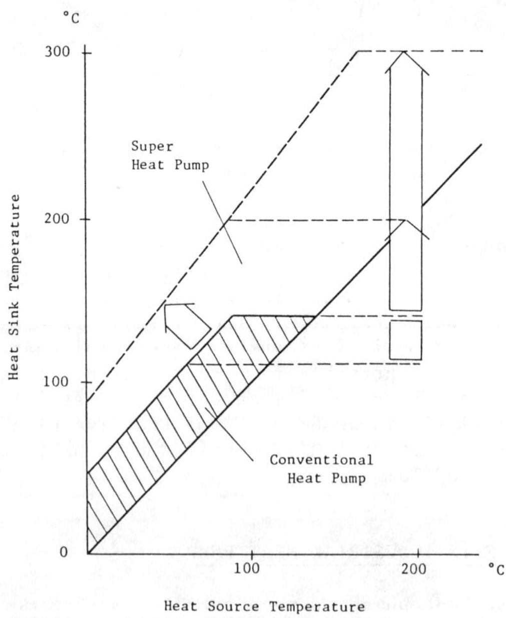

Fig. 6. Working temperature range target of Super Heat Pump.

temperatures of 150 to 300°C as shown in Fig. 7. The component developments for this project are:

- New working fluids, including fluid mixtures.
- High performance compressor and systems.
- High performance heat exchangers.

In the development of chemical heat storage and heat pump systems, it is aimed at getting high temperature systems of up to 200°C. In the total system study it is intended to develop plant simulation technology and to evaluate dynamic performance of Super Heat Pumps under typical plant operating condition.

CONCLUSIONS

The future market potential of industrial heat pump is estimated to be enormous. However in order to bring it into reality, there seems to be some barriers in front of us:

- Technical development of high performance, high temperature and high temperature lift heat pumps at reduced cost.
- Improvement of economical competitiveness by reducing installation costs and by matching heat source and heat sink.
- Energy environment particularly relative energy prices plays an important role for the economical competitiveness of heat pumps. However the trend of energy prices, specifically the oil price in the coming years is difficult to predict.

Therefore for the time being, government's support for the implementation of industrial heat pumps is required along with the support for the development of advanced heat pumps.

REFERENCES

1. F. Moser and H. Schnitzer, "Large Industrial Heat Pumps Present Potential and Outlook", Proceedings of IEA Heat Pump Conference, Graz, Austria, May 1984, PP.221-234.

2. C.L. Lopez-Casicedo, "Factors Affecting the Utilization of Heat Pumps in Industry", Proceedings of IEA Heat Pump Conference, Graz, Austria, May 1984, PP.235-248.

3. E. Pedersen, "Heat Pumps in Danish Industry-Present Status, Possible Applications and Future Prospects", Proceedings of IEA Heat Pump Conference, Graz, Austria, May 1984, PP.249-260.

4. J. Berghmans, Heat Pump Systems Applied in Industry, Final Report IEA ANNEX III, 1985.

5. The Development of Super Heat Pump Energy Accumulation System, NEDO (New Energy Development Organization) Tokyo, Japan, 1986.

Chapter 23

Future Prospects for Industrial Heat Pumps in North America

Richard C. Niess and Joel S. Gilbert

ABSTRACT: Low fuel oil and natural gas prices in North America up through the mid-1970's inhibited industrial heat pump development and application. Following the oil and gas crisis and oil embargo, both industry and government looked at heat pumps with a moderate development effort in closed cycle heat pumps operating up to the 80 C to 104 C output range and open cycle (mechanical vapor compression) heat pumps into a slightly higher temperature range.

Of the key industries having potential heat pump applications -food processing, chemical & petrochemical, textiles, metal fabrication, and pulp & paper - relatively few made token installations and one (textiles) ignored this emerging energy alternative. Only wastewater treatment plants in the more northerly climates installed a significant number of heat pumps because of the external driver of government incentives.

Today, this situation is changing with a much higher level of interest being shown by all interested segments - potential industrial users, heat pump manufacturers, utilities and public utility commissions. This presentation will describe the industrial heat pump applications made in the United States and Canada and discuss why the interest level has sharply increased, despite current lower oil and gas prices and rising electric rates.

HEAT PUMPS: Prospects in Heat Pump Technology and Marketing (Proceedings of the 1987 IEA Heat Pump Conference), Edited by Kay H. Zimmerman

INTRODUCTION

The large-scale application of heat pumps in North America had its beginning with the first known installation of an air-to-air heat pump in Reading, Pennsylvania, in 1932. Since that time, the air-to-air heat pump market has grown until it now represents a significant portion of the year-round air conditioning installations in residential and small commercial projects. A major consumer attraction is the heat pump's ability to supply both cooling and heating at a small installed cost difference over cooling-only systems and at a reasonable price addition over the predominant heating systems. This contrasts to other major areas, such as much of Europe, where summer mechanical cooling is not deemed needed and many heating systems do not consider a cooling function other than through ventilation. No market for "heating-only" heat pumps existed prior to 1973. Also, few large heat pumps of any type were installed, since the price of energy (gas, oil and electric) was too low to encourage owner investment.

Then came the oil/gas crisis of 1973, spearheaded by the oil embargo that completely revolutionized the energy pricing and availability structure world-wide. One major effect was the sudden interest in waste heat recovery, heat exchange and heat pumping, which could conserve energy and lower both consumption and total cost. At the same time, solar and geothermal resources were studied, however the motivating drivers were not sustained and heat recovery and heat pumping emerged as a more acceptable way to use these alternative energy sources.

Since 1973, industrial heat pump development and commercialization has occurred in North America. However for several reasons interest and acceptance fell below levels that attract major investment in marketing and product enhancement.

HEAT PUMP CYCLE DEVELOPMENT

Basic developments in North America have largely centered on four cycle types[1]. Mechanical vapor recompression, as shown in Figure 1, is one cycle. The most common use is steam compression in industrial facilities, where owning and operating costs make it attractive to use hot wastes or to flash hot condensate to make sub-atmospheric steam or to use low-pressure waste steam. These sources are compressed to higher pressure steam for reuse in the plant. Correcting a plant steam imbalance is another use; this occurs where there is an excess of low pressure steam and a need for more steam at a higher pressure. Economics are often favorable when low compression ratios are involved.

The second cycle, shown in Figure 2, using the addition of an evaporator/condenser, finds use in the food processing, chemical, and petrochemical industries for evaporation, concentration, crystallization, distillation and other thermal separation processes. This cycle is particularly effective where there is a small boiling point difference and the necessary temperature rise by compressing the vapor is not more than 50 C. It also offers a good alternative to multiple-effect evaporators.

The third cycle, shown in Figure 3, is the waste-heat driven Rankine cycle. This application is not yet widely used, due to the

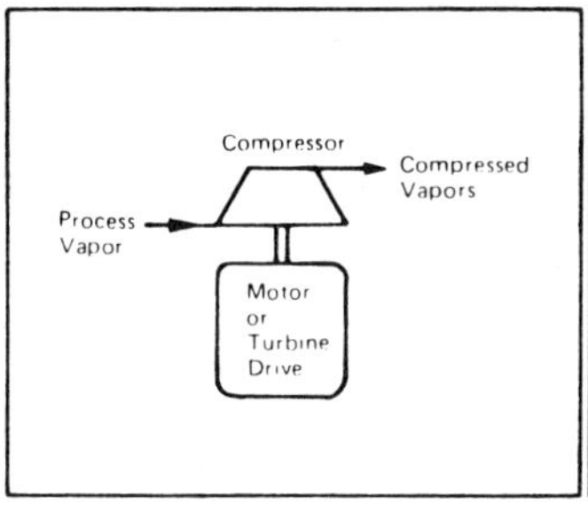

Figure 1. Open Vapor Recompression Cycle

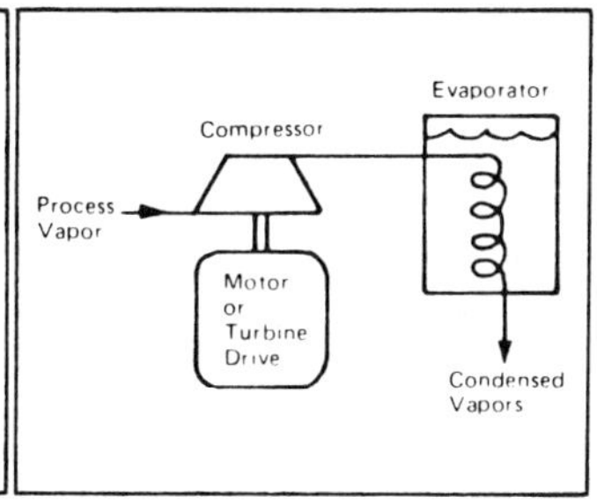

Figure 2. Vapor Recompression Cycle with Evaporator

high capital cost and difficulty in matching heat source, heat sink, and compressed vapor needs. A variation, in the form of an air or Brayton cycle has been developed; one demonstration installation has recently been started, condensing industrial solvents from the exhaust air of drying ovens.

Figure 3. Waste Heat Driven Rankin Cycle

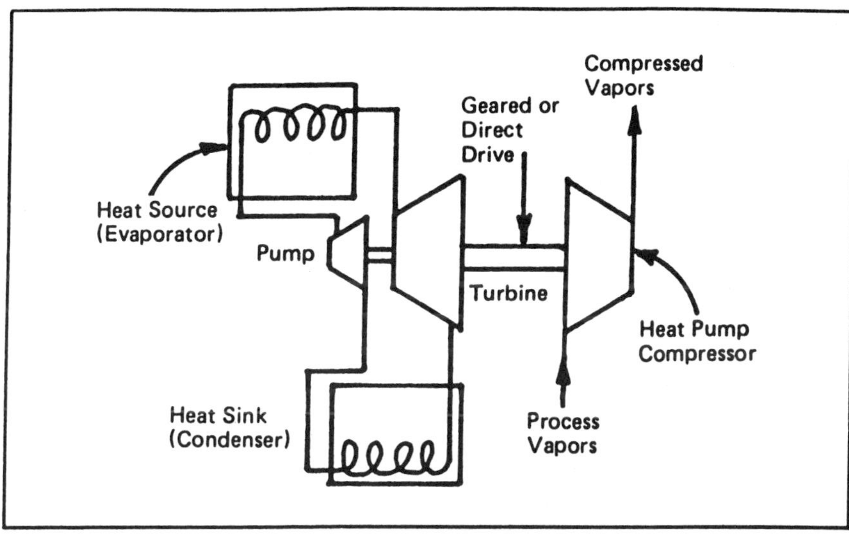

The fourth cycle developed, shown in Figure 4, is the closed cycle vapor compression heat pump. Its genesis was the reciprocating-, screw-, and centrifugal-compressor based water chillers which are extensively used throughout North America. Using readily available fluorocarbon refrigerants as the heat pump working fluid, this cycle is the most commonly used because of its wide application opportunities in both process and space conditioning opportunities.

Figure 4. Closed Vapor Compression Cycle

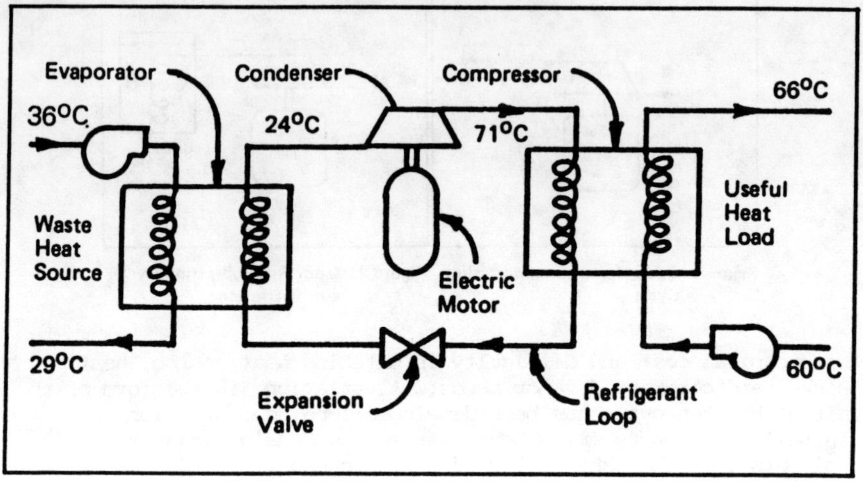

Despite all the development effort by governmental agencies, manufacturers and industrial users, the growth and acceptance has lagged expectations for the North American market.

WHY A PAST LACK OF MARKET GROWTH?

There are a number of reasons why the market for industrial heat pumps has not achieved the growth rates expected at the time of development funding. Some are compelling, some are valid and some simply provide a convenient excuse to explain away the situation.

Initially, electric rates were holding steady or even falling slightly while oil and gas prices were experiencing double-digit escalation. It then follows that growth projections, made on the assumption (wrong) that this condition would continue, were often excessive. When oil and gas came into plentiful supply and prices started down while electricity started to rise, many computer-modeled projections went awry.

These conditions were often used to explain away the situation and the most compelling reasons overlooked. In our opinion the lack of growth stemmed from these three reasons:

1. Lack of a driving influence.
2. Electric utility industry disarray.
3. Considering research as an answer to marketing.

All of these distill into a lack of significant marketing. The concept that, if you build a better mouse-trap the world will beat a path to your door, is very prevalent in this case. In today's market place the choice selection is so overwhelming that only the best marketers usually succeed.

People and industries do something because of some external motivation. This can be as simple as needing more boiler capacity or as exotic as a "get-even" mentality or the perception that electric rates will continue to rise dramatically. Economics, contrary to most belief, are used to justify a project and are rarely the prime driving force.

The electric utility industry in the United States and Canada enjoyed a series of decades where every addition to the generation mix lowered the cost of providing electricity and management could do no wrong until:

-the oil embargoes created a fuel supply problem (not a cost problem). The cost became a problem because the supply was limited. Today, OPEC is trying to control the price without being able to control the supply...and it isn't working.

-the consequences of Three Mile Island, snail darters, and the stream of unemployed or disgruntled nuclear engineers, environmentalists and other interveners turned the public utility commissions into circuses.

-the coal industry capitalized on the rise in oil and gas prices to increase their take.

-interest rates and inflation of that past decade, coupled with skyrocketing construction costs associated with changes in work in progress as a result of the intervenor's efforts, had the inevitable financial shock when the bills finally had to be paid.

As a result, the electric utility industry dismantled their marketing departments, renamed the effort as conservation, and did all they could to help their customers reduce their consumption of, and dependence upon, electricity. For the last ten years, marketing was a word unheard in the electric utility industry.

Marketing of industrial heat pumps took a back seat everywhere. The manufacturers felt they couldn't or shouldn't do the bulk of it as they perceived they didn't have that much to gain. In North America, the major firms considered heat recovery and heat pumps to be a small part of their business (2 to 3 %), to have limited growth potential and to take an disproportionate amount of selling effort in comparison to their main stream product lines. In addition, their sales forces are almost exclusively compensated by commission, yielding little incentive to promote this equipment. Naturally the sales forces have paid scant attention to these products.

Process engineers were too busy with new process developments and available opportunities to focus on heat pumps. Besides, there weren't any sale people around to push them into looking at this new technology.

The electric industry associations and the governmental agencies became enamored with research and didn't want to fund anything that sounded like marketing.

In summary, there has been no driver and no constituency or following.

WHAT HAS CHANGED?

The one event that is causing a major reversal of the past experience is an old technology that has attracted much recent attention, particularly in the United States. That technology is **cogeneration**.

Cogeneration, as we know it today, originated in Europe and spread to North America in the late 1880's[2]. As industrial growth occurred, the need for power increased. Because prices were high and the utility distribution network was not complete or fully reliable, industries began to generate their own electricity on-site; much of this power cogenerated with the systems producing both process steam and electricity.

As the utility industry matured, by the 1920's cogeneration started to decline, accounting for only 15% of generating capacity by 1950. By 1970 it had declined to about 4%. Spurred by the oil and gas crisis, the problems of the nuclear industry, a declining gas market and rising supplies, the gas industry has mounted an extensive campaign to revive cogeneration in the North American markets.

As a result, the U.S. Congress passed the Public Utility Regulatory Act (PURPA) in 1978 and amended it in 1982. Small power producers and cogenerators which meet certain standards are eligible for special incentives. Electric utilities are required to purchase available electric energy from them at "avoided cost" rates and must provide them with electric service at just and reasonable rates. They are also exempt from most of the regulations to which the electric utilities are subject.

The paradox is that the most fertile market areas for the cogenerator project developers are those areas where the utilities have excess capacity. This occurs because this excess capacity is normally a result of a new large central power plant being brought on-stream with the resultant raise in the electric rate base. With these higher electricity prices, the utility finds itself in difficulty competing with today's low gas and oil prices.

As a result of the current "fuel war", the electric utility is rising to the occasion and are starting to mount a major marketing effort to combat cogeneration. The thrust of this campaign is to offer their industrial customers an array of electro-thermal alternatives, with the heat pump being the focus. Other elements will include heat exchange, thermal storage and incentive rates, particularly in the off-peak hours.

The driver, cogeneration, is in place and the electric utility industry has the marketing prowess to make this campaign succeed. Once they get rolling the heat pump equipment manufacturers will see the opportunities and will add to the effort.

WHAT ELSE HAS CHANGED?

Industry today has, for the most part, made most of the low-cost changes in their energy effort and are now looking at the more capital intensive areas. Spurred by the cogeneration developers and with the utilities seeing that they are given proposals on alternative technologies, industrials are taking a close look at their heat pump opportunities. They are also more interested in and understanding of the concepts of waste heat recovery and are looking at competing technologies over a longer range.

After the last few decades of price fluctuations in all the energy forms, industrials are perceiving that the current low price scenario for oil and gas could soon come to an end. They are enjoying it while it lasts but are not counting on it over the long range. They are also perceiving that the price of electricity could once again resume its downward trend as the major construction programs are completed and the rate base adjusted.

Additionally, the legislative constraints on most utilities are being eased. Consequently, new utility capacity will trend toward more efficient designs such as combined cycle plants. New advances in coal gasification and indirect firing will further improve utility competitiveness. Meanwhile, the oil-producing countries are forecast to maintain attempts to keep oil prices up.

All of these trends today and in the foreseeable future indicate that industrial heat pump prospects are in the ascendancy.

KEY INDUSTRIES

Various assessments have indicated that the key opportunities for industrial (process) heat pumps in North America are:

- Food processing
- Textile
- Pulp and paper
- Chemical & petrochemical
- General Manufacturing (including automotive)
- Wastewater treatment plants

There are numerous applications outside these industry segments, but these seem to offer the greatest total potential and repeatability, although all projects are quite site-specific and this is not likely to change.

Food Processing

The process heat pump opportunities in this segment include both closed and open cycle designs. For closed cycle heat pumps, this industry offers a unique combination of using large quantities of moderately warm (60 C) water for processing and cleanup while having a ready source of waste heat currently thrown away by the refrigeration plant condensers. Reclamation of some of this heat is making increasing economic sense to the plant operators. For open cycle designs, a number of food processes involve evaporation.

The major concentration of potential appears to be in the those areas subsequently listed.

Poultry processing and meat packing

Recovery of waste heat from the refrigeration and ice-making operations can be put to economical use to provide process hot water needed for scalding, washing and cleanup. This heat can be reclaimed either by directly condensing the plant refrigerant in the heat pump condenser or by a separate recovery condenser connected by a water loop to the heat pump. The entire system can be integrated with heat recovery of overflow hot water and thermal storage to offer maximum benefits.

Poultry and meat product rendering is an open cycle opportunity discussed later.

Fluid milk processing

Similar applications to the poultry industry also occur here with the additional use of hot water for milk pasteurizing. Open cycle heat pumps are an attractive alternative to steam use in single or multiple effect evaporators in condensed milk and cheese production.

Rendering

Trends in edible and inedible rendering are to the continuous cooker. A more efficient alternative involves the open cycle recompression of the water vapor driven off from the feedstock and uses this recompressed vapor to provide the heat needed to evaporate this water. This heat pump approach can reduce the energy requirement by up to 70% compared to conventional rendering methods.

This method does not necessitate additional vapor condenser capacity, decouples the process from the weather and allows completely automated operation. Lower cooker pressure and temperature gives potentially higher product quality. The design allows for operation with boiler steam, as well as recompressed steam, for increased availability and control of electric demand during peak periods.

Breweries

Waste heat from the plant refrigeration system can be reclaimed by heat pumping to make process hot water for pasteurizing the beers as well as for other process and cleanup needs.

In the barley malting process, a heat pump can be used in lieu of the conventional approach using hot air from a furnace or from steam coils. Using the exhaust air as a source, a heat pump can provide the needed hot air for the needed drying.

Soft drink bottling

The principal opportunity in a bottling plant is to reclaim the heat rejected by the refrigeration system, using a heat pump to make hot water for the bottle or can warmer, cleanup and for winter space heating, if desired.

Textile Industry

This industry includes weaving mills, knitting mills, textile finishing, floor coverings,and yarn and thread mills. These mills use huge quantities of process water heat for washing and bleaching operations and discharge this heated water to waste. The heat pump opportunity is in combining effective heat exchange with closed cycle heat pumps to reclaim the waste heat and produce the needed process hot water.

Pulp and Paper Industry

Canada has led the way in North America for the application of large industrial heat pumps, both open and closed cycle, in the pulp and paper industry with the United States gradually increasing their acceptance of this technology. The two most prevalent opportunities are vapor recompression evaporators to preconcentrate black liquor, and wastewater heat recovery to preheat wash and boiler feedwater. Both systems offer production and economic benefits and are increasingly being installed here.

Additionally, new concepts are being investigated as the industry becomes more competitive. These include using heat pumps to upgrade heat from stonewood grinders and mechanical pulping to generate dryer steam. For example, drying requires about 4GJ of steam per tonne of newsprint[5] and numerous waste heat sources can be used with heat pumping to economically deliver this energy.

Lumber drying

Heat pump dehumidification dryers are gaining rapid acceptance in North America. The sawmillers and furniture manufacturers are increasingly installing this equipment because of the superior quality of the dried lumber compared to the conventional steam drying kilns; savings in operating cost is not the motivation. Most of these dryers are built in Canada from European designs.

Chemical and Petrochemical Process Industries

Up to 80% of the energy costs can be saved when distillation columns equipped with open cycle heat pumps (vapor recompression) are used in the production of chemical and petrochemical intermediates and in thermal separation processes. Heat pumps have, in the past, been disregarded by most of this industry. This is now changing in North America as the industry focusses on energy improvements. The market is primarily developing in distillation, evaporation, separation and other process areas where latent heat exchanges occur. The recent increased sophistication in the computer heat exchange network analyses and the "pinch technology" approach are highlighting heat exchange and heat pump opportunities for the process engineers. Major producers are now taking a keen interest in this area in their struggle to remain competitive world-wide.

Conventional distillation columns are among the biggest energy consumers in the plant. Many of the possibilities for saving energy by improvement of conventional columns have already been exploited, including the use of coupled columns. Further reduction is achievable only by redesign of column internals and recovering the heat thrown away to ambient by column overheads. Fortunately, these go hand-in-hand.

With the vapor recompression heat pump, this heat is taken from the overhead and transferred to the reboiler by the vapor itself, being compressed to the higher temperature level to give off this heat in the reboiler by condensation. It has the added advantages of decoupling the process from the weather and for being interruptible for electric demand control purposes, with the boiler steam picking up the load without process disruption.

There are many separation processes which have temperature lifts that make heat pumping economically attractive (up to 50 C), even at today's low fossil fuel prices. Many retrofit applications can result in paybacks of 1 to 3 years. Alternatives that include heat pumping between intermediate heater and coolers, to the column feed, and between adjacent columns are being investigated.

Improvements in column packing designs in recent years have also added to the heat pump opportunities as the pressure drop across the separating column is reduced, which in turn reduces the volumetric flow rate and the pressure ratio, making the heat pump even more economical.

In the United States, significant new regulations are possible concerning industrial plant emissions. Consequently, in addition to the energy conservation benefits, electric-driven heat pumps may provide emission reduction that can be used to offset emissions from a planned modification or expansion of the facility.

General Manufacturing Industries

There are numerous processes in these generalized industries that use relatively low temperature (under 120 C) heat, either in the form of hot water of low-pressure steam. Many of these processes start with boiler steam at higher pressures, which is then throttled down to the using level to reduce steam distribution piping costs or as routine past practice.

Many of these steam end-use requirements can be met in whole or in part by reclaiming waste heat being discharged locally, using heat pumps and heat exchange.

Open cycle heat pump applications

The major applications for the open cycle heat pump are:

- -Steam recompression to correct a plant steam imbalance.
- -Concentration of watery solutions to reclaim or reuse valuable constituents, such as plating solutions.
- -Evaporation of plant effluents to reduce disposal problems.
- -Generation of process steam from high-temperature wastewater.

Where the open cycle heat pump is applied to an evaporation process, it offers additional advantages over other forms of evaporation, including:

- -Decouples the process from the weather, since the barometric condenser is only used for standby.
- -Usually increases the capacity of the evaporator in a retrofit project.
- -Reduces the required steam flow, which may be a significant factor in a plant with little or no boiler reserve capacity.
- -Reduces or eliminates the need for a significant flow of cooling water.

Closed cycle heat pump applications

Waste heat sources for these heat pumps are numerous and include cooling tower or mill water used to pick up heat from air compressors, billet heaters, injection molders, electric welders, refrigeration and air conditioning, dynamometers, extruders and the like. Other sources include refrigeration equipment rejecting heat directly to evaporative condensers, warm water effluents, vent steam, and hot exhaust vapors being directly vented or sent to air-cooled or barometric condensers.

The principal applications for the closed cycle heat pump in these industries are:

- Evaporation or concentration of heat-sensitive or highly corrosive solutions in the 20 C to 50 C range, without the use of steam or cooling water.
- Production of hot rinse water or phosphatizing solutions for parts preparation and washing.
- Solution heating and cooling in electroplating.
- Vapor degreasers.
- Process or boiler feedwater heating
- Hot water for space heating, showers, cleanup, etc.
- Environmental control in paint spray booths.
- Plastic products injection molding.

With energy costs being an important component of product cost, these plants are focussing their attention on these thermal alternatives with heat pumps taking an ever-larger share of the selection. This will increase even more as the electric utilities mount marketing campaigns to make it more economical for their customers to use the heat pump alternative to cogeneration.

Waste Water Treatment Plants

In this category, a significant number of waste treatment plants in the colder climate areas of North America have installed large closed cycle water-to-water heat pumps to provide both the process and space heat needed, using the plant effluent as the heat source. This saved the gas generated by the digesters for year-round use in engine-driven pumps and generators[3]. The driving influence that caused this relatively high level of acceptance was government incentives.

In the United States, the Clean Water Act and subsequent amendments provided federal funds for those treatment plant projects that used innovative energy conservation designs. With these incentives, municipalities demanded that their design engineers look at the available alternatives and heat pumps emerged as the most economical in the more northerly climates where space heating of the plant was a significant part of the heating load.

With the huge heat source at relatively favorable temperature for heat pumping (>10 C), some of these plants may be expanded to provide other services in the arena of district heating.

District Heating

Political influences, historically low energy prices, and other factors have inhibited the growth of municipal district heating systems in North America, in contrast to their popularity in Europe and Scandinavia in particular. Commercial district heating was first applied here in 1877. Many large northern cities have, or had, small systems, usually with steam distribution. While a number of such systems were put in service decades ago, their number have been declining and only one (St. Paul, Minnesota) has been constructed in recent years. This does not include large, single-owner building complexes, such as universities, military installations, industrial complexes, and medical centers, which are exceptions.

District heating systems based on or using heat pumps have had rapid growth in recent years in Scandinavia[4]. In our opinion, this concept will have limited acceptance in North America, largely due to the political barriers and interveners. We predict that some large waste treatment plants located close enough to urban areas may be expanded to provide heat pump based hot water distribution district heating. In a few cases, municipal refuse fueled cogeneration plants may be utilized in the same manner but, in relation to the apparent potential, the growth of this market in North America will be minuscule.

Summary

Industry in North America is examining all of their plant operations with a heavy emphasis on cost reduction and competitiveness. The high cost plants are being closed and integrated into the low cost plants with major disruptions in the areas affected. Utility companies are now recognizing they must do all they can to make the plants in their service territory the low cost producer and their contribution is in the energy area. They don't want cogeneration with its built-in bias against future electrification and see heat pumps and other such alternatives as heat exchange and thermal storage as their major weapon in the "fuel war".

With electric utility marketing coming out of hiding and with the driver of cogeneration, the future prospects for industrial heat pumps in North America are bright indeed.

REFERENCES

1. J. S. Gilbert, "Industrial Heat Pumps," Plant Engineering, 74-77, November 23, 1983.

2. G. Lovin, Cogeneration World, March/April 1986.

3. R. C. Niess, "Effluents and Energy Economics," Water/ Engineering & Management, 52-58, August 1981.

4. ASEA-STAL, "Reference List-large heat pump plants," December 1985.

5. Potential for Electric Heat Pumps in the Canadian Newsprint Industry, Canadian Electric Association, July 1983.

Section V

Advances in Thermally Activated Heat Pumps

Chapter 24

Developments in Gas-Fired Absorption Heat Pumps in North America

Robert C. DeVault

ABSTRACT: Gas fired absorption heat pumps for space conditioning (heating and air conditioning) are being developed by several U.S. organizations. Single-effect cycle absorption heat pumps for residential applications were developed in the 1970's, but proved to be uneconomic and were not manufactured. Advanced cycle absorption heat pumps with substantially higher efficiencies are now being developed, and have the potential to be economically attractive to consumers.

BACKGROUND

Residential and commercial buildings account for 37% of the total energy used in the United States primarily for heating, cooling, and hot water. Natural gas and oil are major fuel sources for building heating, and almost all air conditioning is done with electricity. Figure 1 shows energy consumed by fuel type in the U.S. for heating and cooling.

Air conditioning is a major energy use in residential and commercial buildings. Currently in the U.S., about 85% of all new houses are air-conditioned, and about 30% of new houses use electric heat pumps for heating and air-conditioning [1].

Figure 2 shows the relative operating fuel cost to the consumer for electric heat pumps, gas furnaces, and gas fired heat pumps for a range of electricity to gas price ratios. As can be

---

*Research sponsored by the Office of Buildings and Community Systems, U.S. Department of Energy, under contract DE-AC05-84OR21400 with Martin Marietta Energy Systems, Inc.

HEAT PUMPS: Prospects in Heat Pump Technology and Marketing (Proceedings of the 1987 IEA Heat Pump Conference), Edited by Kay H. Zimmerman

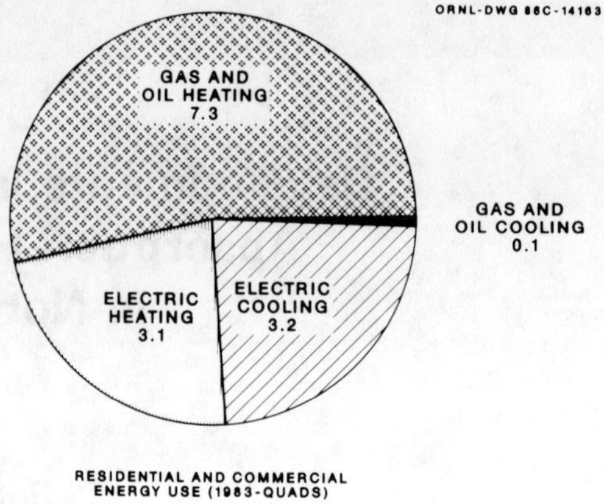

Figure 1. Natural gas and oil are major fuel sources for building heating.

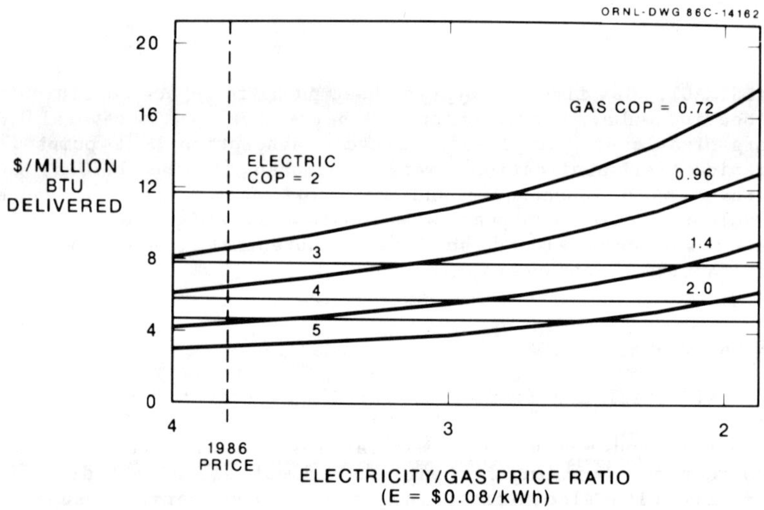

Figure 2. Advanced absorption heat pumps will cost less to operate than electric heat pumps.

seen, with modestly efficient gas fired heat pumps (coefficient of performance (COP) > 1.4), heating fuel costs should be lower than electric heat pumps for any credible electric to gas price ratio.

The U.S. Department of Energy (DOE) and the Gas Research Institute (GRI) have been supporting research and development for absorption heat pump systems since the late 1970's. The goals for development of gas-fired absorption heat pumps are to produce equipment that will save primary energy compared to existing technologies, be economical to the customer to purchase and operate, and also provide for the use of natural gas for air-conditioning as an alternative to electricity.

PROJECT SUMMARIES

Initial projects, first funded in the 1970's and now completed, were for the development of prototype single-effect absorption heat pumps for residential applications.

Arkla Industries, a manufacturer of ammonia-water air conditioners, developed a single-effect ammonia-water heat pump for heating only with steady state COPs of 1.25 at 8.3°C (47°F) and 1.12 at -8.3°C (17°F) [2]. These performance levels are fuel based efficiencies including flue losses using the higher heating value of natural gas.

Allied Corporation, under sponsorship of DOE, the American Gas Association (AGA), and GRI developed a single-effect heat pump for both heating and cooling for residential applications, using a fluorocarbon refrigerant and an organic absorbent. Prototype heat pumps were developed to achieve a heating COP of 1.2 at 8.3°C (47°F) and a cooling COP of 0.5 at 35°C (95°F) [3]. The prototype development used R-123a (1,2-dichloro 1,2,2-trifluoroethane) for the refrigerant and ETFE (ethyl-tetrahydro furfuryl ether) for the absorbent. Further refinement of the Allied prototype, including a condensing flue, showed potential for about a 1.5 heating COP and a 0.6 cooling COP.

Both the Arkla and Allied heat pump prototypes were successfully operated and achieved their performance goals. However, the single-effect performance levels were not sufficient to offer reasonable economic incentive for customer acceptance, so neither machine was further developed or manufactured.

Columbia Gas Service Corporation started the development of a residential size double-effect absorption heat pump using a proprietary absorption fluid pair in the mid-1970's. Their performance target was a heating COP of 1.55 and a cooling COP of 0.80. In 1980, GRI joined Columbia Gas in funding development of this double-effect concept. In 1984, Columbia Gas reached their original performance goals with laboratory breadboard prototype equipment [4]. Recently, in October 1986, they have operated a packaged prototype machine and demonstrated gas-fired COPs of 1.7 in heating and 0.85 in cooling, exceeding the original 1980 program goal [5].

Based on the achieved performance results of the Arkla and Allied single-effect cycle prototypes, and the need to provide economic payback to customers of less than three years against conventional heating and air conditioning systems, the DOE absorption heat pump program goals were re-evaluated in late 1981. The decision was made to evaluate advanced absorption cycles capable of producing higher efficiency heat pumps than conventional single-effect cycles while still using existing absorption fluid combinations. Higher efficiency levels were needed, especially for air-conditioning operation, in order to save enough money in fuel costs to offset the assumed high initial cost of absorption machines.

In 1982, the DOE absorption program, managed by Oak Ridge National Laboratory's (ORNL) Building Equipment Research Program, issued a Request for Proposal (RFP) for the development of advanced absorption cycles with performance goals of COPs of 1.6 in heating and 0.7 cooling, as compared with current state-of-the-art single-effect values of 1.2 and 0.5. The RFP was distributed to more than 50 potential bidders within the U.S. Three contractors were selected for the work under the advanced cycles program, and work was started in late 1982 and early 1983. The three contractors are Carrier Corporation, the Trane Company, and Phillips Engineering Company.

The work in the advanced absorption cycles program is divided into three phases. Phase I is the analytical evaluation of advanced cycles, with a preferred advanced cycle being selected for further development. Phase II is the development and testing of a laboratory breadboard prototype of the selected advanced cycle. Phase III is the development and testing of a packaged proof-of-concept prototype after successful completion of Phase II.

The goal of the Phase I selection of a preferred advanced cycle was not necessarily to pick the highest efficiency cycle, but rather to select cycles with a combination of features, such as high efficiency, low manufacturing cost, high inherent reliability, etc., that appeared to offer the most likely path to a manufactured product by the 1990's.

In Phase I analysis work, Phillips Engineering examined a variety of cycles including: (1) double-effect cycles, (2) resorber augmented cycle, (3) generator-absorber heat exchange (GAX) cycle, (4) double-effect regenerative cycle, (5) variable-effect cycle, and (6) two-stage generator-absorber heat exchange cycle [6]. Phillips evaluated each cycle on the basis of (1) potential efficiency, (2) heat exchanger surface requirements, and (3) ability to operate over a broad range of temperatures. The GAX cycle was selected as having the best combination of high potential efficiency and low manufacturing cost for residential applications. Both ammonia-water and ammonia-lithium bromide-water were selected as candidate absorption fluid systems for this cycle.

The Trane Company and Carrier Corporation each evaluated advanced cycles for commercial applications. The Trane Company conducted a Phase I analysis similar to Phillips Engineering and also selected a GAX cycle, but with additional features and internal heat exchange for higher efficiency [7]. Economic analysis predicts customer payback of less than 3 years in commercial buildings while simultaneously giving a better utility load profile for both gas and electric utilities.

Carrier's study of advanced cycles resulted in selection of a dual-loop cycle based on the use of methylamine-lithium bromide-water in the lower (temperature) loop and lithium bromide-water in the upper loop [8]. The DOE has been granted six patents based on the Carrier Phase I work. Figure 3 shows a schematic of the Carrier dual loop heat pump.

Each of these advanced cycles selected by Trane, Carrier, and Phillips offers potential for a rated heating COP in the range

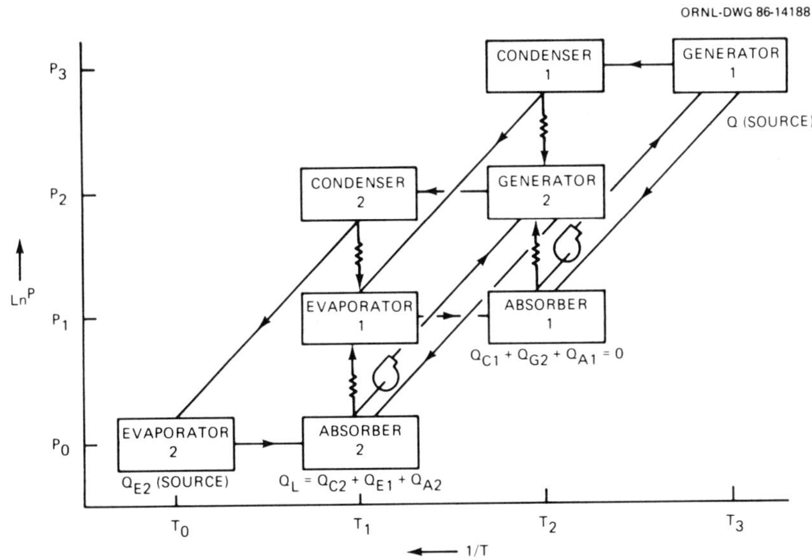

Figure 3. The Carrier dual loop cycle uses different fluids in each loop for best performance.

of 1.8 to 2.0 and cooling COP in the range of 0.8 to 1.2, significantly exceeding the original program goals set in 1982.

Phase II of the projects, the laboratory breadboard demonstration of the cycles identified as most promising, is underway. During 1985, Phillips Engineering successfully operated a proof-of-principle breadboard prototype of a residential-sized unit with performance exceeding the original advanced cycle RFP goals. The GAX cycle breadboard achieved COPs of over 1.8 in heating at 8.3°C (47°F) and 0.8 in cooling at 35°C (95°F). An independent cost analysis of the Phillips Engineering prototype concept has predicted a customer-installed cost equal to existing mid-line gas furnace/electric air-conditioner combinations or to electric heat pumps, indicating the potential for advanced absorption cycle residential heat pumps to become a major consumer product in the future.

Trane expects to demonstrate proof-of-principle with a laboratory breadboard prototype model of a commercial-sized unit in 1987.

Carrier is currently scheduled to build and test two separate laboratory breadboard prototypes. The first prototype, a single-stage commercial-sized absorption heat pump capable of below-freezing operation is scheduled for operation in 1987. The second prototype using the dual-loop concept is scheduled for testing in 1988.

In mid-1983, GRI funded a dual-cycle advanced absorption residential size heat pump concept proposed by Battelle-Columbus. The objective is to develop a residential-size gas heat pump with 1 1/2 to 3 tons cooling capacity and COPs of 1.80 heating and 0.94 cooling [4]. During 1985, Battelle operated an integrated breadboard of the dual-cycle absorption heat pump concept and achieved the target COPs [9]. Phase II work to design, build, and test a series of packaged prototypes for field testing is currently underway, with a manufacturing participant. Large scale field evaluation of the Battelle dual-cycle advanced absorption heat pump is scheduled to start June 1, 1987, with subsequent commercialization projected for early 1989.

The American Society of Heating, Refrigeration, and Air Conditioning Engineers (ASHRAE) is sponsoring absorption supporting research on heat and mass transfer in absorbers to address absorber performance at previously unexplored operating conditions applicable to advance cycle systems [10].

DOE has also sponsored the development of a modular user-oriented absorption cycle computer model for use in evaluating advanced absorption cycles for heat pumps and heat transformers. The basic model has been completed and is available for use on main-frame computers and for IBM compatible personal computers [11].

The Institute of Gas Technology (IGT), under DOE sponsorship, has undertaken the cataloging of absorption fluids properties data and references for all available U.S. data [12] and also for available foreign data [13].

In support of the advanced heat pump work, ORNL has developed a unique instrument using fiber optics to measure precisely the refrigerant concentration in the refrigerant-absorbent mixture in an operating absorption device [10]. This fiber-optic refrigeration cycle monitor measures the refractive index of fluids. Because most refrigerant-absorption fluids are binary mixtures, the refractive index of the fluid is directly related to the concentration of its individual components. In tests with $LiBr/H_2O$, the instrument has demonstrated the ability to measure local refrigerant concentration to an accuracy of 0.1% inside an operating absorber. Combining measurements of local concentration and temperature allows on-line monitoring and control of the refrigeration system, as well as a detailed study of heat and mass transfer processes occurring within absorption systems. This new measurement capability should help in the design of new absorption systems for increased operating efficiency.

Figure 4 shows primary energy consumption in heating and cooling for existing technologies, the prior single-effect cycles and the current advanced cycle concepts. Based on recent progress in the DOE and GRI programs, advanced absorption cycle heat pumps show the potential for significant energy savings with acceptable consumer economics for both residential and commercial applications.

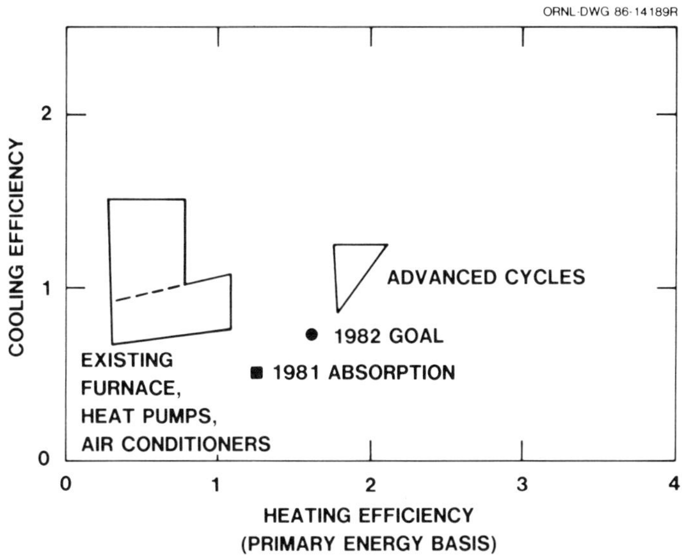

Figure 4. Advanced cycles have performance potential beyond current technology.

Note: Assumes electricity delivered to customer at 33% efficiency including generation and distribution losses. Heating performance based on 47°F standard steady state rating for all technologies, cooling performance based on 95°F ambient steady state rating for all technologies.

Notes:

All performance figures are based on high heating value of natural gas.

Studies of corrosion ratios aid absorption fluids thermal stability in advanced absorption cycles as included in each project listed.

REFERENCES

1. "1986 Statistical Panorama," *The Air Conditioning, Heating and Refrigeration News*, April 7, 1986.

2. R. H. Merrick and D. Kuhlenschmidt, *Engineering Development and Demonstration of an Ammonia-Water Absorption Heat Pump*," prepared for ORNL by Arkla Industries, Inc., ORNL/Sub/80-61603/1 (in press).

3. K. P. Murphy, et al., *Development of a Residential Gas-Fired Absorption Heat Pump - Physical and Thermodynamic Properties of R123a/ETFE, System Development and Testing and Economic Analysis*, prepared for ORNL by Allied Corporation, ORNL/Sub/79-24610/4, August 1985.

4. A. R. Maret, "The GRI Gas Heat Pump Program - An Update," *Proceedings of the DOE/ORNL Heat Pump Conference*, December 11-13, 1984, Washington, DC.

5. Private communication with Columbia Gas, October 1986.

6. B. A. Phillips, "Advanced Residential Absorption Heat Pumps Cycle Analyses - Construction of a Test Model," *Proceedings of the DOE/ORNL Heat Pump Conference*, December 11-13, 1984, Washington, DC.

7. F. C. Hayes, "Evaluation of Advanced Design Concepts for Absorption Heat Pumps," *Proceedings of the DOE/ORNL Heat Pump Conference*, December 11-13, 1984, Washington, DC.

8. R. C. Reimann, "Advanced Absorption Heat Pump Cycles," *Proceedings of the DOE/ORNL Heat Pump Conference*, December 11-13, 1984, Washington, DC.

9. D. A. Ball and W. T. Hanna, *Development of the Battelle Absorption Heat Pump for Residential Application - Phase I Final Report*, October 1985.

10. H. Perez-Blanco and R. Radermacher, "Absorption: an Update," *ASHRAE Journal*, November 1986, pp. 25-26.

11. G. Grossman and E. Michelson, "A Modular Computer Simulation of Absorption Systems," *ASHRAE Trans.*, V. 91, Pt. 2, 1985.

12. R. A. Macriss and T. S. Zawacki, *Absorption Fluids Data Survey-Final Report on U.S. Data*, ORNL/Sub/84-47989/1, 1986.

13. R. A. Macriss, J. M. Gutroj, and T. S. Zawacki, *Absorption Fluids Data Survey-Final Report on Foreign Data*, ORNL/Sub/84-47989/2 (in press).

14. D. T. Bostick et al., "Real-Time in-situ Refractometer for Concentration Measurements in Absorption Machines," *ASHRAE Transactions*, PO-86-04, no. 3, 1986.

Chapter 25

Developments in Utilization Technology for Absorption Heat Pumps in Japan

Shigekichi Kurosawa

ABSTRACT: The absorption heat pump is one of promising means to promote efficient use of heat energy and thus achieve energy saving. It is capable of recovering low temperature waste heat, which would otherwise be discharged to the air or water, in efficient way and converting it to high temperature heat energy which can be utilized for various purposes. In this respect the absorption heat pump is considered as an optimal heat recovery technology. Commercial development of the absorption heat pump has been carried out in the past ten years.

The Agency of Industrial Science and Technology under the Ministry of International Trade and Industry started a research and development project on waste heat utilization technology and system in 1976, and selected the absorption heat pump as one of major subjects. This project was incorporated into Moonlight Project in 1978. A small prototype to heat hot spring water was developed in the same year and a large prototype in 1980.

Today, the absorption heat pump is used for air-conditioning and heating, producing hot water and preheating boiler feed water in various types of facility including office buildings, computer centers, public bathhouses, district cooling and heating systems, and industrial plants.

This paper discusses present state of absorption heat pumps and their applications for commercial purpose as well as future outlook for utilization technology for them.

HEAT PUMPS: Prospects in Heat Pump Technology and Marketing (Proceedings of the 1987 IEA Heat Pump Conference), Edited by Kay H. Zimmerman Printed in U.S.A.

1. COMMERCIAL PRODUCTION OF ABSORPTION HEAT PUMPS

At present, six manufacturers of gas-fired absorption water chiller-heater are manufacturing absorption heat pumps in Japan. The absorption heat pumps currently manufactured are divided into two types: Type I has a capacity of 80 - 10000 kW, and Type II of 170 - 5000 kW. Type II absorption heat pump is used as a heat transformer to produce high temperature water or steam at industrial plants. 10 units have been manufactured up to this date.

On the other hand, Type I absorption heat pumps have been commercially developed in 1979 and approximately 140 units have been shipped up to 1984. (Table 1)

This paper discusses Type I absorption heat pump.

Table 1 Shipment of Absorption Heat Pump

Year	Type I absorption heat pump		
	Total	Domestic market	Export
1980	6	6	0
1981	40	40	0
1982	31	31	0
1983	30	28	2
1984	22	19	3

2. TECHNOLOGY SIDE OF WASTE HEAT RECOVERY

A Heat Driven Absorption Type Heat Pump owes its name to the absorption cycle that requires no driven force from any motive power source but is driven by heat energy. The heat pump, therefore, consumes little power energy, has a merit of low level vibration and noise, and permits easy operation and maintenance.

This heat pump uses lithium bromide as the working fluid and water as the refrigerant. The latent heat of vaporization of refrigerant has a great influence on the coefficient of performance (COP) of a refrigerator. So, the large latent heat of water, the chemical stability of lithium bromide, and the operation under the vacuum condition give the heat pump excellent features such as energy saving and safety.

This heat pump consists of an evaporator, absorber, condenser and generator.

The absorption cycle is operated in either cooling mode or heat pump operation mode.

In the cooling mode, the generator is heated by high temperature exhaust gas or hot waste water, that is, waste heat is utilized directly. As for application in buildings, there is an alternative: to recover the exhaust gas from gas engines or gas turbines, or to utilize the waste heat contained in it by making hot water in the heat exchanger. Solar heated water could be used as the heat source if the site conditions allow it.

In the heat pump operation mode, the waste heat is recovered as the heat source of the evaporator, and the role to heat the generator is taken by gas combustion as the fuel. Hot waste drain from a hotel, hospital, or bathhouse and the heat obtained from the cooling water are used for space heating, hot water supply or reheating during intermediate season between winter and summer in office building. With a double bandle type condenser, the hot water from the condenser can be used for reheating in intermediate seasons or for heating water supply during the summer season.

The absorption cycle is assumed to be the most prosperous of all heat recovery methods because it finds wide application in combination with a co-generation system or as a chiller-heater as stated above.

3. TYPES AND FEATURES OF ABSORPTION HEAT PUMP

3.1 Absorption Heat Pump Cycle

Fig. 1 shows an absorption heat pump cycle.

The refrigerant drips in the evaporator and evaporates obtaining heat from the heat source water. That is, the heat recovery from waste heat takes place in the evaporator.

The refrigerant vapor produced in the evaporator is absorbed into the strong solution that is dripping in the absorber. The heat liberated in the absorber heats the water flowing in the coil (first heating process).

The weak solution diluted in the absorber is heated in the generator by city gas to liberate the refrigerant vapor. The solution is concentrated in this process.

The refrigerant vapor generated in the generator then enters into the condenser where it is deprived of its heat by the water feeded from the absorber and condenses into liquid. The heat of condensation of the refrigerant vapor heats further the water heated in the absorber (second heating process).

The refrigerant condensate returns to the evaporator and deprive the heat source water.

Thus, this cycle produces hot water continuously.

3.2 Types of Absorption Heat Pump

The absorption heat pumps used for commercial purposes can be divided into three types which means heat reclaim type, double bundle type and economizer type (Fig. 2).

3.2.1 Heat Reclaim Type Heat Pump

The heat reclaim type heat pump is further divided into single effect type and double effect type. Both types supply heat from the absorber and condenser. The heat reclaim type is employed for most of commercial absorption heat pumps.

In this type of heat pump, if an amount of heat recovered in the evaporator (Q_E) is sufficient, input to the generator is controlled according to hot water outlet temperatures. Upper limit of hot water temperature is 95°C for the single effect cycle and 50°C for the double effect cycle. The latter is set low since pressure in the high temperature generator is maintained below atmospheric pressure.

3.2.2 Double Bundle Type Heat Pump

The double bundle type supplies heat from the condenser (Q_C), and consists of a side arm condenser and a main condenser (for hot water heat exchanger). When hot water load is small, surplus heat will be discharged to a cooling tower through a side arm condenser to maintain heat balance. In terms of relation to heating load (Q_H), the double bundle type is conventionally used when $Q_E > Q_H$ and the heat reclaim type when $Q_E \leqq Q_H$. In practice, however, the double bundle type is mainly used as a secondary system to supply chilled water for air-conditioning. Also, because of relatively low temperature of hot water outlet, the double bundle type is introduced as a simplified system to preheat the feed water.

The double effect type can extract heat which is approximately 50% of heating value of heat source water (Q_E).

3.2.3 Economizer Type Heat Pump

In the economizer type heat pump, hot water is produced in the single effect cycle on the hot water heat exchanger, while chilled water is produced in the double effect cycle on the evaporator side.

Generally, refrigerant vapor produced in the high temperature generator is used as a heat source, and hot water is produced in the hot water exchanger which is connected to the high temperature generator. Then the refrigerant condensates are returned to the evaporator to produce chilled water. This process forms the single effect absorption heat pump cycle. Upper limit of hot water outlet temperature is approximately 70°C.

3.3 Feature

The gas-fired absorption heat pump shows COP of 1.3 or greater, resulting in energy saving of 40% or greater compared to the boilers.

Also, this type has advantages in low noise and vibration, ease in operation, no need for qualification in operation (not subject to any regulation because of vacuum operation), excellent reliability and safety, and compact design.

4. BASIC DESIGN CONCEPT AND PROCEDURE

The high efficiency of heat pump is important in attaining high economy. At the same time, economic efficiency of the system as a whole is governed by its operation rate. In the absorption heat pump, difference in temperature between the heat source water side and hot water side affect COP less than other types of heat pump do. Thus, it is important for the user to select optimal temperature condition to maximize the operation rate. One way of achieving the high operation rate is to use the evaporator side for air-conditioning.

4.1 Range of Temperature Rise

Generally, temperature range usable as the heat source water is between 10 - 50°C. Temperatures of heat source water (Q_E), input (Q_G), and hot water ($Q_A + Q_C$) may be set on the basis of the following relationship:

$$\Delta T_1 = C \times \Delta T_2$$

Whereas; ΔT_1 : Difference temperature in temperature between the input and hot water outlet. (°C)

ΔT_2 : Difference in temperature between the hot water outlet and the heat source water. (°C)

C : Proportional constant. Value of C is generally 2.

4.2 Heat source Water

Heat source water should be selected in consideration of the following factors:

- The heat source water should be available in sufficient quantities. If not, supplemental heating will be required, eventually the economic efficiency will decrease.
- The heat source should be located near the place of use.
- The heat source water should not be corrosive and should not contain contaminant.
- Temperature variation of heat source water should be small. There should be no risk of freezing.

4.3 Water Quality and Quantity

Since steel pipes are principally used for heat transmission, water circulated in the pipes should satisfy the water quality standard for cooling water established by the Japan Refrigeration and Airconditioning Industry Association so as to prevent corrosion and scaling from developing in the pipes. However, considering the fact that waste water is often used as the heat source water, cupronickel pipes, stainless steel pipes or titan pipes may be used where required, or the heat exchanger may be provided to supply clean water to the heat pump. Furthermore, flactuation in flow rate of heat source water and hot water cannot be avoided sometimes, although constant flow rate is preferrable, so that measures should be taken to maintain at least 50% of flow rate all the time.

4.4 Heat of Heat Source Water and Heating Load

Except where a large amount of heat in the heat source water is available, heating load and amount of heat in the heat source water do not balance in ordinary cases. In such case, the following principles may be applied:

- If a sufficient amount of heat in the heat source water is available all the time, heat balance needs not to be considered, e.g., recovery of heat from river water or air.

- If the sufficient amount of heat is not available compared to load requirement, a supplemental heat source should be provided.

- If heat surplus and deficit occur during the day, a heat storage tank should be provided to increase the rate of utilization.

5. APPLICATIONS OF ABSORPTION HEAT PUMP

Application of absorption heat pump system in office buildings is summarized in Table 2.

5.1 Application to Hot Water Supply

Fig. 3 shows the small-sized type for the energy-saving hot water supplying system installed in Tokyo Branch Office of Tokyo Gas Company to reclaim heat from waste hot water of the bathroom. The gas-fired absorption heat pump has hot water output of 81 kW (single effect cycle, COP: 1.3, high calorific value). The waste hot water exchanger exchanges heat between the waste hot water and heat source water. The feed water is heated through the hot water supply heat exchanger. The boiler operation mode starts to operate automatically when waste hot water is in short supply. As the hot water output of the absorption heat pump drops to 60% of the rated capacity during the boiler operation, the flow rate is regulated by a control valve in the water feed piping system so as to maintain the hot water temperature above 50°C.

The absorption heat pump used in this system functions as a boiler, designed to supply hot water efficiently, if an amount of heat in the heat source water is not sufficient. This dual function has an advantage over the operating system in which heat deficiency in the heat source water is supplemented by a back-up gas boiler while stopping heat pump operation, which has a low heat recovery rate and requires relatively complex system configuration. (Table 3)

Table 3 Specification for Gas-fired Absorption Heat Pump

Operation mode		Heat pump operation	Boiler operation
Heat source water temperature	°C	27 → 13	-
Hot water temperature	°C	35 → 55	43.6 → 55
Heat source water input	kW	36	-
Hot water output	kW	81	47
Gas consumption	kW	62	
Outer dimension (L x W x H)	mm	930 x 1140 x 2120	

The 160 kW type is also available.

Table 2 Application of Gas-fired Absorption Heat Pump

System	Outline	System flow	COP (excluding fuel efficiency)	Example
Heat pump type for hot water supply	The system for producing hot water of a fixed temperature by reclaiming waste hot water.	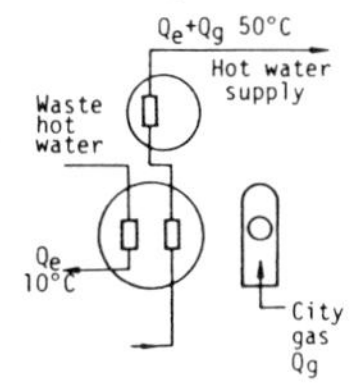	COP $= \frac{Q_a + Q_c}{Q_g}$ = 1.75 (City gas single effect cycle)	°Bathhouse °Swimming pool
Double bundle condenser type	The system for producing hot water by reclaiming the heat contained in used cooling water of air-conditioning. This system uses a condenser of double bundle type.	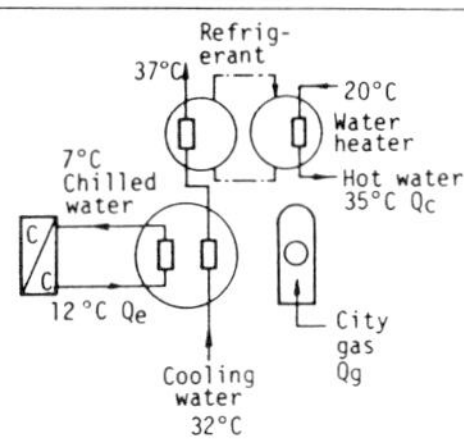	COP $= \frac{Q_e + Q_c}{Q_g}$ = 1.2 ∿ 2.1 (Double effect cycle)	°Office building °Community center
Heat reclaim type	The system for space heating by reclaiming the waste heat from the absorber and condenser of an air-conditioner. This system is used where there is a cooling load even in the winter, e.g., air-conditioning of a computer room. This system is operated in the single effect cycle when the required hot water outlet temperature is high, and in the double effect cycle when the air-conditioner is running in the summer.	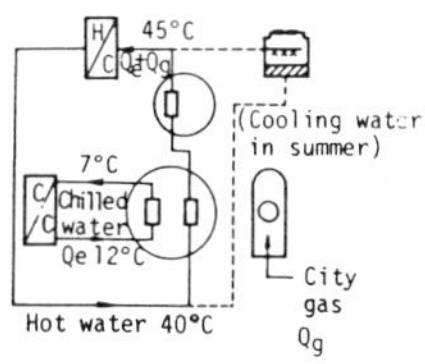	All waste hot water is utilized. COP $= \frac{Q_e + Q_a + Q_c}{Q_g}$ = 2.5 (Single effect cycle) = 3.4 (Double effect cycle)	°Office building °College laboratory
Heat reclaim type combined system	This system consists of two gas-fired absorption water chiller-heaters combined. One water heater-chiller is run in the heat pump mode when there are simultaneously cooling and heating loads. The heat reclaimed from the used cooling water during air-conditioning operation of one water heater-chiller is utilized for space heating by another machine running in the double effect cycle heat pump operation mode. The combination allows the evaporator temperature to be higher than the single machine and the heat pump to provide hot water of sufficient temperature even in the double effect cycle. In the summer, the two water chiller-heaters are separated and run in the ordinal cooling mode independently.	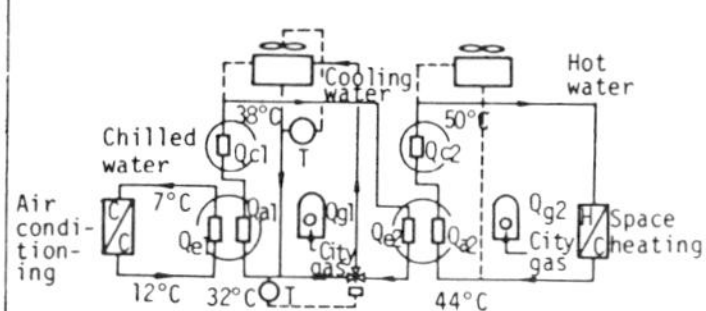	All waste hot water is utilized. COP $= \frac{Q_{e1} + Q_{a2} + Q_{c2}}{Q_{g1} + Q_{g2}}$ = 1.9 (Double effect cycle)	°Computer center
Economizer type	This system uses a concurrent hot and chilled water supplying gas-fired water chiller-heater and is used where there exist both cooling and heating loads. The refrigerant condensate obtained in the hot water exchanger is conducted into the evaporator. This adds the chilled water output equivalent to the hot water load.	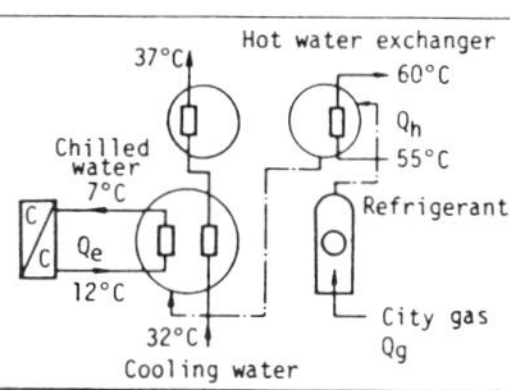	$Q_e = Q_h$ COP $= \frac{Q_e + Q_h}{Q_g - Q_h}$ = 2.4 (Double effect cycle)	° Office building

5.2 Application to Sauna Bath

In this system, an oil-fired boiler which is conventionally used to supply steam, hot water, and space heating is eliminated, and a part of hot water requirement is satisfied by the gas-fired absorption heat pump. Waste hot water at 35°C is circulated through the absorption heat pump to recover the heat.

The system control method is shown in Fig. 4. The hot water outlet temperature is set at 75°C, which is maintained by proportional control of gas flow rate control valves.

Design criteria for the heat pump is: Heat source water inlet temperature at 33°C; and temperature of hot water inlet at 10°C during the winter and 30°C during the summer. Difference in temperature between heat source water inlet and outlet is generally 5 - 15°C in consideration of evaporator design. If the temperature difference is set at 15°C, the temperature of heat source water outlet will be lowered to 18°C. This means, 700 kW is recovered from the heat source water, and the cycle efficiency is 1.67. Thus, hot water output is 2.5 times of heating value obtained from the heat source water, or 1750 kW. With this heat reclaim system employed, system efficiency of the absorption heat pump is 130%, considerable increase over 77% for the previous system using an oil-fired boiler. As a result, energy consumption is expected to decrease by 45%.

5.3 Application to Airconditioning of Computer Center (1)

In the computer center building of Tokyo Gas, the steam heated absorption heat pump recovers heat from cooling water in the steam heated double effect absorption refrigerator which handles cooling load generated by computers.

The recovered heat is used for heating the fan coil unit system in an attempt to save energy consumption in the winter. Furthermore, surplus heat generated from load variation in the fan coil unit system is used to preheat water for hot water supply.

The system flow sheet is shown in Fig. 5.

COP during the daytime was approximately 1.7, and daily mean COP including the nighttime was 1.63.

5.4 Application to Airconditioning of Computer Center (2)

In the computer center of an electric appliance manufacturer, heat generated in the computer room is recovered by the absorption heat pump (Table 4) for heating during the winter.

The computer room is air-conditioned by a water-cooled package type airconditioning system throughout the year. Its cooling water at 32°C is circulated to the absorption heat pump to produce hot water at 60°C which is supplied to the general office heating system.

During the summer, the heat pump is shifted to a double effect absorption water chiller-heater to carry out cooling operation.

Cooling load for the entire building is 3500 kW, including 1900 kW for the computer room, and heating load is 1160 kW. (Fig. 6)

Table 4 Specification for Absorption Heat Pump

Operation mode		Heat pump mode	Cooling mode
Heat source water input	kW	473	-
Heating output	kW	1300	-
Cooling output	kW	-	670
Temperatures of heat source water and chilled water	°C	30 → 25	14.5 → 7
Temperatures of hot water and cooling water	°C	52.5 → 60	32 → 38
Gas input	kW	1050	720
Outer dimensions (L x W x H)	mm	4640 x 2960 x 2850	

5.5 Recovery of Waste Heat from Laboratory

A university's laboratory is developing a laser system to be used for research on solid state physics. This project requires very clean air and constant temperature/humidity. Waste heat is recovered from the laboratory through airconditioning and is used for heating during the winter. The heat reclaim system uses a gas-fired absorption water chiller-heater (dedicated to cooling operation) and a gas-fired absorption heat pump.

As shown in Fig. 7, hot water is produced from a cooling water circuit in the absorption heat pump and is circulated through an auxiliary heat exchanger to stabilize the hot water outlet temperature.

5.6 Application of Double Bundle Condenser Type

The gas-fired absorption heat pump is used in Cultural Center for reheating during intermediate seasons. Its specifications are shown in Table 5.

Hot water for reheating during cooling operation, under the same temperature conditions as cooling water, is produced by the double bundle condenser and is supplied to hot water coils in the airhandling unit. The hot water control system is designed to maintain a constant outlet temperature of heat exchanger for reheating.

As shown in Fig. 8, refrigerant drained in the condenser is regulated in accordance with detected outlet temperature of hot water for reheating. This way, heat transmission area on cooling water coils is varied to control heat transmission in the heat exchanger for reheating. This means, by giving a priority to controlling of the heat exchanger for reheating, heat discharged to the cooling water circuit can be controlled in accordance with reheating load.

Table 5 Specification

Operation mode		Cooling	Reheating	Heating
Capacity	kW	1055	523	1047
Flow rate	L/min	3020	1150	3000
Temperature	°C	12 → 7	32 → 38.5	55 → 60
Gas consumption	kW	1343 (Summer)		1407 (Winter)

5.7 Application of Triple Bundle Condenser Type

A branch office of Tokyo Gas has water loads of a cooking class and employee's bath. The waste heat from the condenser of the gas-fired absorption water chiller-heater is recovered as surplus heat.

In general, refrigerant vapor generated in the high temperature generator is used by the low temperature generator as a heat source in the second stage of concentration process. In this case, the feed water from the cistern is circulated through the condenser to the refrigerant drain waste heat separator for increasing the temperature of feed water.

The cycle flow is shown in Fig. 9.

5.8 Application to District Cooling/Heating System (Cooling-water Waste-heat Recovery Type Heat-pump System in Shinjuku District Cooling/Heating Center)

In the Shinjuku District Cooling/Heating Center, a steam absorption type heat pump (1400 kW) is installed in order to recover some of the waste heat discharged from the steam-turbine-driven (condensate type) centrifugal refrigerator to the cooling tower. (Fig. 10) The boiler feed-water is preheated, using the cooling water as the heat source. As a result of installing this heat pump, the energy consumption is reduced. The savings amount to approximately 14 million yen a year, after deducting the operation control expenses (electricity charges and periodical repair expenses). This is based on an average COP of 1.50 for January through September.

Various devices are used to increase the load rate and operation rate of the heat recovery system. To increase the load rate, flow rate cascade control is applied to make sure that the heating hot water flows to the boiler preferentially. To increase the operation rate, the line connection can be changed by remote control, so that the heat recovery system is efficiently operated according to the combination of refrigerator and boiler operations in the plant. 100% operation rate throughout the year is set as the target.

The amount of input steam is regulated by cascade control. The temperature is kept between the boiler feed-water intermediate temperature and the temperature of the heat source water inlet, in order to prevent crystallization of the solution during times of low heat source water temperature.

In actual operation during the winter, the load was low at 40 to 50%, but the COP was 1.60, recording data that exceeded expectations (heat source water temperature is 22.5°C to 21°C, and boiler feed-water temperature is 51°C to 67°C).

With the absorption heat pump, cascade control of feed-water temperature is carried out according to the following flow.

6. FUTURE OUTLOOK

Since the absorption heat pump has been developed on the basis of airconditioning technology, there are several technical problems before producing hot water usable for a wide range of purpose from a low temperature heat source. For instance, Type I heat pump has limitation in hot water outlet temperature. Also, its utilization technology is still to be improved in some respects. Nevertheless, application of the absorption heat pump is not limited to airconditioning. There are many cases identified to be suitable for industrial use in consideration of amount and quality of waste heat source. Industrial use has advantage in terms of pay back period because the operating time is more than twice of airconditioning. At the same time, however, design consideration should be given to a variety of requirements for temperature condition of waste heat, type and quality of water among industries, as well as back-up requirement to take into account durability in long-time continuous operation.

Also, operational constraints such as freezing in coils on the heat recovery side and crystalization of absorbent solution in case of low hot water inlet temperature.

If air heat source can be used by the absorption heat pump, the above problems will be solved to expand its scope of application considerably.

Thus, as the first step, we are developing an air cooling type gas-fired absorption water chiller-heater using water LiBr combination, and its commercial development in 1988 is targetted.

The absorption heat pump has cleared the first development stage, and now is expected to be applied to a wide range fields through research and development on utilization technology based on its features.

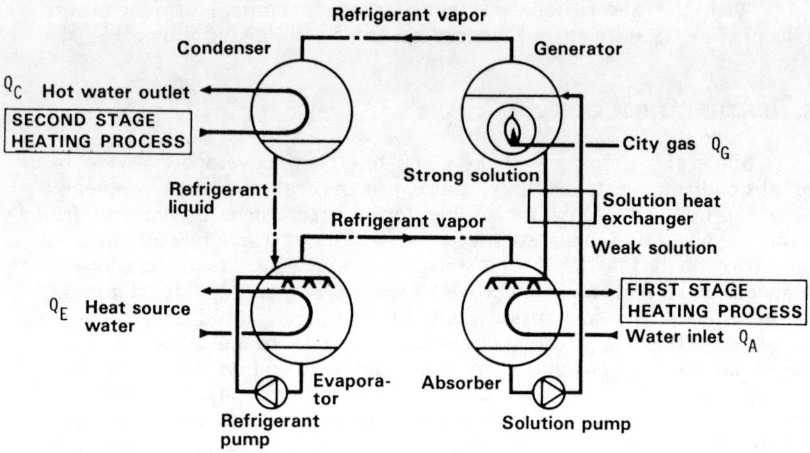

Fig. 1 Absorption Heat Pump Cycle

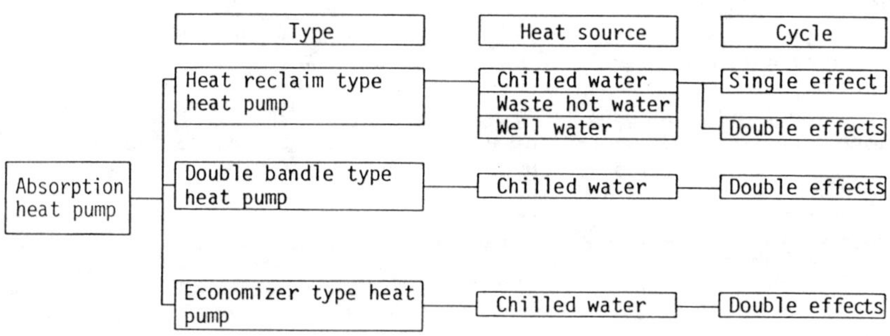

Fig. 2 Type of Absorption Heat Pump

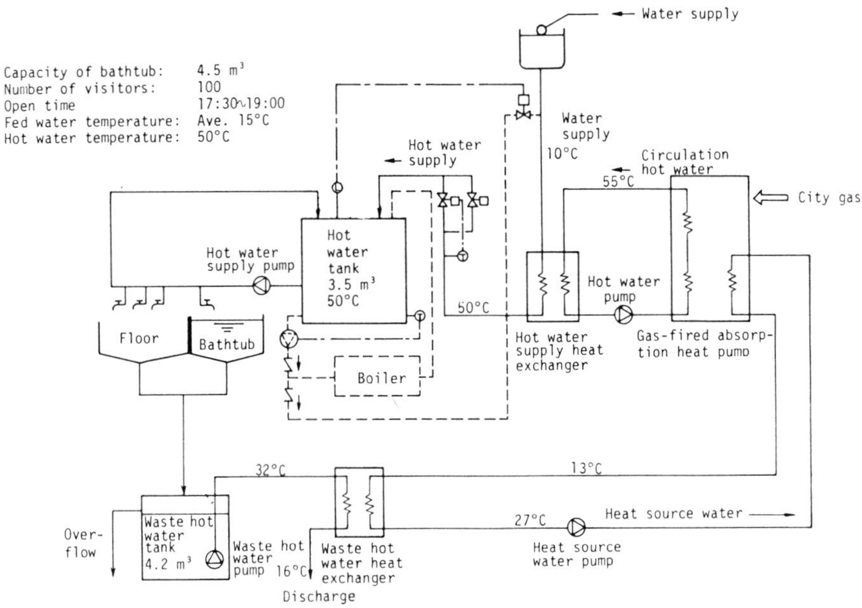

Fig. 3 Field Test Flow Sheet

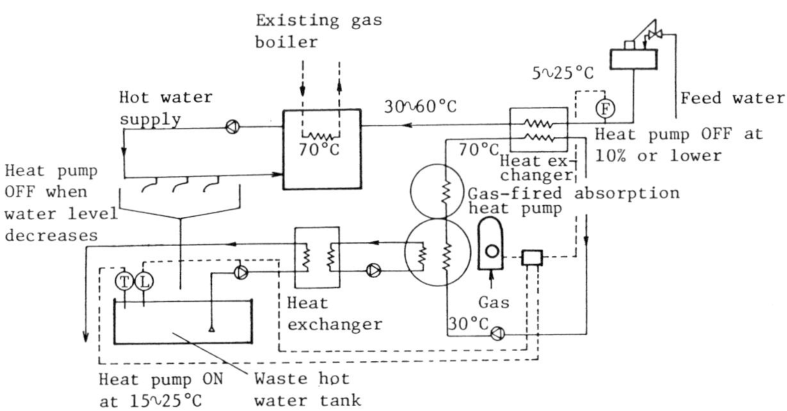

Fig. 4 Flow Sheet for Sauna Bath System

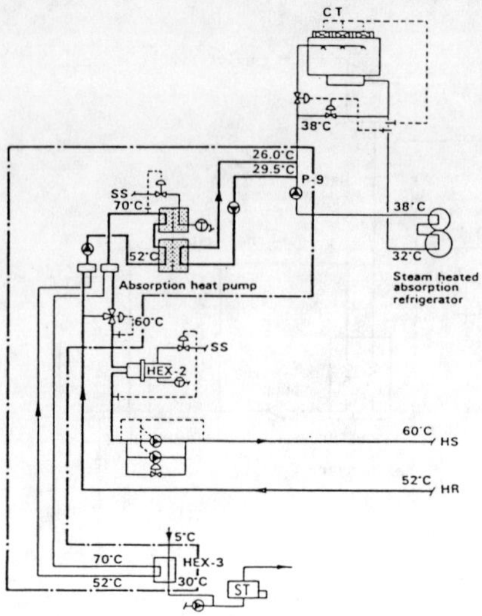

Fig. 5 System Flow

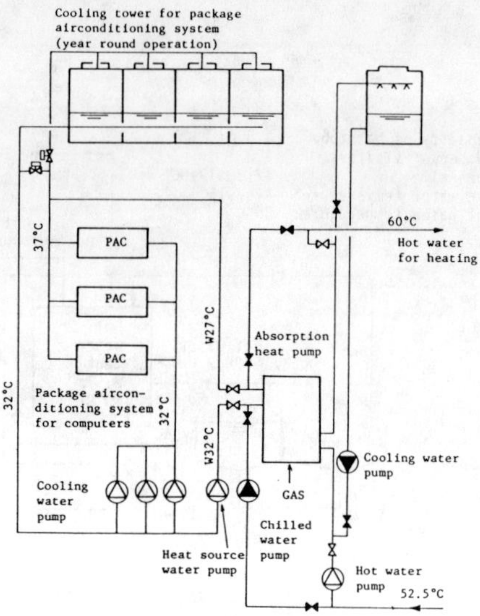

Fig. 6 Flow Sheet (Heat Pump Mode)

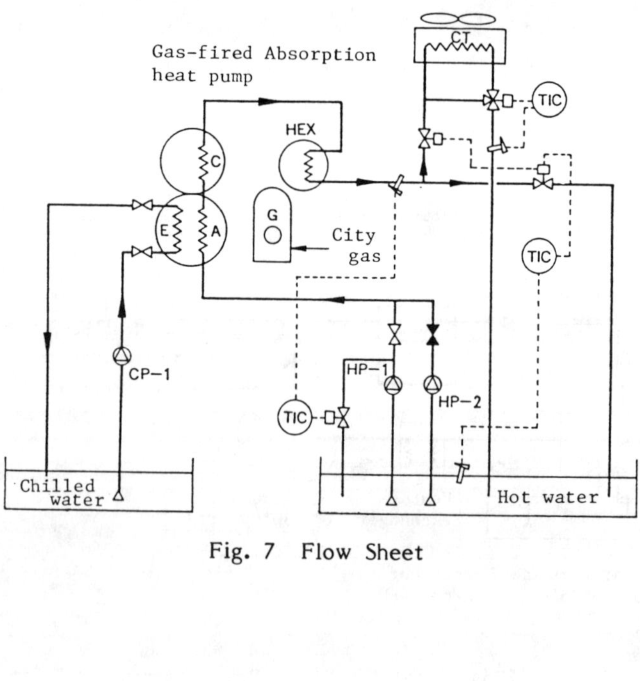

Fig. 7 Flow Sheet

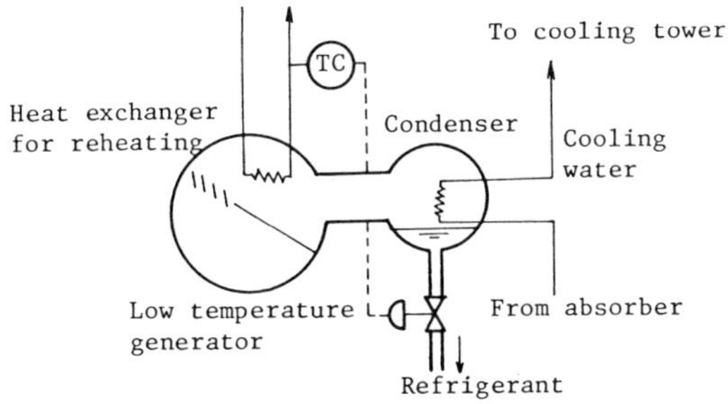

Fig. 8 Principle to Control Hot Water for Reheating

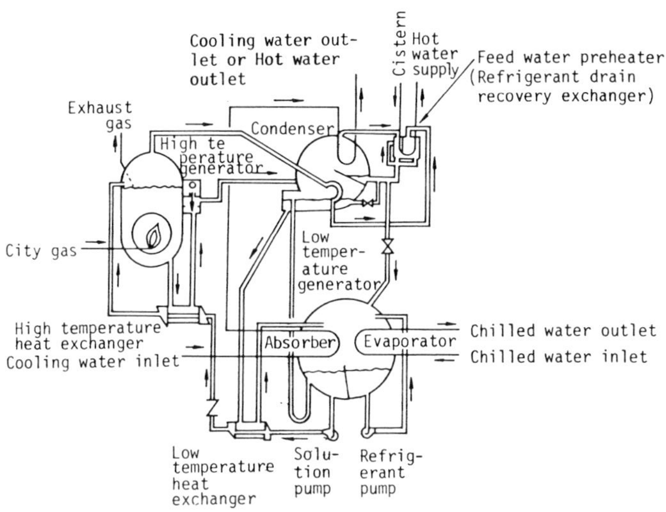

Fig. 9 Cycle Flow Sheet of Triple Bundle Condenser Type

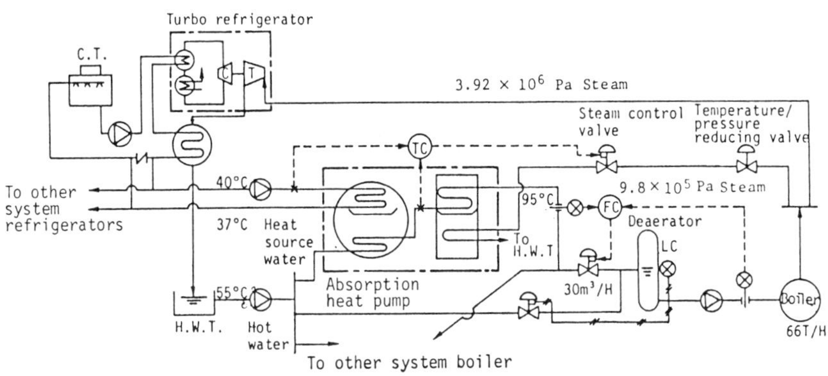

Fig. 10 System Flow Sheet

Chapter 26

What Needs To Be Known About Fluid Pairs To Determine Heat Ratios of Absorber Heat Pumps and Heat Transformers

Georg Alefeld

ABSTRACT: The heat ratios for absorber heatpumps and heat transformers are usually calculated by enthalpy balances. Rather accurate and elaborate enthalpy-concentration diagrams are required to determine the heat ratios.

Starting from First and Second Law from the very beginning, simple analytic equations for the heat ratios have been derived. A small set of nondimensional fluid parameters is sufficient to predict the heat ratios quantitatively. The analysis yields figures of merit for processes and for fluid pairs. The concept may be helpful in the search and evaluation of fluid combinations.

INTRODUCTION /1/, /2/

Heat ratios of absorber heatpumps and heat transformers are, in general, calculated by enthalpy balances in enthalpy concentration charts. Although only First Law is explicitly used, these heat ratios never violate Second Law requirements. The reason is, that the Second Law is implicitly taken into account by using experimental enthalpy values of real fluids. The heat ratios are determined by the differences of large numbers which, therefore, must have higher accuracy than desired for the heat ratios. It is unsatisfactory that Second Law, which is independent of fluid properties, must be recalculated for each case considered, needing an extensive amount of experimental data.

HEAT PUMPS: Prospects in Heat Pump Technology and Marketing (Proceedings of the 1987 IEA Heat Pump Conference). Edited by Kay H. Zimmerman Chelsea, Michigan 48118. Printed in U.S.A.

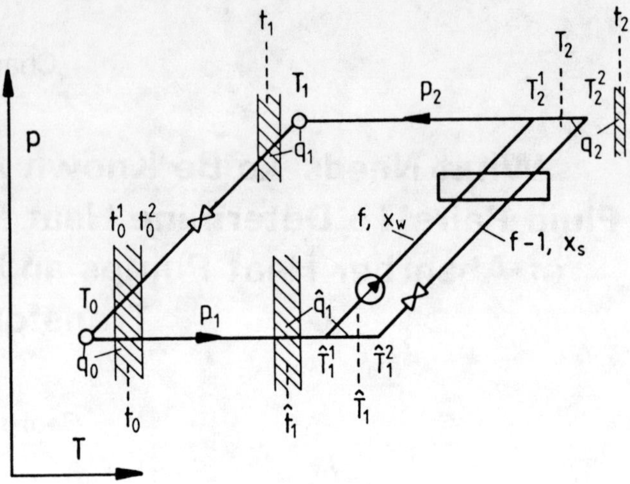

Figure 1: Temperatures and Pressures for an Absorption Heatpump

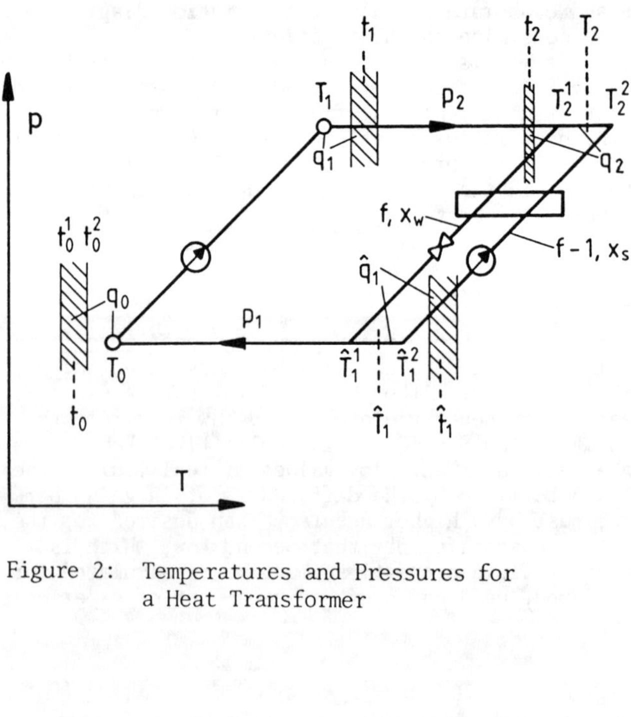

Figure 2: Temperatures and Pressures for a Heat Transformer

In this presentation Second Law is explicitly used from the very beginning. The accuracy required for the fluid data and their amount is thus widely reduced. For new fluid pairs or mixtures rather fast estimates for heat ratios are possible.

FIRST AND SECOND LAW FOR THE ABSORBER REFRIGERATOR

Figure 1 shows temperatures and pressures for the four exchange units of an absorber heatpump. First and Second Law can be written as /2/:

$$\text{I. Law:} \quad q_2 - q_1 - \hat{q}_1 + q_0 - q_r + w = 0 \tag{1}$$

$$\text{II. Law:} \quad -\frac{q_2}{t_2} + \frac{q_1}{t_1} + \frac{q_1}{\hat{t}_1} - \frac{q_0}{t_0} + \frac{q_r}{t_r} = \sum_1^n \delta s_i \tag{2}$$

with q_i, $\delta_i \geq 0$ (for q_i see Figure 1)

w = pumping power
q_r = heat of rectification
t_r = external temperature at which heat of rectification is removed
$\Sigma\delta s_i$ = sum over all entropy producing process steps including heat transfer to the external heat carrier fluids

The quantities q_i , w, δs_i are actually fluxes and refer to one mass unit of circulating working fluid.

The temperatures t_i are exactly defined averages over the temperature spread $t_i^2 - t_i^1$ of the heat carrier fluids:

$$\frac{1}{t_i} = \frac{1}{q_i}\int_{t_i^1}^{t_i^2} \frac{c_i dt}{t} \qquad q_i = \int_{t_i^1}^{t_i^2} c_i dt \tag{3}$$

For most cases the middle temperature is a good approximation (with an accuracy better than 1%!):

$$t_i \approx \frac{1}{2}(t_i^1 + t_i^2) \tag{4}$$

From now on the pumping power w and the heat of rectification q_2 will be neglected which is possible for the pair $H_2O/LiBr$. Furthermore it will be assumed that: $t_1 = \hat{t}_1$ and $T_1 = \hat{T}_1$. For more details see /1/.

The following entropy producing steps can be identified:

δs_1 = desuperheating of the fluid between T_2 and T_1
δs_2 = cooling of the fluid from T_1 to T_0 (throttling)
δs_3 = heating of the fluid from T_0 to $\hat{T}_1$

δs_4 = heating or cooling of the weak (in respect to absorbent) solution to T_2^1

δs_5 = cooling or heating of the strong (in respect to absorbent) solution to $\hat{T}_1^2$

δs_6 = entropy generation in the solution heat exchanger

$\left.\begin{matrix}\delta s_7\\ \delta s_8\\ \delta s_9\\ \delta s_{10}\end{matrix}\right\}$ = heat transfer between heat carrier fluids and working fluids

δs_{11} = all the rest, e.g. irreversibilities of the pump, pressure drops, heat conduction, etc. (δs_{11} will be neglected in the following)

Eliminating $q_1 + \hat{q}_1$ from Equation (1) and (2) yields the following equation for the heat q_2 and the efficiency $\eta = q_0/q_2$ for refrigeration:

$$q_2 = q_0 \frac{t_1 - t_0}{t_0} \Big/ \frac{t_2 - t_1}{t_2} + \frac{t_1 t_2}{t_2 - t_1} \sum_1^{10} \delta s_i \qquad (5)$$

$$\eta = \eta_r(t) \Big/ \left(1 + \frac{t_0 t_1}{t_1 - t_0} \sum_1^{10} \delta s_i / q_0\right) \qquad (6)$$

with the efficiency $\eta^r(t)$ for a reversible machine, operating between t_2, t_1 and t_0:

$$\eta^r(t) = \frac{t_2 - t_1}{t_2} \Big/ \frac{t_1 - t_0}{t_0} \qquad (7)$$

The quality of the process (effectiveness), including heat transfer, can be measured by the following ratio:

$$g(t) = \eta / \eta^r(t) = 1 \Big/ \left(1 + \frac{t_0 t_1}{t_1 - t_0} \sum_1^{10} \delta s_i / q_0\right) \qquad (6a)$$

The entropy, generated in the heat transfer, is the sum of four terms:

$$\sum_7^{10} \delta s_i = q_2\left(\frac{1}{T_2} - \frac{1}{t_2}\right) + q_1\left(\frac{1}{t_1} - \frac{1}{T_1}\right) + \hat{q}_1\left(\frac{1}{\hat{t}_1} - \frac{1}{\hat{T}_1}\right) + q_0\left(\frac{1}{T_0} - \frac{1}{t_0}\right) \qquad (8)$$

This is an exact relation if, for the internal temperatures $\hat{T}_1$ and T_2, the analogous entropic averages as shown in Equation (3), but over internal heat distributions, are used. Again for most cases the middle temperatures are good approximations, even if heat production is not completely homogeneous over the temperature interval, e.g. over $\hat{T}_1^2 - \hat{T}_1^1$. Inserting Equation (8) into Equation (2) and eliminating $q_1 + \hat{q}_1$, now the following set of equations holds:

$$q_2 = q_0 \frac{T_1 - T_0}{T_0} \Bigg/ \frac{T_2 - T_1}{T_2} + \frac{T_1 T_2}{T_2 - T_1} \sum_1^6 \delta s_i \tag{9}$$

$$\eta = \eta^r(T) \Bigg/ \left(1 + \frac{T_0 T_1}{T_1 - T_0} \frac{\sum_1^6 \delta s_i}{q_0} \right) \tag{10}$$

with the efficiency $\eta^r(T)$ of a reversible machine, operating between the internal temperature levels T_2, T_1 and T_0:

$$\eta^r(T) = \frac{T_2 - T_1}{T_2} \Bigg/ \frac{T_1 - T_0}{T_0} = \left(\frac{1}{T_1} - \frac{1}{T_2}\right) \Bigg/ \left(\frac{1}{T_0} - \frac{1}{T_1}\right) \tag{11}$$

The value of η from Equation (10) is identical to that of Equation (6).

Out of the three temperatures T_0, T_1 and T_2 only two can be chosen independently, the third is determined by the individual properties of the solution field. Equation (11) can be expressed by the slopes of the isosteres for the pure fluid $\bar{r}$ and the middle value of the slopes for the strong and weak solution $\bar{r} + 1$.

$$\eta^r(T) = \frac{\bar{r}}{\bar{r}+1} = \frac{1}{1+1/\bar{r}} \tag{12}$$

Only far from the critical point is the slope $\bar{r}$ identical with the heat of vapourization r. This is the case for H_2O around room temperature.

Equation (9) and (5) can be interpreted as follows: The first term represents the minimum amount of driving heat q_2, required for a given refrigeration power q_0. Due to irreversibilities the required heat input is increased proportional to the entropy production. It should be noted that the proportionality factor $T_1 T_2/(T_2 - T_1)$ differs from that which one finds in calculating exergy losses. From Equation (10) a quality factor (an effectiveness) for the process, now not including heat transfer, can be defined as:

$$g(T) = \eta \Big/ \overset{r}{\eta}(T) = 1 \Big/ \left(1 + \frac{T_0\ T_1}{T_1 - T_0} \frac{\sum_1^6 \delta s_i}{q_0}\right) \tag{13}$$

(No ambiguous temperature t_u of the ambient appears in this equations, as it does in an exergy analysis!)

The entropy increase δs_1 is determined by the following integrals:

$$\delta s_1 = \frac{1}{T_1} \int_{T_1}^{T_2} c_p\, dT - \int_{T_1}^{T_2} \frac{c_p\, d}{T} \tag{14}$$

Taking the average for c_p and expanding to lowest order in $T_2 - T_1$ one gets:

$$\delta s_1 = c_p \left(\frac{T_2 - T_1}{T_1} - \ln \frac{T_2}{T_1} \right) = \frac{c_p}{2} \frac{(T_2 - T_1)^2}{T_1\ T_2} + \ldots \tag{15}$$

Similarly δs_2 to δs_6 can be calculated /1/:

$$\delta s_2 = \frac{c}{2} \frac{(T_1 - T_0)^2}{T_0\ T_1} + \ldots \tag{16}$$

$$\delta s_3 = \frac{c_p}{2} \frac{(\hat{T}_1 - T_0)^2}{T_0\ \hat{T}_1} + \ldots \tag{17}$$

where c, c_p = specific heat capacity of the liquid and gaseous working fluid.

For the total entropy, created in the solution cycle, one finds after a lengthy calculation, in spite of the many possible cases, a common simple expression (18)

$$\sum_4^6 \delta s_i = \frac{1}{2}\left[f\, c_w + (f-1)(1-2\,\eta_s) c_s\right] \frac{(T_2 - \hat{T}_1)^2}{T_2\ \hat{T}_1} + \eta_s (f-1) c_s \frac{(\hat{T}_1^2 - \hat{T}_1^1)^2}{\hat{T}_1\ T_2}$$

where η_s = efficiency of the solution heat exchanger, f = circulation ratio, c_s, c_w = specific heat capacity of the strong and weak solution.

The last term in Equation (18) is small, compared to the first one and may be neglected in most cases.

The refrigeration power per unit mass after throttling is given by

$$q_0 = r\,(1 - \frac{c(T_1 - T_0)}{r}) \tag{19}$$

Inserting Equations (15) to (19) into Equations (9) and (10) yields the final equations for the total heat input q_2 and the efficiency η of the refrigerator.

The denominator in Equation (10), which is a correction to the reversible (Carnot) efficiency, is a simple function of a set of nondimensional fluid parameters listed in Table 1. The first two parameters are characteristic of the working fluid only. For the third and forth parameter the specific heat capacities of the absorber fluid must be known. With the fifth parameter the Carnot efficiency can be calculated without specifying the three temperatures T_0, T_1, T_2 explicitly. Only the ratio of the slopes of vapour pressure lines (or the temperature differences $T_2 - T_1$ and $T_1 - T_0$ and one temperature) must be known with the same accuracy as required for the heat ratios. Since all the parameters listed in Table 1 are causing corrections of the order of 10% or less (see Table 2), an accuracy of an order of magnitude less is sufficient. If the pumping power must be considered, a further fluid parameter is needed /3/. The absolute pressures, at which the machine operates, is only relevant for this parameter.

For the Tables 1 - 4 the following input data have been used: Fluid pair: $H_2O/LiBr$; $c = 4{,}19$ kJ/kgK, $c_p = 1{,}90$ kJ/kgK, $r = 2490$ kJ/kg, $T_0 = 2°C$, $T_1 = 38°C$, $T_2 = 80°C$, $x_s = 0{,}595$, $\hat{x}_w = 0{,}570$, $c_s = 1{,}97$ kJ/kgK, $c_w = 2{,}09$ kJ/kgK, $f = 16$, $T_1^{\prime} - \hat{T}_1^{\prime} = 6$ K, $\eta_s = 0{,}8$, $|T_i - t_i| = 5$ K.

Table 1. Nondimensional fluid parameters for $H_2O/LiBr$

1	2	3	4	5	6
cT_1/r	c_pT_1/r	$(f-1)c_sT_1/r$	$f\,c_wT_1/r$	$1/r$	pump parameter
0,52	0,24	3,74	4,18	0,10	0,0

Table 2 shows that the additional heat input caused by the irreversibilities δs_1, δs_2 and δs_3 amounts only to about 5% of the total heat input. This holds for H_2O as working fluids and changes appreciably for NH_3 or R 22. The largest term results from the solution circuit. Again this term strongly increases for other pairs.

Table 2. Heat input q_2 per refrigeration power q_0 for a refrigerator or heat pump (HP: Equation 9, η_s = 0,8) and: Heat output q_2 per heat output q_0 for a heat transformer (HT: Equation 29, η_s = 0,8)

	$\frac{q_2}{q_0}$	$\frac{1}{\eta^r(T)}$	$\frac{T_1 T_2}{T_2 - T_1} \frac{\delta s_1}{q_0}$	$\frac{T_1 T_2}{T_2 - T_1} \frac{\delta s_2}{q_0}$	$\frac{T_1 T_2}{T_2 - T_1} \frac{\delta s_3}{q_0}$	$\frac{T_1 T_2}{T_2 - T_1} \frac{\delta s_4 + \delta s_5 + \delta s_6}{q_0}$
HP	1,32 (120%)	1,10 (100%)	0,017 (1,5%)	0,036 (3,3%)	0,016 (1,4%)	0,152 (13,8%)
HT	0,90 (81,8%)	1,10 (100%)	0,016 (1,4%)	0,033 (3,0%)	0,015 (1,3%)	0,139 (12,5%)

Table 3. Heat ratios q_0/q_2 and quality factors g (effectiveness) for an absorption refrigerator (Equations (6), (6a), (7), (10), (11), (12), $|T_i - t_i|$ = 5 K)

	$\eta^r(T)$	η_s = 0,8	η_s = 1,0	η_s = 0,0	η^r (t)
$\eta = q_0/q_2$	0,91	0,76	0,83	0,58	1,56
g(T)	---	83%	91%	64%	---
g(t)	---	49%	53%	37%	---

In Table 3 the reversible heat ratios and three real heat ratios for different heat exchanger efficiencies η_s are listed. The reversible value 0,91 is below 1,0 due to the increased slope of the vapour pressure lines with LiBr. The value 0,76 is typical for a heat exchanger efficiency of 0,8. Even for η_s = 0, the heat ratio stays as high as 0,58, which is only the case for H_2O/LiBr. Without heat transfer the effectiveness is rather high, i.e. the heat ratio reaches about 80% of the Carnot value.

THE COP AND EFFECTIVENESS FOR THE ABSORBER HEATPUMP

Using the relation

$$\eta_{HP} = 1 + \eta \tag{20}$$

the COP and a quality factor (effectiveness) of the absorber heatpump can easily be calculated as:

$$\eta_{HP} = \frac{\dfrac{T_2 - T_0}{T_2}}{\dfrac{T_1 - T_0}{T_1}} \; \frac{1 + \dfrac{T_0\, T_2}{T_2 - T_0} \sum_1^6 \delta s_i \,/\, q_0}{1 + \dfrac{T_0\, T_1}{T_1 - T_0} \sum_1^6 \delta s_i \,/\, q_0} \tag{21}$$

with the COP for the completely reversible heatpump

$$\eta^r_{HP} = \frac{T_2 - T_0}{T_2} \Bigg/ \frac{T_1 - T_0}{T_1} \tag{22}$$

and an effectiveness

$$g_{HP}(T) = \eta_{HP} / \eta^r_{HP} = (1 + \eta) \,/\, (1 + \eta^r(T)) \tag{23}$$

As a quality factor for the fluid one can use g_{HP} (T) with the best value for the solution heat exchanger efficiency. Table 2 holds for the heatpumps as well. In Table 3 the first line must be increased by one. The quality factors for the heatpump are thus higher, e.g. 1,76/1,91 = 92% instead of 83%. In contrast to refrigerators for heatpumps the increased amount of heat due to irreversibilities is not lost, but recovered at T_1 as useful product.

FIRST AND SECOND LAW FOR THE HEAT TRANSFORMER

According to Figure 2, First and Second Law of an absorption heat transformer are identical to Equation (1) and (2) (ignoring q_r and w!) except that the sign for the heat fluxes q_i is reversed:

I. Law: $$-q_2 + q_1 + \hat{q}_1 - q_0 = 0 \tag{24}$$

II. Law: $$\frac{q_2}{t_2} - \frac{q_1}{t_1} - \frac{\hat{q}_1}{t_1} + \frac{q_0}{t_0} = \sum_1^n \delta s_i \tag{25}$$

with $q_i, \delta s_i \geq 0$.

The entropy production in heat transfer is now given by (26)

$$\sum_{7}^{10} \delta s_1 = q_2\left(\frac{1}{t_2} - \frac{1}{T_2}\right) + q_1\left(\frac{1}{T_1} - \frac{1}{t_1}\right) + \hat{q}_1\left(\frac{1}{\hat{T}_1} - \frac{1}{\hat{t}_1}\right) + q_0\left(\frac{1}{t_0} - \frac{1}{T_0}\right)$$

With the expansion to the lowest order in $T_2 - \hat{T}_1$ or $T_1 - T_0$ respectively, the entropy production for the heat transformer δs_1 to δs_6 is identical to that of the heatpump /1/. Thus Equation (15) to (18) can be used in Equation (25).

The heat q_0 rejected at the condenser is in contrast to Equation (19) larger than r, namely:

$$q_0 = r\left(1 + \frac{c_p\,(\hat{T}_1 - T_0)}{r}\right) \tag{27}$$

Equations (24) and (25) can now be solved for q_2 and $\eta_T = q_2/(q_1 + \hat{q}_1)$ eliminating q_0. Yet the results are more transparent and can be compared more easily with those of the heatpump if first, the heat ratio q_2/q_0 is determined, as has been done for the refrigerator.

The efficiency of the heat transformer can then be calculated with the equation:

$$\eta_T = \frac{q_2/q_0}{1 + q_2/q_0} = \frac{1}{1 + q_0/q_2} \tag{28}$$

Inserting Equation (26) into Equation (25) and eliminating $q_1 + \hat{q}_1$ yields:

$$q_2 = q_0 \;\frac{T_1 - T_0}{T_0} \Bigg/ \frac{T_2 - T_1}{T_2} - \frac{T_1\,T_2}{T_2 - T_1}\sum_{1}^{6} \delta s_1 \tag{29}$$

Equation (29) is identical with Equation (9) except for the differing value of q_0 (Equation (27)) and the minus sign of the last term. Clearly for a heat transformer irreversibilities will reduce the heat output q_2.

Combining Equation (29) with Equation (28), one finds for the efficiency of the heat transformer:

$$\eta_T = \left(\frac{T_1 - T_0}{T_1} \Bigg/ \frac{T_2 - T_0}{T_2}\right) \frac{1 - \dfrac{T_0\,T_1}{T_1 - T_0}\sum_{1}^{6} \delta s_i \,/q_0}{1 - \dfrac{T_0\,T_2}{T_2 - T_0}\sum_{1}^{6} \delta s_i \,/q_0} \tag{30}$$

This equation is the reciprocal of the heatpump efficiency Equation (21), but with minus signs in front of the sum for the entropy increases.

The completely reversible heat transformer has the efficiency

$$\eta_T^r = \frac{T_1 - T_0}{T_0} \Big/ \frac{T_2 - T_0}{T_2} = \left(\frac{1}{T_0} - \frac{1}{T_1}\right) \Big/ \left(\frac{1}{T_0} - \frac{1}{T_2}\right) \tag{31}$$

The quality or effectiveness of the process can be measured by the following ratio:

$$g_T(T) = \eta_T \Big/ \eta_T^r = \frac{1 - \dfrac{T_0\ T_1}{T_1 - T_0} \sum_1^6 \delta s_i / q_0}{1 - \dfrac{T_0\ T_2}{T_2 - T_0} \sum_1^6 \delta s_i / q_0} \tag{32}$$

The analogous equation holds for the effectiveness $g_T(t)$, including heat transfer replacing the internal temperatures T_i by the external temperatures t_i and including the heat transfer in the sum of the irreversibilities

$$g_T(t) = \eta_T \Big/ \eta_T^r(t) \tag{33}$$

The fluid pair by itself may be characterized by g(T) yet taking for the heat exchanger efficiency η_s in Equation (18) the best ($\eta_s = 1$) or the worst ($\eta_s = 0$) value.

The reversible efficiency η_T^r can be expressed by the slopes of the isosteres in vapour pressure diagrams:

$$\eta_T = \frac{1}{1 + \dfrac{\bar{r}}{\bar{r}+1}} = \frac{1 + 1/\bar{r}}{2 + 1/\bar{r}} \tag{34}$$

For the heat transformer the same nondimensional parameters, as listed in Table 1, can be used.

In Table 2 the second line shows, how much the heat output q_2, relatively to the heat output q_0 is reduced by irreversibilities. The small differences in the third to sixth column between heatpump (HP) and heat transformer (HT) result from different values of q_0, namely $q_0(HP)/q_0(HT) = 0{,}915$. For the heat transformer the sum of column 3 to 6 is subtracted from column 2 to yield column 1.

In Table 4 heat ratios of a heat transformer for three heat exchanger efficiencies and for reversible processes as well as quality factors are listed.

Table 4. Heat ratios $q_2/(q_1 + \hat{q}_1)$ and quality factors g (effectiveness) for a heat transformer (Equations (26), (27), (28), (29), $|T_i - t_i| = 5$ K).

	η_T^r (T)	$\eta_s = 0{,}8$	$\eta_s = 1{,}0$	$\eta_s = 0{,}0$	η_T^r (t)
$\eta_T = \frac{q_2}{(q_1+\hat{q}_1)}$	52,5%	47,6%	50,1%	34,3%	60,0%
g_T(T)	---	90,7%	95,5%	65,3%	---
g_T(t)	---	79,5%	83,5%	55,5%	---

Such high values as η_T = 47 to 48% and about 90% of Carnot efficiency have been reached in laboratory prototyps /4/ as well as in large commercial plants /5/.

CONCLUSION

Starting from First and Second Law, very simple formulae for the heat ratios (COPs) of absorber refrigerators, - heatpumps and - heat transformers have been derived. These formulae are particular useful if only a limited amount of fluid data or data with limited accuracy are available.

As a by-product a Second Law analysis is presented, which does not have the problems, well-known for an exergy analysis /6/. In such an analysis exergy losses are calculated. The exergy is the work which could be gained from heat in a reversible power station working against the lower temperature level t_u. The exergy losses are proportional to the entropy production with t_u as the proportionality factor. Therefore, the choice of t_u leaves an ambiguity which is unsatisfactory. The effectiveness of a machine, defined as the real efficiency, in respect to that of a reversible machine, should not depend on an arbitrary chosen or accidentally existing temperature t_u and thus vary with day and night or with the season.

Of more relevance than the problem of how much potential work is lost in a process step are, to our opinion, the following questions: how much additional heat at T_2 must be supplied due to irreversibilities? - Or: how much is the heat output at T_2 reduced due to irreversibilities?

The answer can be found in the Equations (9) or (29) of this paper. These additional heat quantities are proportional to $\Sigma \delta s_i$, as in the exergy loss. Yet, if these heat quantities are converted into exergies, one finds that the factors by which the entropy increase must be multiplied, can qualitatively and quantitatively differ widely from t_u. In the analysis of this paper only process temperatures enter into the equations for the efficiency or effectiveness. The equations are independent from the ambient.

REFERENCES

/1/ This presentation is an extract of a more detailed paper by the author on: "The Heat Ratios of Absorber Heatpumps and Heat Transformers from First and Second Law", submitted for publication.

/2/ It is recommendable to read the introduction of the authors paper "What Needs to Be Known about Working Fluids to Calculate COPs?" in these proceedings.

/3/ See e.g. D.L. Hodgett, "New Working Pairs for Absorption Processes", Proceedings of a Workshop in Berlin (1982) Editor: Swedish Ccuncil for Building Research.

/4/ P. Riesch et al., "Partload Behaviour of an Absorption Heat Transformer", to be presented at the Third Intern. Symposium on the Large Scale Application of Heat Pumps, Oxford, March 1987.

/5/ Data sheets by Sanyo, Hitachi Zosen or GEA

/6/ e.g. W. Fratscher and J. Beyer, "Stand und Tendenzen bei der Anwendung und Weiterentwicklung des Exergiebegriffs", Chem., Technik 33, pp. 1-10 (1981)

Chapter 27

Development of Gas-Engine Heat Pumps in Japan

Tadashi Fukuda

ABSTRACT: Energy saving is the matter of the first priority in Japan, because Japan is heavily dependent for fuel supply on imports. One of the most promising method for energy saving can be said the development of gas engine heat pump system. Osaka Gas has been devoted to R&D on gas engine heat pumps cooperated with other Gas Companies for a long time.

The way of development consists of two streams. One is the development of reasonable priced package gas engine heat pump air conditioner. The other is of gas engine heat pump cooling/heating and hot water supply system.

This paper describes the development of both gas engine heat pump systems.

1. INTRODUCTION

Japan, with her little energy resources, had to face two oil crises occurred in 1973 and 1979 and was pressed for countermeasures with all-out nationwide efforts since the greater feeling of risk prevailed in this country than any other advanced countries. Since then, two policies have been promoted actively as drastic measures to get out of the constitution dependent of imported energy, the oil in particular.

HEAT PUMPS: Prospects in Heat Pump Technology and Marketing (Proceedings of the 1987 IEA Heat Pump Conference), Edited by Kay H. Zimmerman

One of those policies is to develop and introduce oil-substituting energy and the other is to push forward the energy saving measures.

Gas companies in Japan attempted to become oil-independent and started importing liquefied natural gas from Alaska (U.S.A.) since 1969 and tried to switch from the conventional supply gas such as gas produced from oil and coal to natural gas supply.

In the Osaka Gas area as of 1986, about 80% of gas consuming households are supplied with natural gas and major gas companies including Tokyo Gas, Osaka Gas, and Toho Gas plan to complete to switch their gas supply to all consumers to the natural gas between 1988 and up to 1992.

The natural gas contains CH_4 as a main component and has excellent anti-knock property as compared with existing city gas containing large amount of H_2, and makes it an optimum fuel for gas engines.

In a system that employs a gas engine as its source of driving force, effective use of the engine exhaust heat by recovery is considered, and it is possible to compose a system that brings forth by far high energy saving effect depending on the degree of the ratio of use such engine waste heat.

Therefore, the development of excellent gas engine heat pump is one of important policies of this country in that such effort conforms to the national policy of oil-independence and energy saving, and related researches and developments have been implemeted to date, mainly by gas companies.

On the other hand, as recent buildings tend to lease their spaces and quarters to tennants, changes are seen of the design concept of air conditioning equipment. Ever since those oil crises, designs of air conditioning equipment used to employ central heat source systems with highly efficient heat source equipment as the primary objective to save the energy of the whole building. As a result of this, new buildings in Osaka area of over 3000 square meters equipped with the central heat source systems account for about 90% share in terms of the capacity made available from gas fired absorption chiller/heater. However, in small and medium size tennant buildings built recently, diversified patterns and easy operation of the system are regarded as important on the part of the user of the air conditioning equipment, and many such buildings tend to adopt individual type air conditioning systems which can easily meet individual needs of users.

For equipment that are applicable to such individual air conditioning systems, direct expansion type electric heat pumps are generally installed.

Since 1982 electric machinery and apparatus makers began marketing electric heat pump air conditioners for business use (generally called the multitype) specifically intended for such applications. They are basically the unit type package air conditioners of 3.7 KW to 22 KW combined with one or more outdoor units equipped with 3.7 KW refrigerant compressors, and one or two indoor units are connected to one compressor. Many of these package air conditioners are adopted by a number of small and medium sized buildings, but the recent trend indicates installations by many buildings of 3000 square meters or over of space capacity. Number of units sold in the Kinki area in 1985 amounted to 1,165 units, a continuous increase of over 40% every year.

Under such circumstances, the development of gas engine heat pumps in Japan was directed toward two main courses of the development of systems seeking after the energy saving capacity and the development of gas engine heat pump air conditioners which can be applied to individual type air conditioning systems. In the development of high energy saving type gas engine heat pumps, primary purpose is to reduce running costs and it is necessary to design so that the engine exhaust heat of each system can be utilized most effectively.

Therefore, it is difficult to standardize the entire system and the equipment become slightly complicated, causing initial costs to increase to some extent as compared with ordinary systems.

On the other hand, in the development of direct expansion type gas engine heat pump air conditioners, the greatest objective is to reduce the initial costs through mass production effect, and yet the development will involve difficulties to make available equipment which have better characteristics in terms of the running cost and functions as compared with the electric heat pump air conditioners already existing in this field.

The following discusses the system which was developed in accordance with respective directivity.

2. HIGH ENERGY SAVING TYPE GAS ENGINE HEAT PUMP SYSTEMS

2.1 Achievements in Operation

The heat pump system of this type was first installed in 1980 for air conditioning of offices, and 105 jobs (137 units) have been completed and are operating in many places in Japan as of the end of March 1986. Installations include offices, plants, hotels, and sports centers mainly for air conditioning or air conditioning and hot water supply, accounting for 84% of all installations.

Table 1. Jobs for installation by buildings

Offices	48
Plants	17
Sports center & pools	13
Hotels	13
Super markets and stores	1
Other facilities	10
Spa buildings	2
Hospitals	1
Total	105 jobs

Table 2. Number of installed units by capacity

Cooling capacity	Number of units
Less than 70 KW (Less than 20 USRT)	26
70 ∿ 175 KW (20 ∿ 50 USRT)	56
175 ∿ 350 KW (50 ∿ 100 USRT)	36
Over 350 KW (Over 100 USRT)	19
Total	137 units

Table 3. Heat pumps by type of heat source

Air heat source	86 jobs
Water heat source	19 jobs

2.2 Example of Installations

The following deals with gas engine heat pump systems for air conditioning hot water supply, and heating swimming pool water of a sports center with warm water swimming pool set up in a shopping center. This system has been in operation since January 1986.

(1) System structure

Table 4. Specifications of main equipment

Gas engine heat pump of water heat recovery type			
Performance	Cooling capacity (cooling mode)	KW	130 (7°C)
	Cooling capacity (cooling mode with hot water supply)	KW	120 (7°C)
	Heating capacity of heat pump	KW	157 (45°C)
	Waste heat recovery from engine	KW	63 (70°C)
Gas engine	Model		Yanmar 6T95LG
	Cylinders (cycle)		6 (4)
	Displacement	cc	4678
	Revolution	rpm	900 ∿ 1800
	Rated output	PS/rpm	30/900 ∿ 60/1800
Compressor	Model		Daikin 6c752
	Cylinders		6
	Displacement	cc	1643
	Refrigerant		R-22
Capacity control		%	100 ∿ 50 - 33 - 0 Engine revolution control + unloader control
Overall dimension (weight)		mm(kg)	L2800 W2200 H1800 (3600)
Unit type gas fired absorption chiller/heater			
Performance	Cooling capacity	KW	70
	Heating capacity	KW	61
Gas fired water heater			
Rated output		KW	349
Storage tank			
Hot water storage capacity		ℓ	3200
Hot water temperature		°C	60

(2) System flow chart

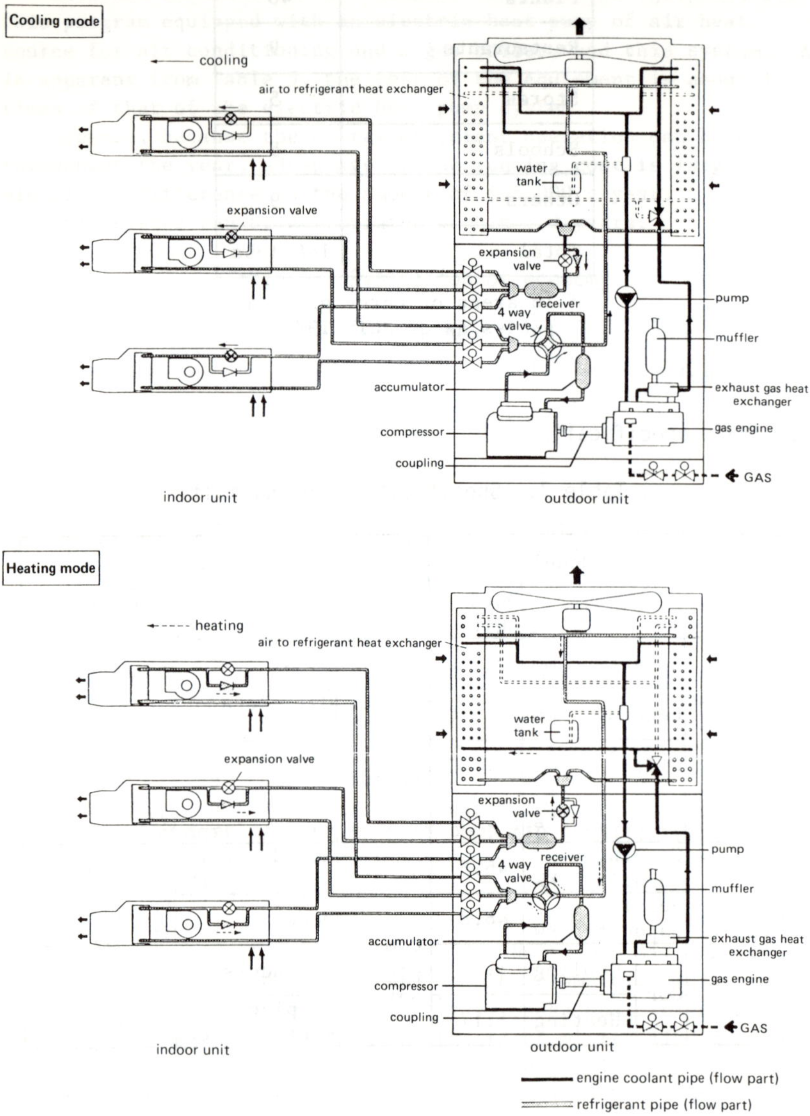

Fig. 3. System flow chart

Three indoor units are connected to one open type compressor which is connected to a gas engine, and each of the indoor unit can be started and stopped. Three types of indoor unit are made available; ceiling suspension type, ceiling cassette type, and ceiling embedded duct connection type, which can be selected depending on applications.

(Refrigerant flow)

The basic refrigerant flow is not at all different from electric heat pump, but has the following features unique to the gasmulti.

i) The refrigerant piping lead into three indoor units are divided inside the outdoor unit into three systems by the solenoid valve to open and close the passage according to the operation conditions of each indoor unit.

ii) Capacity of the refrigerant is controlled by controlling the gas engine revolution.

iii) An expansion valve is used for the pressure reducing device of refrigerant.

(Exhaust heat recovery flow)

The engine exhaust heat recovered from the cooling water line is simply released into the air during cooling. While during heating, the water heated by engine exhaust heat flows through the piping arranged alternately with the refrigerant piping on the surface of the air heat exchanger and the air heat exchanger is heated by the exhaust heat. By this operation, the following large merits are available.

i) Since the refrigerant is indirectly heated, high COP can be obtained.

ii) The air heat exchanger is freed from frosting and the defrosting operation became unnecessary.

As is apparent from the heat balance shown in Fig. 4, the effect of such exhaust heat recovery becomes greater as the outdoor temperature loweres.

(3) Economic efficiency

The price of the unit is still high as compared with that of the electric heat pump. Structure-wise, we cannot hope lower price than that of the electric type, but the price difference should be narrowed as the technical developments, are promoted and the number of sold units increased. The fuel cost was calculated based on 450 hours of cooling and 350 hours of heating for the operation hours equivalent to the total load. Maintenance cost of the gasmulti includes the personnel expenses and cost of periodic replacement parts required for twice-yearly periodic inspection.

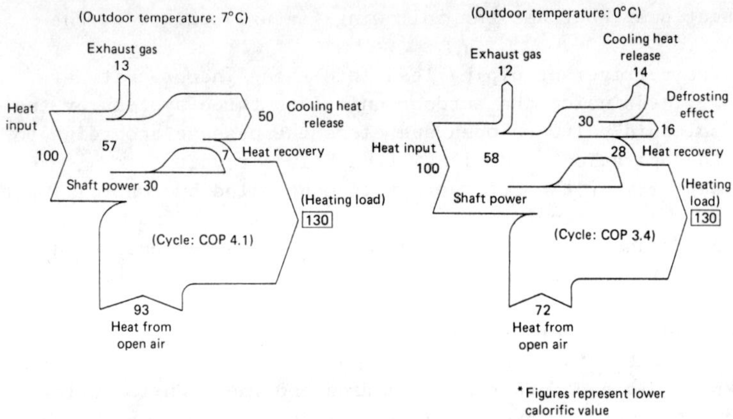

Fig. 4. Heat balance during heating

Table 8. Evaluation of economy

	Gas multi	Motor driven heat pump
Fuel cost	\$ 2,700	\$ 4,600
Maintenance cost	\$ 600	–
Total of running cost	\$ 3,300	\$ 4,600
Difference	–	\$ 1,300
Price of unit	\$28,100	\$25,800
Difference	\$ 2,300	–
Pay-back period	1.8 years	–

Rate of translation \$1 = ¥160

(4) Features of gasmulti (Summary)

i) A multitype which can control three indoor units individually.
ii) Revolution control of gas engine provides high partial load efficiency.
iii) Utilization of gas engine exhaust heat
  ° High COP
  ° Defrosting is not necessary.
iv) Low running cost
v) Power consumption can be reduced.

3.3 Other Development Plans

Development of small gas engine heat pump for business and home use is under way. Currently, protype equipments of capacities listed in Table 9 are being field tested. One of them are scheduled for commercialization in the spring of 1987.

Table 9. Planned gas engine heat pump models

	(1)	(2)	(3)	(4)	(5)	(6)
Cooling capacity KW	6.3	6.3	4.1	8.1	14	17
Heating capacity KW	9.4	9.3	5.8	9.3	16	21
Hot water supplying capacity KW	10	9.3	-	-	-	-
Number of indoor units	3	3	1	1	1	2

4. CONCLUSION

This paper described the two directions of the trend in the development of the gas engine heat pump in Japan; one is the development of gas engine heat pump system of high energy saving type and the other is the commercialization of gas engine heat pump air conditioner of production type.

However, this does not mean to divide the development methods into two directions and it is natural that related costs should be reduced by standardization and energy saving efforts made as much as possible in the future development of either of these types.

Now that the high energy saving performance and low fuel cost of the gas engine heat pump has been verified, the purpose of the future development should be placed to reduce the initial cost, expand the product variations, simplify the maintenance, and improve the reliability. We believe that demands will be increased as pursue these purposes.

REFERENCE

1. H. Kuroda, et al., No. II Summary of Gas Engine & Gas Turbin Systems, Japan Gas Association Technological Development Commitee, Tokyo, 1986

Chapter 28

Gas-Engine-Driven Heat Pumps — Design, Components, Experience, Layout, and Economic Feasibility

Wilhelm Struck and Franz Hirschbichler

SUMMARY: A design drawing is used to illustrate the M A N concept of gas engine heat pump modules. The coefficient of performance of the heat pump module is determined mainly by the following components: engine, compressor, condenser, evaporator and refrigerant. The state of the art of the components currently available is described. The optimal heat pump capacity to be installed is obtained with the aid of economic feasibility computations. Apart from capital investment costs, the main factor influencing the economic feasibility of the heat pump is the energy savings. This can be found, with the aid of a computer programme, through simulation of annual operation in heating mode. It is also shown how the economic feasibility computation is made with the aid of a computer programme.

1. INTRODUCTION

With increasing awareness of the limited nature of the earth's resources and with the increasing environmental consciousness in the western industrialized countries, heat pump technology is finding ever more acceptance. Using gas engine driven heat pumps for residential heating primary energy savings of approx. 50 % can be achieved compared to heating systems with conventional boilers. The noxious emissions of gas engines can be kept on a very low level using either catalytic converters or lean burn combustion principles. For reasons of environmental protection gas engine driven heat pumps have reached a market share of 95 % of the total market volume of IC-engine driven heat pumps in the Federal Republic of Germany. With over 550 units installed yet, the gas engine driven heat pumps can be considered as beeing established. Initial problems with maintenance and operation are now solved sufficiently.

HEAT PUMPS: Prospects in Heat Pump Technology and Marketing (Proceedings of the 1987 IEA Heat Pump Conference), Edited by Kay H. Zimmerman © 1987 Lewis Publishers, Inc., Chelsea, Michigan 48118. Printed in U.S.A.

2. DESIGN FEATURES OF M A N STANDARD HEAT PUMPS

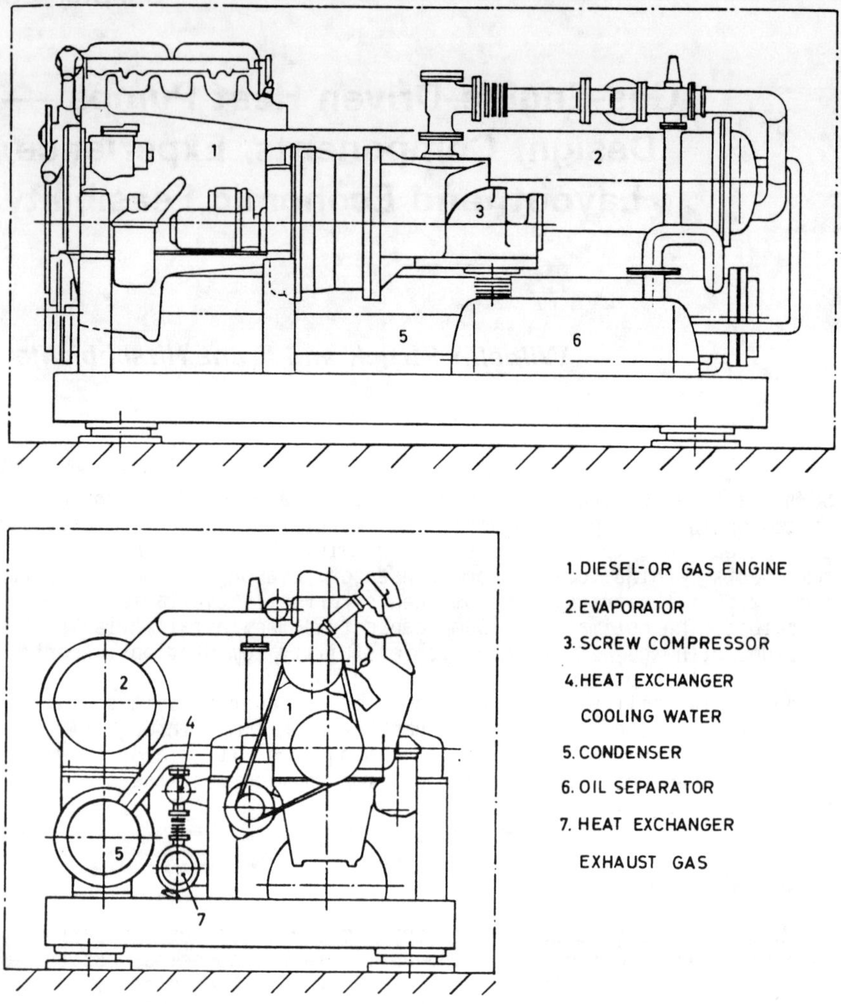

Fig. 1.

All the heat pump's main components are mounted on a common base frame. This allows virtually all the piping and wiring to be fitted at the manufacturer's plant, which yields the advantages of high assembly quality, reduced on-site installation work, and easier implementation of quality assurance measures. The compressors are rigidly flanged to the engines. The irregular torque of the internal combustion engine is not transmitted to the compressor via the base frame but through a torsionally rigid clutch housing. The engine-compressor unit is however flexibly mounted on the base frame.

This keeps the base frame and the components mounted thereon as free as possible from the vibrations of the engine-compressor unit, thus minimizing the danger of leaks, fractured piping or other vibration-induced damage. This construction also obviates alignment problems between engine crankshaft and compressor shaft. The compact construction also means that under normal circumstances there are no problems in transport and setting up.

3. MAIN COMPONENTS OF THE HEAT PUMP

The gas engine heat pump comprises a number of components, all of which affect the heat pump's operating performance.
Since the main components, i.e. engine, compressor, evaporator and condenser, as well as the refrigerant, exert the greatest influence on the heat pump process, these components are discussed below.

3.1 The engine

4 and 6 cylinder inline engines and V-12 engines are used with M A N standard heat pumps. The 6 and 12-cylinder engines can be supplied in naturally aspirated or in turbocharged versions.

In order to comply with the European and US emission regulations M A N standard gas heat engines can be supplied as rich burn engine with three-way catalytic converter and as lean-burn engine. To see the relationship between output, efficiency and noxious emissions the values measured for the two engine types E 2842 E and E 2842 LE were plotted as a function of excess air ratio (Figure 2.). As can be seen from the chart, the lean-burn concept led to lower output in the case of the naturally aspirated engine (E 2842 LE). This can be counteracted by turbocharging with intercooling (E 2842 LE).

The one concept is not necessarily superior to the other. The lean-burn concept makes it possible to comply with the emission regulations in West Germany and also with those in the USA. But the most rigorous reduction of noxious emissions is achieved in rich-burn operation with three-way catalytic converter. Among the advantages of the lean-burn engine is that this concept enables noxious emissions to be kept correspondingly low even when fuels containing constituents which would damage the catalytic converter are used.

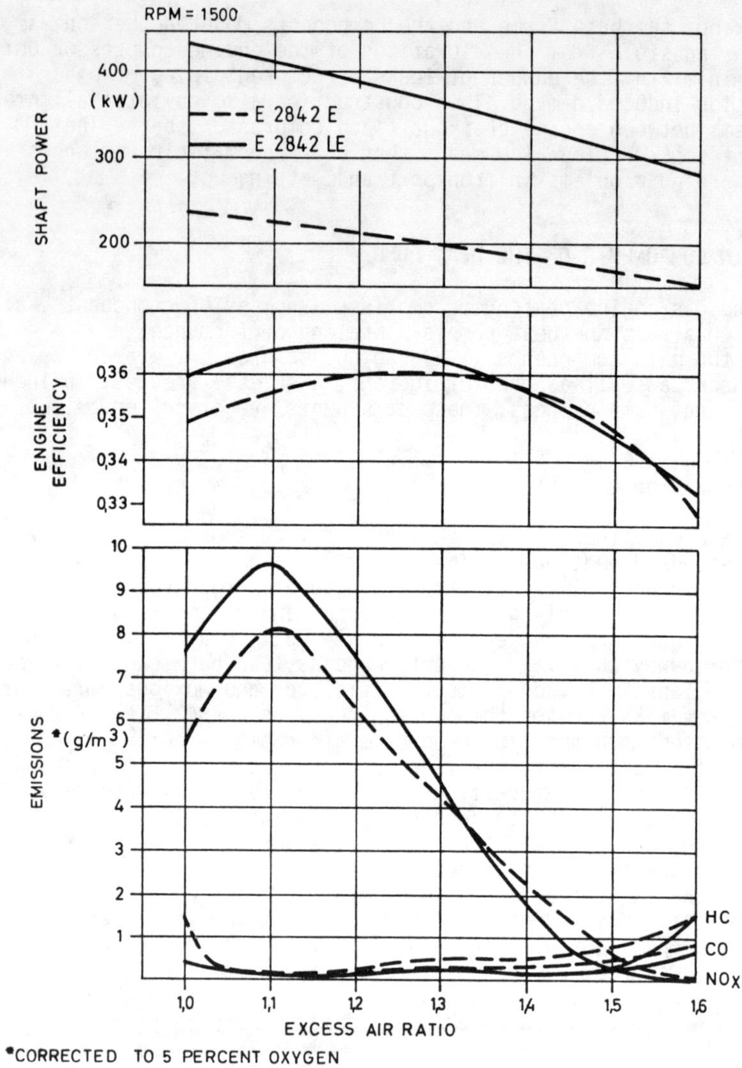

Fig. 2.

3.2 The compressor

Reciprocating compressors, screw compressors and centrifugal compressors are mainly used in compression type heat pumps. Whereas centrifugal compressors are restricted to heat pumps of high rating, screw compressors and reciprocating compressors are largely used for medium and low ratings. Originally M A N heat pumps were fitted with both reciprocating and screw type compressors. After many years' experience with both types, screw compressors are now used almost exclusively from M A N's CF and CR series.

The screw compressors in the CR series can be used with all commercially available refrigerants. The maximum permissible operating pressure for CR compressors is 25 bar. Used in conjunction with the refrigerant R 12 this corresponds to a condensation temperature of 85°C. If, however, it is possible to get on with lower heat pump flow temperatures, as used in many residential heating systems, M A N heat pumps are fitted with the much cheaper and simpler screw type compressors of the CF series. These are air compressors modified for heat pump operation. Maximum operating pressure of the CF compressor is 14 bar. Thus if R 12 is used the unit can be operated at a condensation temperature of 57°C. Unlike the CR compressors the rotors are running in ball bearings. The lubrication system, with oil feed only at the bearings, has been kept very simple. Oil circulation is maintained by the pressure gradient present in the compressor, therefore there is no need for oil pumps.

Figure 3 and Figure 4 show the operating data for the screw type compressors of the CF series predominantly installed in M A N standard heat pumps.

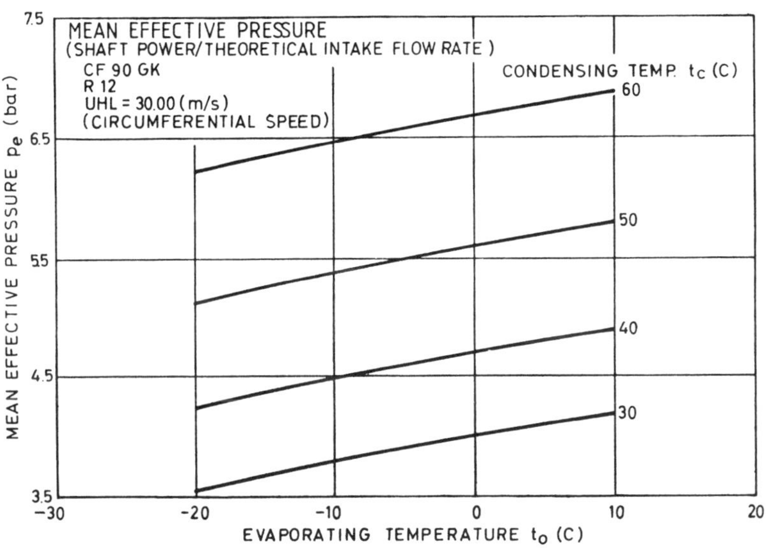

Fig. 3.

$p_e = \frac{P_e}{\dot{V}_{th}}$ ... mean effective pressure

$COP = \frac{\dot{Q}_o}{P_e}$ ... coefficient of performance

P_e ... shaft power

$\dot{V}_{th}$ ... theoretical intake flow rate

$\dot{Q}_o$ ... refrigerating capacity

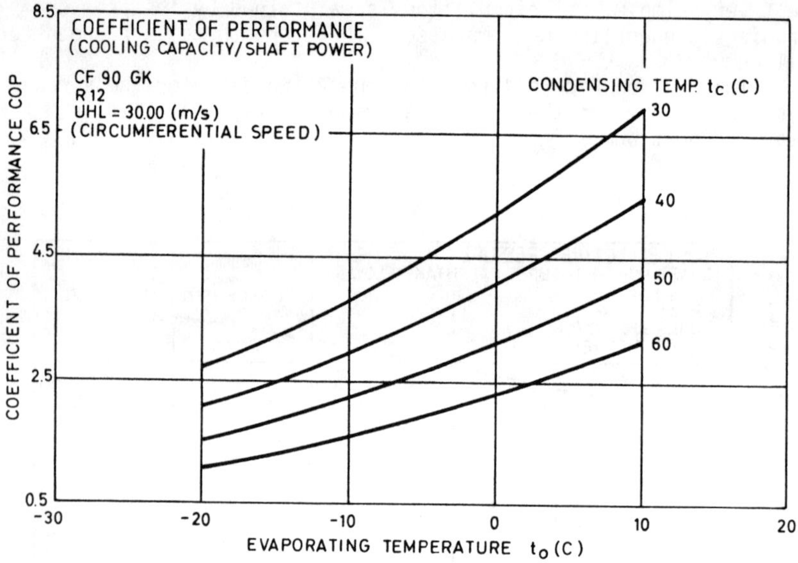

Fig. 4.

In the case of CF screw compressors part-load regulation is carried out solely by variation of speed. Figure 5 shows the part-load performance of a speed governed CF screw compressor operated with R 12. Used in this example is a CF 90 GK.

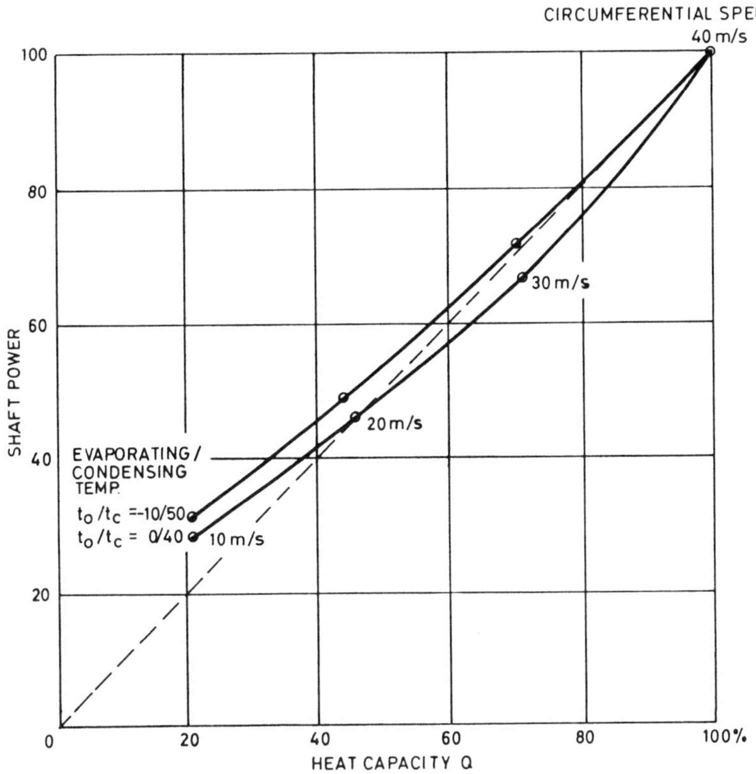

Fig. 5.

3.3 The heat exchangers

The heat exchangers with the greatest influence on the process data of heat pumps are evaporator and condenser. There are many manufacturers of such equipment, and as the market for refrigerating plants and heat pumps is very varied, there are many different types. Here we will only deal with those most frequently installed in heat pumps.

3.3.1 The condenser

Two types of condenser which are frequently to be found in heat pumps are the double-shell tube condenser (coaxial condenser) and the multitubular surface condenser. Whereas coaxial condensers are mainly used in small heat pumps, multitubular surface condensers are fitted in those of medium and large capacity. In this apparatus, which is relatively simple in design and operation, the refrigerant flows through the shell and condenses on the outer surface of the tube, through which heating water flows. The multitubular surface condensers customarily used by M A N in its heat pumps are designed for mean logarithmic temperature differences of 5 - 7 K.

3.3.2 The evaporator

Depending on the anergy source used in the heat pump a distinction is made between air-to-refrigerant and water-(or brine) to-refrigerant evaporators. A distinction is made between dry type and flooded evaporators depending on the evaporation process. The dry type evaporators mainly used in the lower and medium rating ranges are characterized by vaporizing refrigerant in the tubes, the heat being applied to the outer surface of the tube in shell and tube evaporators (water- or brine-to-refrigerant evaporators), or in ribbed-tube or lamellar-tube evaporators. The dry evaporator not only allows considerable superheating of the refrigerant vapour at the outlet but also relatively problem-free removal from the evaporator of the oil transported in the refrigerant. For larger ratings flooded vaporization, mainly through water- or brine-to-refrigerant evaporators of shell and tube construction, are used. In this application the refrigerant evaporates on the outer surface of the tube, like in the multitubular surface condenser. The refrigerant vapour is led off via a vapour dome in which the co-transported fluid droplets are separated. A special suction device is needed to remove the oil in the refrigerant. The water- or brine-to-refrigerant evaporators customarily used in M A N standard heat pumps are designed for gradients of 4 - 6 K, the air-to-refrigerant evaporators for gradients of 8 - 10 K. Gradient as used here means the difference between air intake temperature and evaporating temperature in the case of the air-to-refrigerant evaporator, or between water (brine) output temperature and evaporating temperature in that of the water-to-refrigerant evaporator.

3.4 The refrigerant

The main criteria in selecting a refrigerant suitable for use in heat pumps are evaporating temperature, volumetric heat output, COP, saturation pressure range and thermal stability.

To enable preliminary selection on the basis of saturation pressure range the refrigerant operating temperature range most frequently specified for residential heating, i.e. - 20°C (air-refrigerant evaporator) to + 70°C (output temperature), is delineated. If subpressure in the refrigerant circuit is to be avoided then, of the customary refrigerants listed in Table 1, only R 12, R 22 and R 500 come into consideration. For conventional refrigerant compressors a maximum operating pressure of 25 bar is permitted. Using the refrigerant R 22 this corresponds to a condensation temperature of 62 °C. To obtain the desired heat pump output temperature of 70 °C the refrigerant R 22 can't be used. If one compares the values for the two refrigerants R 12 and R 500 obtained in the theoretical comparison process (Figure 6, Table 1), R 500 shows a greater volumetric heat output and R 12 a greater COP for the annual heating cycle. Computations for the annual heating cycle with M A N gas engine heat pumps show almost equivalent results in respect of energy savings for both the refrigerant R 12 and R 500. Since the longterm behaviour of the mixture R 500 and oil is not yet fully known, M A N uses the refrigerant R 12 in its standard heat pumps.

THEORETICAL COMPARSION PROCESS

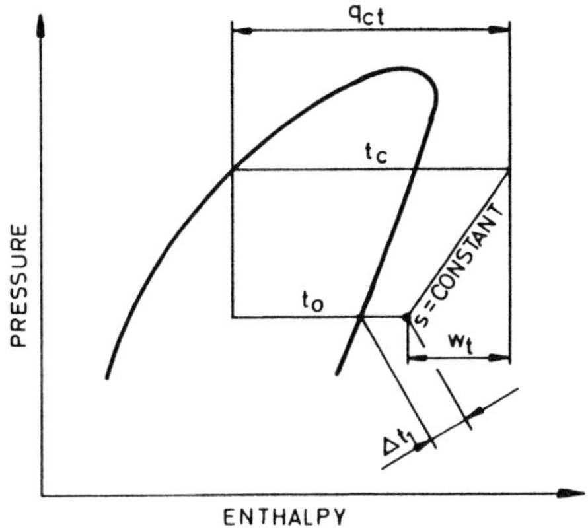

t_o	EVAPORATING TEMPERATURE
t_c	CONDENSING TEMPERATURE
Δt_1	SUCTION GAS SUPERHEAT
q_{ct}	SPECIFIC HEAT CAPACITY
q_{cvt}	VOLUMETRIC HEAT CAPACITY
v_1	SPECIFIC VOLUME AT COMPRESSOR INTAKE
w_t	SPECIFIC WORK OF COMPRESSION
s	SPECIFIC ENTROPY

$$q_{cvt} = \frac{q_{ct}}{v_1} \qquad COP = \frac{q_{ct}}{w_t}$$

Fig. 6.

Table 1.

REFRIGERANT	SATURATION PRESSURE (bar) -20°C	70°C	t_o=-20°C, t_c = 70°C, Δt_1 = 10 K q_{cvt} (kJ/m³)	COP	THERMAL STABILITY IN CONJUNCTION WITH LUBRICATION OIL UP TO
R11	0.16	4.12	181.4	3.01	130°C
R12	1.51	18.96	1103.1	2.63	150°C
R12B1	0.52	8.69	474	2.87	120°C
R22	2.45	29.83	1879.5	2.63	150°C
R114	0.37	7.43	282.8	2.43	160°C
R500	1.77	22.41	1289.8	2.6	150°C

4. SELECTION OF OPTIMAL HEAT PUMP VALUES

The optimal performance data for a heat pump can only be defined after thorough investigation of the heat pump's economic feasibility for a particular application. The two main values determining the economic feasibility of a heat pump installation are its operating data (energy savings) and the capital investment required.

4.1 Calculation of heat pump performance data

To calculate the economic feasibility of using a heat pump it is necessary to know its operational behaviour (performance data) when integrated into the heating system over the entire heating period. To enable such simulations to be carried out a computer programme was drawn up. This computer programme incorporates the performance characteristics of internal combustion engine and compressor, and the corresponding characteristics of the heat exchangers, under all operating conditions. Since the operating conditions for the heat pump are greatly influenced by the heating system concept, this interaction of heat pump and heating system was also incorporated in the computer programme. Besides the relationship between ambient air temperature and heat capacity and the water temperatures of the heating system, the computer programme also takes into consideration the frequency distribution of the ambient air temperature and thus the meteorological conditions for various climate zones in the Federal Republic of Germany. In this way not only can the heat pump performance data at design point be calculated but also the energy savings over an annual heating cycle through heat pump operation compared to a conventional boiler (Example see Figure 7).

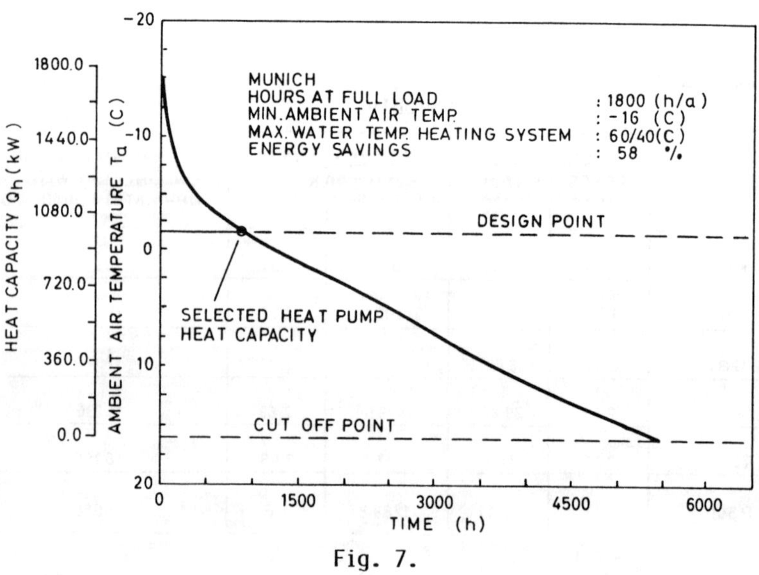

Fig. 7.

4.2 The economic feasibility calculation

To enable optimal selection of heat pumps for use in different heating systems and also to be able to give our customers expert advice on the financial aspects, a computer programme for the calculation of economic feasibility was drawn up. To make the results of the computer calculations as clear as possible and to be able to present them to the customer in an easily comprehensible manner the net present value (NPV) method was used. Under this method the net present value of the difference between the costs for the comparison system (heating system with heat pump) and those for the reference system (conventional boiler) is calculated for each year of the period under consideration. Besides calculating the amortization period for the comparison system the net present value at the end of the system's service life is ascertained, since this figure is also needed to evaluate the system's economic feasibility. State aid, such as capital investment subsidies, preferential depreciation or other grants and allowances, is also included in the computation. When carrying out sensitivity analyses, calculi of variation for the capital investment required for the comparison system, for increases in energy prices, and for the theoretical interest rate can also be performed with the aid of this computer programme, and the results then plotted out. Figure 8 shows the results of a calculus of variation for the increase in price for natural gas.

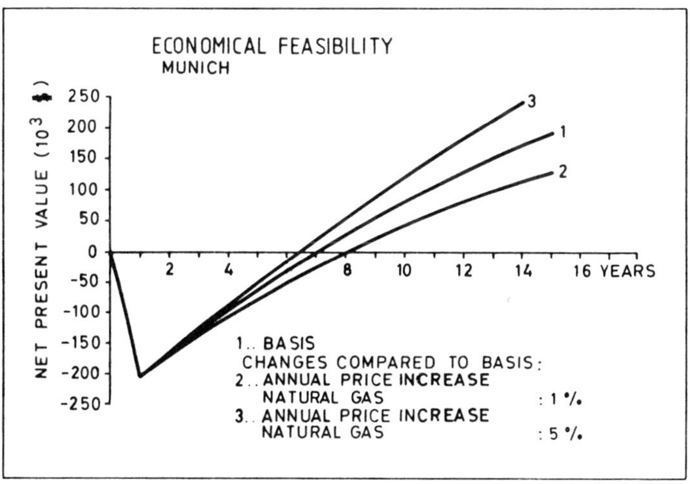

Fig. 8.

Besides the operating data given in Figure 7 the input data listed in Table 2 were also used in this calculation.

Table 2.

Heat capacity of heat pump at design point	980 kW
Capital investment for heat pump, including peripheral equipment	288,000 $
Service life	15 a
Natural gas price	0.18 $/m^3
Capital investment grant	7.5 %
Theoretical interest rate	8 %
Annual repair costs	2,880 $
Annual maintenance and operation costs	4,320 $
Annual energy savings compared to reference system (conventional boiler)	58 %
Annual cost increases for:	
natural gas	3 %
repairs	3 %
maintenance and operation	4 %

Chapter 29

Development of a Kinematic Stirling-Engine-Driven Heat Pump System

Russell E. Monahan

ABSTRACT: A gas-fired, Stirling engine-driven heat pump has been projected to provide coefficients of performance (COP's) of 1.8 for heating at 8°C (47°F), and 1.1 for cooling at 35°C (95°F), resulting in significant energy savings. Progress in the development of the V160 engine for this application is presented and a commercialization schedule is outlined. Recent engine and system developments and technical accomplishments are detailed, including the environmental laboratory testing of a 35 kW (10-ton) breadboard prototype heat pump system and the accumulation of over 120,000 engine and test rig hours.

INTRODUCTION

Heat pumps driven by electric motors have been strongly accepted for both residential and commercial applications. For many years the gas industry has recognized the need for a gas-fired alternative to the electric heat pump. The development and commercialization of such a product is expected to ensure growth of the gas industry's customer base.

The Stirling engine lends itself well to the application as a heat pump prime mover because of its low noise and vibration, high torque at low speeds and high inherent thermodynamic efficiency. In addition, most waste heat is transferred to the engine coolant, resulting in low exhaust gas temperatures. This high percentage of useable heat permits domestic hot water heating and space conditioning that normally require supplemental electric resistance heaters. Also, the Stirling engine needs very little maintenance and can operate for long periods of time without service.

HEAT PUMPS: Prospects in Heat Pump Technology and Marketing (Proceedings of the 1987 IEA Heat Pump Conference), Edited by Kay H. Zimmerman

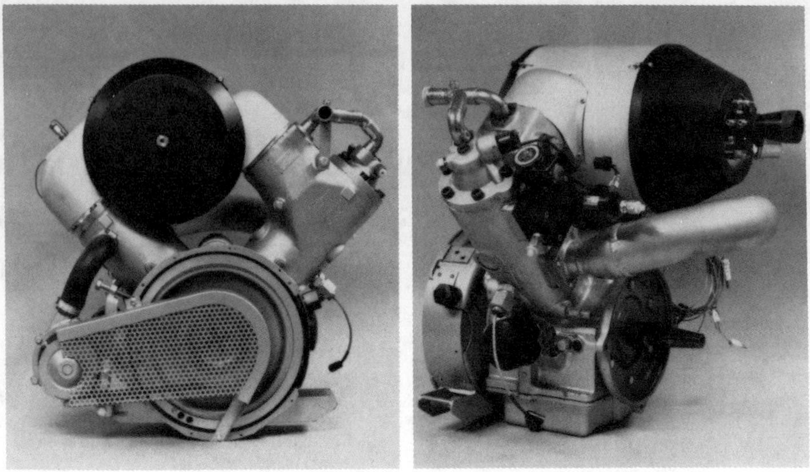

Figure 1. The V160 Stirling Engine

The V160 is a 90°V, two cylinder, kinematic, single-acting Stirling engine as shown in Figure 1. With the working gas at a pressure of 15 MPa (2175 psi) and temperature of 700°C (1292°F), the 160cc swept volume produces a peak 20 kW (26.81 hp) power at 3000 rpm. The efficiency exceeds 30% (LHV) at approximately 1000 rpm and 78 N-m (58 ft-lbs) torque. Typical engine characteristics are shown in Figure 2.

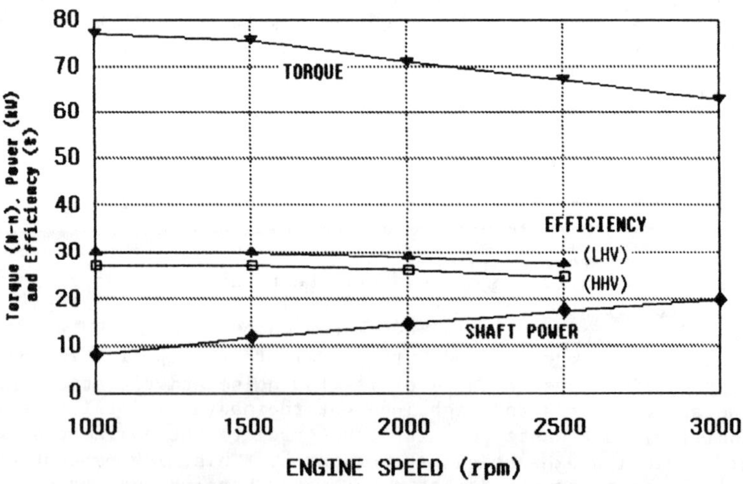

720°C Hydrogen Working Fluid, 30°C Coolant

Figure 2. V160 Performance Characteristics.

Originally developed for generator set applications, the V160 engine was adapted for the natural gas-fired heat pump prime mover through the replacement of the liquid fuel system with an ejector/venturi arrangement. Engine control software was modified to provide speed modulation. A number of component design changes have resulted in reduction in package cost and increases in durability and performance.

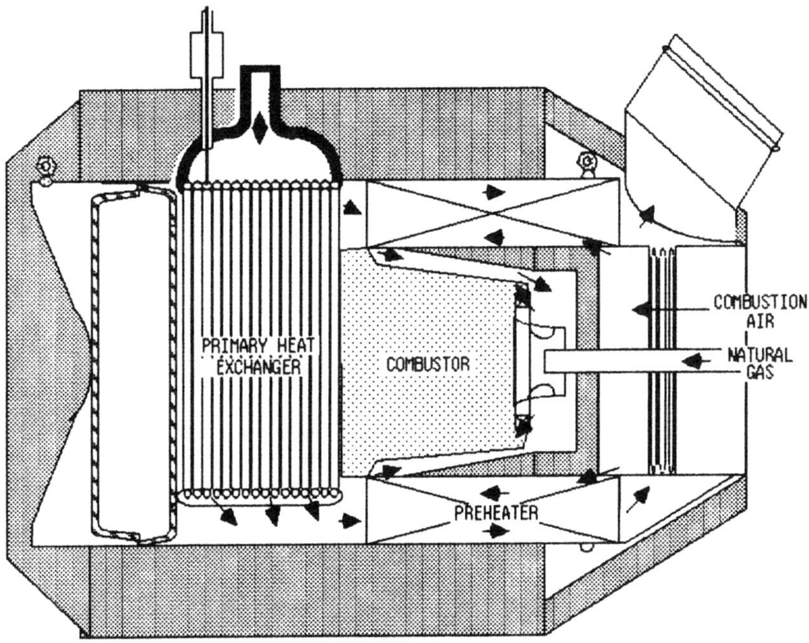

Figure 3. V160 Heating System.

The V160 heater head (shown in Figure 3) allows exhaust gases to preheat incoming air before it enters the combustion chamber, offering a substantial overall increase in efficiency. Combustion products pass through the primary heat exchanger, transferring energy to the working fluid as it passes through the tubes

The moving parts in the V160 drive system are enclosed and cannot be contaminated by the outside air or the fuel in the combustion chamber. The pistons and piston rings operate dry in the pressurized working gas system, separate from the lubricating oil system. This unique combination results in a cost-competitive and reliable product.

The engine's microprocessor control unit (MCU) monitors safety functions and regulates the two major control loops, speed/load and heater temperature. Engine speed and load are functions of the working gas pressure, which is controlled by means of solenoid valves and a small compressor; the higher the pressure, the more power output. Heater temperature is sensed through a thermocouple and air is delivered by a speed modulated combustion blower; the more air and fuel inducted, the more thermal energy released to the working fluid.

Over 120,000 hours of testing was performed on V160 engines and test rigs. One engine, operating on digester gas at a sewage plant, exceeded 20,000 hours of operation with little maintenance. Stirling Power Systems continues to develop component and system designs that will increase reliability and efficiency and reduce manufactured cost.

System Development

A concerted computer modelling effort preceded the assembly of the first V160-driven prototype heat pump system. A thorough engine mapping exercise provided complete speed, power, torque and fuel consumption data for this simulation. Using well documented characteristics of the components of the baseline electric heat pump, several candidate compressors were evaluated by "mating" with the engine data. Based on the calculated seasonal performance, a variable displacement (two or four operational cylinders) compressor was selected.

Assembly of the breadboard prototype was carried out at the Borg-Warner Research Center. For ease of testing and construction, a split system heat pump was chosen as the basis of the breadboard system. A system schematic is shown in Figure 4. Two recirculating air handling test loops were used in the preliminary evaluation of the system. These test results showed good agreement with the computer model.

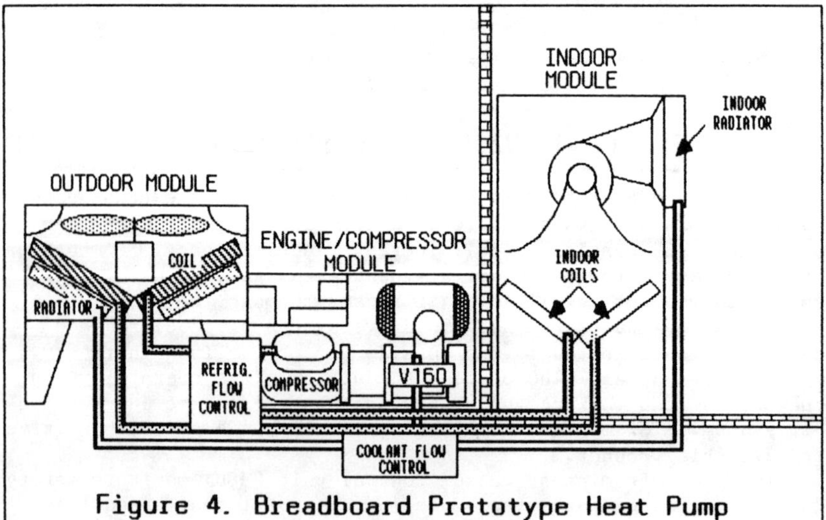

Figure 4. Breadboard Prototype Heat Pump System Schematic

System Testing

Full scale environmental laboratory testing of the breadboard prototype system was carried out in the first quarter of 1986. Without standardized test procedures for engine-driven heat pump systems, a program plan was outlined before testing and revised as the evaluation proceeded. All cooling mode tests were run with indoor conditions of 20°C (67°F) wet bulb and 27°C (80°F) dry bulb temperatures. All heating mode tests were made with the indoor temperature at 21°C (70°F) dry bulb. Tables 1 and 2 summarize the test results for cooling and heating modes, respectively.

Table 1. Cooling mode test results.

	DB/WB Temps		Engine Data				System Data			
Run No.	ID°C	OD°C	Speed (rpm)	Torque (N-m)	Power (kW)	Eff. (%)	No. Cyl.	Capacity (kW)	COP's Thermal	COP's Cost
1	27/20	35	1700	66.7	11.86	25.8	4	32.8	.713	.541
2	27/20	28	2000	64.3	13.5	25.7	4	37.0	.706	.550
3	27/20	35	2700	39.5	11.19	20.5	2	27.3	.500	.393
4	27/20	28	2900	38.5	11.71	20.8	2	29.8	.530	.417
5	27/20	35	1000	60.1	6.29	26.6	4	23.6	.996	.629
6	27/20	28	1000	55.2	5.78	26.5	4	25.7	1.17	.723
7	27/20	20	1000	47.5	4.97	26.8	4	26.8	1.44	.825
8	27/20	28	1000	23.9	2.50	16.5	2	15.3	1.01	.532
9	27/20	20	1000	16.5	1.73	13.7	2	15.3	1.22	.578

Table 2. Heating mode test results.

	DB/WB Temps		Engine Data				System Data					
Run No.	ID°C	OD°C	Speed (rpm)	Torque (N-m)	Power (kW)	Eff. (%)	No. Cyl.	Capacity (kW)	Q. to Rad. (kW)	% Total by Rad.	COP's Thermal	COP's Cost
10	21	8/16	2050	58.2	12.46	24.1	4	65.1	26.5	40.8	1.26	.979
11	21	-8/-9	2400	46.9	11.79	25.5	4	48.9	23.6	48.3	1.08	.798
12	21	-21/-21	3000	41.8	13.13	22.2	4	50.3	32.4	64.5	.850	.677
13	21	-22/-22	2400	42.4	10.67	28.3	4	35.1	18.7	53.3	.934	.674
14	21	8/16	2700	33.6	9.47	27.4	2	56.0	24.4	43.6	1.16	.889
15	21	-8/-9	3000	26.6	8.36	19.7	2	39.1	23.6	60.4	.939	.682
16	21	-22/-22	3000	41.8	13.13	22.2	2	50.3	32.4	64.5	.850	.677
17	21	8/16	1000	47.6	4.98	25.7	4	33.4	9.0	26.9	1.72	.991
18	21	-8/-9	1000	40.9	4.51	25.1	4	23.9	9.4	39.3	1.33	.744
19	21	8/16	1000	21.0	2.20	17.3	2	20.7	6.2	30.1	1.63	.778
20	21	-8/-9	1050	21.3	2.34	17.5	2	15.1	5.8	38.4	1.13	.547

To determine the 35°C (95°F) outdoor cooling mode rating point, the engine was run to its standard maximum working gas pressure of 15 MPa (2175 psi) and was able to achieve a speed of 1700 rpm. At this condition, the measured net cooling capacity was 32.8 kW (111,857 Btu/hr or 9.3 tons). Subsequent testing with higher engine output showed performance degradation indicating refrigerant side limitations caused by flow restrictions.

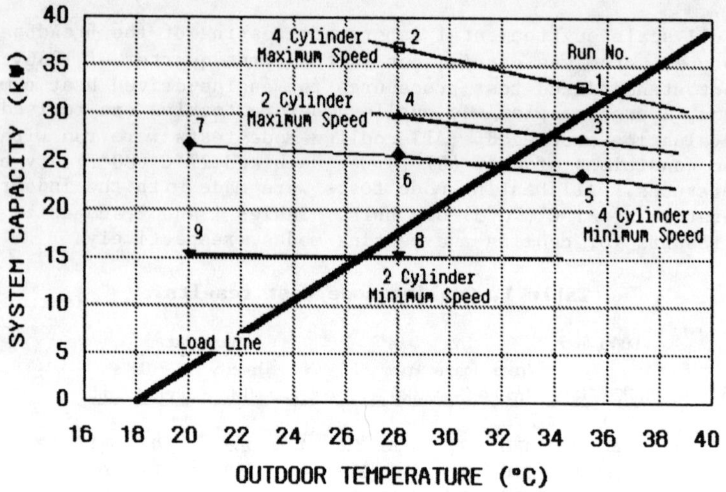

Figure 5. Cooling Mode Test Results.

Figure 5 shows the steady state cooling mode test points with the load line for the system's seasonal performance evaluation as outlined in proposed ARI standard 210/240 for variable speed electrically driven heat pumps. Table 1 shows the data corresponding to each run number. Two experimental speed and outdoor temperature combinations, at maximum and minimum engine conditions for both two and four cylinder active compressor states were chosen. This allowed interpolation of load matching points along the load line.

Thus, with the compressor being driven at full power in the four cylinder mode, test points 1 and 2 define the system maximum capacity. In contrast, with the compressor operating at 1000 rpm in the two cylinder mode, test points 8 and 9 show the minimum system operational level. Cycling of the system will occur at outdoor temperatures less than 32°C (89°F) for four cylinder operation and 27°C (81°F) for two cylinders. Cooling load matching is possible through engine speed modulation above these temperatures to the system capacity at approximately 36°C (97°F).

Thermal COP's for cooling mode operation range from .706 to 1.44 for four cylinder operation, with lower results for the two cylinder mode.

Figure 6 shows the maximum and minimum design heat requirement (DHR) load lines as determined using the procedure outlined in the appendix of ARI standard 240. These represent the two extremes in the application of 35 kW (10 ton) equipment with most buildings having requirements closer to the minimum DHR line. Table 2 shows the data corresponding to each run number.

The maximum system capacity exceeded 64.5 kW (220,000 Btu/hr) at 8°C (47°F), with four active compressor cylinders, 2050 rpm and the maximum engine working gas pressure of 15 MPa (2175 psi). At -21°C (-5°F), the breadboard system's output was over 49.8 kW

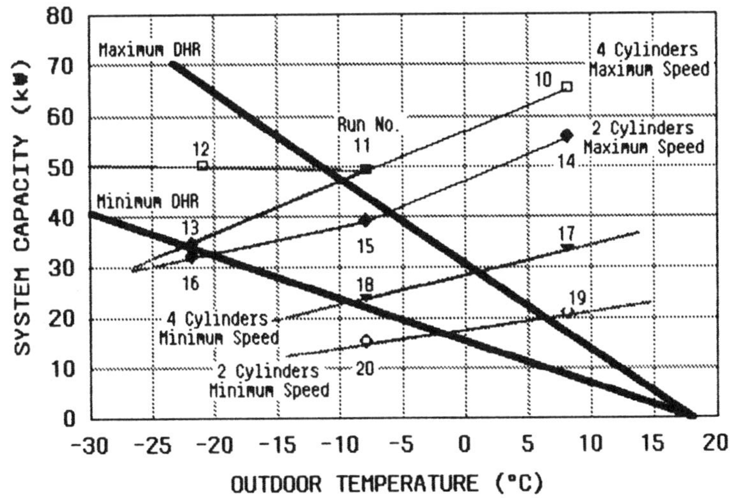

Figure 6. Heating Mode Test Results.

(170,000 Btu/hr), considerably higher than the standard electric-driven system which requires substantial resistance heating for outdoor temperatures lower than -8°C (17°F) along the minimum DHR line.

Due to lower refrigerant densities at the compressor inlet, maximum available engine speed increased as outdoor temperature dropped. Therefore, for minimum DHR applications, load matching would be possible through engine speed modulation from minimum capacity (approximately -2°C (28°F) for two cylinders and -9°C (15°F) for four cylinder compressor operation) down to outdoor temperatures near -34°C (-30°F).

With the minimum four cylinder capacity at 8°C (47°F) outdoor temperature, the system thermal COP is 1.72. Even at -21°C (-6°F), the thermal COP is .934, which is comparable to the highest efficiency gas furnaces. As shown in Table 2, engine coolant heat can be used for additional space conditioning and provides a substantial portion of the total system capacity, especially at lower outdoor temperatures.

Seasonal Performance

To allow the comparison of electric-driven and gas-fired engine-driven heat pump systems, the "Cost COP" was defined as:

$$\text{Cost COP} = \frac{\text{Net Cooling or Heating Capacity}}{\text{Gas Used} + \frac{E}{G} \times K_c \times \text{Electricity Used}} \quad [1]$$

where E/G = Electric/Gas price ratio (3.65 for this study)
K_c = Energy conversion factor (if needed).

This index weighs the use of electricity, relative to gas consumption by the consumer's electric/gas price ratio and provides a figure proportional to the operating cost of the heat pump.

Using climatic data for Chicago and Dallas, a comparison of the seasonal operating costs of the breadboard engine-driven and standard electric systems was undertaken. A computer program was developed to compute the seasonal performance for both systems. This involved the use of "temperature bins" for which the system Cost COP was determined. By combining these with the anticipated number of hours of operation in each bin for both geographic locations, seasonal operating costs were calculated.

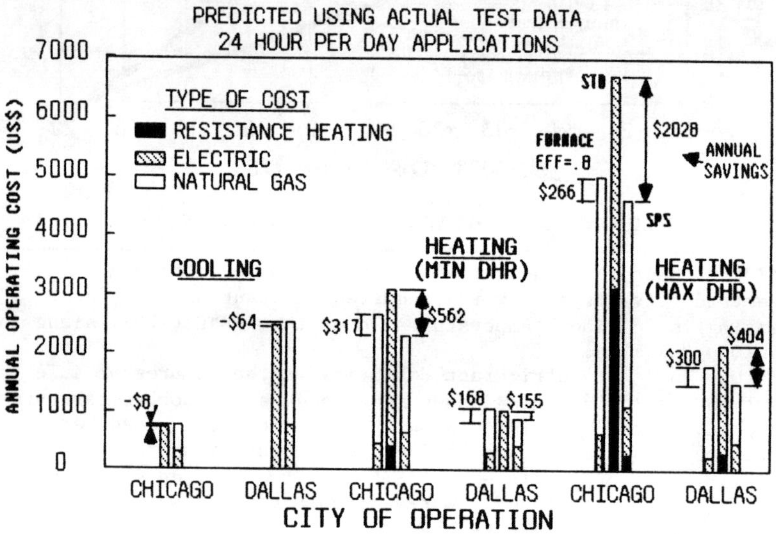

Figure 7. Annual Operating Costs.

Figure 7 and Table 3 show the comparison of these costs for both heat pump systems and a gas furnace. Compared to the electric heat pump and assuming a cyclic degradation factor of .25 for both systems, the V160-driven heat pump will save a consumer in Chicago between $562 and $2020 in annual heating costs while cooling costs are equivalent.

Table 3. Operating cost comparison.

V160-Driven Heat Pump vs. Standard Electrical vs. Gas Furnace

	Cost COPs								Annual Operating Costs							
Operating Mode	V160		Std.		Furnace		V160/Std.		Std.		V160		Furnace		V160/Std. Savings	
	Chi.	Dall.	Chi.	Dall.	Chi.	Dall.	Chi.	Dall.	Chi.	Dall.	Chi.	Dall.	Chi.	Dall.	Chi.	Dall.
Heating-MinDHR	.760	.793	.607	.669	.665	.659	1.252	1.185	2792	995	2230	840	2547	1008	562	155
Heating-MaxDHR	.770	.863	.527	.684	.726	.723	1.461	1.262	6428	1944	4400	1540	4666	1840	2028	404
Cooling	.645	.631	.651	.648	NA	NA	.991	.974	826	2404	834	2468	NA	NA	-8	-64

CONCLUSIONS

The test program and seasonal performance analysis have indicated a number of interesting points. Even though system cycling would be reduced, two cylinder compressor operation is not cost effective because the higher engine efficiency at the larger four cylinder loads more than compensates for the increased cycling losses. At very low ambient temperatures, a decrease in engine shaft efficiency actually increases system performance as the thermal energy delivered to the engine coolant provides more useable heat than can be obtained through the refrigerant system. Maintenance and service costs are expected to be determining factors in the economic viability of any engine-driven heat pump system; because of the Stirling engine's reliability and long life, the V160 is a prime candidate for this application. Future changes in the prices of electricity and natural gas would significantly affect the cost analysis; however only an increase in the electric-to-gas price ratio is foreseen for the near future.

Future Plans

Tests of the breadboard heat pump indicate the need for several design changes to the system. Refrigerant line restrictions must be reduced to bring up the cooling mode capacity, indoor and outdoor coil designs must be improved to increase efficiency and parasitic electrical loads (fans, pumps and blowers) must be decreased in order to lower operating costs. A second generation V160-driven heat pump system will be assembled and evaluated during early 1987 and will incorporate these design changes.

With the successful completion of an engine durability demonstration program in late 1986, field experimental testing of the V160-driven system is being scheduled for the summer of 1987. Preproduction manufacturing of the engine is now underway and units are being prepared for market seeding in a variety of applications. Full scale commercialization is expected by 1991.

ACKNOWLEDGEMENTS

1. K. Kountz, J. Glaw, and J. Rider, "Performance Test of Experimental SPS/Champion 10 Ton Heat Pump", Borg-Warner Research Center Project 4552, July 1986.

2. L. Johansson, J. Agno, and W. Percival, "The Advanced Gas-Fired Stirling Engine as a Heat Pump Drive", Stirling Power Systems, SAE Paper 860881, May 1986.

CONCLUSIONS

[illegible]

[illegible]

[illegible]

Section VI

Heat Pumps for District Heating Applications

Chapter 30

Heat Pumps in Block Centrals

Rolf Westerlund

ABSTRACT: In Sweden existed 1984 8.400 block-centrals consuming 1,9 million m^3 oil per year. Despite that a large share of these will eventually be connected to district heating is the market potential for heatpumps large. Electric driven heatpumps are in Sweden the best solution for the most small and medium size block-centrals.

INTRODUCTION

In 1984 there existed some 8.400 block-centrals in the non-industrial sector in Sweden. They consumed approximately 1,9 million m^3 (500.000 US gallons) oil per year.

In the same year the district heating consumed some 1,2 million m^3. In the district heating only 43 % of the energy output was produced by oil.

In block-centrals are close to 100 % of the heat production produced by oil. Therefore the block-central is an important market for equipment which can reduce the oil consumption and for many reasons are heatpumps well suited for this market. Since 1984 the number of block-centrals have decreased due to the connection to the district heating system. The total number in 1986 is estimated to be 8.000.

HEAT PUMPS: Prospects in Heat Pump Technology and Marketing (Proceedings of the 1987 IEA Heat Pump Conference), Edited by Kay H. Zimmerman

TECHNICAL DESCRIPTION OF BLOCKCENTRALS

A block-central in this context is defined as a plant producing hot water for heating and sanitary hot water for more than one building. A block-central is often owned by an organization which ownes a large part of the connected buildings.

Water temperatures up to 120°C (250 F) are used. Steam is rarely used in Sweden for heat production. In some hospital are centralized produced steam in use but the general trend is to substitute the steam with hot water sterilizers or by decentralized electric boilers for sterilization.

The industry uses steam more frequently but are not included in this study.

The size of the block-centrals can be from 300 kW up to 30 MW. They can mainly be divided into three different catagories depending on the size. The smallest block-centrals have only one sub-central placed in the boiler room. The largest have the same technical standard as district heating systems. The medium size block-central has several sub-centrals but of a lower standard than the district heating.

Small size block-centrals

The majority of the block-centals which consumes less than 200 m^3 of oil per year have only one sub-central. This sub-central is installed in the boiler room and the energy is distributed to the connected buildings through a four pipe system. Two of the pipes are used for heating and the other two for sanitary hot water, see figure 1 below.

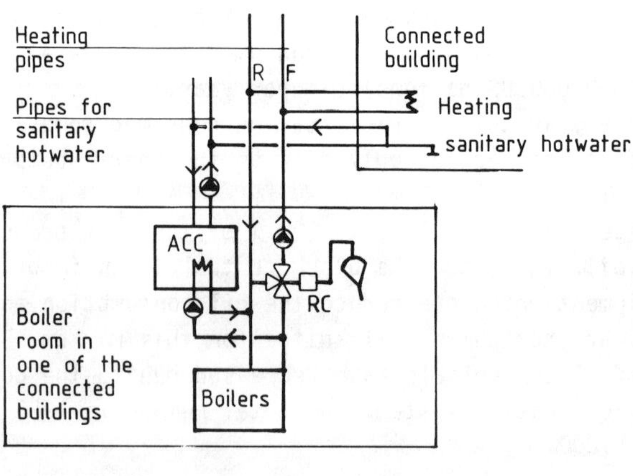

Figure 1. Small size block-centrals

The water temperature in the distribution system is low. The forward temperature for heating is at the most 50-60°C. In the beginning of the heating season is the forward temperature 30-35°C. The sanitary hot water temperatures are with small variations between 50-60°C throughout the year. The return temperature is about 10-15°C lower than the forward temperature. The flow rate is constant.

Sanitary hot water is normally produced and stored in an accumulator in the boiler room. The accumulator is in most cases rather small and with poor installation and stratification.

In Sweden there are approximately 5,800 small block-centrals consuming more than 550.000 m^3 of oil per year.

Medium size block-centrals

The medium size block-centals consumes between 200 and 2.000 m^3 of oil per year. The boilers are installed in a separate building or at the ground level of one of the connected buildings.

A medium size block central has more than one sub-central but the primary system is not separated from the secondary system with a heat-exchanger. Energy is distributed with a two pipe system, here called the primary system. Sanitary hot water and water for the heating system is produced in the sub-centrals, see figure 2.

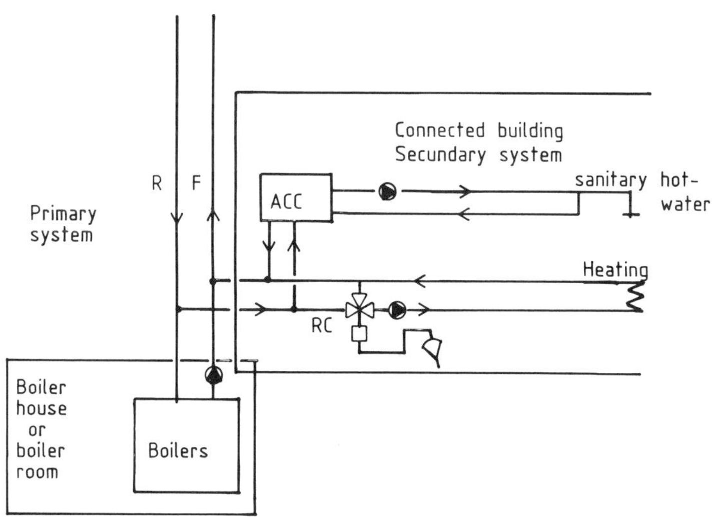

Figure 2. Medium size block-centrals - installations

In the primary system is the flow rate constant and the forward temperature is set at a fixed level for a long period. Boiler-water is used in the primary system which dictates the use of high temperatures. The forward temperatures in the primary system is often 90-110°C. The return temperature is often higher than 70°C. In the secondary system the temperature levels are the same as in the small size block-centrals, see figure 3.

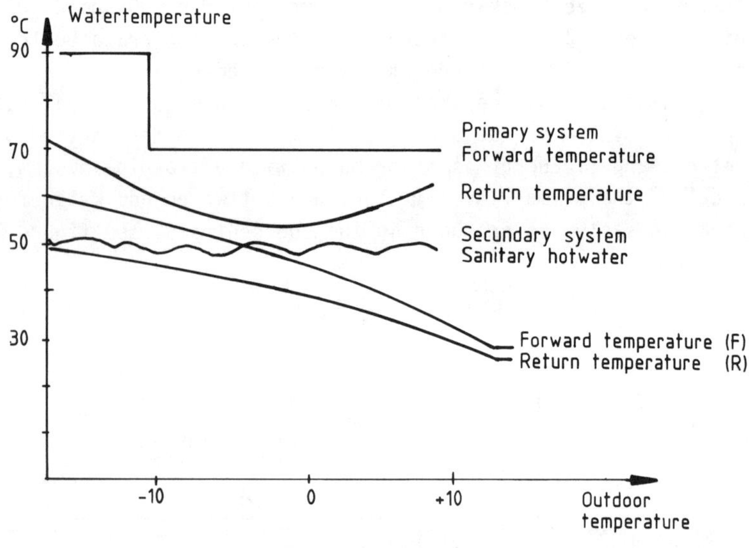

Figure 3. Medium size block-centrals - watertemperature

In the sub-central sanitary hot water is produced in a small accumulator and water of the correct temperature for the heating system is created. In some medium size block-centrals exists a mixture of sub-centrals. Some sub-centrals have heat-exchangers and others have not.

There is a total of 2.500 medium size block-centrals in Sweden. They consume 1,000,000 m³ of oil per year.

Large block-centrals

Large block-centrals have sub-centrals with heat-exchangers and are more or less operated in the same way as the district heating systems. The sub-centrals are in somewhat simpler then district heating and the operation strategic is also simpler.

The forward temperature is between 80-120°C and the return temperature between 70-90°C. The flow rate is often constant in the primary system. In Sweden exists about 100 large block-centrals which consumes some 250.000 m^3 of oil per year.

HEAT PUMPS IN BLOCK-CENTRALS

The most common way to install heatpumps in block-centrals is to place them in the boiler room or in a separate building nearby. Installation in the sub-centrals can be done if exhaust air will be used as a heat source.

In block-centrals heatpumps has to compete with the continuing use of oil, wood-chips, coal or other fuels. In Sweden was installation of electric boilers an interesting alternative in the years between 1982-1985.

A study in 1984 showed that the technical and economical possibilities for heatpumps are best in small och medium size block-centrals. In large size block-centrals are installation of boilers for wood-chips often a more economical alternative.

In small and medium size block-centrals can normally only more expensive refined types of solid fuels be used. At the same time the problems associated with heatpump installation is smaller in small and medium size block-centrals, e.g lower temperature levels, less demand of heat source, than in larger ones.

The installation of heatpumps in block-centrals requires some modifications to the distribution system to secure good working conditions. Regardless of the size of the block-central must all reasonable measures which reduces the temperature level in the system be carried out. Fast variations of the load can be reduced by installing well functional accumulators and control equipment.

Reduction of the temperature level in the secondary heating system is carried out by different energy saving measures such as airtight isolation and balance of the heating system. Replacement of some radiators is also often necessary.

Reduction of the sanitary hot water under 50°C at the taping point is not feasible. To reduce that the temperature drops in the pipes, water circulation must be installed. This also reduces the water consumption.

The possibility of introducing the above mentioned measures are much bigger in block-centrals than in district heating systems. This is due to the fact that in block-centrals the heat production and the connected buildings are owned by the same organisation.

In medium and large size block-centrals the temperature level in the distribution pipes must also be reduced. This is done by using new controlling equipment and by protecting the boilers from cold water. The sub-centrals must be rebuilt to cope with the lower water temperatures. A reduction of the temperature in the distribution system will reduce the heat losses. Normally are the heat loss between 5-20 % depending on the pipe size, length and insulation standard.

In the small size block-centrals the accumulators must be installed in the boiler room, as there are no sub-centrals. In the larger block-centrals the accumulators can be installed in a sub-central, but normally they are also installed in the boiler room.

The costs of these measures can represent as much as 30 % of the total cost of the installation.

Figure 4 shows the average cost of installation of heatpumps and boilers for solid fuel, in existing block-centrals. This figure assumes that no particular problems occurs during the installation.

If an existing boiler can be converted to solid fuel, the cost will be reduced by 20-30 %.

In Sweden all commercial heatpumps are driven by electricity. Diesel-driven heatpumps cannot compete today due to the low price of electricity. If the price of electricity would increase it is still doubtful if the diesel-driven heatpumps can compete with a diesel-driven co-generation plant, as shown in figure 5.

Figure 5 shows which system is the most economical in a large size block-central taking into consideration oil and electricity prices, assuming that the existing boilers are worn-out.

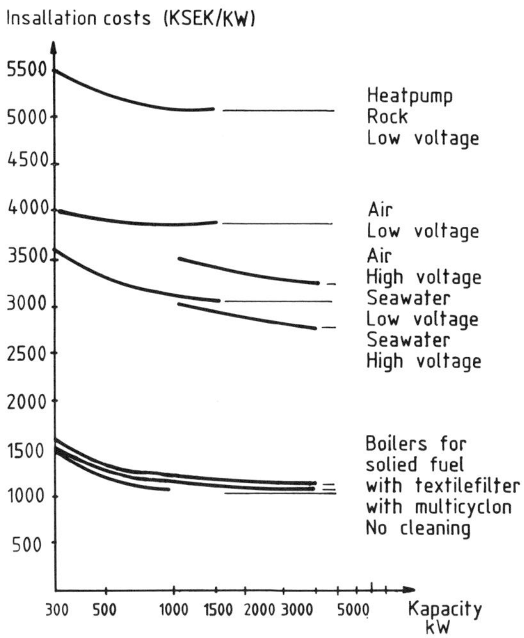

Figure 4. Installation costs

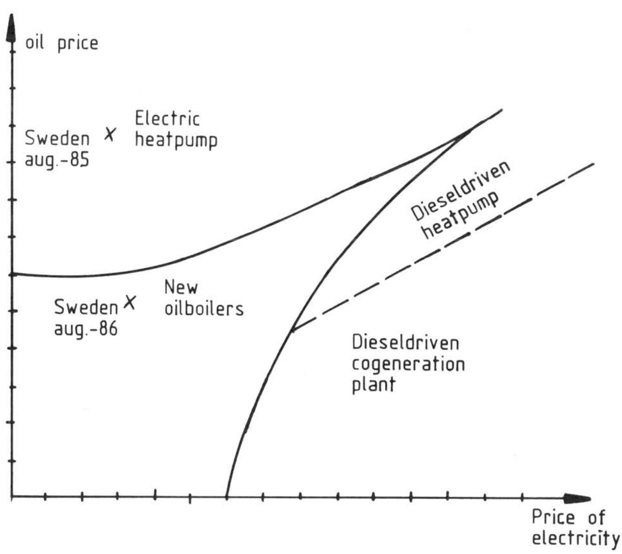

Figure 5. Economical potential for different installations

MARKET POTENTIAL

In block-centrals heatpumps will compete with continuing use of oil, wood-chips or simular fuels or connection to district heating system. Which alternative that is chosen depends upon many factors. In many block-centrals is the installation of heatpumps not possible due to lack of space in the boiler room, its surroundings or the lack of suitable heat source. For others are installation of wood-chip boilers or connection to district heating system more suitable. The continuing use of oil is in some cases a good solution.

In Sweden approximately 50 % of the block-centrals will be connected to district heating. Experience shows that it is very difficult to compete with district heating in district heated areas even if another alternative are more economical.

Table 1 shows the estimated long-term market for different sizes of block-centrals. Block-centrals which can install several different systems are supposed to install the most economical alternative. Oil and electricity prices calculated as in August 1985. The drop of oil prices between August 1985 - August 1986 has reduced the economical potential for heatpumps. Unless the present oil prices are conserved for many years, the calculated potential for heatpumps in block-central are valied.

Increased prices of electricity could reduce the potential as shown in figure 5 and create a market for small diesel-driven co-generation plants.

Table 1. Market potential for heatpumps and other systems in block centrals

Block-central size	Plants today	Convertion to District heating	heat-pumps	wood-chips	Continuing use of oil
Small	5,800	2,900	1,700	400	800
Medium	2,500	1,200	600	600	100
Large	100	50	10	40	0

Chapter 31

Heat Pumps for District Applications: Operating Experience in Japan

Takanori Chiba

ABSTRACT:

In Hikarigaoka Park Town, a 15-minutes drive from the heart of Tokyo, Heat pump system for District H & C utilizing urban waste heat was adopted. As demand for load gradually increased since the initial operations in March 1983, an additional heat pump with a capacity of 3.6 G cal/H utilizing a Heating-Tower was constructed in October 1985 and is now in full operation.

This thesis relates (1) operating experience since 1983 & some measures taken for the operations, (2) characteristics of predicted operations by the Heat pump with Heating-Tower, and the results of the operations.

This thesis also relates (3) another DHC project based on the above system, Heat pump with Heating-Tower. The final capacity of system is about 15 Gcal/H with Heat storage tank 7,200 m^3. It introduces outlines of the project and cost trade-off studies of several cases for the DHC which is now under construction to cover 5.3-hectare Hibiya area located in the center of Tokyo, Japan.

HEAT PUMPS: Prospects in Heat Pump Technology and Marketing (Proceedings of the 1987 IEA Heat Pump Conference), Edited by Kay H. Zimmerman Chelsea, Michigan 48118. Printed in U.S.A.

SYSTEM AND FEATURES OF HIKARIGAOKA PARK TOWN

Hikarigaoka Park Town is a new town with a total area of 186 ha and it is designed to accommodate 40,000 people by 1988. In the town, the heating, cooling and hot water are supplied with a heat pump system. This uses the urban waste heat and operation was started in March, 1983.

The waste heat condensation from the incinerating plant and waste heat from cooling of the extra-high voltage underground cable is effectively used for this system. The structure of the system is the center plant with a heat storage tank of large capacity and the substations which are decentralized.

The equipment for the plant is installed in accordance with progressive stages in order to avoid unreasonable investment in advance.

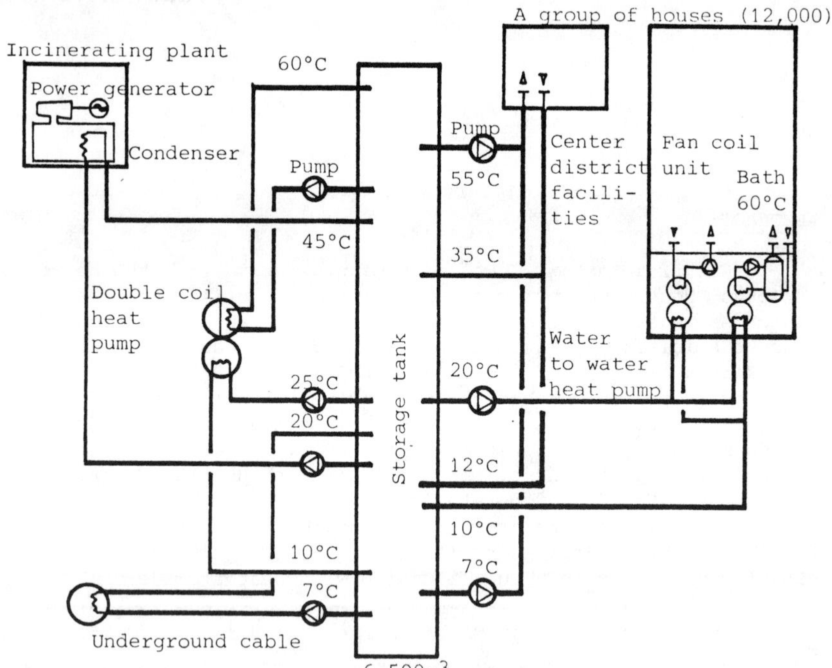

Fig.-1.Schematic diagram of the total DHC system.

Operating Experience

The district heating and cooling (DHC) has been in operation for four years, since the start of supply in 1983. During this period, in order to coordinate the increase of heating and hot water supply load for the residential zone and the cooling load for commercial equipment, the heat source for the center plant and substations has been increased.

The heat pump with Heating Tower is utilized as the primary heat source of the center plant, and the number of water-to-water heat pumps for the substations was increased.

During this period, volume control system was introduced to the substation in order to deal with the maximum load in the winter time.

The increase of substations and the installation of control valves

To maintain operation with the increase of substations from 15 (as of 1983) to 58, it became more important to consider to maintain the temperature difference and flow balance of water from the center plant to retain high efficiency operation.

For this purpose, control valves were installed at the substations at the end of lines from the center plant.

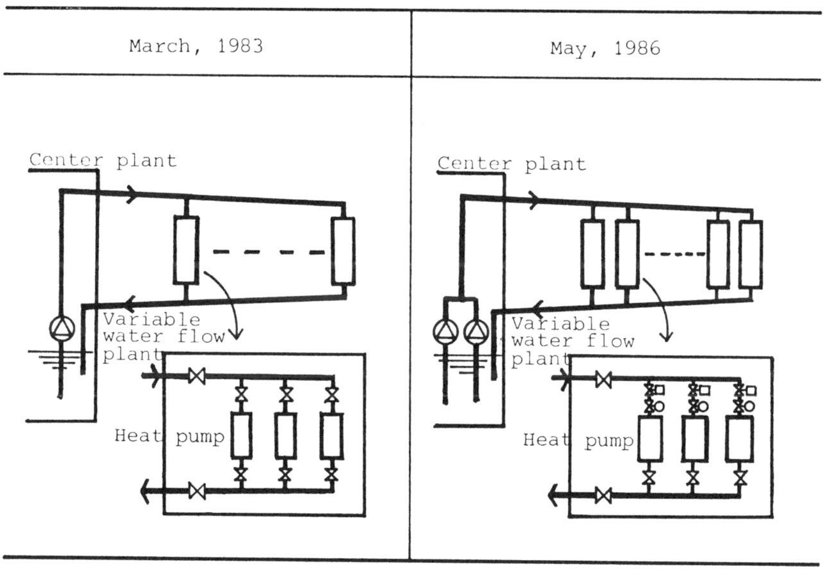

Constant volume valve   Solenoid valve

Fig.-2.The installation of control valves.

The change of coefficient of performance (C.O.P.) of the heat pumps installed in the subplant is shown in Fig.-3.

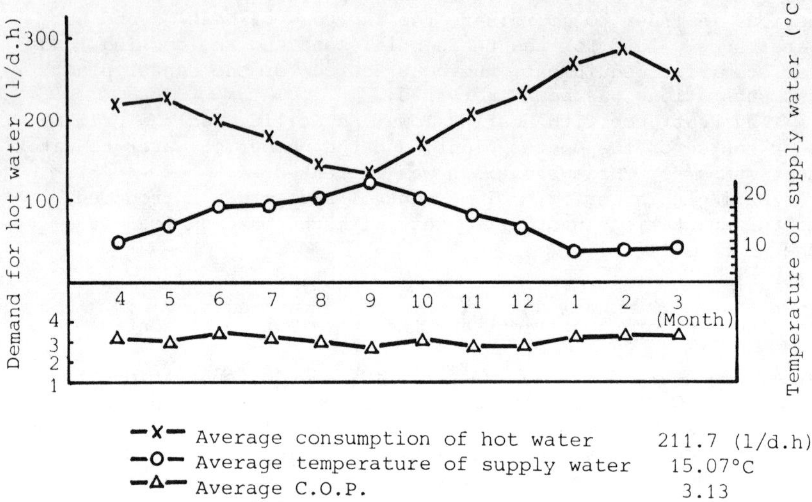

Fig.-3.Seasonal value of hot water supply and COP in A substation.

Increase of heat source for the center plant

In keeping with the increase of the load, two 615^{RT} heat pumps with Heating-Tower were installed in the center plant in January, 1986.

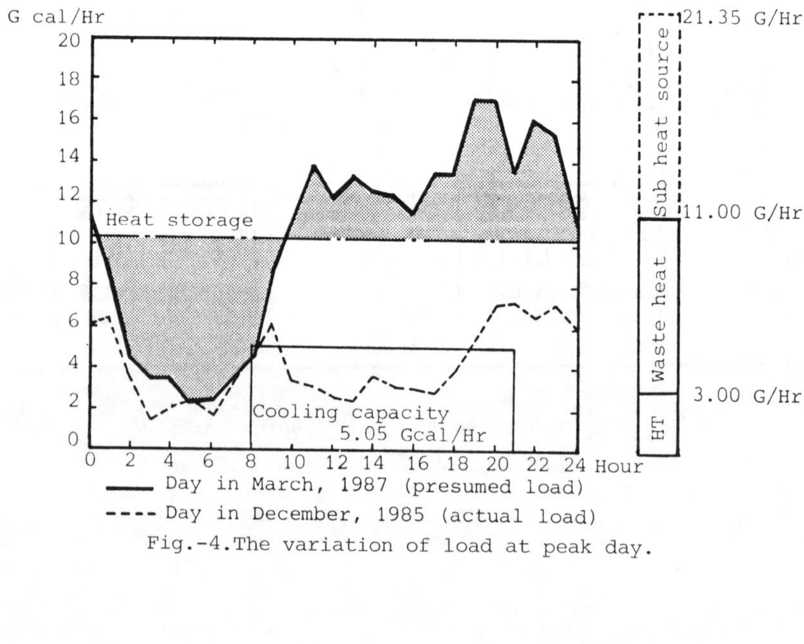

Fig.-4.The variation of load at peak day.

Heat Pump With Heating Tower

Design specification

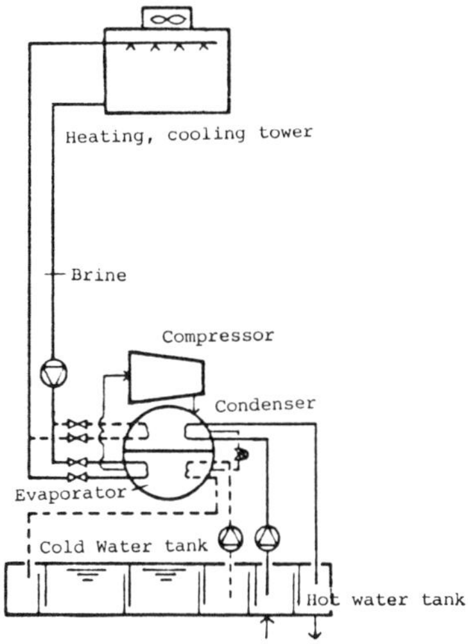

Fig.-5.The skeleton

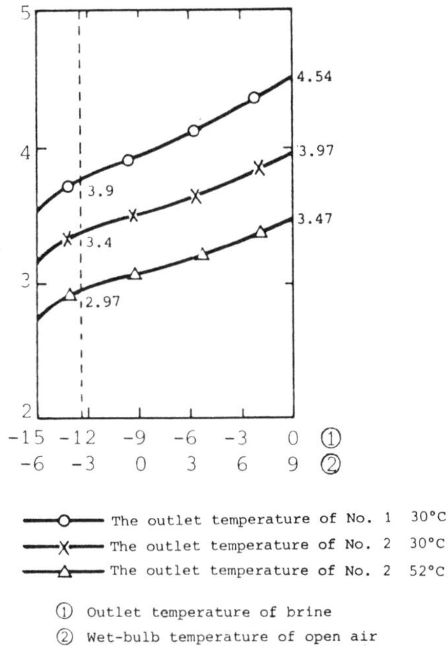

Fig.-6-Design COP.

Table 1. Specification

Items		Unit	Cooling	Heating (52°C)	Heating (30°C)
Cooling capacity		Kcal/Hr	1,860,000	--	--
Heating capacity		Kcal/Hr	--	1,500,000	1,500,000
Range of control		%	100 - 30	100 - 30	100 - 30
Cold	Variation	--	W	E	E
	Temperature	°C	15 → 5	-7.5 → -11.0	-7.2 → -11.0
Hot	Variation	--	W	W	W
	Temperature	°C	31.5 → 36.4	37.1 → 52	15.2 → 30
Main compressor		KW	460	251	233
Booster compressor		KW	--	335	279
Refrigerant		--	R-11		

W: Clear water E: Ethyleneglyclo 35 wt%

Installation of heat pump with heating tower

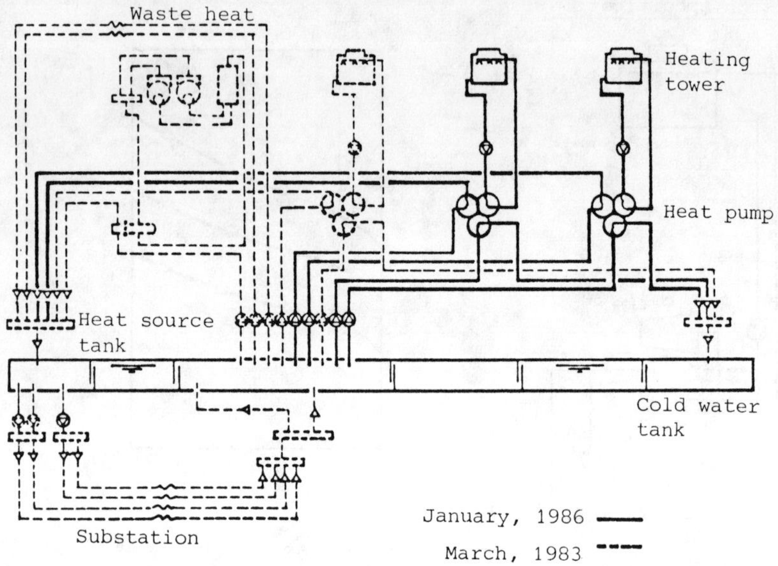

Fig.-7.Skeleton of center plant.

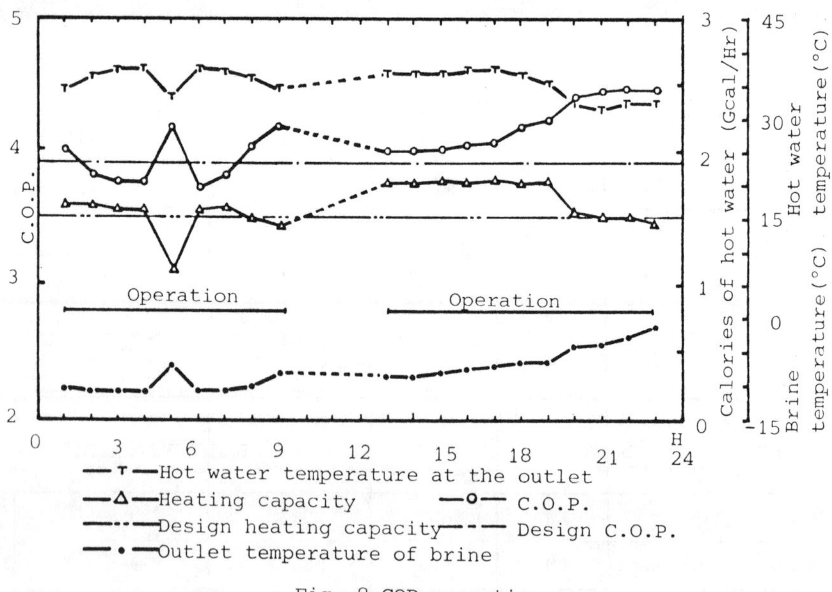

Fig.-8.COP operation

THE PLAN OF DHC AT HIBIYA AREA

DHC In Japan In Recent Years

The majority of DHC in Japan have been the type mainly used for heating in a cold district or the urban type which supplies steam and cold water using gas as its main heat source. However, in recent years, the number of systems using the urban waste heat or with the heat pump as its heat source which possesses a thermal storage tank, such as ice storage, is increasing. Examples from recent years and an example from the Hibiya area are shown.

Table 2. Current example of DHC Japan

Area	Area Size ha	Heat load	Cooling calories Gcal/h	Heating calories Gcal/h	Fuel	Supply stars
Shibaura	9.5	Offices	18.28	18.4	Gas	Feb. 1984
Ginza	2.5	Department stores, offices	4.3	2.1	Elect-ricity	Arp. 1984
Nishi Shinjuku	4.7	Hospitals, hotels, offices	8.4	25.5	Gas	Sept. 1984
Nishi Ikebukuro	12.6	Offices, hotels	3.7	7.1	Gas	June 1985
Akasaka, Roppongi	7.9	Offices, hotels, residents, halls	21.02	19.4	Gas	Apr. 1986
Hibiya	5.3	Offices, shops, theaters	11.6	6.7	Elect-ricity	Sept. 1987

Outline Of Hibiya Area

As the D.H.C. plan for the area where the redevelopment of the existing city is to be carried out, the hot water and cold water supply system using an electric heat pump was adapted, although the steam and supply system using a gas source has been used in the nearby area.

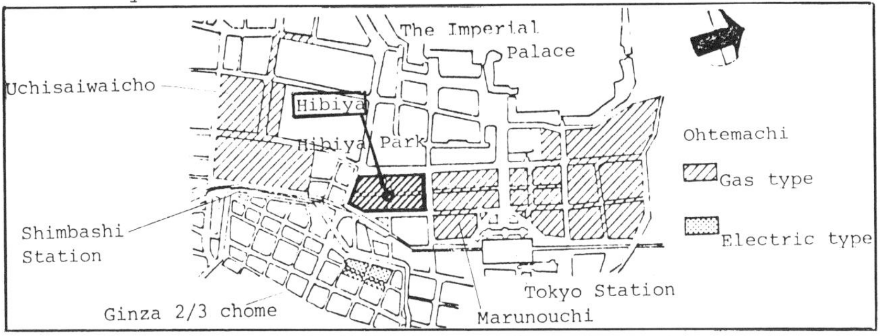

Fig.-9.The map of Hibiya surrounding the project site.

The capacity of loads and the features of the district

Table-3. The capacity of loads

		Step I	Step II
Total subjected floor area m^2		86,200	296,000
Plant load	Cooling Gcal/H	7.1	19.1
	Heating Gcal/H	5.5	15.1
Annual energy value	Cooling Gcal/Y	9,270	17,900
	Heating Gcal/Y	3,520	11,430
Size of plant	Area m^2	1,500	2,000
	TST* m^3	5,800	7,100
	Heat source RT	1,300	4,000

Note) TST* = Thermal Storage Tank

The D.H.C. with the electric heat pump system was introduced to this area for the following reasons:

1. The Tokyo Metropolis intended to promote the district heating and cooling.

2. The off peak rate system of electric fee was adopted.

3. The capacity of the heat source was reduced with the thermal storage tank by utilizing the structure of the building.

4. The heat supply line would be shortened as much as possible.

5. The capacity of the system could be reduced by the accumulated efficiency of the required load of individual buildings.

6. The use of urban waste heat could be used.

7. The waste heat recovery of the heating and cooling load would be carried out.

The Outline Of The Cooling/Heating Plant System In Hibiya

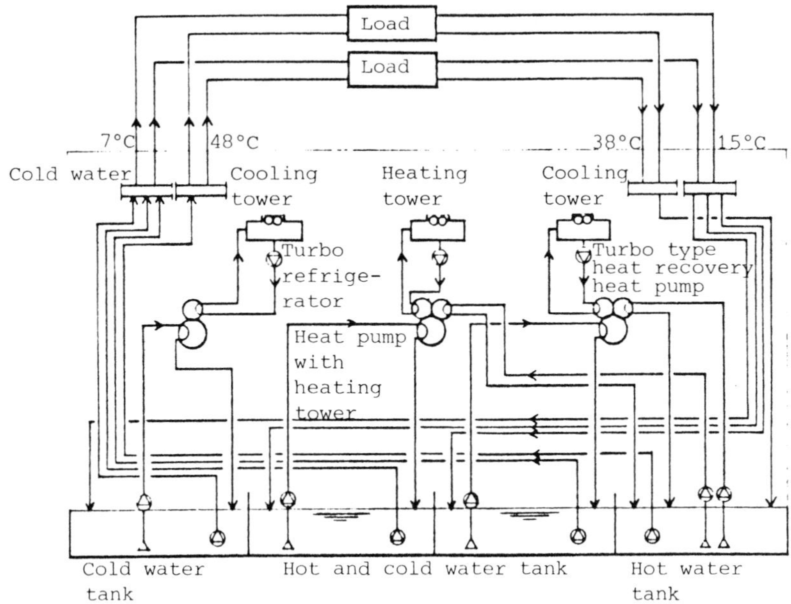

Fig.-10. System flow.

Table 4. Outline of equipment

1. Ability of main equipment

Type of equipment	Quan-tity	Specification
Turbo type heat re-covery heat pump	1	Heating 1.04 Gcal/h Cooling 1.82 Gcal/h
Turbo type heat re-covery heat pump	1	Heating 2.46 Gcal/h Cooling 1.97 Gcal/h
Heat pump with heat-ing tower	3	Heating 1.6 Gcal/h Cooling 2.1 Gcal/h
Electrical turbo refrige-ration machine	1	Cooling 3.61 Gcal/h

2. Capacity of thermal Storage tank

Hot and cold water tank	4,800 m^3
Cold water tank	1,200 m^3
Hot water tank	1,100 m^3

The Trial Calculation Of The Heat Unit Cost

The result of trial calculation

Trial calculation results of the heat unit cost of DHC compared with the individual system is shown in the graph below.

Table 5. Unit cost of D.H.C.

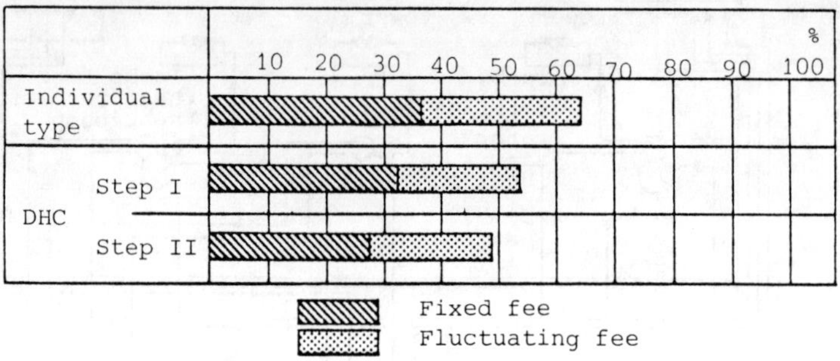

Adoption of a 7,200m^3 thermal storage tank

Table 6. Thermal storage tank

	Cooling operation	Heating operation
Amount of heat storage during night	40.8 Gcal	50.15 Gcal
Translated rate during night	21.2%	32.6 %
Reduction Equipment	7 Gcal/H (36.6%)	6 Gcal/H (44.1%)
Load pattern	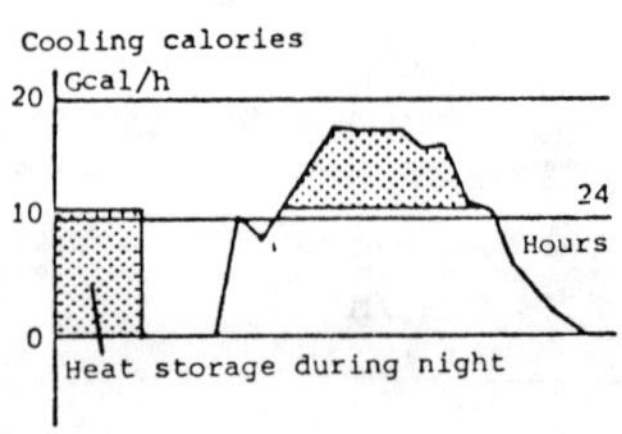	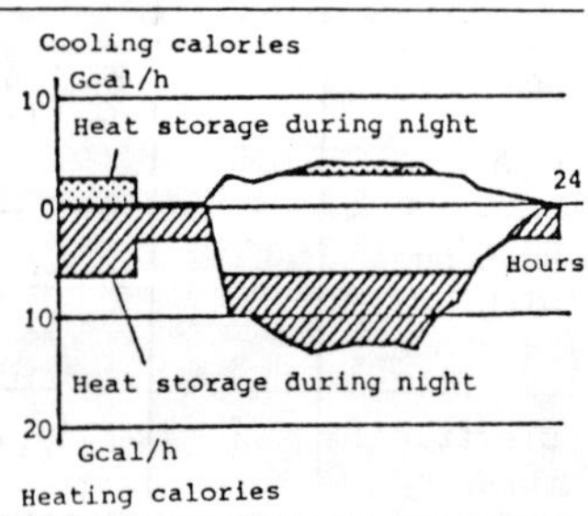

The Detailed Investigation Of Heat Pump With Heating Tower

The features of the heating tower

In addition to the merits of the existing heat pump, the heat pump (with heating tower) offers the following advantages.

1. C.O.P. improves and the running cost decreases.
   Cooling 3.1 → 4.2 (35% increase)
   Heating 3.0 → 3.7 (23% increase)
2. The selection of space for the plant is not unreasonably restricted and the heat pump is very compact.
   The size of the heat pump is 56% and the air-to-air heat exchanger is 45% of the existing types.
3. High temperature (55°C) can be produced economically.
4. The initial cost can be reduced.
   US$1,750/RT → US$1,310/RT (25% reduction)
5. The pump is not vulnerable to frost.
   A qualified specialist is not required since the pump is not regulated under the law related to gas operations.

The problem of the heating tower which should be solved.

1. The control of brine density requires maintenance.
   In case of snow or rain, the percentage of water increases and the density dilutes.

The method of density control adopted for this project

1. The scope of density control
   The minimum density to prevent freezing, 35% (wt) is required since the lowest temperature using brine is -12°C. The lowest denstiy of brine was chosen to keep 40% (wt) within 5% of the allowance.
2. The method of density control -- ① The adoption of brine absorptive tank.
   The variation of the amount of water due to dilution and increase causes the increase and decrease of water value within the system of circulation. In order to maintain the normal operation, a water storage tank of approximately 50m^3 capacity is installed and maintains the density automatically.
3. The method of density control -- ② Dry operation of the heating tower is adopted.
   Because the heat pump intended to produce heat, circulation of diluted brine of normal temperature and discharge of the water into the air is done by operating fans and a circulation pump. However, in order to carry out this operation, some requirements, such as air humidity less than 60%, have to be satisfied.
4. The method of density control -- ③ The sub-gradation system.
   The sub-gradation system for emergency use is now under development in order to cope with the situation in case heat pump operation is required when the condition of ambient air humidity is not suitable for normal operating conditions.

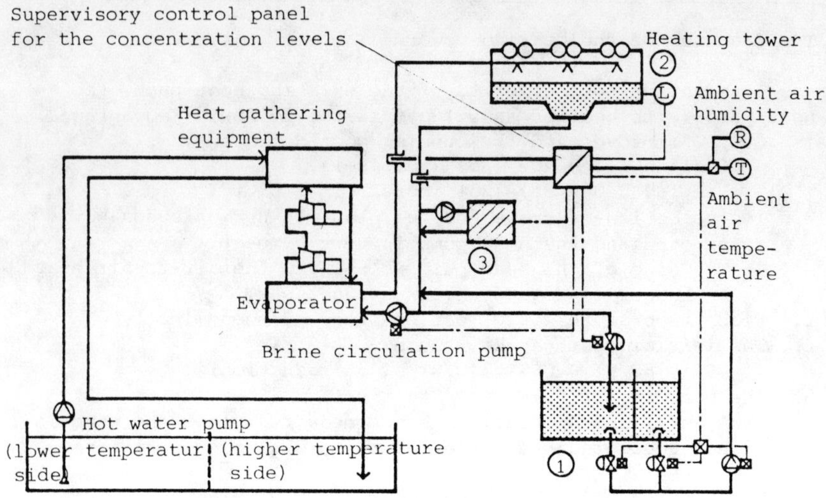

Fig.-11.Heat pump system with heating tower
density control system.

Chapter 32

Operating Experience in Sweden

Lars E. Åstrand

During the past five years heat pumps for district heating have made a remarkable break-through in Sweden. There are a number of reasons for this, but basically there are two factors of great importance. First, district heating is widely spread in Sweden. More than 50 per cent of all multi family housing is served by district heating. Secondly, electricity has been available at low cost due to the fact that hydro and nuclear are the main sources. In 1985 of the total electricity production 53 per cent was hydro and 42 per cent was nuclear. Of the remainder 4 per cent was produced from wood residues from the pump and paper industry and coal in combined heat and power production. Only one per cent was condensing power, coal and oil.

The price background is shown in the **figure 1** which demonstrates how oil and electricity prices, taxes included, have varied since 1978. In the figure also the energy cost advantage of a heat pump compared with oil firing is shown assuming a COP of the heat pump of 3 and an effenciency with oil firing of 90 per cent. If the heat pump replaces oil it will be operated on base load with a load factory exceeding 0.6.

HEAT PUMPS: Prospects in Heat Pump Technology and Marketing (Proceedings of the 1987 IEA Heat Pump Conference), Edited by Kay H. Zimmerman © 1987 Lewis Publishers, Inc., Chelsea, Michigan 48118. Printed in U.S.A.

This gives for e.g. 1982 an advantage of 0.6 x 8 760 x 0,13 = 680 SEK/kW, year which should be compared with installation costs depending on circumstances of 1 - 2 000 SEK/kW in 1982 prices. It is evident that pay-back times in the order of two years were possible which explains the very rapid development.

OIL AND ELECTRICITY PRICES

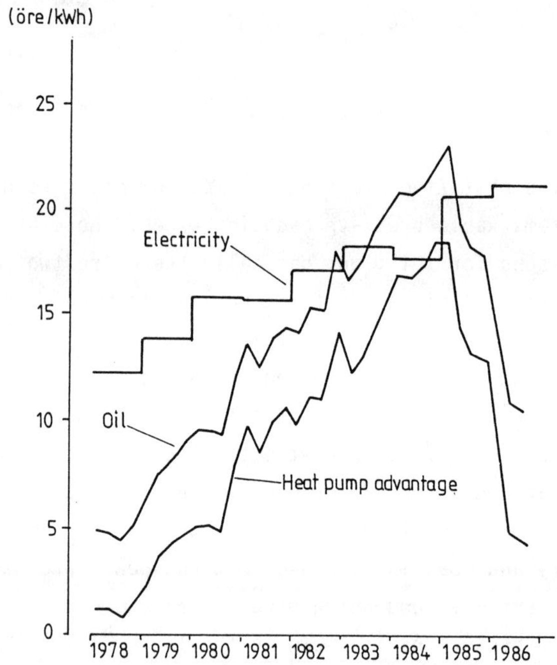

Fig. 1.

Figure 2 shows installed capacity per the end of each year since 1981. The success of heat pump operation is of course depending upon the availability of a heat source available continuously through the year at an approximately even temperature. It must also be reasonably close to the district heating network so that connection costs can be kept low. In every city there exists such a heat source at the waste water treatment plant. The effluent from this plant has a fairly constant temperature and flow. 52 per cent of the installed capacity uses such a soures. Other sources are industrial waste heat 14 per cent, sea or lake water 27 per cent, ground water including geothermal applications 4 per cent and air 1 per cent.

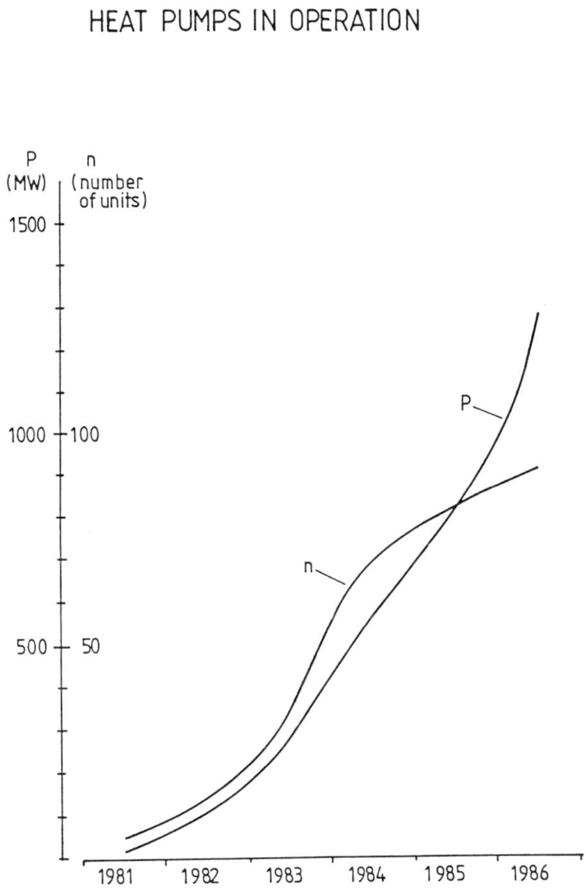

Fig. 2.

Of a total of 120 units larger than 1 MW, almost all are compression type pumps. Only two units are absorption heat pumps. Of the compressions pumps most units smaller than 10 MW have screw compressors, whereas the larger machines have turbo compressors. Unit size of the larger machines have been 13-15 MW or 25 MW. The largest plant has a total capacity of 100 MW. A typical schematic diagram for a large compression pump is shown in **figure 3**. The evaporator is typically a shell and tube heat exchanger with the heat source on the inside of the tubes. For lower temperature heat sources, such as lake and sea water, a spray-type evaporator is used.

COMPRESSION HEAT PUMP

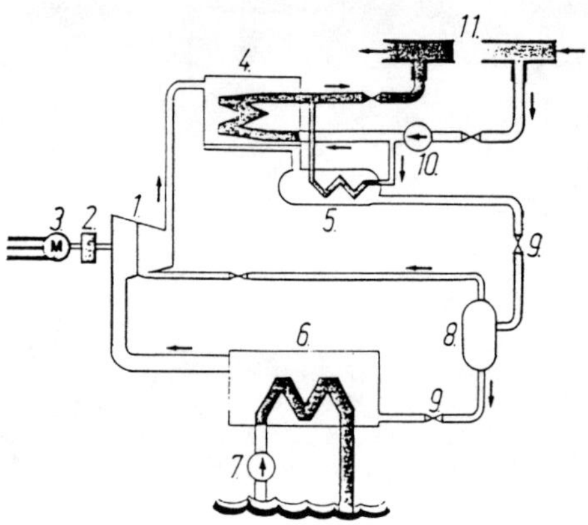

1 COMPRESSOR	7 WASTE WATER PUMP
2 GEAR	8 FLASH BOX
3 ELECTRIC MOTOR	9 CONTROL VALVES
4 CONDENSER	10 DISTRICT HEATING PUMP
5 CONDENSATE COOLER	11 RETURN PIPE OF DISTRICT HEATING NET
6 EVAPORATOR	

Fig. 3.

Recently also two absorption heat pumps have been taken into operation, one with 7 MW output and the other with 50 MW. A schematic diagram of the latter is shown in **figure 4**.

ABSORPTION HEAT PUMP

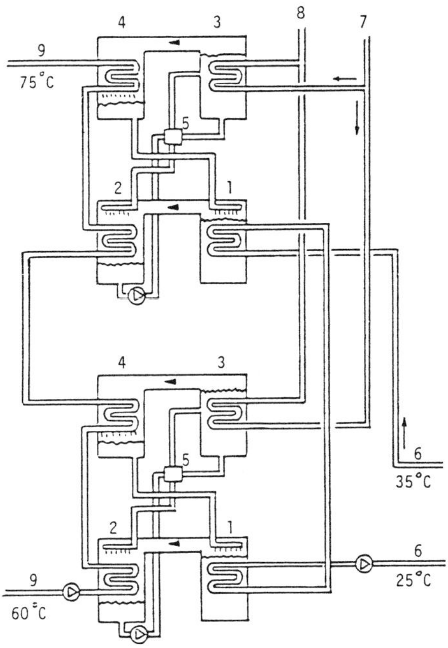

1 EVAPORATOR
2 ABSORBER
3 REGENERATOR
4 CONDENSER
5 HEAT EXCHANGER
6 WASTE WATER
7 STEAM
8 STEAM CONDENSATE
9 RETURN PIPE OF DISTRICT HEATING NET

Fig. 4.

Operating experience will be reported under the headings availability, COP, leakage and operating problems.

The availability has in general been very high. During 95 average availability for turbo compressor units was 90.7 per cent and for screw compressor units 86.5 per cent. Availability is here defined as operating time divided by the sum of operating time and forced outages, e.g. outside of planned outages. The median value was higher, 93 per cent for turbo compressors and 98.3 for screw compressors. From the data one can expect an availability well above 90 per cent after the initial operating period. The largest part of the unavailability in the plants that run well, have mainly been due to outside disturbances, such as high district heating temperature and interruptions in the electrical supply.

The average COP was for turbo compressors 3.09, for the unit itself and 2.91 for the plant and for screw compressors 2.69 and 2.40 respectively.

Leakage of cooling medium measured as the need for refilling has been just below 10 per cent of the freon content per year, most of which is due to slow leakage, i.e. through shaft seals valve spindles etc. Some of the leakage may be due to the working fluid dissolving in the lubricating oil.

As is evident from the preceding the operating experience has been very good. The most common causes of disturbance within the plant have been faults in the auxiliary equipment and control equipment. Disturbances from outside the plant have come from insufficient supply of waste heat or too low temperature of the waste heat, too high temperatures in the district heating systems and interruptions in electricity supply. The heat pump as a source for district heating has thus shown itself to be a very reliable and economic piece of equipment.

Section VII

Economics, Marketing, and Promotion

Chapter 33

The International Heat Pump Council: A Vital Step Toward Successful Heat Pump Marketing

Werner Hochegger

ABSTRACT: The International Heat Pump Council was established on September 18th, 1986 at a workshop in Graz, Austria on approaches to the European heat pump market by a group of heat pump experts, associated with the International Energy Agency research and development program as well as industry and utilities. The membership includes top-level representatives of governments, power utilities and industry from a number of countries. They rely primarily on the IEA Heat Pump Center for the scientific and technical information that is one of the bases of the marketing and promotion work, and work closely with the Center. Dissemination and coordination of information to the participants in the council will serve to increase product quality, installer reliability and consumer awarness.

INTRODUCTION

After the onset of the first oil crisis, the HP became one of the most important energy-saving technologies. Today, in the aftermath of Chernobyl, it still retains that distinction, in spite of the drop in the price of oil. This tremendous technology has been fully developed for commercial application for more than 15 years; thus the question arises as to why the HP market is currently declining instead of growing vigorously. In many countries, individual groups have

HEAT PUMPS: Prospects in Heat Pump Technology and Marketing (Proceedings of the 1987 IEA Heat Pump Conference), Edited by Kay H. Zimmerman

gone to great lengths to wake this sleeping energy giant and put it on its feet, but in most cases no real lasting success has been achieved. If we look at the last decade, nothing has really changed in spite of all the attempts to help the HP to a breakthrough. All the current activities, if continued, have no real chance of revitalizing the market in the future either.

PROBLEMS IN HEAT PUMP MARKETING

Why is this? The main reasons that the HP has failed to achieve the position in the heating market that it deserves are:

1. Lack of communication and coordination on the part of those groups involved in HP marketing, and
2. Wrong approaches and wrong equipment for a given job.

INFLUENCES ON HEAT PUMP MARKET

With regard to the first problem, let us consider the four groups on four different levels than influence the HP market. These are:

1. Government (national, regional, local)
2. Power utilities
3. Industry
4. Installers

Every measure and decision taken by one of these four groups affects the entire HP situation. Government policies have a considerable influence on market development.

Power utilities long blocked development of the HP market owing to misconceptions about the competitive threat of the HP to direct electrical heating, and adverse effects on the power network. Progress has been made in correcting these misconceptions and the power utilities are coming to appreciate the very large market they can help to open up with HPs for existing and new buildings.

The role of industry and installers in the HP market in general leads us to the second problem mentioned above: wrong approaches and wrong equipment for a given job. Bad decisions in this area, probably more than anything else, have created the bad image heat pumps suffer from today. This situation, however, has at least been recognized, and can also be corrected. First, the markets for the various types of HPs must be clearly defined. The main part of Europe is a heating-only market for air/water systems for small residential buildings, and this is therefore a most important kind of equipment that should be promoted if everyone, down to the consumer, is to be happy.

HEAT PUMP MANUFACTURERS

Another problem is that HPs are made by two kinds of manufacturers, each with specific problems. The HP manufacturers from the air-conditioning branch have to become accustomed to making heating-only products that perform well at low outdoor temperatures. The second group of manufacturers, the boilermakers, has to learn to make equipment with a refrigeration cycle. And both manufacturers and installers have to come to realize that a HP is not simply a different kind of boiler or air-conditioner, but a system in its own right that requires very careful design, selection and installation.

EQUIPMENT RATINGS

A further major probelm is the inadequacy of current equipment ratings. More complete and descriptive ratings will be necessary to improve system design and performance, leading in turn to satisfaction of all involved, down to the consumer.

LACK OF COORDINATION

These, then are the groups and problems that come to bear on the international HP market. So far, the four groups - government, power utilities, industry and installers - have each in their own way spent a lot of time, money and effort to develop the HP market, but none of these groups, working alone and without coordination and cooperation with the others, has managed to produce a significant and long-term positive effect. There are of course exeptions, where very good local trends have developed, but these tend to be limited in size and duration.

INTERNATIONAL PROMOTION AND MARKETING

The only answer to this problem appears to be the establishment of an association for international HP promotion and marketing. The arguments for the establishment of such an organization are as follows:

- A mass market is needed in Europe, North America and Japan.
- Markets that once flourished, such as France and Sweden in Europe, but have since declined, must be revitalized.

- International collaboration offers an efficient means of increasing competence.
- Other products and technologies have profited from international lobbies; the same should be true for the HP.
- An international organization will serve to optimize the total system of power production and end use for space heating and domestic hot water.
- The organization will be able to combat barriers and opposition to a technology that protects the environment and conserves energy.
- Policy makers will be provided with joint and coordinated information.
- Existing infrastructure and competence will be utilized before they are lost.
- Heat pump marketing efforts will be coordinated at a high international level.

All of which sounds very good in theory, but one must ask who will do the work, assume the responsibility, provide the competence, and pay for this effort.

HEAT PUMP CHARTER

A major positive step was taken on September 18, 1986, in Graz, Austria at the International Workshop on the European HP Market, at the Energiesparhaus Graz, with the ratification of a Heat Pump Charter to establish an International HP Council. The charter was formulated by a group of HP experts associated with the International Energy Agency research and development program, as well as industry and utility activities addressing technologies and marketing.

This charter recognizes a new energy paradigm for the next quarter century, a transition period away from dependence on oil and energy systems that are harmful to the environment. The approach is more evolutionary than revolutionary and seeks a change toward clean, safe, reliable, comfortable and economic energy. Advanced HPs have a critical role in this transition.

RATIONALE FOR NEW ORGANIZATION

HPs permit use of abundant low-temperature heat sources, which otherwise could not be used practically, by increasing the temperature level. The HP therefore provides the means of using otherwise untapped natural heat sources or wasted heat from buildings and industrial processes. A concerted offert is needed to realize the widespread use of HPs. If mature and reliable HP technologies and proven user know-how emerge, HPs provide a solution that can be put into practice. If this is not the case, this HP application will be lost for many years to come. Successful application of HPs can only be achieved if steps are taken to ensure that reliable and readly understood information is available to assist the prospective designers, installers and purchasers of HP systems. The currently high initial costs of HPs must also be reduced. Even at present, however, lower energy demands lead to reduction of investments for energy supply systems that offset the extra costs for efficient systems. The environmental benefits to society of HP use must also be considered. The conclusions from the heat pump position papers prepared by the 13th World Energy Conference Experts Group on HPs were endorsed by the signatories to the Charter as important and valid guidelines for future activites.

KEY ISSUES

The Charter signatories are concerned that, in the aftermath of the collapse of the oil price and the Chernobyl accident, widespread use of HPs will not be realized, with consequent loss of energy conservation, and environmental and national economic benefits that these systems can provide. Failure to continue development of HP application will also result in loss of the established HP competence and capability in industry, utilities and governmental agencies. The required promotional activities directed both at consumers and governments exceed those which may be reasonably expected of utilities, particularly in the present economic environment. Moreover, related facets of the problem, such as involvment in building design, exceed the traditional roles of utilities. Finally, the required international coordination, product development and policy making necessitates involvement on the part of manufacturers, governments and others, in addition to utilities.

Recommended configurations and design methods for HP systems should be developed and disseminated. Reliable performance and economic data must be made available both for products and installed systems.

At present, there is insufficient information available to properly appraise the (European) HP market. A variety of basic information is available but there is a need for further analysis.

Current HP sales levels are not sufficient to present an attractive investment opportunity to manufacturers or to achieve low unit costs. Fragmentation of the market by requirements of individual countries and the variety of HP types further complicate the situation. Means must be found to develop a market volume in Europe in the hundreds of thousands rather than thousands of units per year.

Heat pump and air-conditioning sales are growing in southern Europe. Low summer electricity prices in northern Europe also provide a possibility for HP sales to grow, particularly in conjunction with the increasing market for air-conditioning.

Efforts are needed to eliminate international barriers to HP trade resulting from restrictive manufacturing or rating standards, and especially from the proliferation of differing codes among individual countries.

Finally, the environmental and national economic benefits of HPs must be evaluated.

Manufacturers would welcome a cohesive system for preparation and dissemination of catalogs to present the various products and components available for HP systems.

Manufacturers, as well as power utilities, also seek better communication regarding HP marketing issues, removal of institutional barriers, and better understanding of the benefits of HPs among national governments, local governments and consumers.

The signatories of the HP Charter believe that HPs are of value to the economy as a whole, as they provide increased primary energy flexibility, increased energy efficiency, reduced environmental emissions and immissions, more efficient energy carriers, and improved national balance of trade.

PROPOSED ACTIONS

The signatories to the Heat Pump Charter encourage the following specific actions:

1. An International HP Council is intended to provide a basic communication program on HPs.
2. To implement the charter and the International HP Council, cooperation with the IEA should be provided.

The HP Charter was signed in draft form on Sept 18, 1986, by 17 representatives of government, utilities and industry from 5 European countries.

When the proposed actions have been taken to implement the work of the International Heat Pump Center, the first order of business of the Council will be a variety of publicity initiatives.

PUBLICITY INITIATIVES

The International Heat Pump Center sees itself as a center to coordinate activities of all those involved inwork. Its first task will be to introduce itself to appropriate governmental agencies, power utilities, manufacturers and industry, planners and installers, to inform them of its activities and the possibilities it offers, and to invite their membership and cooperation in the International Heat Pump Center.

Cooperation is here understood to mean the providing of information on current activities of a given organization, especially those of a marketing and/or promotional nature. This material will then be collected and summarized in a periodical, tentatively entitled "Council News", to be published quarterly in several languages. A mailing list including key people from governments
power utilities
industry
installers
press
will also be compiled.

The publicity initiatives themselves will concentrate on developing promotion materials and presenting them to important organizations and individuals in positions to influence HP prospects. Position papers and marketing plans will form the basis of the International Heat Pump Centers publicity materials. Promotional acitivities will be tailored as follows for each target group:

1. National and local governments

   - prepare briefings and brochures showing the public benefits of heat pumps.
   - identify and visit key officials to deliver briefings.

2. Utilities

   - Develop briefing explaining benefits of additional HP load.
   - Develop briefing explaining favorable environmental effects of increased HP penetration
   - Develop briefing explaining potential benefits of large HPs in district heating systems.
   - Identify and visit the key utilities in each important national market.

3. Consumers, architects, builders

   - Identify and document goodsystem types for each application of major interest.
   - Identify benefits to the consumer and to the environment resulting from the installation of HPs.

- Develop list of products and installers meeting cost/performance and quality guidelines.
- Develop such advertising, brochures and publicity material as required.
- Develop and implement strategy for contacting target groups (e.g. mailing, advertising).

4. Producers, distributors, installers

   - Announce the establishment of the International Heat Pump Center, explain its benefits and solicit contributions.
   - Develop mailing list for this target group in each country.
   - Obtain participation in developing HP system cost/performance guidelines.

These target groups are viewed from an international perspective. In addition, the IHPC will seek cooperation with each important national market to work with local industry to improve the quality of HP installations and service. Specifically, this will involve:

1. Development of a mailing list for manufacturers, distributors, consulting engineers and installers.
2. Circulation of the announcement of the formation of the IHPC to the mailing list, explaining the IHPC program and inviting participation.
3. Follow-up with telephone calls and visits to the most important installers; exlanation of procedure for obtaining certification.
4. Development of certification and training procedures for installers and consulting engineers.
5. Development of a list of certified installers and trainers for use by consumer.
6. Monitoring of field results, to be compiled in a list of successful installations.
7. Monitoring of new product offerings, to be included in a list of certified products.
8. Assistance to installers, consumers, etc., with specific questions as required.

The International Heat Pump Center is organized as an independent body, supported by funding from governments, utilities, HP manufacturers, distributors and installers. The manager of the International Heat Pump Center will at his discretion engage HP and marketing experts to carry out the actual work of the Council. Close cooperation with the IEA Heat Pump Center is intended. The activities of the International Heat Pump Center will be reviewed and approved by a board of directors made up of representatives from the funding organizations.

CONCLUSIONS

We are confident that this comprehensive strategy covers all the important aspects of the complex problem of how to promote and market HP technology in such a way as to put it in the position it deserves in today's energy-conscious world. This overall strategy is, however, sufficiently flexible and finely tuned to be able to accomodate adjustments at any time to suit specific circumstances or changes in the general situation.

HP technology is ready and waiting for much broader commercial application than has so far been the case in most of the world. Suitable marketing and promotion measures to bridge the gap between the promise and reality of this technology are now available with the International Heat Pump Center.

Chapter 34

Factors Affecting the Marketing of Heat Pumps in North America

Ted C. Gilles

ABSTRACT: Electric heat pumps continue to achieve greater market share in new residential construction in North America with over 30% of the new single family homes constructed in the United States in 1985 being so equipped. Electric heat pumps, particularly add-on types used as an alternate to cooling only air conditioning with fossil fuel furnaces, provide attractive year round revenue for electric utilities. Many home owners are attracted to heat pumps because of the convenience of electric heat. However, the final buying decision is generally an economic one, and higher fossil fuel costs and higher efficiency heat pumps are making them more competitive and therefore attractive to those making the choice. A simplified chart method for comparing the economics of various residential comfort heating systems is a principal part of this paper. Continuing, and in some cases, accelerating research and development by HVAC equipment manufacturers will cause even more efficient and reliable heat pumps to be introduced into the market in the not too distant future.

INTRODUCTION

Electric heat pumps of all types, including air and water source and ground coupled systems, are widely used in residential and commercial applications in North America [1]. In fact, approximately 30% of all new single family dwellings constructed in the U. S. in 1985 were equipped with electric heat pumps. There were over 800,000 unitary (less than 65,000 Btu/h) air source heat pumps sold in the U. S. in 1985. It has been estimated that approximately 70% of these were used in new construction while the rest were replacement; some for conventional cooling only air conditioning. [2]

HEAT PUMPS: Prospects in Heat Pump Technology and Marketing (Proceedings of the 1987 IEA Heat Pump Conference),
Edited by Kay H. Zimmerman

SELECTION DECISION FACTORS

Owner-user Decision Factors

The buying decision to select a heat pump is affected by many factors, not the least of which is the operating characteristics in relationship to the heat loss profile of the structure. Figure 1 compares the heat pump performance profile to the heating and cooling profile of a residence over the range of incurred outdoor temperatures. The principal characteristic is that at temperatures below the balance point between the structure heat loss and the capacity of the air source heat pump, supplementary heat must be provided from a source other than the heat pump. This particular aspect of air source heat pumps has the greatest influence on the selection of the type of heat pump and its application during the buying decision process.

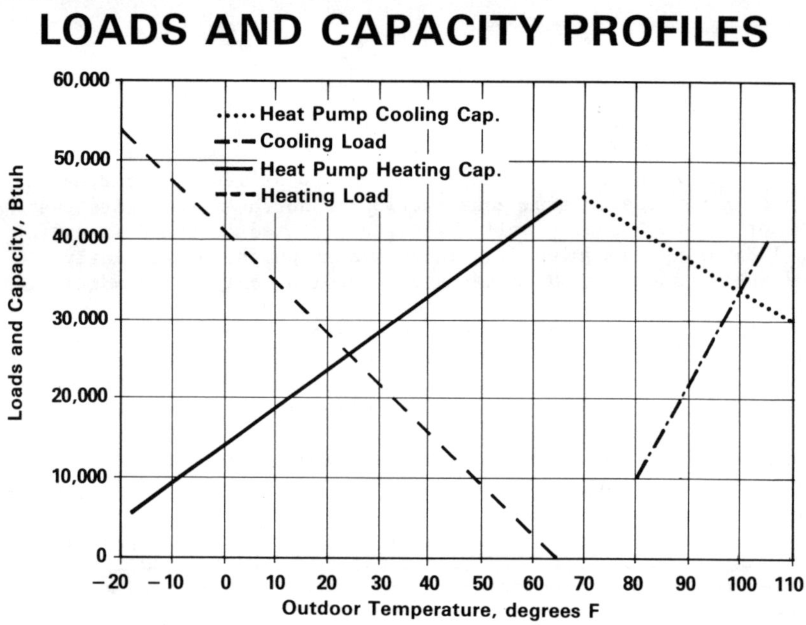

Figure 1.

Factors that influence the buying decision for a heat pump include:

- Is the availability of air conditioning important to the owner?
- Is natural gas not available as heating fuel?
- Are electric rates more attractive for heating than natural gas?
- Is a unique source of relatively constant temperature water such as from a lake or underground available inexpensively for a water source heat pump?

Heating only heat pumps have never achieved a significant level of acceptance in North America because nearly all geographical locations have some time of the year when cooling with air conditioning is desirable. Therefore, any installation of electric powered air conditioning is a defacto candidate for an electric heat pump.

If natural gas is not available for the source of heating energy, then a heat pump is often selected because of the concern of many owners of the questions regarding safety of the storage of No. 2 heating oil or liquefied petroleum gas (LPG). This decision is tempered by the economic aspects of the respective systems for the owner. Comparative system economics will be discussed in detail in following paragraphs of this paper.

ELECTRIC UTILITY SYSTEM FACTORS

Using a computer-assisted analysis [3], the impact of different residential systems upon the electric utility can be compared. The following table shows that the annual load factor of the various systems expressed as "kWh/kW" can be anything from 603 to 2,559 hours in some locations.

Table 1. Computer-assisted Comparison of Electric Powered Systems

	Bismarck, ND	Cincinnati	Atlanta	Houston
Air Conditioning				
Annual kWh	2410	3971	5971	9349
kW	4	4	4	4
kWh/kW	603	993	1493	2338
Elec. Heat & A/C				
Annual kWh	38522	18813	19322	18448
kW	20	13	12	9
kWh/kW	1926	1447	1610	2050
Conv. Heat Pump				
Annual kWh	26075	13016	11221	14755
kW	20	13	12	9
kWh/kW	1304	1001	935	1639
Add-on Heat Pump				
Annual kWh	12797	11445	10915	12547
kW	5	5	5	5
kWh/kW	2559	2289	2183	2509

As can be seen from Table 1 above, conventional air source electric heat pumps with direct element resistance supplementary heat but with less annual kWh provide the electric utility with less kWh for revenue per kW than a system using only direct element resistance heaters and air conditioning in all of the four geographical areas. This presents a particular problem to electric utilities with winter peaks. In extremely severe cold weather, the heat pump and the supplementary strip heating will impose a demand on the system nearly equal to the demand of a conventional electric heat furnace. However, the Coefficient of Performance (COP) of the heat pump will reduce the kWh of energy use, and the accompanying revenue, during all of the milder periods of the heating season. The winter peak for the supplementary heaters with the heat pump is no problem for the summer peaking electric utility and produces revenue kWh during the heating season without a need for additional generation and transmission capacity investment. The greater desirability of add-on heat pumps, those using a source of supplementary heating other than direct element resistance heaters, such as a gas or oil furnace, and with the indoor coil in the usual position of an air conditioning evaporator, is evident for all geographical areas.

Add-on Heat Pumps

Add-on electric heat pumps present a particularly attractive marketing opportunity for electric utilities for numerous reasons including the following [2]:

1. With over 18,000,000 central air conditioning systems presently installed with various types of forced air furnaces in single family residences, over 1,000,000 a year become candidates for change out for either maintenance or operating cost economic reasons.
2. They will add significant amounts of annual revenue kWh without capital investment for generation and transmission capacity.
3. In most cases they are significantly less expensive for the owner to install than alternate low kW demand systems such as ground water source heat pump systems.

HEAT PUMP FEATURES

Electric heat pumps have relatively apparent differences when compared to other types of residential heating systems. These differences are often presented, or at least perceived by the prospective owner, as benefits that enhance the desirability of electric heat pumps. These electric heat pump characteristics that set them out as being different from other residential systems, particularly those that use fossil fuel, include the following:

1. The implied safety of no combustion of fuel and eliminating the risk of a structural fire.
2. The elimination of the need for fuel storage.
3. No need for a flue venting system.
4. No fuel piping system.

OWNER ECONOMICS

As desirable as these characteristics may be to the prospective owner, the final heating system choice is most often made based on comparative economics of the candidate systems. As previously stated, most geographical areas of North America have weather conditions that make the availability of air conditioning an attractive feature of the residential comfort system. Since the air conditioning mode performance of most air source heat pumps is comparable with equivalent cooling only units, the operating cost economics are generally determined by the heating mode operation compared to other candidate systems.

Universal Heating Operating Cost Chart

All unitary electric heat pumps used for residential applications are tested and rated based on the United States Department of Energy (DOE) procedures [4]. The heating season performance criteria of any heat pump is based on the Heating Season Performance Factor (HSPF). The HSPF of a particular type, capacity and efficiency heat pump is determined by the detailed DOE tests. The Air-Conditioning and Refrigeration Institute (ARI) publishes a directory that is updated twice each year that lists the capacity and HSPF of most of the unitary heat pumps sold in North America. Although manufacturers test and rate their heat pumps for each of the prescribed DOE Regions shown in Figure 2, only the HSPF for Region IV is listed in the ARI Directory.

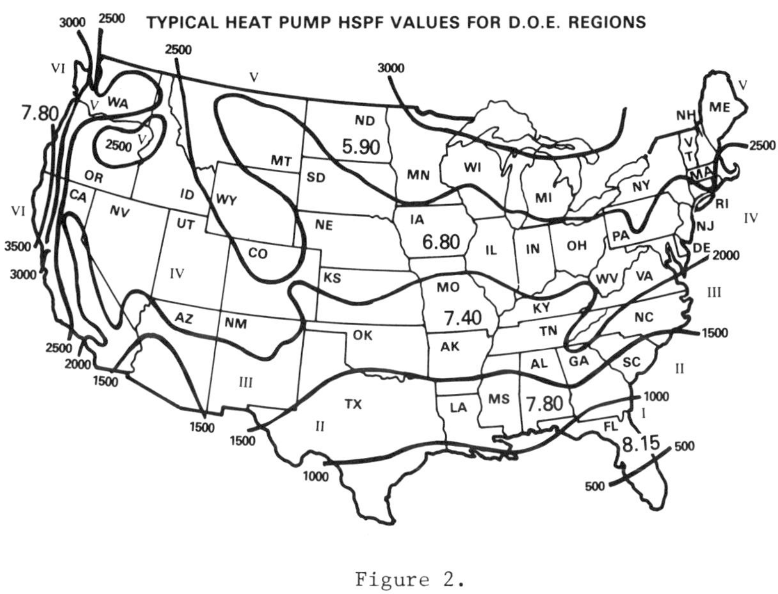

Figure 2.

The representative comparative operating costs of various electric heat pump systems can be evaluated against fuel-fired heating systems using the chart shown as Figure 3. The chart accommodates the following variables:

1. Any climate for any geographical location can be determined by varying the heat pump HSPF to reflect the local climate.
2. Any heat pump performance level can be considered also by varying the HSPF.

3. Any electric power and fuel cost can be compared.
4. Simultaneously compares electric and fuel-fired systems for a specific installation.
5. Evaluates on a proportional basis any annual heating energy requirement (100,000,000 Btu/year delivered to the conditioned space is shown).
6. Indicated difference between five year systems operating costs is equivalent to five year simple payback on first cost difference.

Electric Heat Pump and Fossil Fuel Systems Operating Costs

Figure 4 shows the comparative operating cost "zones" of various residential heating systems in use in North America today. Electric heat pumps have typical HSPF's ranging from 7.0 for good quality air source units to over 11.0 for water source units using relatively constant temperature underground or lake water. The "zone" shown covers a power cost range of $0.05 to $0.08 per kWh. Similarly, fuel-fired heating systems have Annual Fuel Utilization Efficiency (AFUE) ratings of 70 to 97%. The fuel costs shown range from slightly less than $5.00 to $7.00 per million Btu. This chart indicates that today's electric heat pump and fuel-fired systems are in general very competitive to each other. The range of five year operating costs for electric heat pump systems is from about $2,800 to $4,500 compared to $2,800 to $4,000 for the fuel fired systems.

Figure 5 shows the possible "zones" of five year operating costs for advanced electric and fuel-fired heat pumps in the heating season. This particular chart would indicate that fuel-fired heat pumps will have a potentially significant operating cost advantage over advanced design electric heat pumps if present design goals for the fuel-fired units are met.

Figure 6 shows the comparative five year operating costs of an 80% AFUE natural gas furnace at $5.50 per million Btu and a 7.0 HSPF air source electric heat pump at $0.06 per kWh. In this particular case, the gas furnace has a five year operating cost advantage of about $1,000. The electric heat pump would have comparable operating costs at $0.045 per kWh. For purposes of keeping the estimating chart simple, the furnace blower power cost is not included. However, neither is any credit taken for the heat added by the blower motor, so there is minimum net impact on the results of the comparison.

Air and Water Source Heat Pump Operating Costs

Figure 7 is another example of how the comparison chart can be used to estimate the operating cost trade-off between various systems. In this case an air source electric heat pump with an HSPF of 7.0 is compared to a water source electric heat pump with an HSPF of 11.0 at $0.06 per kWh. The five year operating cost difference is approximately $1,600 (4,300 - 2,700). If five year simple payback is the purchase decision criteria, the water source system could justify $1,600 additional installed cost to cover the cost of such things as a ground water supply and disposal system.

Add-on Heat Pump Computer-assisted Economic Analysis

Figure 8 is an example of a computer-assisted more complex economic analysis [3] as required for an add-on heat pump with a gas-fired furnace compared to a stand alone gas furnace with the same AFUE rating. The results of this study show that if a furnace and heat pump cost $600 more to install than the same furnace with an air conditioner with the same cooling SEER (Seasonal Energy Efficiency Ratio) as the heat pump, the owner would realize a 10 year return on investment (ROI) of 34.78% before tax.

HEAT PUMP RELIABILITY AND MAINTENANCE

Occasionally, another determining consideration for selecting an electric heat pump is the question of long term reliability. This question can be evaluated from at least two vantage points:

1. Two recent studies conducted by the Electric Power Research Institute (EPRI) have shown that the service life of electric heat pumps is 15 years in Chicago and over 20 years in Alabama [5].
2. Five year extended protection plan maintenance programs are available from a major insurance underwriter at a cost of about $100 per ton.

CONCLUSIONS

Electric heat pumps, particularly air source types, have matured technologically and continue to achieve greater acceptance in the residential and commercial sectors in North America. The established reliability of air source heat pumps has basically reduced the buying decision to one of the overall cost effectiveness between first cost and operating cost compared to alternate comfort conditioning systems. Manufacturers of electric heat pumps continue to further enhance the cost effectiveness with the development of new designs using variable speed and capacity methods and sub-zoning and will likely introduce new compressors, including scroll and advanced concept reciprocating types, in the not too distant future.

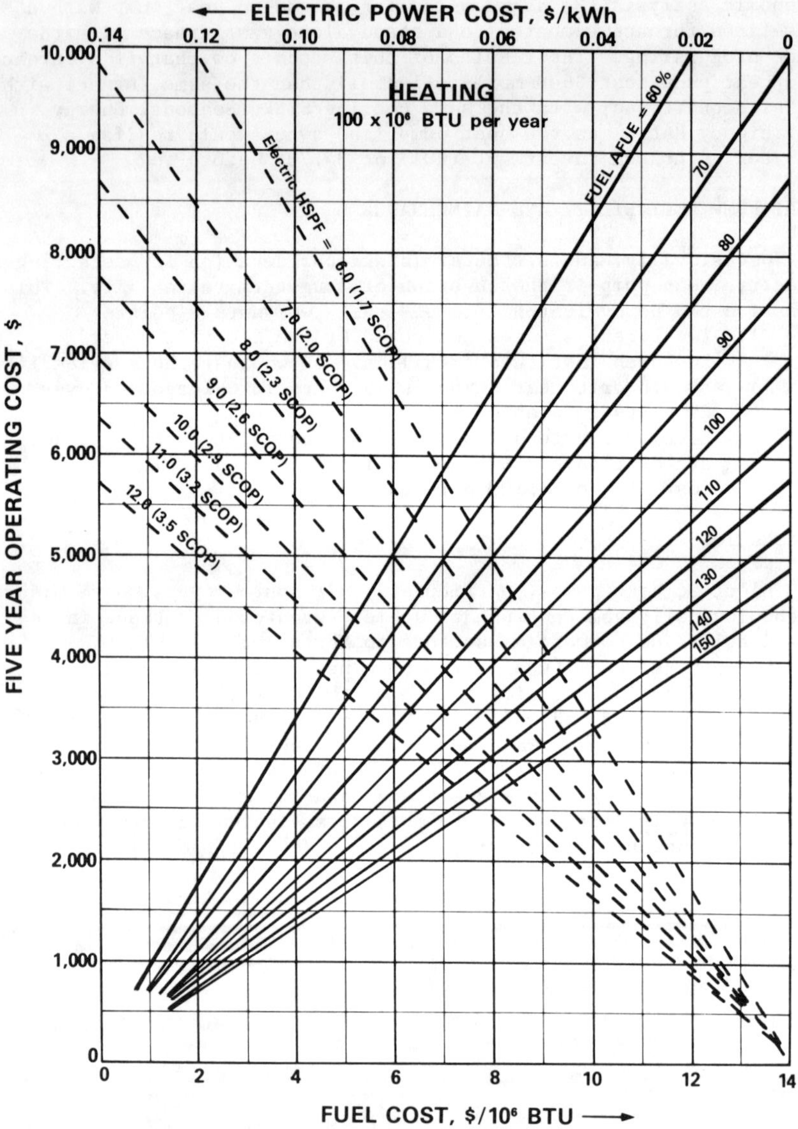
ELECTRIC POWER COST, $/kWh
0.14
0.12
0.10
0.08
0.06
0.04
0.02
0
HEATING
100 x 10[6] BTU per year
Electric HSPF = 6.0 (1.7 SCOP)
7.0 (2.0 SCOP)
8.0 (2.3 SCOP)
9.0 (2.6 SCOP)
10.0 (2.9 SCOP)
11.0 (3.2 SCOP)
12.0 (3.5 SCOP)
FUEL AFUE = 60%
70
80
90
100
110
120
130
140
150
FIVE YEAR OPERATING COST, $
10,000
9,000
8,000
7,000
6,000
5,000
4,000
3,000
2,000
1,000
0
0
2
4
6
8
10
12
14
FUEL COST, $/10[6] BTU

Figure 3.

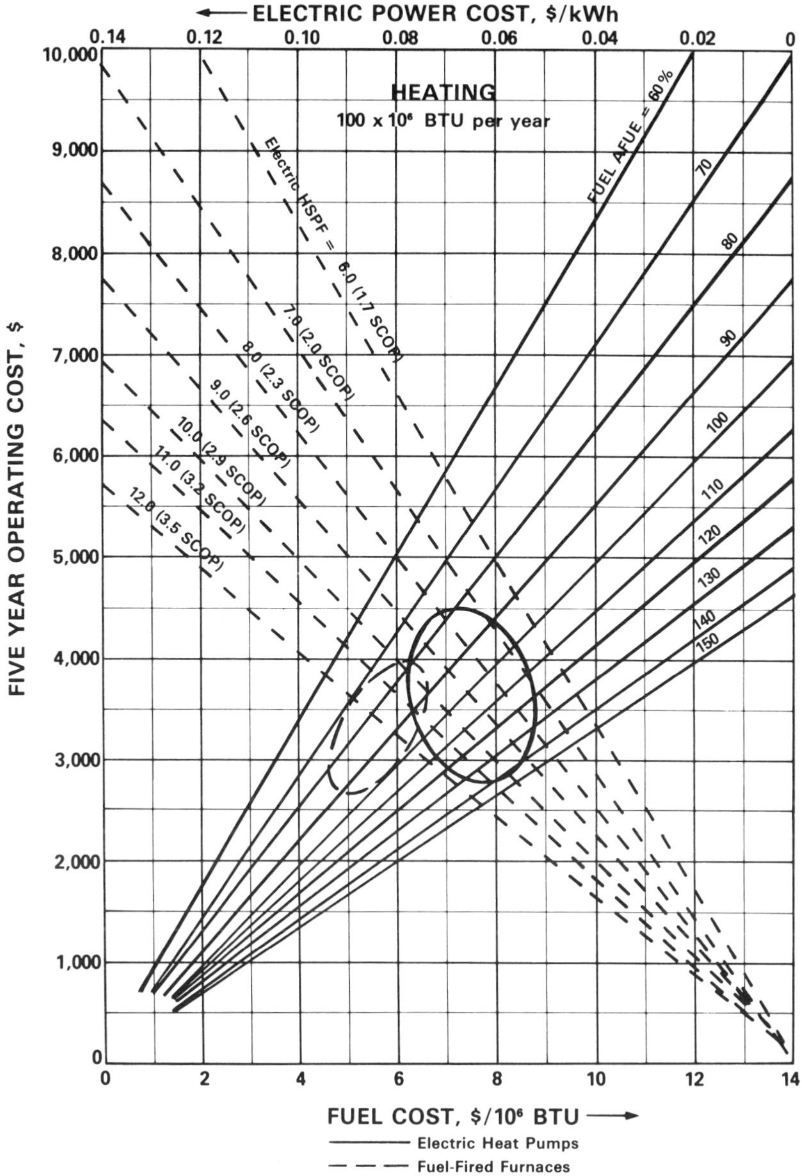
ELECTRIC POWER COST, $/kWh
0.14
0.12
0.10
0.08
0.06
0.04
0.02
0
HEATING
100 x 10⁶ BTU per year
Electric HSPF = 6.0 (1.7 SCOP)
7.0 (2.0 SCOP)
8.0 (2.3 SCOP)
9.0 (2.6 SCOP)
10.0 (2.9 SCOP)
11.0 (3.2 SCOP)
12.0 (3.5 SCOP)
FUEL AFUE = 60%
70
80
90
100
110
120
130
140
150
FIVE YEAR OPERATING COST, $
10,000
9,000
8,000
7,000
6,000
5,000
4,000
3,000
2,000
1,000
0
0
2
4
6
8
10
12
14
FUEL COST, $/10⁶ BTU
Electric Heat Pumps
Fuel-Fired Furnaces

Figure 4.

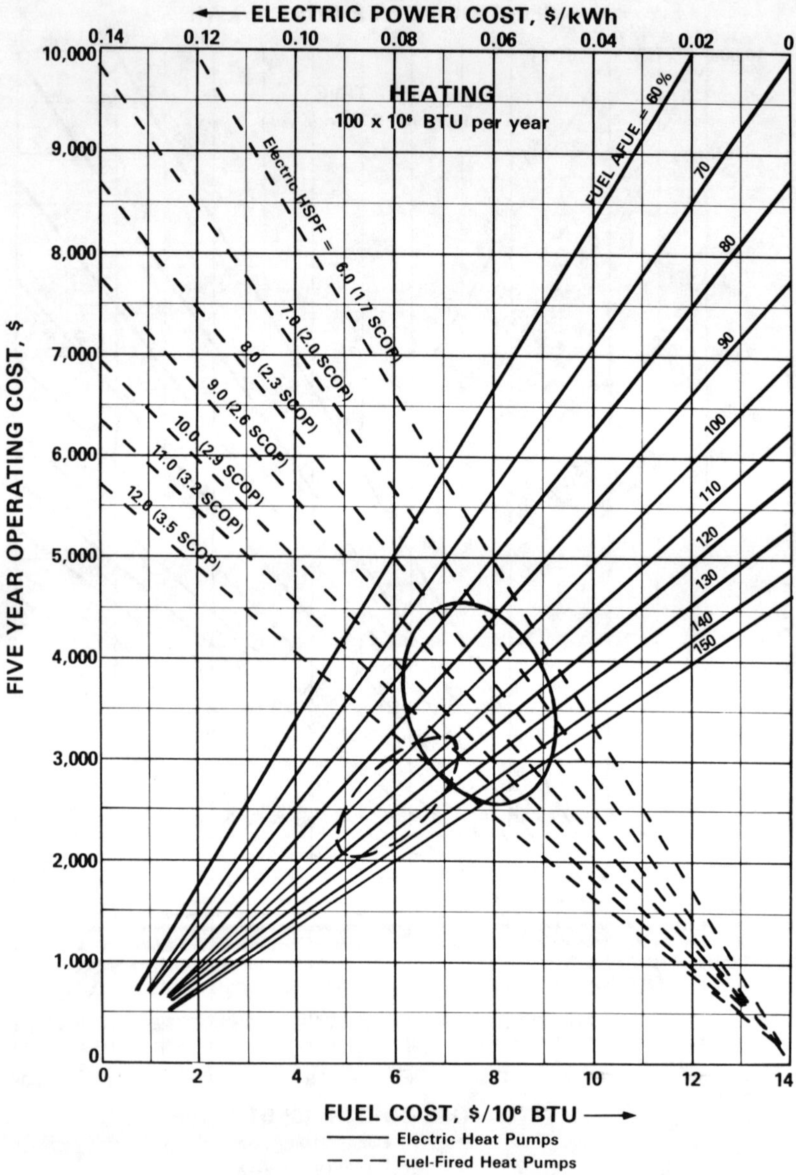

← ELECTRIC POWER COST, $/kWh
0.14
0.12
0.10
0.08
0.06
0.04
0.02
0
HEATING
100 x 10⁶ BTU per year
Electric HSPF = 6.0 (1.7 SCOP)
7.0 (2.0 SCOP)
8.0 (2.3 SCOP)
9.0 (2.6 SCOP)
10.0 (2.9 SCOP)
11.0 (3.2 SCOP)
12.0 (3.5 SCOP)
FUEL AFUE = 60%
70
80
90
100
110
120
130
140
150
FIVE YEAR OPERATING COST, $
10,000
9,000
8,000
7,000
6,000
5,000
4,000
3,000
2,000
1,000
0
0
2
4
6
8
10
12
14
FUEL COST, $/10⁶ BTU →
Electric Heat Pumps
Fuel-Fired Heat Pumps

Figure 5.

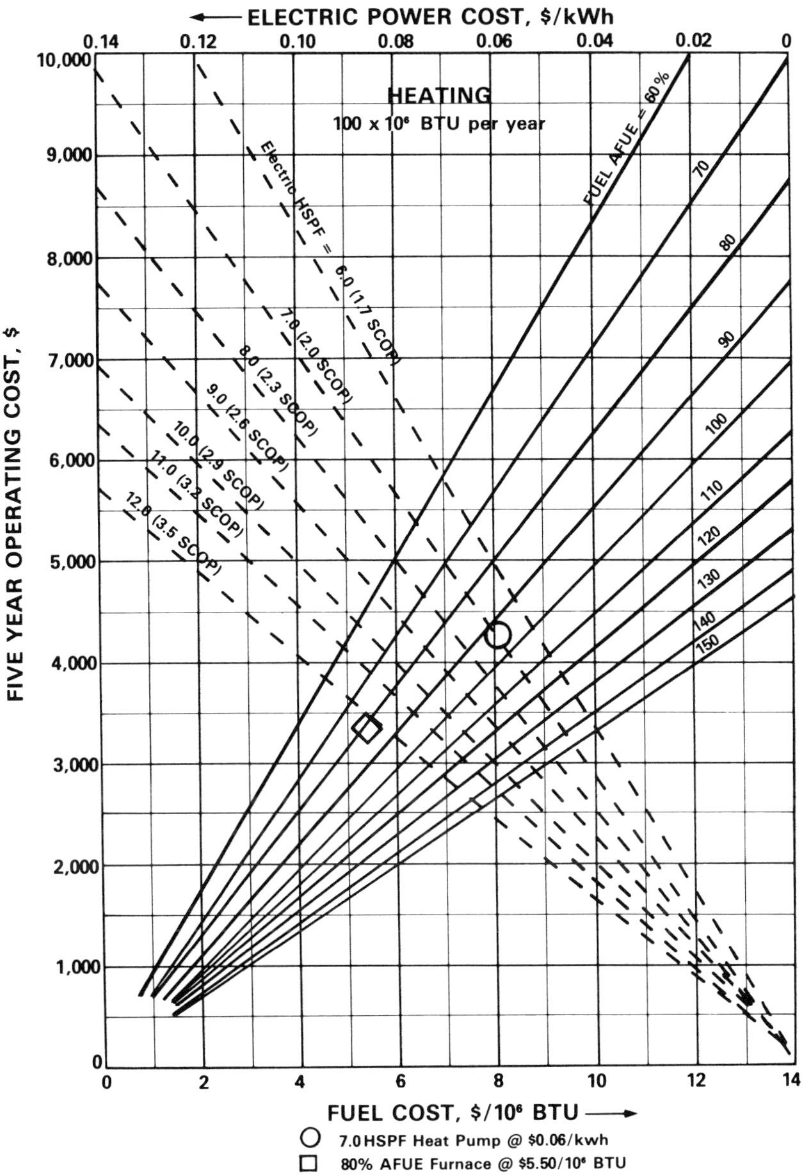
ELECTRIC POWER COST, $/kWh
0.14
0.12
0.10
0.08
0.06
0.04
0.02
0
HEATING
100 x 10⁶ BTU per year
Electric HSPF = 6.0 (1.7 SCOP)
7.0 (2.0 SCOP)
8.0 (2.3 SCOP)
9.0 (2.6 SCOP)
10.0 (2.9 SCOP)
11.0 (3.2 SCOP)
12.0 (3.5 SCOP)
FUEL AFUE = 60%
70
80
90
100
110
120
130
140
150
FIVE YEAR OPERATING COST, $
10,000
9,000
8,000
7,000
6,000
5,000
4,000
3,000
2,000
1,000
0
0
2
4
6
8
10
12
14
FUEL COST, $/10⁶ BTU
7.0HSPF Heat Pump @ $0.06/kwh
80% AFUE Furnace @ $5.50/10⁶ BTU

Figure 6.

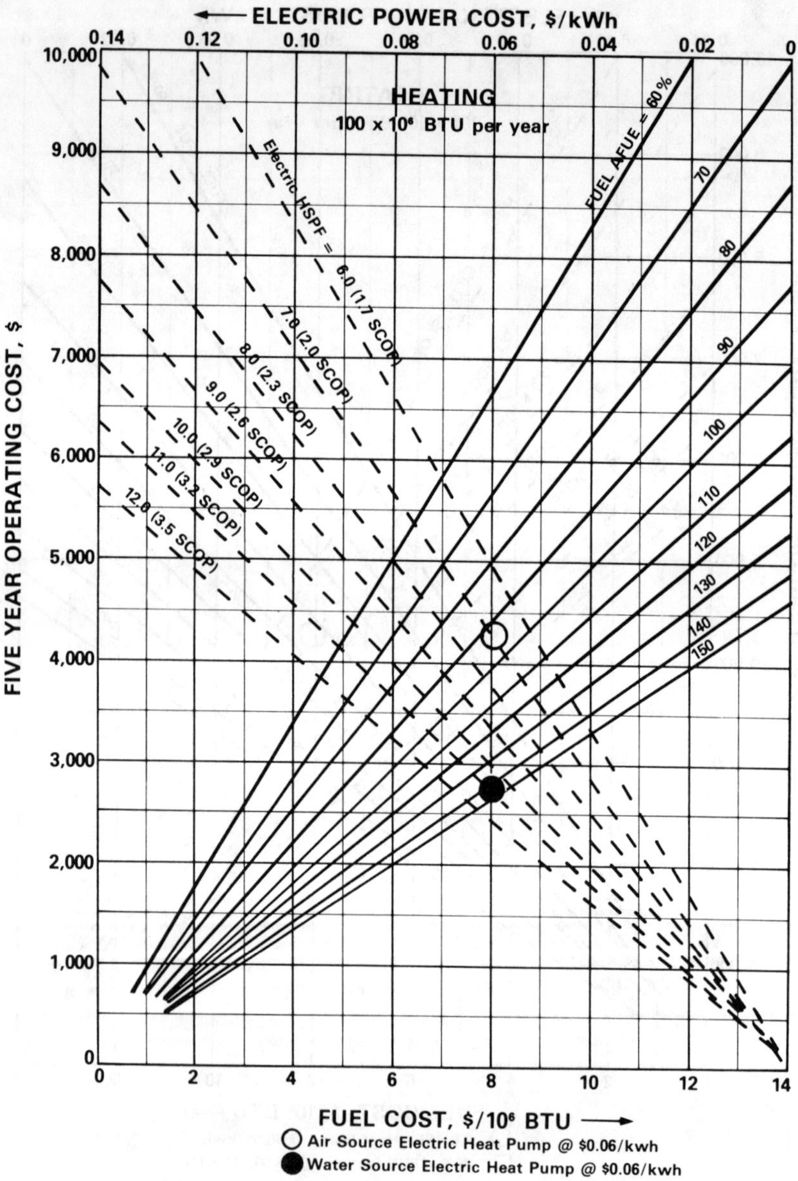

ELECTRIC POWER COST, $/kWh
0.14
0.12
0.10
0.08
0.06
0.04
0.02
0
HEATING
100 x 10⁶ BTU per year
Electric HSPF = 6.0 (1.7 SCOP)
7.0 (2.0 SCOP)
8.0 (2.3 SCOP)
9.0 (2.6 SCOP)
10.0 (2.9 SCOP)
11.0 (3.2 SCOP)
12.0 (3.5 SCOP)
FUEL AFUE = 60%
70
80
90
100
110
120
130
140
150
FIVE YEAR OPERATING COST, $
10,000
9,000
8,000
7,000
6,000
5,000
4,000
3,000
2,000
1,000
0
2
4
6
8
10
12
14
FUEL COST, $/10⁶ BTU
Air Source Electric Heat Pump @ $0.06/kwh
Water Source Electric Heat Pump @ $0.06/kwh

Figure 7.

LENNOX ENERGY SAVINGS INVESTMENT ANALYSIS

LOCATION **Sioux City**

COST DIFFERENCE OF Add-on Heat Pump OVER Gas Furnace	=	$ 600.00
DOWN PAYMENT DIFFERENCE	=	$ 0.00
INTEREST RATE	=	17.0%
LOAN LENGTH (MONTHS)	=	60
MONTHLY PAYMENT DIFFERENCE	=	$ 14.91
ANALYSIS LENGTH (YEARS)	=	10
TAX BRACKET	=	30%
DESIGN HEATING LOAD (BTU's/HR)	=	60,000
HEATING LOAD HOURS	=	2575
Add-on Heat Pump		
HEATING DESIGN (OUTSIDE) TEMPERATURE	=	−7
SYSTEM 1 LOW TEMPERATURE CAPACITY	=	18,800
SYSTEM 1 LOW TEMPERATURE COP	=	1.8
SYSTEM 1 HIGH TEMPERATURE CAPACITY	=	35,200
SYSTEM 1 HIGH TEMPERATURE COP	=	2.8
WINTER ELECTRIC RATE ($/KWH)	=	$ 0.055
WINTER ELECTRIC ESCALATION RATE	=	5.0%
GAS RATE ($/THERM)	=	$ 0.550
GAS ESCALATION RATE	=	5.0%
A.F.U.E. RATING	=	65.0
WINTER ELECTRIC USE (KWH)	=	12,963.7
GAS BTU'S (MILLIONS)	=	34.7
FIRST YEAR ENERGY COST	=	$ 904.01
Gas Furnace		
A.F.U.E. RATING	=	65.0
GAS BTU'S (MILLIONS)	=	183.0
FIRST YEAR ENERGY COST	=	$ 1,006.63

	HVAC INVESTMENT			ALTERNATE INVESTMENT		
YR	ENERGY SAVING	TAX SAVING	NET SAVING	BEGINNING BALANCE	INTEREST @ ROI%	DEPOSIT (−) WITHDRAWN (+)
0			0			0
1	103	29	−48	0	0	−48
2	108	24	−47	48	12	−47
3	113	19	−47	106	26	−47
4	119	12	−48	179	44	−48
5	125	5	−50	271	66	−50
6	131	0	131	386	94	131
7	138	0	138	349	85	138
8	144	0	144	297	72	144
9	152	0	152	225	55	152
10	159	0	159	128	31	159
	1291	88	484			484

RETURN ON INVESTMENT (ROI) PERCENTAGES

AFTER TAX ROI (%)	24.34
BEFORE TAX ROI (%)	34.78

Energy savings plus tax saving minus annual loan payment equals NET savings. The above comparison shows that your investment in a Add-on Heat Pump as compared to a Gas Furnace at a $600.00 difference will provide the same annual cash return to you over a period of 10 years as if you had invested the down payment plus any subsequent deposits at the indicated percent ROI in an alternate financial opportunity. Your initial investment of $ 0.00 was recouped and earned an additional $484.41 for you.

Figure 8.

REFERENCES

1. James M. Calm, "Electric Heat Pump Trends in the United States," Proceedings of International Energy Agency Conference, May, 1984, Graz, Austria.

2. Ted C. Gilles, "Evaluating Heat Pump Supplemental Heat Alternatives for Residential and Commercial Applications," Proceedings of Electric Power Research Institute Heat Pump Conference, April, 1986, Kansas City, Missouri.

3. Lennox Industries Inc., "Lennox LOGIC Computer-Aided HVAC Analysis Programs," Form No. LOGB-852-L8, Dallas, Texas.

4. Test Procedures for Central Air Conditioners, Including Heat Pumps, Federal Register, pp. 76700-76723, Dec. 27, 1979.

5. Electric Power Research Institute, "Heat Pump Reliability," Form No. EU3004, Palo Alto, California.

Chapter 35

How to Raise the Competitive Position of Heat Pumps in Air-Conditioning Systems of Large Office Buildings

Kihachiro Kubota and Kimio Morino

ABSTRACT: Heat pump, combination of boiler and chiller, and combined boiler and absorption refrigerator are widely used as heat generators in air-conditioning system of modern office buildings in Japan.

The past record shows that heat pump's competitive position is higher than other equipment where total floor space of the building is under 5000 m^2 but it becomes lower when the total floor space is over 5000 m^2. This is mostly because the price of heat pump is higher than that of other heat generators. To make heat pumps more economical in air-conditioning systems for large office buildings, it is necessary to reduce both the price and the running costs of heat pumps.

This paper discusses the economical position of heat pumps among other popular heat generators in the air-conditioning systems of large buildings and shows some actual experiences on the effective use of heat pumps in large office buildings.

INTRODUCTION

Heat pumps are widely used as a heat generator in the air-conditioning systems of office buildings in Japan. The past record shows that 48% of office buildings have installed electrical driven heat pumps. In particular the air-to-air unitary heat pump is very popular in small buildings where the total floor space is less than 5000 m^2 and has an economic advantage over other heat generators.

The share of heat pumps, however, goes down as the office building becomes larger. This tendency is due to the heat pump's price and its running costs which increase as the heat pump grows larger. So it is necessary both to reduce the price of heat pumps and to study effective applications to improve their competitive position in the case of large buildings.

HEAT PUMPS: Prospects in Heat Pump Technology and Marketing (Proceedings of the 1987 IEA Heat Pump Conference), Edited by Kay H. Zimmerman © 1987 Lewis Publishers, Inc., Chelsea, Michigan 48118. Printed in U.S.A.

This paper discusses the economical merits of heat pumps in air-conditioning systems with the heat generator's share and the computer simulation of air-conditioning systems. And this paper shows some actual experiences on the effective use of heat pumps in the air-conditioning systems of large office buildings.

THE COMPETITIVE POSITION OF HEAT PUMPS IN THE AIR-CONDITIONING SYSTEMS FOR JAPANESE OFFICE BUILDINGS

Popular Heat Generators

There are four popular heat generators in air-conditioning systems for modern Japanese office buildings as follows:

A. Unitary Heat Pump (UHP)
B. Heat Pump for a central heat generator (CHP)
C. Combined Absorption Refrigerator and Gas-fired Boiler (AR-GB)
D. Chiller and Gas- or Oil-fired Boiler (C-GB)

Usually UHP is installed in the perimeter of office spaces and CHP, AR-GB or C-GB are installed in a central plant of building.

Share of Heat Pumps in Office Buildings

Figure 1 shows the share of four popular heat generators in four hundred air-conditioning systems selected from among buildings which were built from 1980 to 1984 [1].

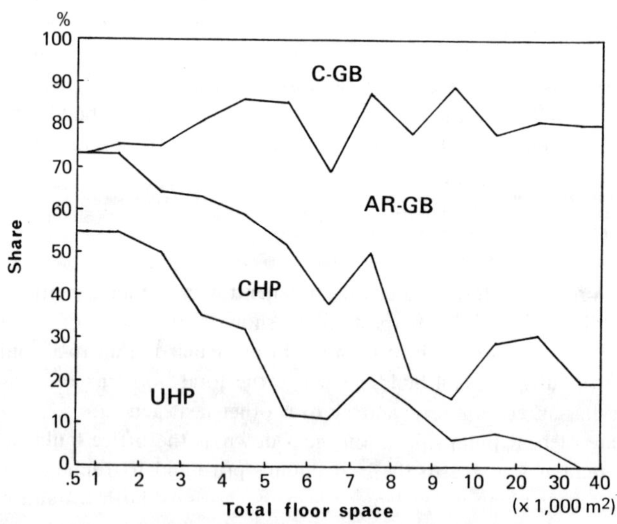

Fig. 1. Share of heat generators.

The following comments can be derived from this figure:

1) The share of Unitary Heat Pumps is larger than other heat generators when the total floor space of buildings is less than 5000 m^2. Adding the share of Central Heat Pumps to it, the total share is over 50%.
2) The share of Unitary Heat Pumps decreases as the total floor space of the building becomes larger than 5000 m^2 and also the total share including Central Heat Pumps is below 50%.
3) When the total floor space is over 8000 m^2, the share of Combined Absorption Refrigerator and Gas-fired Boiler is far higher than that of heat pumps.

The circumstances stated above mean that the competitive position of heat pumps is still low in large buildings.

Unitary Heat Pumps have merits of economy in small buildings as follows:

1) The price is low because it is a ready-made device.
2) Installation, operation and maintenance are easy, so their costs are low.
3) The installation space is small as it is compactly built in a unitary air-conditioner of small size, so the office space can be used effectively.

However these merits diminish in large buildings because many units are needed as the building becomes larger. This increases the maintenance costs and the problems of controlling a number of units.

From this point of view it is usually planned to install a large heat pump as a central heat generator. But such plans are often refused by building owners because the heat pump's price and running costs are high. Usually a large heat pump is custom-made machine and its price per unit of output is about twice as much as a ready-made one at the present time in Japan. In addition, the cost of electric power in Japan is higher than that of gas and oil in commercial use nowadays.

POSSIBILITY OF RISING HEAT PUMP'S POSITION IN LARGE BUILDING

Trials of Rising Heat Pump's Position

Buildings which installed heat pumps as a heat generator in air-conditioning systems appeared in the 1970's in Japan. Since that time such buildings were built in large numbers. The main reasons for using of heat pumps in the air-conditioning systems were to prevent of air pollution in cities and also to achieve energy savings.

Some of these buildings installed a big water tank for heat storage in order to improve the heat pump's effectiveness on cost. As the power rate in the night is cheaper than that in the daytime in Japan, the heat pump is operated by using cheap power in the night and the heat produced by it for heating and cooling the office space is accumulated in a heat storage tank in this case. This method offered not only lower heat production costs but also a decrease in the price of the heat pump owing to the reduction of heat pump capacity. At the same time, this system increases the initial investment and the pump motive power for circulating the hot or cold water in the system. Though there are cost disadvantage factors for heat pumps in this system, this offered more advantage to heat pumps than in a system without the heat storage tank.

Some new buildings built recently have a heat recovery system or use solar energy as shown later, and have had good results.

Cost Evaluation of Heat Pumps in Air-conditioning Systems

For finding the heat pump's situation on cost in large buildings, we evaluated the initial costs and the running costs of an air-conditioning system model with different heat generators by a computer simulation program.

Conditions in cost simulation

The simulation study was done for four popular heat generators as followers:

1) Central Heat Pump System (CHP)
2) Central Heat Pump System with Heat Storage Tank (CHP-ST)
3) Combined Absorption Refrigerator and Gas-fired Boiler (AR-GB)
4) Chiller and Gas-fired Boiler (C-GB)

The specifications of each heat generator are shown Table 1.

Simulation conditions were as follows:

a. The building is in Tokyo and air-conditioning service space is 8000 m^2 (Figure 2).
b. The annual heating and cooling load are 31 Mcal/m^2 and 44 Mcal/m^2, respectively.
c. The air-distribution systems are single duct constant volume system and fan-coil system.

Table 1. Specifications of heat generators

	Heat Generator		Output	Input
CHP	Air-to-water Screw Heat Pump			
		cooling	220 USRT	231 kw
		heating	650 Mcal/h	268 kw
CHP-ST	Air-to-Water Screw Heat Pump			
		cooling	86 USRT	102 kw
		heating	217 Mcal/h	101 kw
AR-GB				
		cooling	220 USRT	60 m^3/h [a]
		heating	665 Mcal/h	70 m^3/h [a]
C-GB	Turbo Refrigerator			
		cooling	220 USRT	188 kw
	Sectional Boiler			
		heating	740 Mcal/h	82 m^3/h [a]

[a] LNG: 11,000 kcal/m^3

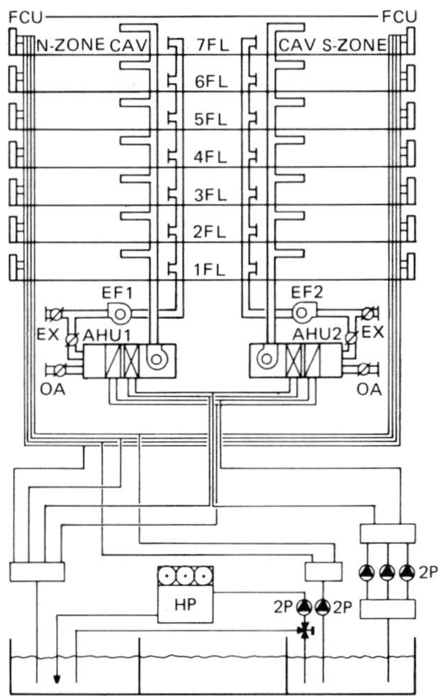

Fig. 2. Air-conditioning system of model building (CHP-ST).

Initial investment

Table 2 shows the initial investment of heat generate equipment including devices such as pumps, controllers and other accessories which are related to heat generator by ratio. The investment for CHP-ST is about 1.3 times as much as AR-GB, and CHP is also more expensive compared with AR-GB and C-GB.

The heat pump price accounts for about 65% in CHP and 30% in CHP-ST of the whole investment. The price of the heat storage tank is half of the whole investment. Compared with CHP, the price of the heat pump in CHP-ST is about half because its capacity can be reduced by the heat storage tank.

Table 2. Initial investment of heat generators

kind of heat generator	CHP	CHP-ST	AR-GB	C-GB
investment cost (ratio)	113	127	100	95

Running cost

Table 3 shows the annual consumption of electric power and gas in each case as estimated by the simulation program.

From this result the running cost (Table 4) can be calculated by referring to the tariffs for electric power and gas given by the electricity and gas companies. Both the electric power rate and gas rate consist of a basic fundamental charge depending on the user's maximum power demand or gas volume contracted with the supply company and the amount charge.

The CHP system is inferior in running cost to AR-GB, but CHP-ST can compete with AR-GB. The running costs of CHP-ST are low because the electric power rate under contract with the electricity company could be decreased by reducing the user's maximum power demand by means of a reduction in heat pump capacity and using cheap power effectively at night.

Table 3. Annual energy consumption

System		CHP	CHP-ST	AR-GB	C-GB
Energy consumption	Electric power [Mwh]	443	468 day: 345 night: 123	267	286
	Gas [$\times 10^3 m^3$]	- - -	- - -	38	21

Table 4. Annual running cost (1000 yen/year)

System	CHP	CHP-ST	AR-GB	C-GB
Electric power	1,947	1,531	970	1,365
Gas	- - - -	- - - -	536	332
Total	1,947	1,531	1,506	1,697
Proportion	129	102	100	113

Table 5. Energy cost (yen) per one Mcal

System	CHP	CHP-ST	AR-GB	C-GB
Electric power	18.4	13.4 day: 14.7 night: 9.7	14.9	19.5
Gas	- - - -	- - - -	13.0	14.4
Total	18.4	13.4	14.1	18.2

The rate of power and gas per one Mcal in each case are shown in Table 5. The rate for CHP-ST is cheapest. The power rate in the night is about 65% of that in the daytime.

How heat pumps raise the competitive position

Viewed from the standpoint of total cost, it seems that the heat pump is at a disadvantage in air-conditioning systems of large buildings as compared with other heat generators at the present time. But it is possible to rise heat pump's position in running cost by use of a heat storage tank. In order to improve the heat pump competitive position in this system it is important to improve the total COP, including devices which are related with the heat pump, though it is necessary of course to make efforts to reduce the heat pump price and improve the COP of the heat pump.

Figure 3 shows the items of energy consumed by heat generator, auxiliary equipment, pumps and air-handling fans. The energy consumption of the heat pump in CHP-ST is low as compared with AR-GB and C-GB. But that of pumps became larger because the heat storage tank was installed. If this energy could be reduced, the CHP-ST could get an advantageous position. A solution to this problem is the use of a heat storage tank such as a tower. The items shown on the right hand in Figure 3 show how this solution can be applied (shown as notation CHP-TST). The energy consumption of the pumps becomes less than in CHP-ST because of pump head reduction. An actual application will be shown later.

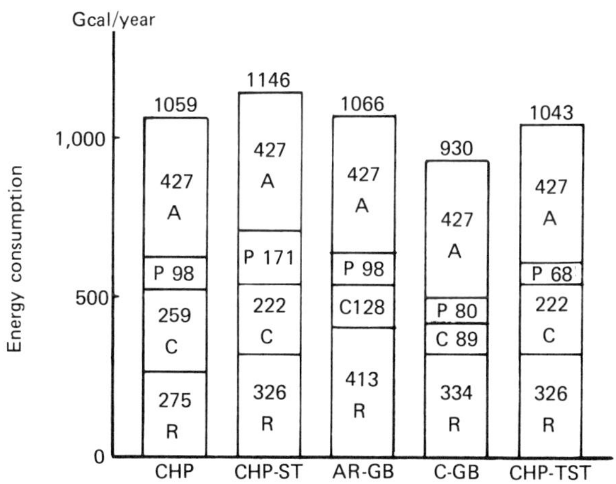

R: Heat generators
C: Fan and pump of cooling tower or fan of air-heat exchanger
P: Primary and secondary water pumps
A: Air-handling fans

Fig. 3. Items of energy consumption.

Actual Experience of Use of Heat Pumps

Chubu Denryoku Okazaki Building [2]
(Owner is Chubu Electric Co., Inc.)

This is one of the most advanced buildings for energy saving. A lot of thought on energy saving was applied to the architectural structures and the air-conditioning system.

One of this building's features is to recover heat from the usual office space, and the computer room. Transferring heat from the exhausted air to the air supplied to the usual office space is done by an air-to-air total heat exchanger. Heat generated in the computer room is recovered by a double-bundle type heat pump (Figure 4). The latter heat recovered is accumulated in heat storage tanks. The heat recovery system is controlled effectively in accordance with indoor and outdoor climatic conditions by computer.

Measurements have shown that the energy used in this building is half that of buildings of conventional design. The initial investment for energy saving is about 100 million yen and about 40 million yen is saved in energy costs annually. It is reported that the initial investment cost will be paid back in four years.

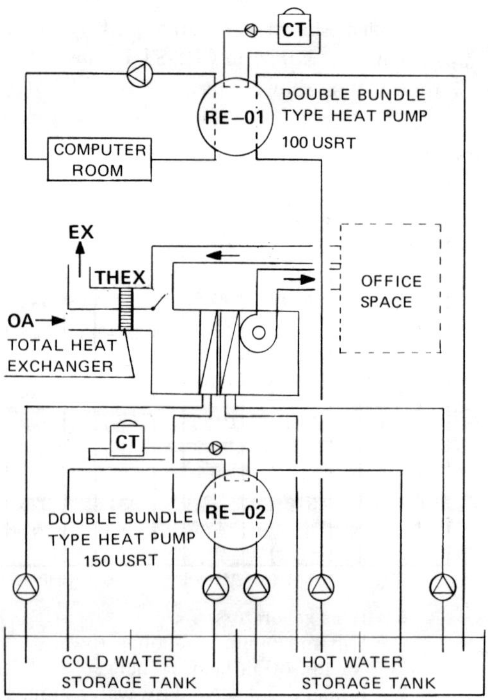

Fig. 4. Heat recovery system.

Kenchiku Kaikan (Owner is Architectural Institute of Japan) [3]

Ready-made unitary heat pumps were installed in the perimeter of this building (Figure 5). They are composed of the closed-loop water source heat pump system and can give either heating or cooling by absorbing or rejecting heat from the closed loop.

Generally, many office buildings require simultaneous heating and cooling during occupancy. Figure 6 shows the actual requirements of simultaneous heating and cooling measured in this building in winter. The hatched areas indicate the recovered heat being rejected from the spaces that require cooling even in winter. The heat recovery in this system contributes 20% of energy saving to the whole heating load in winter.

The seasonal COP of the heat pumps in this system is 3.6.

Further more this building has solar collectors to assist heat pumps, supplying about 16% of the heating load in winter.

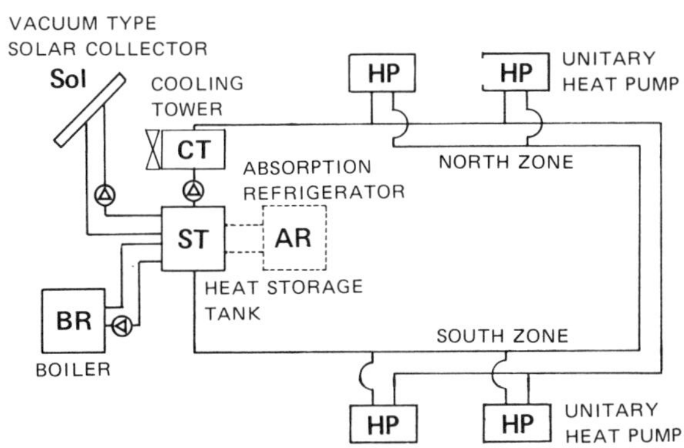

Fig. 5. **Diagram of heat recovery system.**

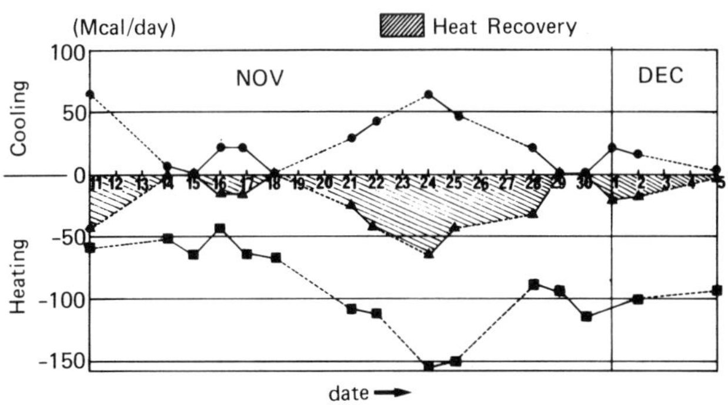

Fig. 6. **Heat recovery in winter season.**

Building in Institute of Shimizu Co., Ltd. [4]

The installation of the heat storage tank is useful to improve the heat pump's ranking and it is needed to reduce the power of the water supply pumps. Because the heat storage tanks used in Japanese buildings are usually like a pool and almost the water system between it and other devices is not closed.

The air-conditioning system of this building has air-source heat pump and heat storage tank like a tower to try the decrease of pump power (Figure 7).

Measurement result shows that the power of the water supply system is about 40% of the power in the case of a standard tank and the cost saving is 1.4 million yen annually.

It is reported that the initial investment of this system would be paid back in about eight years as compared with no heat storage tank heat pump system.

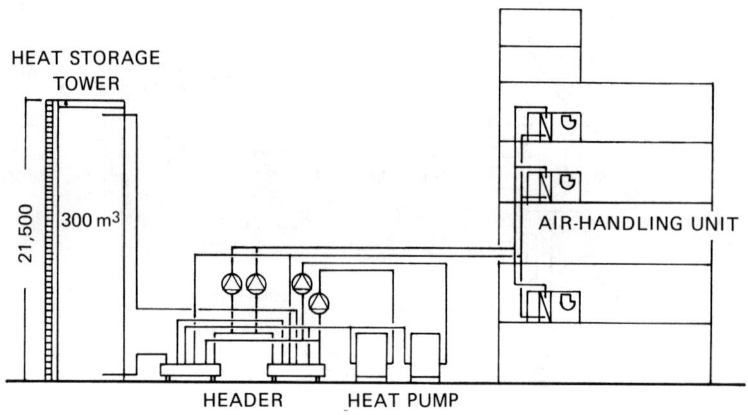

Fig. 7. Tower type heat storage tank system.

REFERENCES

1. "Completion Facility List," *Air-conditioning and Sanitary Engineering,* SHASEJ, 1982–1986
2. M. Sakurai et al., "HVAC System of Chubu Denryoku Okazaki Building," *Air-conditioning and Sanitary Engineering*, SHASEJ, 13–38, Vol.55, No.12, 1981
3. M. Sakurai et al., "Study on energy consumption in new KENCHIKU-KAIKAN (part 1 – 5)," *Summary of technical meeting of Architectural Institute of Japan*, 1982–1985
4. Y. Konishi et al., Development of the Vertical Heat Accumulator, *Report of the Research Institute of Shimizu Construction Co., Ltd.*, 93–103, Vol.37, April 1983

Chapter 36

European Market for Small Electric Heat Pumps: Economic Attractiveness of Heat Pumps and Short-Term Market Prospects

Jean-Francis Harris

ABSTRACT : In most European countries with heavy dependence on oil imports, the market for space heating only heat pump systems developed rapidly and peaked between 1982 and 1984 to decline equally rapidly. While the escalation and subsequent drop in oil prices has had a considerable impact on the development of the heat pump market, significant marketing misjudgements have been equally crucial factors in the fall-off in sales. Even with a high oil price scenario and a significant decrease in equipment price, it is doubtful that the heat pump market for space heating only applications will increase substantially. However niche markets for both space heating and cooling applications are developing in the Southern part of Europe. This paper analyses the different market segments for heat pumps in France, Sweden and F.R.G. and provides both qualitative and quantitative sale estimates in the short term in those segments.

ECONOMIC ATTRACTIVENESS OF HEAT PUMPS

In all market segments, except those of the "early adopter", the overiding criteria for the successful penetration of heating or cooling applications is the consumer's perception of the cost-effectiveness of the equipment. Even in the residential market, customers are above all concerned with low first investment costs (capital + installation) and to a certain extent reduced short term operating costs. This fundamental problem was rapidly identified by most of the European governments and utilities. Until the market attained sufficient dimensions to allow the benefits of mass production, states and utilities

HEAT PUMPS: Prospects in Heat Pump Technology and Marketing (Proceedings of the 1987 IEA Heat Pump Conference). Edited by Kay H. Zimmerman

compensated for additional costs of heat pumps through the use of financial aids to end-users. In addition, some electric utilities offered preferential electric tariffs for heat pumps.

The detailed analysis of the French, Swedish and German markets in the following paragraphs highlights the impact of economic factors on the development of the heat pump market. However price is not the only factor driving the development of a market. Identification of appropriate segments, communication and distribution are other importants factors. Chapter 2 will show the impact of those factors through examples of marketing misjudgements. Finally chapter 3 will identify potential markets for heat pumps within the next five years.

Price Of Energy Sources

As shown in Exhibit 1, domestic energy prices to end users vary significantly from one country to another, in particular for electricity. Prices for light heating oil until mid-1985 were almost the same in European countries (2.7 cents/kWh), except in France where taxes on this form of energy are fairly high. Natural gas is particularly cheap in the Netherlands and in the United Kingdom (2 cents/kWh) where it is an indigenous energy source. Abundant cheap gas in those two countries was a natural barrier for the development of heat pumps. Electricity prices vary from 3.1 cents in Sweden to 9.8 cents in the Netherlands. A large quantity of hydroelectricity enables Sweden to offer one of the lowest electricity prices in Europe, second only to Norway.

Exhibit 1 : Domestic energy prices to end-users (mid 1985) (a)

U.S. Cents/kWh	F.R.G.	France	Netherlands	Sweden	United Kingdom
. Light heating oil	2.7	3.2	2.7	2.7	2.7
. Natural gas	2.9	3.1	2.0	-	2.0
. Electricity					
- average	-	7.0	9.8	3.1	4.2
- peak	6.2 (b)	6.5	-	-	7.4
- off-peak	3.8	3.7	-	-	2.4
- Electricity/oil	2.3	2.2 (c)	3.6	1.15	1.6
- Electricity/gas	2.4	2.3	4.9	-	2.1

(a) energy prices by mid 1985 before decline in oil prices
(b) preferential tariff for heat pumps
(c) 1.25 with "effacement jour de pointe" tariff

The lack of specific electricity tariffs for space heating in the Netherlands explains the absence of electric heating in this country. The major UK electric utility (CEGB) has set up an attractive off-peak electricity tariff for electric heating system with night storage in order to respond to the competition of natural gas. As shown in Exhibit 1, the German utilities also offer an attractive electric off-peak tariff for heating systems with night storage. The same utilities did not offer, however, an attractive tariff for heat pumps. Although France has the lowest electricity tariffs within the Common Market, these tariffs are not very favorable to domestic end users. Electricite de France recently introduced a new electric tariff called "effacement jour de pointe". In order to obtain this tariff, end users will have to switch from electricity to another energy source during peak hours. With this particular tariff end users had the benefit of a favourable electricity/oil ratio. However the recent slide in oil prices (2.2 c/kWh for light heating oil in mid 86) makes electricity unattractive in all European countries even with the use of efficient electric heating systems.

The French Heat Pump Market

Electricite de France's promotional campagn in 1977-1979 stimulated a rapid growth of the French heat pump market (see Exhibit 2).

Exhibit 2 : Heat Pump Sales in France, F.R.G. and Sweden

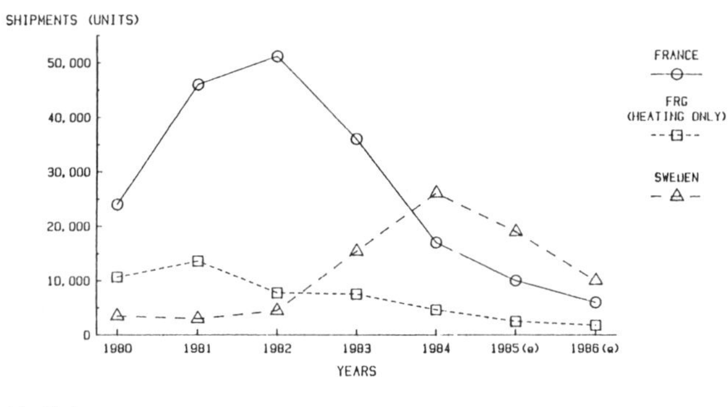

During this period E.D.F. promoted the installation of small electric air-to-air heat pumps in new well insulated single family houses, using exhaust air as the heat source and with electric resistance as a back-up system. However the poor seasonal performance factor of such systems, the insufficient qualifications of mechanical contractors who installed air distribution systems and the attemps of builders to minimise construction costs have led E.D.F. to discontinue its efforts in the direction of this market. In the early 1980's, E.D.F. and the French government oriented their efforts in the direction of existing single family houses equipped with an oil boiler. About 3 million single family houses were equipped with such systems and were situated far from the natural gas distribution system. After three years of demonstration and diffusion programmes,

E.D.F. started the Perche i campaign (electricity heat pump added to an existing boiler). Households received a $ 400 subsidy for the installation of a bivalent heat pump system. Sales of heat pumps peaked rapidly at 50 000 units. Campaign objectives were to reach sales of about 100 000 units/year by 1990. However as shown in Exhibit 3, electric heat pumps were not very attractive for households consuming 3 cubic meters of oil per year.

Exhibit 3 : Economic Attractiveness of Heat Pumps in an Existing Single Family Residence (a)

	Perche i Heat Pump (b)	Perche i Heat Pump (c)	Diesel Heat Pump
Annual consumption after retrofit (d) (kWh)			
. oil	7,870	7,870	17,120
. electricity	7,400	7,400	220
Annual operations cost after retrofit (FF)			
. oil	1,694	1,694	2,680
. electricity	3,142	2,134	100
. maintenance	840	840	1,600
. installed cost (FF)	25,000	25,000	33,000
. annual savings (FF)	2,344	3,352	2,640
. pay-back	10	7	13

(a) with 1982 energy prices
(b) with current peak/off peak tariff, SPF : 2.6
(c) with estimated "effacement jour de pointe" tariff
(d) consumption of light heating oil : 34,000 kWh

Rapid growth in the market was due to purchases of "early adopteres" and households with larger oil consumption. Despite the possibility to obtain a subsidy for the installation of a heat pump, the initial price of the system was seen as a major barrier to the development of the market. Then, the French government in coordination with E.D.F. developed a new strategy in order to deliver to the final consumer equipment with a maximum installed price and a 5 year guarantee. However some marketing misjudgements, in particular at the distribution level, led to poor results even when the price of oil was still high. In spite of a vigorous promotion campaign (TV advertising, etc.) sales of heat pumps declined rapidly even before the oil price dropped (see Exhibit 6). Estimates for 1986 show sales equivalent to those before the second oil crisis.

The German Heat Pump Market

During the 1970's a large number of new residences were equipped with night storage electric heating systems. These systems were promoted through preferential electric tariffs by electric utilities to smooth the demand curve. Except through this heating system, electricity was not competitive with fossil fuel. With the second oil crisis and rapid escalation in the oil price, electric air to water heat pumps seemed to be attractive. After a rapid increase of the market until 1981 (see Exhibit 2), the market for space heating only heat pumps declined equally rapidly. There were three major reasons in the poor development of this market :

- as shown in Exhibit 4, the price of equipment was very high. Only high income households could afford such equipment. This market segment was rapidly saturated. Important efforts have been conducted to reduce the installed cost of equipments by 30 %. However by the time that heat pump manufacturers were able to achieve this cost reduction, oil prices were declining rapidly.
- electric utilities were not able to offer an attractive electricity tariff in connection with heat pumps. When they realised the necessity to offer an attractive tariff, oil prices dropped.
- as opposed to Sweden or France, the German government did not offer a direct subsidy for the installation of a heat pump, but a tax deduction over 10 years of the installation cost. A direct subsidy might have been a more attractive formula.

Exhibit 4 : Economic Attractiveness of Heat Pumps in a Two Family House (25 kW) (a)

	Air to Water Electric Heat Pump	Gas Absorption Heat Pump	Gas Boiler
. Seasonal Performance Factor (%)	300	128	78
. Annual Fuel Consumption (kWh)	28,240	37,760	56,850
. Annual Fuel cost (D.M.)	3,050	2,565	4,065
. Maintenance Cost (D.M.)	500	400	200
. Installed Cost (d)	48,000	43,000	30,000
. Pay Back Period	12 (c)	9 (b)	-

(a) with 1982 energy prices ; needs : 40,000 kWh
(b) relative to a gas boiler
(c) relative to an oil boiler
(d) including distribution

F.R.G. has been the only European country with a fair development of heat pumps for hot water. Sales for hot water heat pumps peaked in 1981 with 35 000 units. Despite the decline in oil prices, this market remains stable with about 25 000 units per year. Several reasons may explain the development of this product. (1) Its initial cost is acceptable for most consumers. (2) The majority of German households are equipped with boilers providing both space heating and hot water. In this case the seasonal performance for hot water is very low and the heat pump fairly attractive. (3) Finally the production of hot water through night storage in combination with electricity is not as developped in F.R.G. as it is in other countries. However boilers are now able to produce hot water with a fairly good performance, thus limiting the future development for hot water heat pumps.

Despite the efforts conducted by manufacturers and electric companies to offer a cheaper heat pump product and a more favourable electric tariff, the space heating only heat pump system is not attractive with current energy prices. As shown in Exhibit 2, sales of heat pumps should not exceed 2 000 units in 1986.

Exhibit 5 : Economic Attractiveness of Heat Pumps in an Existing Single Family House (a)

	Ground Source Heat Pumps (b) Monovalent	Ground Source Heat Pump (c) Bivalent	Air to Water Heat Pump Bivalent
. Seasonal Performance Factor (%)	2.3	2.85	2.2
. Annual Fuel Consumption (kWh)	15,052	17,800	33,089
. Annual Fuel cost (Sek)	3,642	4,930	7,262
. Maintenance Cost (Sek)	1,400	1,400	1,000
. Installed Cost (Sek)	60,000	55,800	25,000
. Pay Back Period (d)	8 (c)	9 (b)	6

(a) with 1982 energy price
(b) horizontal earth heat pump in clay + sand ground
(c) vertical earth heat pump in clay + solar assistance
(d) relative to an oil boiler

The Swedish Heat Pump Market

The Swedish market for heat pumps developed less rapidly than in the other countries : until 1981 heat pump sales were dominated by ground source systems. As shown in Exhibit 5, the initial cost of a ground source heat pump is fairly high and despite an attractive electricity tariff and loans with preferential interest rates, the pay-back remained unattractive. Sales of heat pumps developed with the introduction of exhaust and outdoor air heat pumps. Both the introduction of cheaper systems and a 15 % investment subsidy had a considerable impact on the heat pump sales. Relative penetration of heat pumps over total sales of heating equipment was the highest in Sweden : heat pumps accounted for 25 % of total sales of heating equipment in Sweden when it was only about 10 % in France in 1982 and 3 % in F.R.G.

Both the drop in oil prices and the discontinuation of the 15 % investment subsidy are equally significant factors in the collapse of the market. Heat pump sales in 1986 are not expected to exceed 10 000 units in 1986.

OTHER FACTORS AFFECTING THE HEAT PUMP MARKET

Most of the efforts conducted by manufacturers, electric utilities and governments were in the direction of a dramatic decrease of the installed cost of the heat pump either through subsidies or through guaranteed installed prices. The Perche GTI program is a good demonstration of how a potentially successful marketing operation turned into a significant failure. As shown in Exhibit 6, the French heat pump market started to decline when oil prices were still continuing to increase and when both the French government and E.D.F. started the Perche GTI campaign.

Exhibit 6 : Evolution of Heat Pump Sales versus oil prices in France

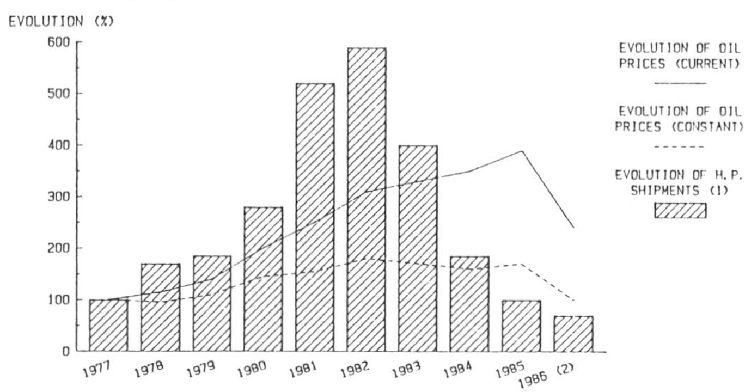

Source : DHYCA, EdF
(1) Base 1977
(2) July 31st, 1986

The objective of this program was to provide customers with a heat pump with a maximum installed price and a 5 year guarantee. Installers worked with a "delivered price/manufacturer price" ratio of about 4 with a conventional system while the Perche GTI had a ratio of about 2.5. As a result, only 25 % of the 20 000 installers and distributors in the network sold GTIs. This demonstrates that the price of a system is not the only key factor. Distribution is also key, especially when 2/3 of final price to end-users is through distribution and installation.

Another key factor is the recognition of real market segments and the evolution of those segments. Although single family houses with an oil system are numerous in Europe (about 8 million units), they do not represent an homogeneous segment. The market of houses with large oil consumption is limited and probably saturated in the same way as the "early" adopters. Most of the households also had to replace their existing boiler thus adding the cost of a new boiler to the cost of the heat pump. Until the end of this decade there will be an important need to replace existing boilers installed about 15-20 years ago. When households have replaced these boilers, they might be less interested in adding a heat pump.

SHORT THERM PROSPECTS FOR HEAT PUMPS IN EUROPE

In all European countries, the market for domestic space heating only heat pumps will continue to decline with the current energy prices. Overall European market for space heating only and hot water heat pumps should not exceed 40 000 to 45 000 units by 1987. Even with a new escalation in oil prices, it is doubtful that the market for heat pumps will increase again. Early adopter's market segment is already saturated and most of the largest oil consumers in single family houses have already a heat pump or conducted other energy saving measures. However there are still niches for heat pumps in various residential and tertiary segments. With the development of heat pump applications in cooling, overall market for heat pumps should be about 100 000 units by 1990.

France

There is little hope in the short term for a new development of the heat pump market in single family houses even in the case of an important escalation in oil prices. Untapped market segments for space heating only heat pumps are in the multi family buildings and in the tertiary sector.

With a price of 3 cents/kWh for light heating oil, the pay back period for an electric heat pump using the Peakday Effacement tariff is about 4 years for a building with 50 apartments. Heat pumps will be attractive only in non gas areas since it is difficult to compete with gas condensing boilers. However both residential and tertiary markets are fairly limited and represent sales of about 500 to 600 units of 150 kW per year.

Small retail outlets, in particular the distribution of high value added products (perfumes, fashion, shoes, etc.) are an opportunity for heating and cooling applications. An important promotion campaign will start in early 1987 in this segment. Sales of equipment should reach rapidly 10 000 to 15 000 units per year. Markets of the same size exist also in Italy and in Spain. It is doubtful that in the short term the new construction will be the next target for space heating and cooling applications, except in very specific segments (i.e. the fitting up of attics in 8 to 10 year old houses). The development of cooling applications in the new construction will rely on Gaz de France success in introducing gas condensing warm air furnaces. In order to compete with direct electric heating systems Gaz de France is currently developing a warm air system. Air may then compete with hydronic distribution and open the new construction market to air conditioning. However, heat pumps, even if they are offering some cooling, will have to compete against a very cheap gas system.

Exhibit 7 : Short Term Evolution of Heat Pumps in Sweden

Heat Source	Distribution	0.3 kW	4-10 kW	11-25 kW	26-100 kW	101-1000 kW
Out-air	Water	Decrease	Drop	Flat	Increase	Increase
Out-air	Air	Increase	Increase	Flat	Flat	Flat
Exhaust.	Water	Flat	Flat	Flat	Flat	Flat
Exhaust.	Air	Flat	-	-	-	-
Ground	Water	-	Drop	Drop	Decrease	Flat
Ground	Air	-	-	-	-	-

Source : Killanan Services Ltd

Sweden

Short term forecasts on the Swedish market show a decline for most of the market segments except for the multi-family building sector and for small air to air heat pumps in single family houses (see Exhibit 7). Other market segments should continue to decline. Unless an important escalation in oil prices or new subsidy schemes, the Swedish market for heat will remain flat and below the 1986 level (10 000 units).

F.R.G.

Despite the effort of German electric companies to offer a

heat pump package at a set price and preferential electric tariffs, the heat pump market will continue to decline for space heating only applications. Sales of hot water heat pumps will maintain the overall heat pump market at about 20 000 units per year for the next three years. Small retail outlets are not yet perceived as a potential market in F.R.G. However, like in France and other European countries with moderate climatic conditions, this segment offers opportunities for cooling applications.

Chapter 37

Experience with Electric Utility Incentive Programs and Market Prospects in Europe

Michael F. Schneeberger

ABSTRACT: Electric utilities incentive programs in Europe are described and analyzed. Due to great differences in the structure of electricity production, in national energy policies and in scientific and technical capacities, inhomogenious situations in heat pump introduction developed in Europe. Based on a description of the present situation of European utilities with details on the structure of electricity production, the specific situation in European countries is analyzed. The outstanding activities of Swedish utilities, triggering utility programs in other countries are described. The programs of Electricité de France (EdF) and of German utilities are compared. The autor resumes an optimistic view on future aspects concerning heat pump utility programs since many possibilities of heat pump applications are not yet discovered and the benefits of this new electricity application not yet understood on a large scale.

INTRODUCTION

"Innovative Utility Programs are the key to Successful Heat Pump Development" has been the editorial of the IEA-HPC Newsletter Vol.3, No.3 from Oct. 1985 (1). Experience has shown, that heat pump installations have increased considerably, if electric utilities have fully supported the application of heat pumps with a broad range of activities. In Europe there are programs of the Swedish State Power Board, EdF, electric utility companies in the Federal Republic of Germany and Austria. Activities included in

HEAT PUMPS: Prospects in Heat Pump Technology and Marketing (Proceedings of the 1987 IEA Heat Pump Conference), Edited by Kay H. Zimmerman

these programs are: Technical consulting, selection of appropriate systems, offers for complete systems, extended guarantees, offers for benefical financing arrangements, and in some cases, low electricity tariffs.

To analyze and compare the different experiences with electric utility incentive programs in Europe, it is important to take into account the different situations of utilities in Europe as concerning structures of organisation, production capacities, price differentials between oil and electricity and energy policy as well as environmental discussion in the respective country. The climatic and geophysical situation, the scientific and industrial capacities of the country as well as the financial potential are decisive factors for the development of heat pump innovation projects utilities are involved.

Due to these reasons, an increasing exchange of informations and experiences developed between European utilities crystallizing in an "Utility Heat Pump Group". This informal group of utility experts of many OECD-countries in heat pump application certainly will spread knowledge and experience to break down many of the barriers to the broad utilization of this benefical technique leading to the protection of our environment and conservation of our resources.

PRESENT SITUATION OF EUROPEAN UTILITIES

The development of the situation of European utilities in the last decade can be resumed in the following general items:

After the first and second oil price shock electric utilities reacted in different ways, depending on their existing structures. The "Oil Substitution" - reduced oil consumption for electricity production - was the main target in many countries. The increase of nuclear energy production and the "renaissance" of coal utilization can be considered as the most important changes in electricity production (see table 1).

This structural changes induced high investment costs increasing the electricity tariffs in periods of declining oil prices. The increasing investments for environmental protection of energy production plants accelerated this development.Countries with important hydropower structures and/or well operating nuclear capacities can stabilize this important price trends.

Due to the nuclear power reactor accident in URSS the production and application of electricity in European countries appear to become a major topic of political discussions in many European countries.

Therefore, the application of electricity, whether in the traditional way by direct or storage heating or in the form of indirect use by heat pump installations appears to become more difficult in the next future.

There is a broad field of heat pump application in tertiary structures and in commercial and industrial projects not yet discovered by many utilities due the lack of information and practical experience. The beneficial impact, both to environmental problems and to utility structures will help to sustain the development.

Table 1. Structure of electricity production in European countries

	Belgium	France	Italy	F.R.G.	U.K.	Sweden	Austria
Production 1984, TWh	52	310	172	372	265	124	43
./.							
Hydro	1	67	45	18	6	69	29
Nuclear	26	182	7	88	47	51	-
Thermal							
coal + lignite	15	45	23	213	121	)	4
fuel	4	8	70	10	87	)4	3
gas	3	4	23	32	3	)	5
others	3	4	4	11	1	-	2
Appr. consumption 1984, kWh per inhabitant	5.300	5.700	3.000	6.000	4.700	15.000	5.700

SPECIFIC SITUATION IN EUROPEAN COUNTRIES (2)

Situated on the western part of the Eurasiatic continent, European countries are situated between 70° and 35° northern latitude. The northern part of Europe is exposed to subpolar climatic conditions, the southern part is influenced by subtropical climatic situations. Western European countries are strongly influenced by the warm Gulf-stream, whereas eastern countries are exposed to continental climatic influence. Due to these variations concerning the most important meteorological parameters - temperature and humidity - the heating and cooling requirements show great variations with a great impact on heat pump development.

Sweden

As one of the leading countries concerning heat pumps development in Europe Sweden can be considered as a heating only area. The rather low prices of electricity, a high standard of industrial development and an important activity of Swedish utilities are important factors for the successful Swedish program. The state owned Swedish State Power Board supported a rather comprehensive R & D program, starting in 1979 and reinforced by the political consequences of the second oil price shock. The State Power Board initiated pilot and demonstration projects, investigated all imaginable heat sources including geothermal and waste heat and realized large size district heat applications. An important participation of Swedish utilities can be found in the training of all partners involved. The information of customers and the realization of excellent technical documentations are further characteristics of the Swedish utility program.

The Swedish program essentially concentrates on electrically driven heat pumps, and up to now about 130.000 heat pumps have been installed. In 1985 about 25.000 installations have been realized. About 90 installations of 1 MW or more increased the total amount of thermal heat pump load in Sweden considerably. The total heat generated per year with this systems is 12 TWh.

Norway

Norway, blessed with energy resources in relation to its population, has very favourable electricity production with 100 % hydropower. There are important activities in research groups as concerning heat pump development, and electricity companies are developing essentially the heat pump installation in industry for evaporation and drying processes. The total power installed is about 100 MW with 500 GW_{th}/year. The thermo-compression plants of 200 MW show an output of 1 TWh/year. Optimistic prognostics are given for the future, based on an integrated effort of research instituts, constructors and utilities.

Finland

Finland's heat pump development is essentially promoted by the Ministry of Trade and Industry and small applications compete with low prices in electricity. IMATRAN VOIMA OY operates heat pumps in district heating systems and much attention is concentrated on industrial applications in drying and evaporating processes.

France

The French electric utility company, "Electricité de France", in accordance with the national energy policy invested considerable effort in the development of heat pumps. Based on scientific research work an important demonstration program has been organized for all types of applications. Special tariffs have been introduced to support the development, the "EJP rate". An important marketing effort has been made after the second oil price shock, with 50.000 annual sales per year in 1982. Due to different reasons, a big drop of sales can be found in the last years in the residential application. The space heating of industrial buildings, low temperature drying, heating of liquids, milk treatment and wood drying appear to become successful areas for heat pumps. Electricity application in agriculture is growing rapidly using heat pumps for greenhouse heating. Presently about

80 installations are in operation in industrial processes. Considerable effort of EdF prepared the terrain for effective penetration of this quite new idea of heat pumps in this field. Growing geothermal energy use, combined with big size heat pump installations has been realized in France with an important promotion activity of the EdF - company.

Federal republic of Germany (FRG)

Compared with France with its centralized structures, the different electric utility companies in the FRG developed different policies concerning heat pumps, depending on the respective electricity production structures. The annual sales of space heating systems are declining to values of 4000 units/year, the sales of hot tap water systems are stabilizing at 30.000 units/year. Promotional efforts have been supported by the big Rheinisch-Westfälisches Elektrizitätswerk together with the German industry for the "Heat pump program 2000". At present, about 50.000 heat pumps are installed in the FRG, with a total electric load of about 350 MW_e. Despite low oil prices German utilities show an increasing interest in new energy conserving applications with heat pumps.

Austria

The Austrian government has supported the heat pump innovation since 1979 by tax reduction. The electric utilities introduced experimental tariffs to help market introduction. At present, about 50.000 installations are operating with about 35.000 hot tap water units. The heating market is declining, the hot tap water application still increasing. The electric utility company of Upper Austria, "Oberösterreichische Kraftwerke AG" (OKA), supported strongly the heat pump innovation in the last years, with important market effects in many application areas. The development of geothermal sources, the use of industrial waste heat, many applications in gastronomy and industry have been realized by energy consulting organized by this utility. The heat pumps installed in the OKA supply area represent about 50 % of the total number of heat pumps in all of Austria, while the OKA supply area itself is inhabited by only 10 % of the Austrian population. An increasing spread of interest in heat pump application can be observed recently in other Austrian utilities.

FUTURE ASPECTS

Based on the experience acquired by electric utility companies in Europe in the last years and the publication of operation results, an increased interest can be observed in this new electricity application field. The integration of electricity demand of heat pump installations in the load management of electric utilities, the increasing environmental discussion and the problems associated with the burning of fossil fuels in highly populated areas will contribute to improve the situation. A broad field of "heat transforming" from low temperature to high temperature levels, for example in geothermal applications, in waste heat of sewage plants and in industrial application appear as important developments in the future. Therefore, the general development of heat pump innovation programs of electric utilities in Europe can be considered as optimistic.

REFERENCES

1. K. Holzapfel, "Innovative Utility Programs are the Key to Successful Heat Pump Development", IEA-Heat Pump Center "News Letter", Vol.3, No.3, October 1985.

2. 13th Congress of the World Energy Conference, Cannes, "National Reports on the Status of Heat Pumps", October 1986.

Chapter 38

Experiences with an Electric Utility Incentive Program —TEPCO's Activities for the Use of Heat Pumps

Katsuhiko Narita

ABSTRACT: The Tokyo Electric Power Company Inc.(TEPCO) acknowledges that he is responsible for energy- and environmental conservation considering his situation; He consumes a large quantity of imported oil and natural gas for power generation.

TEPCO's activities of promoting the use of heat pumps have started since 1960 for this reason, and accelerated in earnest since 1977 when our Energy Conservation Center was established in TEPCO.

This papers states the TEPCO's heat pump promoting activities programms and experiences learned.

INTRODUCTION: TEPCO is one of 9 power utilities in Japan, and supplies electric power to more than 20,000,000 customers in the central Japan,where Tokyo Metropolice is situated. TEPCO's share to national KWH sales reaches approximately one third and the peak demand exceeds 37 GW in 1985.

The electric resources for power generation in TEPCO has been changed with the times ; in early times hydro power, and then thermal power and now nuclear power is going to be dominant. But fossil fuels such as imported oil and liquified natural gas are consumed yet a great deal affecting some impacts to our circumstances in spite of strenuous efforts for prevention of pollution and contamination, and the investment for the prevention reaches more than 20 % of the total.

Heat pump technology contributes to environmental conservation as well as the efficient use of energy resources, and TEPCO is convinced of the great importance of the efficient use of energy for our small islands where a huge amount of energy is thickly onsumed regardless of no energy resources.

HEAT PUMPS: Prospects in Heat Pump Technology and Marketing (Proceedings of the 1987 IEA Heat Pump Conference). Edited by Kay H. Zimmerman © 1987 Lewis Publishers, Inc., Chelsea, Michigan 48118. Printed in U.S.A.

For these reasons TEPCO promotes the use of heat pumps through various programms, which are introduced in this papers and experiences learned through the activities.

The theory of a heat pump is introduced to Japan in 1926. Since then heat pump technology has been applied to many kinds of installations. Most of them were used to commercial buildings utilizing groundwater as a heat source.

In 1950's small air source heat pumps were introduced, but unsuccessful because of falling down heating performance in cold days and noise problem in operation.

TEPCO decided to build GUMMA branch with heat pumps utilising groundwater as a heat source in 1960, which was great success,and designed some branches with heat pumps succeedingly.But groungwater heat pumps are limmited in the application site in spite of their excellent and stable performance, and TEPCO started to develop a large scale airsource heat pump in 1960. The first airsource heat pump of 100 RT with 2 turbo compressors in series was operated in 1965, and was noticed some problems still unsolved. It was the oil price crises that spotlighted a heat pump technology again and the energy coservation center selected the technology to be the most important and vital for energy consevation and started active promoting activities of heat pumps based on the past experiences.

Heat pump promoting activities today are focused on the following four fields; research & development, demonstaration of heat pump installatios in his own facilities, recomendations & consultations of heat pumps to the public, promotion of district heating & cooling by heat pumps as his own business.

1. Research & development activities

D & D activities of heat pumps are managed in R & D center and marketing department. R & D center is responsible for fundamental and long term programm,and marketing department for applicational and short & medium term programm. Most of the programm are carried on in cooperation with related industries. The yearly expenditure for the programm amounts to $10 million.

1.1 Marketing research for a heat pump

—research the diversificated customers' needs and potentiality of heat pump market in residential, commercial and industrial fields.

—analysis & evaluation of new heat pump technology and new products newly brought on the market.

—exchange market information and customers' needs with customers and the related manufacturers.

1.2 Research & development

—develop new heat pumps and their applications, and improve the performance of existing heat pumps in cooperation with related manufacturers.

—research and development of various heat sources for heat pumps including experimental applications.

—develop automated designing and evaluating system for heat pump applications.

2. Demonstration of heat pump systems

2.1 Demonstration of heatpump installation in TEPCO's facilities

—all of TEPCO's branches and facilities are going to be equipped with heat pumps, in some cases newly developed heat pump systems for experimental purposes.

—TEPCO's head office has been heated and cooled by heat pumps utilising waste heat from a transformer station (630 MVA) in electrical network.

—heat pumps & heat pump systems are always demonstrated in some service stations and housing exhibition spaces with explanatory guides and periodiclly in certain events.

2.2 Other demonstrations

—heat pumps are often demonstrated in certain shows sponsored by TEPCO, for example, Japan Housing Show, and Energy Saving Equipment Show.

—in some conferences related and journals of related associations actual results and experiences of heat pumps are reported as scientic reports and, in some cases, as advertisements.

There are about 300 buildings & facilities that belong to TEPCO and in these few years they shall be fully equipped with heat pumps.

3. Consultaions & recommendations

More than 100 engineers devote themselves to consultation on heat pumps open to the public, and often work for making recommendations for national energy policy, and some preparatory works related.

A training center was established in 1984, where training of these consultants is implemented, in some cases engineers who belong to other companies or organisations are also invited to the training and cooperated experimental applications in the training facilities. We find cooperated experiments of these kinds are very effective for new heat pump applications in industrial fields.

3.1 Consultations on heat pumps

—consultations on heat pumps and heat pump applications with building owners and planners.

—introductions of successful installations of heat pumps experienced by not only our own but also general sector.

—consultant engineers often receive public consultations at the standing demonstration spaces and event halls.

These consulting activities are limitted to planning stage taking account of the business of professionals, and often handed over to some reliable professionals on customers' request.

3.2 Recommendations for the use of heat pumps

—recommendations for the use of heat pumps in various fields to conserve energy resources and environments whenever appropriate to the occassions.

—preparatory working to accommodate the favorable situation for heat pump market.

The heat pump technology center of Japan is established this August and most of these activities are supposed to succeed.

4. Promotion of district heating & cooling by heat pumps as his own business

Tokyo is the one of the most rapidly changing cities in the world and urban development and redevelopment are undertaken in many places. Taking the advantage of the opportunity, district heating & cooling systems by heat pumps are going to be introduced to the sites. District heating & cooling systems requires previous investments and consensus among building owners related. Tokyo Metropolitan Government surpports and recommends district & cooling systems in viewpoint of prevention from air pollution.

Some examples of district heating & cooling sites are HIKARIGAOKA Park Town (New Town Project, 12,000 houses, heat pump dispersed system), GINZA Shopping district,and HIBIYA district.

These projects are promoted as TEPCO's own business and about 30 sales engineers and staffs are engaging in persuading related customers to acceptance of the system, and planning, designing, constructing, and managing the system.

Abovementioned are the briefing of TEPCO's incentive programm for heat pump promotion and other electric utilities in Japan are developing the action programms of the same kind.

Lessons learned through the activities

Small heat pumps for residential use are sold in the number of 1.8 million units and medium size heat pumps for small offices and commercial buildings 0.6 million units every year. This success of heat pumps in Japan results from the followings; mild climate, enduser oriented products, motives or incentives for efficient use of energy caused by shortage of energy resources and prevention from air pollution, and others.

A heat pump works most efficiently and economically when heating and cooling load exist together, and thanks to mild climate in Japan most of buildings require to be cooled in summer and to be heated in winter. Development of heat recovery system in space heating accelerate improvement of heating performance.

Small heat pump room units are widely used in Japan and this fact leads to mass production of the standardized heat pump units, and results in inexpensive and reliable units. Generally speaking newly developed products are apt to be expensive and have certain unstable performance to be improved. Therefore they are generally used in business fields where good maintenance is expected. And when the function and performance of the product are accepted, the product begins to be a public product,for the first time, which is inexpensive, reliable, and easy to use. The improvement can be made through mass production.

In 1950's Tokyo has suffered from various pollutions and contaminations caused by industrial processes and urban activities. District heating & cooling systems are recommended to introduce for prevention from air pollution caused by combustion of fossil fuels for space heating. In 1970's energy saving systems were strongly expected to introduce, and then district heating & cooling system by heat pumps has been spotlighted.

Heat pumps and heat pump technology are promising because of the excellent contrbution for energy and environmental conservation.

Section VIII

Government's Role in Energy Conservation

Chapter 39

Government Energy Policy in Japan

Takehiko Shimura

Synopsis

Japan is one of the world's leading consumers of energy. Unfortunately, it lacks domestic energy resouces. So, it is quite necessaly to take effective measures for energy concervation. In 1978, the Ministry of International Trade and Industry (MITI) established a national project for R&D on energy conservation named "Moonlight Project" as one of these measures. In this project, R&D on a high efficiency gas turbine, electric power storage systems using advanced batteries, fuel cells, Stirling engines, and on heat storable heat pumps have been jointly undertaken by national research laboratries, private enterprises, and universities. Research on heat storable heat pumps started in 1984 and will continue over a period of eight years until 1991. The total budget allocated by the Japanese Government for this area of R&D is estimated to total 10 billion yen (US$ 50 million).

1. Introduction

To promote the energy concervation policy, any legal measures should be taken to rationalize the energy utilization as well as to develop furtherance and enlightment activities for saving energy. At the same time, the technical developments for inproving efficiency and waste heat utilization in every sector of energy conversation, transport, storage and consumption will play a very significant role on energy conservation.

The promotion of the R&D program should be performed systematically and efficiently by combining appropriate R&D approaches and systems. From such a point of view, the Agency of Industrial Science and Technology in MITI started in 1978 the

HEAT PUMPS: Prospects in Heat Pump Technology and Marketing (Proceedings of the 1987 IEA Heat Pump Conference), Edited by Kay H. Zimmerman

"Moonlight Project" which promotes systhetically and systematically the R&Ds for energy conservation in mutual cooperation among national research laboratories, private enterprises, and universities involved.

The Moonlight project has been in progress, based on the following six mainstays:

The first is called the "Large-scale R&D projects for energy conservation", which have to be conducted in cooperation of various organizations such as national research laboratories, private enterprises,universities, etc.,because of requiring too much money and time for a private enterprise to carry out the energy conservation project. At present, five large-scale projects are going on a high efficiency gas turbine, electric power storage systems using advanced batteries, fuel cells, Stirling engines, and on heat storable heat pumps(super heat pupm energy accummlation system).

The second is named the "Leading and Basic technical development for energy conservation", which have been conducted at national research laboratories to provide technological seeds for future energy conservation.

The third is to give the financial assistance to private enterprises. Cincretely, subsidies are provided for private enterprises to develop more efficient houshold appliance such as electric refrigerator, etc., on the basis of the competitive developments among several companies.

The fourth is an investigation on the evaluation method of energy conservations technologies.

The fifth is the promotion of energy conservation through standardization.

The sixth is in an international research cooperation related to energy conservation technology.

2. Large-scale R&D projects for energy conservation

2.1. A high efficiency gas turbine

In this project, R&D is being conducted to achieve an combined cycle thermal efficiency higher than 55%, where a reheat-type gas turbine system with low emissions is operated to drive a steam turbine system by using a exhaust recovery boiler. By the end of fiscal year 1983, a gas turbine pilot plant with overall thermal efficiency of approximately 50% is to be developed, with integrating the most sdvanced existing technologies. The demonstration test are to be conducted concerning reliability, efficiency, etc.

2.2. Electric power storage systems using advanced batteries

A 11 year R&D project has been conducted on advanced battery electric power storage system since 1980.

Advanced batteries under development are four types: sodium-sulfur, Zinc-chloride, zinc-bromide and redox flow batteries. Battery characteristics and development tasks were scrutinized with 1kW class batteries in 1983. The development of 10kW class advanced batteries is underway since 1984. The study on the system technology has been also carried out to establish the means for connecting the battery energy storage system to an AC power utility network. In 1986, 1000kW class power storage system with the improved, large lead-acid batteries is scheduled to be applied to practical test.

2.3. Fuel cell power generation technology

There are four types of fuel cells currently under R&D in the Moonlight project; phospholic acid, alkaline, molten carbonate and solid oxide fuel cell.

As for the phosphoric acid fuel cell, which is thought to have the closest chance of commercial acceptabillity, two types of the fuel cell systems are being developed; low pressure type and high pressure type. Base on the component technology study performed by FY 83, the construction of two 1000kW plants, one in each type, has been started in FY 84 with the target of operating them in FY 86. A 1kW alkaline fuel cell stack was made and tested in FY 84. As for the molten carbonate fuel cell, based on the fundamental study performed in FY 83, development of 10kW class cell stacks has been started in FY 84. The solid oxide fuel cell, which may be the most efficient in the future, is being studied on basic problems.

2.4. Stirling engine

Engine designs, component technologies, concerning seal device, heat exchanger, burner system, etc. and trial manufacturing/ testing of engines are to be conducted to establish the practical Stirling engine system in terms of efficiency, durability, reliability, etc.

3. Super heat pump energy accumulation system

3.1. The object of this project

This project aimes R&D on heat pumps of very high performance and thermal storage systems for conversion of excess electric power at night. R&D's of this project are composed of i) fundamental research on refrigerant, chemical reactions and materials, ii) advanced compressor-driven heat pumps, iii) chemical heat storage system and iv) composition of total systems.

The target of the system in this project are:

for total system;

Application		Output Temperature	Energy Efficiency
A.C. of Office Buildings	Heating	45°C	4.5
	Cooling	7	5.3
District A.C.	Heating	45	4.5
	Cooling	7	5.3
	Hot Water Supply	85	6
Industrial Processes	Heating from Waste heat	150	>2.3
		300	>2.3

for compressor-driven heat pumps;

	High Performance Type			High Temperature Type	
	For Heating Only Type	For Heating & Cooling Type		low Temp. Heat Source	High Temp. Heat Source
		Heating	Cooling		
Output Temp.	85°C	45°C	7°C	~150°C	~300°C
C.O.P.	8	6	7	3	3

for chemical heat storage systems;

	High Temperature Storage	Low Temperature Storage
Output Temperature	~200°C	10°C
Capacity of Storage	50 Kcal/kg (material)	30 Kcal/kg (material)
Heat Recovery Ratio	>75%	>75%

The target of compressor-driven heat pumps in this project are plotted and compared with the performance of present heat pumps.

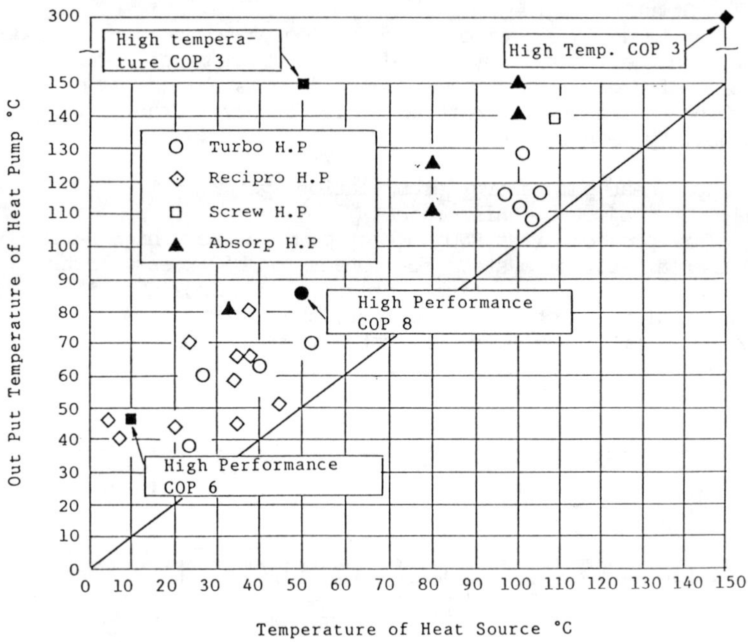

3.2. Program of the project

This project was started in 1984 and will be contiued for eight fiscal years until 1991. Total budget for this project is estimated to be 10 billion Yen (US$ 50 million). As shown in table 1, the program will be proceeded through three stages: for Fy 1984-1986 R&D is concentrated to compressors, heat exchangers, refrigerants etc. for the compressor-driven heat pumps, and to the choice of reaction cycles, appropriate materials etc. for chemical heat strage systems. In Fy 1987-1988, bench-plant of 100 kW class heat pumps and 10,000 kcal class chemical heat strage systems will be tested. In Fy 1987-1991, pilot systems will be tested which

Table 1 Super Heat Pump Energy Accumulation System R & D Schedule

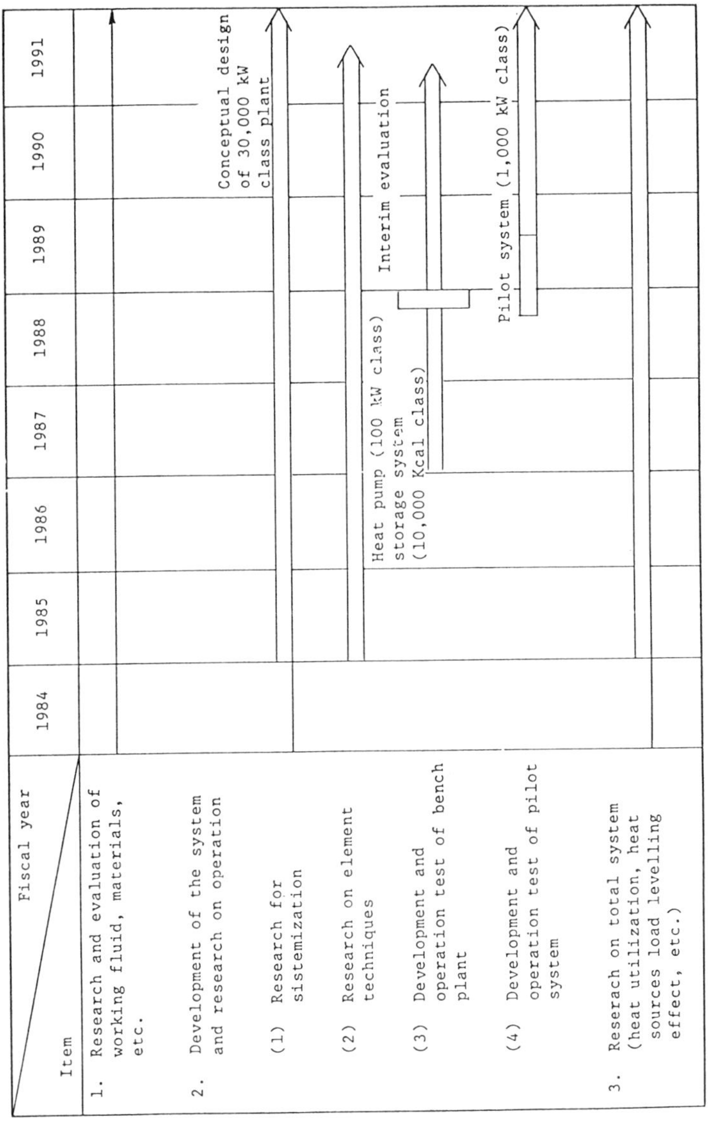

are combined systems and 1000kW class capacities.

"Research and evaluation of working fluid, materials, etc." was started in October 1984 by four national research labolatorys. "Development of the system and research on operation" and "Research on total system" are conducted by New Energy Development Organization, which subcontracts part of the R&D to private enterprises.

3.3. Compressor-driven heat pumps to be developed

3.3.1. A high performance heat pump of COP 8 for heating

A very high performance of COP 8 for the case of output temperature 85 °C from heat source 50°C is expected. This system is shown scematically in the flow diagram.

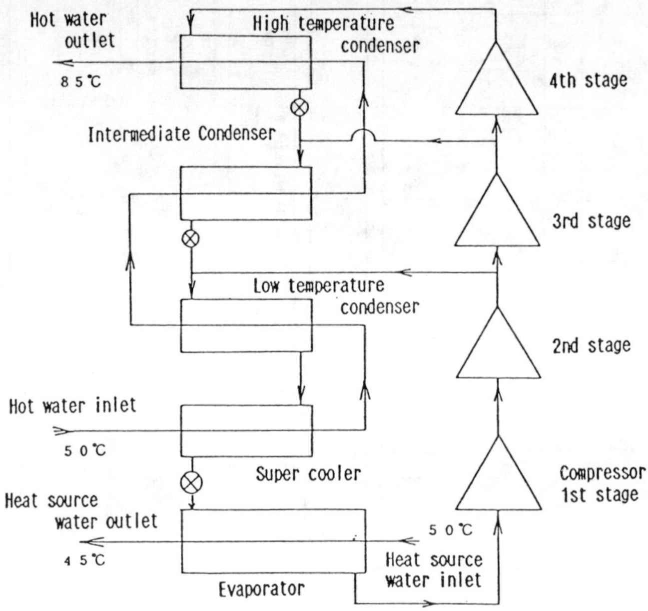

Improved 4 staged turbocompressor hermatically sealed with electric motor will be developed. Reasonable temperature differences will be achieved between hot water, heated by counter flow in three staged condensers, and refrigerant in each condensers by choosing proper mixture of non-azeotropic freons.

Researches on following thema are studied; thermal properties of refrigerant mixtures, thermodinamic analysis for reasonable choice of non-azeotropic mixture, multiple staged compression, condensation and counter flow heat exchangers, cooling methods of hermetic sealed motor, and lubricants etc..

3.3.2. A high performance heat pump of COP 6 (heating) and COP 7 (cooling)

A high performance heat pump for air conditioning is the target. High performance will be achieved by employing two economizers for supercooling the refrigerants, at the same time, by injection liquid-vapour mixture of refrigerants into appropriate intermediate stage of screw compressor. Researches on the following thema are studied; the selection of non-azeotropic mixture of refrigerants for this system, the cooling method by liquid injection, heat exchangers etc..

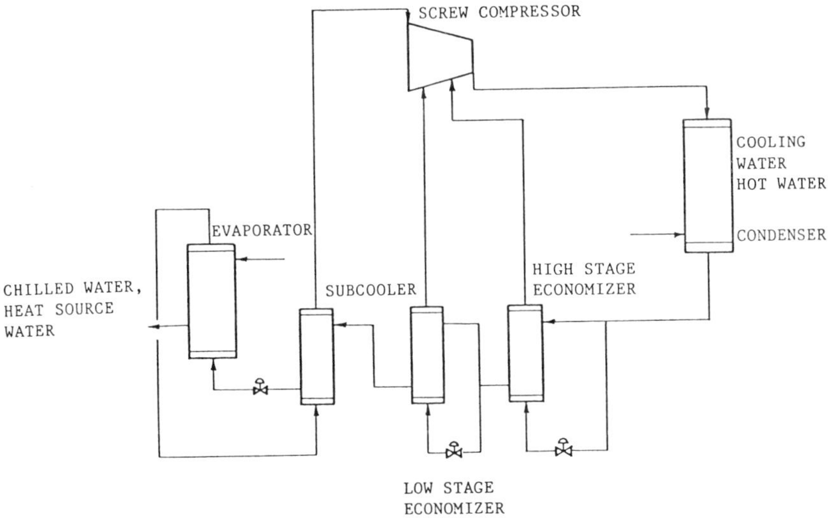

3.3.3. A high temperature output (150°C) heat pump

To attain pumping up of 100°C, two screw compressor-driven heat pumps with screw expanders for recovering power is cascaded.

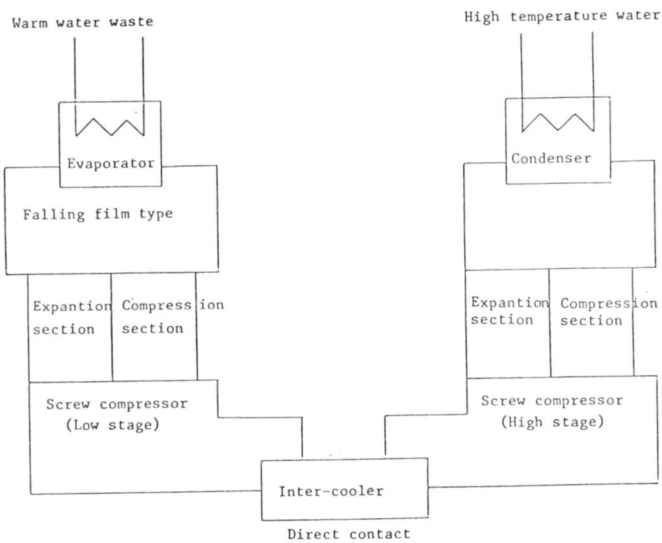

Researches on following thema are studied; high temperature refrigerants which can be operated higher than 150°C, symmetrically arranged new screw compressors to reduce the thrust force, etc..

3.3.4. A high temperature output (300°C) heat pump

This system employing H_2O as refrigerant is shown schematically in the following figure.

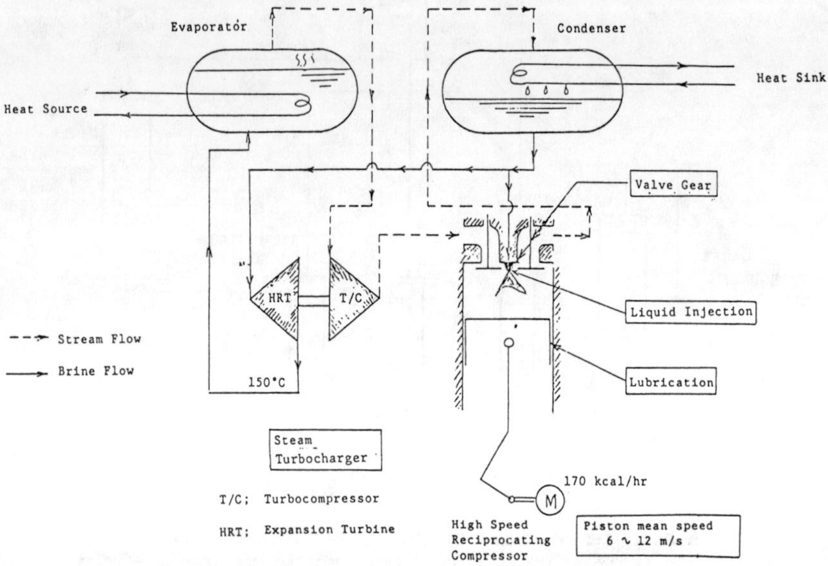

This system is composed of a high pressure condenser, an exaust steam turbine-driven steam supercharger, a high speed reciprocating steam compressor etc..
Researches on the following thema are studied; a high speed compressor with water injection, a steam turbine, etc..

3.4. Energy storage system using chemical reaction to be developed

3.4.1. The reactions for high temperature heat storage system

i) Hydration-dehydration reaction

$$CaBr_2.2H_2O \longleftrightarrow CaBr_2.H_2O + H_2O$$

ii) Ammoniation reactions

$$NiCl_2{\cdot}6NH_3 \longleftrightarrow NiCl_2{\cdot}2NH_3 + 4NH_3$$

$$NaSCN{\cdot}3.5NH_3 \longleftrightarrow NaSCN{\cdot}2.5NH_3 + NH_3$$

iii) Concentration storage

Heat is stored as difference of concentration potential of the solution.

$$TFE{\cdot}E181 \longleftrightarrow TFE + E181$$

TFE: trifluoroethanol

3.4.2. The reactions for low temperature heat storage systems

i) Concentration storage

Heat is stored as difference of concentration potential of the solution of alkaline metal halides in the water.

ii) Heat storage using gas clathrates

Small crystals of fluoromethan, such as R11 or R12, and water called gas clathrates form when a fluoromethan is mixed into water at 3-10°C. This clathrate has a large latent heat of cristalization equivalent to that of water.

Chapter 40

State Support to the Heat Pump Industry: An Assessment of Governmental Policy in the Residential Sector of Sweden, France, and Germany

Torbjorn Bostroem, Jean-Francis Harris, and N. Haskins

ABSTRACT : Since the mid 1970s, European governments, within the objectives of overall energy policy considerations, have implemented regulatory and pricing mechanisms designed to facilitate the penetration of energy efficient heating technologies. These mechanisms have been translated to the building sector through a wide variety of special incentive schemes aimed at supporting research and development, production, domestic consumption and international cooperation and trade. In this paper we analyse the policies of three European governments : Sweden, France and Germany, to bring about the rapid transfer of heat pump technology to the residential heating market.

INTRODUCTION

Since the mid 1970s, European governments, within the objectives of overall energy policy considerations, have implemented regulatory and pricing mechanisms designed to facilitate the penetration of energy efficient heating technologies. These mechanisms have been translated to the building sector through a wide variety of special incentive schemes (subsidies, grants, tax deductions) aimed at supporting research and development, production, domestic consumption and (to a lesser extent) international cooperation and trade.

In this paper we analyse the policies of three European governments : Sweden, France and Germany, to bring about the

HEAT PUMPS: Prospects in Heat Pump Technology and Marketing (Proceedings of the 1987 IEA Heat Pump Conference). Edited by Kay H. Zimmerman

rapid transfer of heat pump technology to the residential heating market. The role of the publicly owned power utilities should not be underestimated, since the combined efforts of state and utilities have been the major impetus in the development of heat pump technology.

Although the broad outlines of energy policy objectives were the same in all three countries i.e., reduction of dependence on imported oil through the development of indigenous energy resources, the role of the state differed significantly according to the structure of the national energy industry. In Sweden where industry is fragmented the state has played a significant role to encourage the development of the heat pump industry. In France where the industry is more concentrated, the state has played a relatively secondary role and the major effort has been provided by the utilities.

Relative penetration of heat pumps is highest in Sweden where the price of electricity has been the most favourable (see Exhibit 1) and where the government has been the most active. While the escalation and subsequent drop in the price of oil has had a considerable impact on the development of the heat pump market, state efforts, particularly at the commercialisation stage, and marketing activities, mainly on the part of the utilities, have been equally crucial factors. In Sweden, for example, the 15 % investment subsidy was a key factor in the development of the market and was equally significant in the collapse of the market when it was discontinued in 1985. In France, EDF's promotional Perche i campaign was extremely effective in bringing about the growth in heat pump sales. However, the follow-up program Perche GTI did not succeed in the long term, largely due to marketing misjudgements.

Exhibit 1 Domestic end-user prices for certain fuels including taxes

U.S. Cents/kWh	F.R.G.	France	Sweden
Light heating oil	2.7	3.2	2.7
Natural gas	2.9	3.1	-
Electricity			
- Average	-	7.0	3.1
- Peak	6.2	6.5	-
- Off-peak	3.8	3.7	-
Electricity/oil	1.4 (1)	2.2 (2)	1.15

1 FF = 0.114 US $; 1 DM = 0.35 US $; 1 SEK = 0.12 US $

(1) Based on off-peak tariff

(2) Based on average tariff. 1.25 with "effacement jour de pointe" tariff.

It is clear that the state has a significant role to play in the emergeance of a new technology by appropriate interventions along the chain of R, D and D, commercialisation and international trade and cooperation and by the elaboration of a coordinated strategy underlying the different policies, in particular when several actors (research bodies, manufacturers, utilities, government agencies) are involved.

EVALUATION OF GOVERNMENT POLICIES

Three key factors distinguish the Swedish efforts in the field of heat pumps from those in France and Germany :

- the scope and level of funding,
- the structure of the delivery system for government funding,
- close cooperation with the public utility.

As can be seen in Exhibit 2, the scope and level of government funding in Sweden far outweighs that of the other two countries. In France where the structure of the energy industry is relatively concentrated, the state monopoly electricity producer, Electricité de France, has been the prime mover in the promotion of heat pumps, whereas in Germany where the electric power industry is very fragmented, the government has played a more important role.

TABLE 2 : Key Government Players in the Development of the Heat Pump Market

Type of Activity	FRG	France	Sweden
R & D	. BMFT		. BFR . STU
Demonstration	. BMFT	. AFME . Direction de la construction	. BFR . STEV
Commercialisation	. Ministry of Finance . County governments	. Ministry of Industry	. STEV National/ . County housing boards

AFME : Agence Française pour la Maîtrise d'Energie
BFR : Swedish Council for Building Research
STU : National Swedish Board for Technical Development
STEV : National Swedish Energy Administration
BMFT : Ministry for Research and Technology

The structure of the delivery system for government

funding also differs in Sweden where there is an intermediate level of independent authorities (see Exhibit 3) responsible for the implementation of government policy. As recipients for public money, these authorities are obliged to define their development strategies within the context of long term program objectives thus leading to a more structured approach. The authorities are also more closely involved with industry than the government ministries. The Swedish Council for Building Research, for example, set up advisory groups involving representatives from industry and research institutes to make recommendations as to product development.

A further striking characteristic is the close cooperation between the Swedish government and the Swedish State Power Board. Although it is not possible to isolate the figure devoted to the heat pump sector, approximately 260 million SEK ($ 30 million) of their own funds has been spent since 1980 on development and demonstration projects as part of its solar energy and heat pump program. In addition to this sum a further 40 million SEK ($ 5 million) from BFR has been devoted to joint projects. By contrast, cooperation between the State and the public utilities has been relatively limited in France where EDF is in a position to pursue its own development activities alone and in Germany where the electricity generating industry is extremely fragmented (over 600 utilities) and comes under the responsibility of the state governments.

Exhibit 3 : Government Expenditure in the Field of Heat Pumps ($ Million)

Type of activity	France	FRG (4)	Sweden (5)
R & D	N/A	18.9 (1)	11.5
Demonstration	N/A	7.4	15.5
Commercialisation	5.5 (e) (3)	NA (2)	139.6
Total	N/A	26.3	166.6

1 SEK = $ 0.12 ; 1 DM = $ 0.41 ; 1 FF = $ 0.11

(e) Estimates

(1) Includes demonstration support for diesel and absorption heat pumps
(2) Tax credits only
(3) Does not include an estimated $ 7 million from EDF's PERCHE campaign
(4) 1875-1985
(5) 1975 - April 1986

There appear to be no mechanisms for formal cooperation between the different government agencies and the other key players in the industry (manufacturers, utilities, research bodies) in any of the three countries. However, informal cooperation appears to be greater in Sweden. This is no doubt a function of the fact that Sweden is a much smaller country with a long tradition of open consultative government.

R, D & D Policy

Clearly the state has an important role to play in orienting the focus of technological research in conjunction with the long-term goals of national energy policy. However, it is not easy to evaluate the success of government policy in this area since, within the framework of national energy policy there are both long-term objectives concerning the development of technology which offers a long-term potential application and requires a sustained support, such as absorption heat pumps, and short-term objectives concerning technologies of more immediate application such as exhaust air electric heat pumps which need a short-term promotional effort in order to bring about rapid market penetration.

In France, state involvement in the funding of research and development has been fairly limited. The role of the AFME in France has been to assist in the development of existing equipment through the funding of demonstration projects rather than to promote fundamental research projects.

In Germany, the Ministry of Research and Technology (BMFT) has been responsible for the funding of both research and demonstration projects in the field of electric and thermally driven heat pumps. The German heat pump industry, which is dominated by a few large manufacturers, was able to finance a large part of their R & D activities. Although this was the case in Sweden for a limited number of large manufacturers, the Swedish heat pump industry is very fragmented (50-60 manufacturers) and characterised by a large number of small manufacturers who did not have the risk capital available to participate in the market development. Government support in the form of experimental building support and prototype demonstration projects has been extremely important to enable the smaller companies to enter the market.

The drawback of a reliance on industry and utility financing is the risk of a bias in R&D efforts towards short-term market driven projects which are likely to produce the most immediate results. In Sweden and Germany government financing has enabled research into the development of absorption heat pumps to continue, even in the current depressed climate. In France, however, where the major impetus has come from industry and the utilities, current research is focusing on more short-term objectives such as the development of small reversible heat pumps for residential application.

In all three countries the state has been instrumental in accelerating the development of heat pump technology, in particular air/air, air/water heat pumps. It is unlikely that industry would have been able to respond so rapidly to market requirements without government assistance, particularly at the development stage. Through the funding of demonstration projects the government has been able to reduce the risk exposure of the parties involved and has enabled them to focus on improving the performance of both systems and components.

Commercialisation Policy

In the residential market, the overiding criteria for the successful penetration of heat pump technology is the consumers' perception of the cost effectiveness of the equipment. The individual customer is above all concerned with low first investment costs (capital + installation costs) and to a certain extent reduced short-term operating costs. Until such time as the market has attained sufficient dimensions to allow the benefits of mass production to come into play, the state can compensate for the additional costs of a new technology over the costs of a more traditional type of equipment through the use of grants or subsidies to the end-user.

Different types of subsidies and marketing incentives have been practised in the three countries (see Exhibit 4) with varying degrees of success.

Exhibit 4 : Marketing Incentives for Heat Pumps in the Residential Sector

Funding Source	France	FRG	Sweden
State	. 400 FF/TOE (1) saved . low interest loans	. 10% of installed cost deducted from taxable income over 10 years	. 15% installed costs (3) . low interest loans
Utility	. FF 3000 bonus(2) . fixed installed price package . favorable electricity tariffs	. low interest loans . fixed price heat pump package . favorable electricity tariffs	

(1) Discontinued
(2) Discontinued in 1982
(3) Discontinued in 1985

The penetration of heat pumps in the total central heating stock has been highest in Sweden (see Exhibit 5) where the government has devoted over 80 % of total government expenditure to market introduction and investment support in the form of direct grants and subsidies. In Sweden, where in the early 1980s the electricity/oil relationship was too low to encourage the use of heat pumps, favouring instead the use of direct electric heating, and where there was little differentiation of electricity tariffs, the subsidies were particularly necessary for the heat pump to be economically attractive particularly in the detached house segment.

Exhibit 5 : Relative Penetration of Heat Pumps in Existing and New Buildings

000 Units	FRG	France	Sweden
Sales of boilers	516	397	91
Sales of direct electric systems	9	155	18
Total	525	552	109
Sales of Heat Pumps	14* (81)	51 (82)	26 (84)
Heat Pumps/ Central Heating Equipment	2.7 %	9.2 %	24 %

* Does not include some 35,000 units for hot water only

The lowest penetration rate of heating only heat pumps is in Germany where the government has favoured the use of tax credits (10 % per year of the installed cost over 10 years). This program has not been successful largely due to the ten-year pay-back period in a market where investment horizons are of a much more limited duration. This example clearly demonstrates the importance of basing the design of government subsidy programs on a sound understanding of the needs of the market place and in particular on consumer behaviour. The use of subsidy programs must also be carefully handled in order to avoid the artificial creation of a market which cannot be sustained in the long term without government assistance.

In Sweden, sales of heat pumps were assisted by the 15 % investment subsidies until 1984. However in that year, sales peaked very dramatically (see Exhibit 6), largely due to market anticipation that the 15 % investment subsidy would be discontinued in 1985. This had a doubly negative effect both overinflating the market (and taxing production capacity) and then abruptly abandoning it. It is generally felt that the investment subsidy should have been phased out gradually in order to retain market confidence in heat pump systems. Unfortunately, the decision to end the investment subsidy also coincided with the drop in oil prices. Thus sales of heat pump which were economically attractive, particularly in the detached house segment, while oil prices were high and the 15 % subsidy was available, plummetted in 1985. In France, the market peaked in 1982 under the stimulus of EDF's PERCHE i program but was unable to sustain the high rate of growth largely due to defects in the follow-up PERCHE GTI program.

Exhibit 6 : Sales of Heat Pumps in FRG, France and Sweden

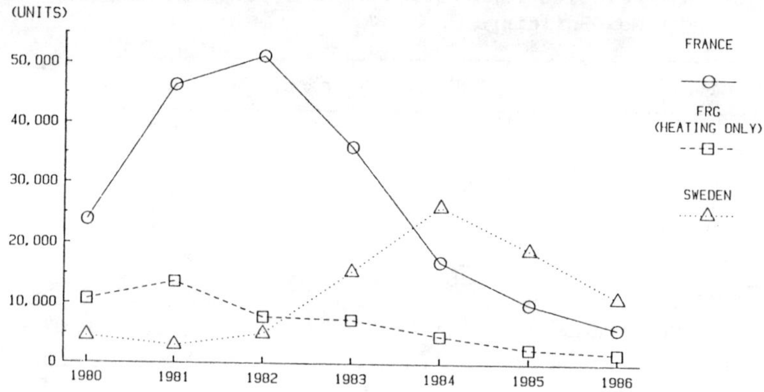

1985 - 1986 ESTIMATES

GROUPE MAC

The PERCHE GTI program, which was launched in June 1982 with the support of the Ministry of Industry, focused on the existing single family house segment. Under the program, the consumer was offered a guaranteed installed price for a variety of heat pumps operating on outside air/water (in the range 1.4-5 Kw) and operating in a bivalent mode (added to existing boiler). The Ministry of Industry provided special loans and also approved the manufacturers. Customers were also eligible for a special tariff "EJP" (or Peakday Effacement) in those zones equipped with remote control (65 % of France in 1983). The particular characteristic of the program, and also its defect was the network of installers and distributors who agreed to maintain set prices and who were responsible for the installation and maintenance of the equipment.

Although the program was extremely successful at the beginning, since over 30,000 heat pumps were sold between June 1982 and June 1983, it was not possible to sustain this degree of sales. The main defect in the program lay with the installers who did not accept the concept of a "maximum installer price". They were used to working with a "delivered price/manufacturer price" ratio of more than 3 with a conventional system while the PERCHE GTI had a ratio of 2.5. As a result, only 25 % of the 20,000 installers and distributors in the network actually sold GTIs.

In Germany, RWE has recently set up a similar program in an attempt to revive the flagging heat pump sales. As can be seen in Exhibit 6 sales of heating only heat pumps began to decline in 1982. While the tax incentive scheme still exists (10 % of the installed cost can be deducted over 10 years from taxable income) it is felt that this incentive is not sufficiently attractive for the individual consumer who is looking for a shorter pay back period.

In the autumn of 1985, RWE launched a program which has a certain similarity with the PERCHE program. This program, however offers the customer a heat pump package (including all the components in a closed system) 6-15 Kw thermal output, air to water bivalent system at a set price. Under an agreement between the manufacturers and the installers the installers' margin on the manufacturers' list price is reduced to 15-20 % (as opposed to 30-40 %). However, only the equipment price is fixed, and the installation cost is freely priced. In addition to the fixed package price, the consumer is able to obtain a loan from RWE up to a maximum of 13,500 DM ($ 4600) at 3 % interest rate to be paid back over four years. A further incentive offered by RWE is a special electricity tariff for bivalent heat pumps. Since manufacturers of air to water bivalent systems have succeeded in reducing production costs by 40 %, RWE's current strategy is to reduce the other elements in the equation : the costs of the components of the total system and the installation costs (two-third of total costs). The scheme is aimed at the "average" consumer in new and existing single and multifamily dwellings where it is essential to have small, efficient and cheap systems. In all three countries it is clear that the installer/distributor network is key to the successful marketing of heat pumps. And it is at this link in the chain that the state can contribute by organising seminars and training courses for contractors, installers, distributors. Both in Sweden and France, state funds have been devoted to such activities but not to a sufficient extent.

Industrialisation Policy

There is little evidence of a coherent industrial policy underlying the development of the heat pump industry, with the exception of France. Government support of the manufacturing industry in Sweden and Germany has been fairly indiscriminate and taken the form of indirect support through the funding of experimental building projects and prototype demonstration schemes. As a result there has been no natural clearing of the market and in both countries there are still about 50-60 manufacturers of heat pumps. In France, by contrast, and consistent with a more interventionist governmental approach, the Ministry of Industry contributed in the early 1980s a total of 100 million FF in the form of direct grants to a small number of selected manufacturers who demonstrated an ability to produce at an industrial level. The objectives of the program were to reach industrial production of two types of heat pump : outside air/water (as a back-up to existing boiler) and water/water in the 5-20 Kw range for single family houses and 50-200 Kw for multi-family buildings. This resulted in the creation of a production capacity capable of responding to the demand created by the EDF PERCHE campaign which was launched in 1981.

Another possible means of clearing the market is through the institution of norms which impose standards on the quality of heat pump equipment. However, the standards imposed in all three countries do not appear to have been sufficiently discriminating to bring about a more concentrated industry structure.

This problem is however, likely to be resolved in the near future as a result of the contraction in the market arising from the drop in oil prices and the phasing out of government incentives. In Sweden it is expected that the number of manufacturers will be reduced to fifteen or twenty, thus creating a more focused supply structure capable of responding efficiently to any potential market upswing.

International Trade and Cooperation Policy

Though all three countries assist in various international cooperative programs under the aegis of the International Energy Agency and the EEC, international trade cooperation has been fairly limited to date. While there is also a certain amount of trade particularly between the European countries it has not yet reached significant proportions. This is largely due to the differing characteristics and requirements of the heat pump markets in each country. While heat pump applications that supply heat and cooling are prevalent in the USA and Japan, there has been little effort to develop this type of heat pump in Europe.

Sweden, who has a long tradition of export trade given its limited domestic market, appears to be the only country to have attempted to set up a specific body to promote international cooperation and trade. The Swedish Export Technology Agency, which was created under the Oil Substitution Fund, provides a forum for manufacturers who are seeking to develop export markets and provides marketing support to those companies. As far as heat pumps are concerned, Sweden's strength lies in the fact that it has followed a niche strategy with two products ground source heat pumps and large heat pumps which should enable it to compete in foreign markets.

CONCLUSIONS AND RECOMMENDATIONS

The government's role in the development of a new technology is to clearly define policy objectives and to ensure the effective translation of these policies to the market place by the provision of funding at strategic points along the chain of R, D&D, commercialisation and industrialisation efforts.

In R&D the government's role is to define long term technological objectives and to provide sustained support to projects which offer a potential long term application such as absorption heat pumps. Short term applications which offer the possibility of a more immediate return will be taken care of by industry and the utilities. While governmental changes are bound to affect R&D policy, an effort should be made to ensure consistency in objectives.

Government policy at all levels must be based on a sound understanding of the needs of the market place. Energy policy considerations should not be allowed to interfere with the basic principles of marketing strategy. Energy efficiency and energy conservation objectives were the driving force behind the development of the heat pump market. This led to a targetting of the mass residential market, a fundamental marketing error. A more effective marketing strategy would have been to follow a niche strategy focussing on specific segments in industry or the public sector, thus creating a secure base from which to expand.

The government has a very powerful public relations role to play in providing a positive and, above all, objective signal of support to the market place. It should not be forgotten that the role of the electric utilities is after all to sell electricity. Subsidies are, however, a double-edged sword and must be handled with care in order to avoid the artificial creation of a market which cannot be sustained in the long term.

The key factor underlying a successful development strategy is that of organisation. The government is in a crucial position to coordinate the activities of the participants in the development of a new market and to avoid the duplication of effort that is bound to occur if there is no underlying strategy. While an interventionist industrial policy is unacceptable in many countries, it should be remembered that an indiscriminate funding strategy is likely to lead to the creation of an unfocused and hence uncompetitive supply structure.

Chapter 41

Impact of Heat Pumps on Energy Resources and the Environment Worldwide

Otto Zellhofer

ABSTRACT: The end-use energy saving potential of heat pumps for space heating versus conventional heating systems for oil substitution factors γ = 3...6 amounts to between 60 and 80 %. Therefrom, the resulting primary energy saving ranges between 20 and 50 %. Thus, the electrically driven heat pump forms an essential instrument for oil substitution, particulary in Austria with its seventy percent share of electricity from hydro power plants which is still to be expanded. Further, oil is substituted by electricity generated from coal, particulary in combined heat and power stations.

Regarding the environment, particularly the electrically driven heat pump shows significant advantages: the reduction of local emissions is 100 % for monovalent heating systems and typically 80 to 90 % for bivalent systems substituting fuel burning heating systems. The heat pump electricity consumption results in no emissions and immissions in case of hydro power plants. However, if the electricity driving the heat pump is - at least partly - generated in fossil-fuel fired power stations, there is an increase

HEAT PUMPS: Prospects in Heat Pump Technology and Marketing (Proceedings of the 1987 IEA Heat Pump Conference), Edited by Kay H. Zimmerman

in power plant emissions. But, especially for urban areas, the overall immission concentration of air pollutants including particulates caused by heat pump operation is lower than the immissions resulting from small, local oil-fired boilers.

1. INTRODUCTION

Today, more than twelve years after the IEA was established and more than ten years after the first IEA R&D projects came into being, it is appropriate to remember the basic aims of IEA. The first aim specified reads as follows: "Co-operation among IEA Participating Countries to reduce excessive dependence on oil through energy conservation, development of alternative energy sources and energy research and development".

The important role that the heat pump is to play in fulfilling the IAE aims has been recognized at an early date:

- The IEA Strategy Report of 1980 /1/ ranked energy conservation in buildings, mainly by the heat pump, as "Priority 1" items.
- The first major international IEA Conference (on New Energy Conversion Technologies and Their Commercializations, Berlin, 1981 /2/) also stressed the importance of the building sector and the need for more R&D and, in particular, for more information.
- A technology review on heat pumps was published by IEA in 1982 /3/.
- And heat pump projects were among the first IEA R&D projects undertaken in the field of end-use energy technology. Within this framework a study on advanced heat pump systems /4/ was undertaken by thirteen countries; the IEA Heat Pump Center, described in another paper of this conference, was established in 1982; and the first IEA Heat Pump Conference was held in Graz in 1984 /5/.

In times of low oil price, the aims of the IEA and the achievements in energy conservation and oil substitution by means of the heat pump should not be forgotten. The role of the heat pump in conserving energy resources and in improving the environment shall therefore be reviewed and policy measures to ensure such positive impacts even in times of low oil prices shall be discussed.

2. IMPACT ON ENERGY RESOURCES

The impact of the heat pump on the energy resources is twofold: The heat pump has a considerable energy conservation potential, and - equally or even more important - it has a large potential for oil substitution or, more general, for the substitution of scarce fossil fuels like distillate fuel oil or natural gas by more abundant or renewable energy sources.

2.1 Energy Conservation

In Table 1, typical data ofelectrically-driven heating-only heat pumps in comparison with an oil-fired boiler are presented. The basic data are similar to the ones used in a paper of the 13th Congress of the WEC /6/.

The amount of end-use energy saved is a function not only of the seasonal performance factor (SPF) of the heat pump, but also of the boiler efficiency, and - in the bivalent case - also of the shares of heat pump and auxiliary boiler in annual heating supply and the change (usually: improvement) of the annual boiler efficiency when switching the boiler from all-year to peaking duty. In the example chosen, the reduction of end-use energy amounts to 57 % for the bivalent, and 80 % for the monovalent case.

In part (b) of Table 1, the argument is carried from secondary energy (electricity, distillate fuel oil) to primary energy. In doing so, the efficiencies of oil supply and processing, and of electricity generation and distribution are to be accounted for. The latter depends on the kind of primary energy used. Results show that there is some primary energy saving even in the bivalent case and 100 % fossil-fuel condensing power generation; the saving is larger for the monovalent case and for pure hydro or partly hydro power generation. It should be noted that the saving will also be larger if combined heat and power (CHP) plants are considered instead of condensing thermal power plants.

2.2 Oil Substitution

End-Use energy is an important figure since it is the energy for which the customer has to pay. On the other hand, primary energy, insofar that it is the sum of different kinds of energy, is only a statistical figure: It consists of parts of quite different energies in terms of scarcity, specific cost, etc., and should therefore be used with discretion.

The qualitative aspect of switching from scarce to more abundant fuels (energy sources) is taken care of by two numbers: The share of scarce fuel (oil, in future perhaps also gas) substituted, and the fuel substitution factor, in particular the factor for oil substitution by electric energy. The latter is defined by the ratio of fuel energy substituted to the (electric) end-use energy necessary for this substitution. In many cases, it may also be expressed as the ratio of annual efficiency (or performance factor) of the electric process to the one of the process to be substituted, i.e. the oil-fired boiler.

Table 1 Energy Conservation and Oil Substition of a Heating-only Heat Pump (Typical Example)

		Oil-fired boiler		Heat pump			
				bivalent (air/water)		monovalent (water/water)	
		El.	Oil	El.	Oil	El.	Oil
(a) End-use Energy							
Share of heat pump and of boiler in annual heat supply	%	-	100	75	.25	100	-
Seasonal performance factor/boiler efficiency	-	-	.7	2.7	.75	3.5	-
End-use energy requirement, percent of delivered heat	%	-	143	28	33	29	-
Free energy used	%	-	-	47	-	71	-
End-use energy, total	%	143		61		29	
End-use energy saved:							
Percent of delivered heat	%	-		82		114	
Percent of end-use energy	%	-		57		80	
(b) Primary Energy							
Efficiency of oil supply and processing,	-			.875			
Efficiency of electricity generation and distribution:							
fossil plants (condensing)	%			.35			
hydro plants	%			.80			
50 % fossil, 50 % hydro	%	1/(.5/.35+0.5/.8)=.49					
Primary energy needs, percent of delivered heat:							
fossil	%	0+163=163		80+38=118		83+0=83	
hydro	%	0+163=163		35+38= 73		36+0=36	
50 % fossil, 50 % hydro	%	0+163=163		57+38= 95		59+0=59	

Table 1 contin. Energy Conservation and Oil Substition of a Heating-only Heat Pump (Typical Example)

		Oil-fired boiler		Heat Pump			
				bivalent (air/water)		monovalent (water/water)	
		El.	Oil	El.	Oil	El.	Oil
(c) Oil Substitution							
Share of oil substituted	%		-		77		100
Oil Substitution factor	-		-		3.9		5.0

In part (c) of Table 1, the oil substitution factor (OSF) for the monovalent case is

$$OSF = 29/143 = 3.5/0.7 = 5.0$$

and for the bivalent case:

$$OSF = (143 - 33)/28 = 3.9$$

In the bivalent case, the simple ratio of the efficiencies cannot be used, since the boiler efficiency is different in the boiler and the bivalent case.

The data of Table 1 are only examples, even if typical ones. In some cases, the SPF of air/water heat pumps may be somewhat lower, and the annual efficiency of modern low-temperature boilers may be somewhat higher. However, only in extreme cases will the OSF be less than 3.0. On the other hand, the boiler efficiency for hot tap water generation outside the heating season will in many cases be of the order of only .25. Assuming an SPF of 2.5 for this application, the OSF then becomes:

$$OSF = 2.5/.25 = 10$$

Even higher oil substitution factors may be achieved in specific industrial applications of heat pumps of the closed-cycle and open-cycle (steam compressor) type.

3. IMPACT ON THE ENVIRONMENT

Recently, the environmental issue has been given much attention for many energy applications. Regarding the heat pump, several approaches are possible /7, 8, 9/. First of all, it should be understood that the question is not whether the operation of a heat pump causes an environmental impact but whether it is larger or smaller relative to the one of another energy system for the same energy service.

Obviously, the <u>local</u> emission is reduced by 100 % if a boiler is replaced by a monovalent heat pump; for a bivalent heat pump, reduction is equal to the share of fuel that is substituted (usually between 70 and 90 %; 77 % according to the example shown in Table 1). The reduction of local emission nowadays is an important argument in deciding on the kind of heating system.

The <u>overall</u> emission of an electric heat pump has to take into account the emissions of the power station supplying the electricity. One approach is to assume the same fuel for the boiler to be fully or partly replaced by the heat pump and for the power station. This approach means that the indirect path of fuel use via electricity and the heat pump is compared to the direct path of using the same fuel locally in the boiler /7/; in part (b) of Table 1 it has been shown that the primary energy requirement will normally be smaller for the heat pump than for the boiler; therefore, also the overall emission will be smaller in the heat pump case. It will be much smaller if the emission control equipment of modern power station is taken into account such as a precipitator for particulates, flue gas desulphurization (FGD) for SO_2, and denitrification (deNOx) equipment for NOx.

The comparison is more difficult if different fuels for the boiler and for the power station are assumed, for instance light fuel oil with a sulphur content of 0.3 % for the boiler and coal for the power plant. In this case, the result will depend on the emission control equipment of the power plant: A coal-fired power plant without FGD and deNOx equipment will generally lead to somewhat higher emissions, a plant with, say, 90 % desulphurization and 80 % denitrification of the flue gas to considerably lower emissions than the oil-fired boiler /8/.

In reality, there will in most cases be a mix of energy sources from which electricity is generated. In particular, in countries like Sweden, Switzerland or Austria where there is an appreciable portion of hydro power, there will be an additional advantage of the electric heat pump due to the lower primary energy requirement (see Table 1) and the emission-free electricity generation in hydro plants /8/.

Regarding heat pumps driven by gas engines or diesel engines, it has been shown that these can environmentally compete with gas- or oil-fired boilers only if the engines are equipped with pollution control devices /9/.

The effects of flue gas emissions on the environment may be grouped into the following categories:

- effects on nature (forest decay),
- health effects on man,
- long-term effects on climate.

The effects on nature may as a first approximation, be assumed to be proportional to the emissions (although is it not yet completely understood which pollutants are the main cause; it may be that the hydrocarbons play the main role and these are emitted from small oil-fired boilers in a far greater amount than from the larger furnace of a power station).

The health effects on man may be assumend to be proportional to the population dose which in turn depends on population distribution and ground-level concentrations (GLC) of a pollutant. Since population distribution and local fuel use are correlated, whereas population distribution and power plant emissions (from high chimneys, far away from urban arears) are not, the population doses, and therefore health effects, of power plant emissions are much smaller than the effects of the same emissions from local fuel use in small furnaces. This ratio is called the transfer factor (TF); its value is about 0.1 for CHP plants in or near cities, and about 0.01 to 0.05 for large power plants /7, 8/. This means that for the health effects on man in urban areas the electric heat pump has an additional advantage by a factor of 10 to 100 vs. the small boiler.

Finally, the long-term effects on climate (greenhouse effect) will depend on the specific CO_2 emission. Here again, the electric heat pump has an advantage, even for the worst case, i.e. if an oil- or gas-fired boiler is compared to a heat pump driven by electricity of which 100 % are generated in a coal-fired power plant. If a ratio of primary energy of 83/163 = 0.51 is assumed (Table 1) and a ratio of fuel-specific CO_2 emission of coal vs. oil or gas of 1.25, the relative CO_2 emission of the heat pumps becomes

$$(0.51)(1.25) = 0.64 < 1$$

Apart from flue gas emissions of the boiler and the power station, there are some other effects of heat pumps on the environment: heat extraction from soil, ground water or surface water, and leakage of brine or working fluid.

Heat extraction from the soil of ground-coupled heat pumps, and from ground or surface water-based heat pumps has been shown to be only of minor consequence to soil and aquatic life if the system is properly designed.

Brine (antifreeze solutions) to extract heat from the soil at temperatures below 0 ^{o}C are mostly based on glycols that are toxic to man and higher animals. Less toxic salts are also available.

Chlorofluorocarbons (CFC) are normally used as working fluids of heat pumps. Recently, the question of CFC release and the possibility of their reducing the ozone concentration in the upper layers of the atmosphere has received attention. A reduction of the absorption of UV radiation in this layer and thus an increase of skin cancer might be the consequence. CFCs are not only used in refrigerators, air conditioners and heat pumps but also as sprays, in foam production, as solvents, for sterilization and fire extinguishing. The release rate had been increasing rapidly up to 1974 when some restrictions on the release began. The effect on the ozone concentration according to present knowledge is not very pronounced, and it is counteracted by the man-made ozone production at lower heights, in particular via NO_x from aircraft. Further research on this topic and on the importance of CFCs as a "greenhouse gas" is under way.

4. STATUS OF HEAT PUMP APPLICATION

Before policy requirements for increased application of the heat pump in the period of low oil price can be formulated, the present statuts of heat pump application is to be investigated. Five major areas can be distingushed:

- heating/cooling and heating/dehumidification applications;

- heating-only applications;
- hot tap water generation;
- district heating applications;
- industrial applications.

The market for heating/cooling application is expanding. Small residential heating/cooling heat pumps are sold in the US at a rate of about 1 million units per year (30 % of all new building) and in Japan at a rate of about 1.8 million units per year (that is more than 50 % of all air-conditioning units). Figures of other countries requiring cooling/dehumidification are not known but may be similar. The use of the heat pump principle in air-conditioned commercial buildings (shops, supermarkets, offices, indoor swimming pools) is also increasing in most countries.

The use of heat pumps for decentralized hot tap water generation has become common in several countries and so far has not been effected too badly by the low oil price. For instance, about 30,000 units are sold annually in Germany, and about 4000 in Austria.

About 90 large heat pumps units with a heat output between 1 MW and 100 MW per unit for district heating systems are in operation in Sweden, yielding a total heat output of about 1000 MW and 5 TWh per year /10/. The rate of deploying further units will, however, decrease, since the most promising sites heave already been used. In other countries (notably Japan, Finland, Romania) large units have also been installed and there is scope for a further increase /10/.

There is a small but steady market for heat pumps in many branches of industry. Widespread use is often limited by the requirement for extremly low pay-back periods /11/.

The important market for small and medium sized heating-only heat pump was most seriously effected by the fall of the oil price. In several countries (France, Germany, Sweden) the market has virtually collapsed, at least as far as the smaller units are concerned. Only in some regions where

- experienced installers,
- well-maintained reference plants,
- initiatives of the utilities,
- tax credits, and
- environmental pressures

exist, was the market preserved. E.g. Upper Austria, the area supplied with electricity from the Oberösterreichische Kraftwerke AG (OKA) is such an area. Anyhow it is the small and medium size heating-only heat pump which requires policy action most urgently.

5. PULIC ACTIONS REQUIRED

At the present time of low oil prices, an energy policy consistent with the long-term interest ofthe country and with the aims of IEA basically requires two goals to be kept in mind:

- continuation of R,D&D work in order to have improved and matured systems available when the economic conditions become more favourable for the heat pump; and
- a minimum level of production and installation of all major heat pump types in order to ensure that know-how and infrastructure are preserved.

The R,D&D work to be carried out in the near future should include the following topics:

- advanced heat pumps of the ambient air/water, exhaust air/water, water(brine)/water and also of the air/air type (including inverter-driven units with low starting current, and packaged units requiring little installation work);
- integration of heat pumps into the heating and ventilating system of new and existing buildings, including control strategies and refined methods of assessment; and
- design and demonstration of optimized energy efficient buildings of which the heat pump is an integral part.

A policy for ensuring a minimum level of heat pump installations will include the following policy instruments:

- Continuation of fiscal measures such as tax credits;
- simplification, where necessary, of the procedure to obtain such tax credits;
- a well-defined energy policy in accordance with the aims of IEA, indentifying the macro-economic and environmental advantages of heat pump use;
- recognition of the heat pump in the environmental legislation;
- use of heat pump in government-owned new and existing buildings;
- stimulation of the use of heat pumps in new buildings financed with public support; and, last but not least,
- clear and consistent information of the public by government, governmental agencies, local governments, and utilities.

REFERENCES

/1/ IEA: A Group Strategy for Energy Research, Development and Demonstration. OECD, Paris, 1980.

/2/ IEA: New Energy Conservation Technologies and Their Commercialization (Proc. Int. Conf. Berlin, April 1981). Springer-Verlag, Berlin, Heidelberg, New York, 1981.

/3/ IEA: Heat Pump Systems - A Technology Review. OECD, Paris, 1982.

/4/ IEA: Common Study on Advanced Heat Pump Systems. Verlag TÜV Rheinland, Köln, 1980.

/4/ IEA (W. Hochegger, ed.): IEA Heat Pump Conference - Current Situation and Future Prospects (Graz, May 1984). Verlag TU Graz, 1984.

/6/ WEC Task Force on Heat Pumps: Heat Pumps for Heating and Cooling. Position Paper, Experts Meeting 2, 13th Congress of the World Energy Conference (Cannes, October 1986).

/7/ WEC Working Group on The Environmental Effects Arising from Electricity Supply and Utilization and the Resulting Costs to the Utility: Electricity Supply and Utilization and the Environment. Position Paper, Working Group Session WG 3, 13th Congress of the World Energy Conference (Cannes, October 1986).

/8/ R. Brandner, et al.: Electric Heat Pumps Result in a Reduction of Air Pollutants. IEA HPC Newsletter Vol. 4, No. 1 (March 1986), p. 5-8.

/9/ M. Meal: Air Pollutant Emission Comparison of Gas and Diesel Engine-Driven Heat Pumps to Oil- and Gas-Fired Boilers. IEA

HPC Newsletter Vol. 4, No. 1 (March 1986), p. 8-11.

/10/ WEC Task Force on Heat Pumps: The Market Development of Heat Pumps. Position Paper, Experts Meeting 2, 13th Congress of the World Energy Conference (Cannes, October 1986).

/11/ F. Moser, H. Schnitzer: Heat Pumps in Industry (Chemical Engineering Monographs, Vol. 20). Elsevier, Amsterdam, 1985.

Section IX

Additions

Chapter 42

Estimating International Heat Pump Use

James M. Calm

ABSTRACT: International heat pump use has grown nearly sevenfold in the last decade. 3.8 million units were installed in 1985, the majority for residential space heating or space heating and cooling, and this quantity is still increasing. Usage for commercial, institutional, and industrial space conditioning is also expanding, as it is for water heating, swimming pool heating and dehumidification, industrial process heating and drying, and district heating. Greatest application is in the highly industrialized and developed countries, particularly those where building air conditioning is common such as Japan and the United States. The trend for space heating is radically different in Europe, where the newer heat pump market is rapidly declining. Uncertainties regarding future energy prices and policies impact heat pump use; the market reaction, most noticeably in Europe, reflects a gamble on low oil prices in the future.

INTRODUCTION

Falling oil, and consequently gas, prices are generally perceived as weakening the near-term market for heat pumps. Compounding this situation, in many countries, are both rising electrical rates and changing policies regarding heat pump promotion; the latter are primarily attributable to governmental reactions to recent nuclear power incidents and the former to new plant amortization. Accurate data and understanding are therefore needed by governmental policy makers, utility planners, and the manufacturing and construction industries to gauge the market impact, identify opportunities, and improve preparations and predictions for the future.

This paper presents the preliminary findings of an attempt to estimate the size and trends of the international heat pump market. Also addressed are the data sources and difficulties encountered to date. A future publication will present a more complete analysis as further data are obtained.

HEAT PUMPS: Prospects in Heat Pump Technology and Marketing (Proceedings of the 1987 IEA Heat Pump Conference), Edited by Kay H. Zimmerman © 1987 Lewis Publishers, Inc., Chelsea, Michigan 48118. Printed in U.S.A.

DATA COLLECTION

The types of data available from individual countries depend on the entity that performs the collection and, not surprisingly, vary considerably. Among the sources of heat pump statistics are:

Recurring census surveys,
One-time or recurring commerce surveys by government agencies,
Manufacturer trade association statistics programs,
Installer trade association statistics programs,
Builder trade association surveys,
Utility trade or research association studies,
One-time or recurring analyses of utility files,
Published and unpublished research surveys, and
Expert estimates.

For the countries examined, every comparison of data from multiple — but independent — sources (the references provide a partial list) yielded conflicting statistics even with, where possible, adjustments for exports and/or imports. Further, although a lesser concern, little consistency exists among the countries or sources for the classification of data. For example, manufacturers and their trade associations tend to categorize statistics by product configurations and capacities. Installers and their trade associations refer, instead, to application characteristics. Builders naturally refer to the type of construction. Utilities generally report statistics only for equipment using the energy form (i.e., electricity or gas) provided and then only within their own service areas.

Some of the differences in data can be attributed to errors in extrapolation from survey samples or from limited service area records to national statistics. In other cases, particularly when multiple trade associations are involved, some conflicts

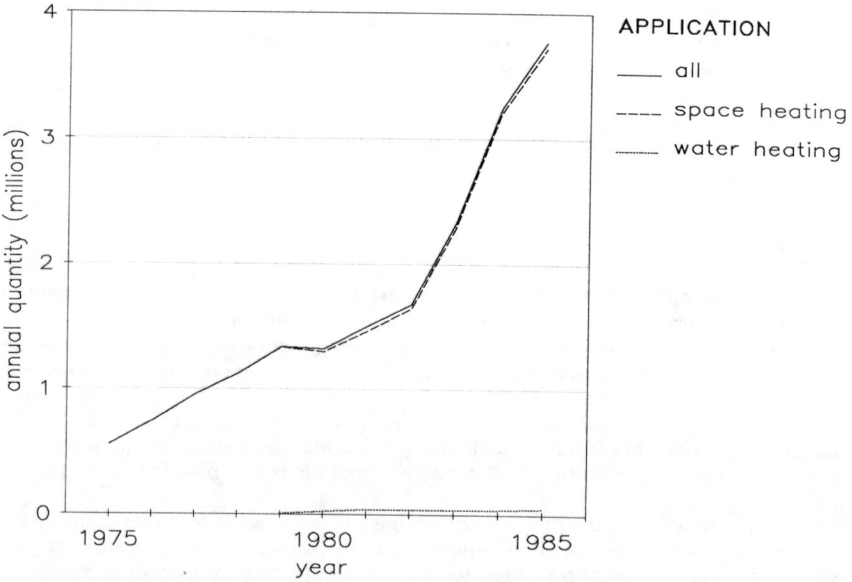

Figure 1. International heat pump use (preliminary data)

result from disparity in membership, often with none of the reporting associations representing all of the firms actually involved.

PRELIMINARY FINDINGS

International heat pump use has, with the exception of 1980, increased fairly steadily for more than a decade as shown in Figure 1. The aberrant year followed the second international oil-price shock, but a more plausible explanation is general economic recession with growth following 1980 — due to the inertial lag — resulting from the impacts of the 1979 price jolt.

Space heating heat pumps clearly dominate the the market.

Space Heating

For Canada, Japan, and the United States, the most common heat pump for space heating is the air-to-air, dual-mode (heating and cooling), electric heat pump. The majority of units in these countries are unitary (consisting of one or more units or modules designed to be used together without specific application engineering) split systems (having a remote outdoor unit connected to the indoor unit via refrigerant lines). These heat pumps are generally monovalent, using electric-resistance heat for supplemental and/or backup needs.

Air-to-water, heating-only heat pumps are the most common type for the European countries. Electric-motor drives are, again, the most common, although a smaller market exists for gas-engine-driven, and to a lesser extent for diesel-driven, heat pumps. By contrast to Canada, Japan, and the United States, the

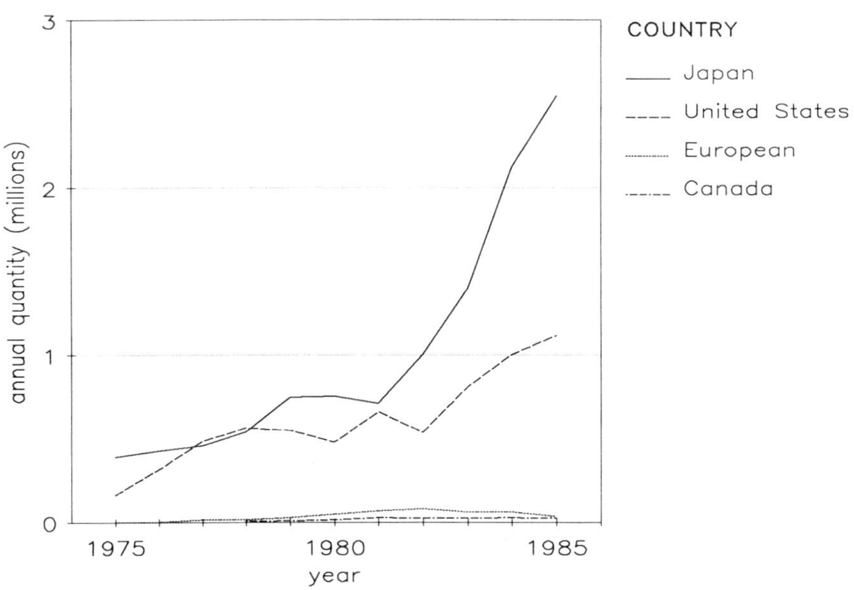

Figure 2. Heat pumps for space heating (preliminary data)

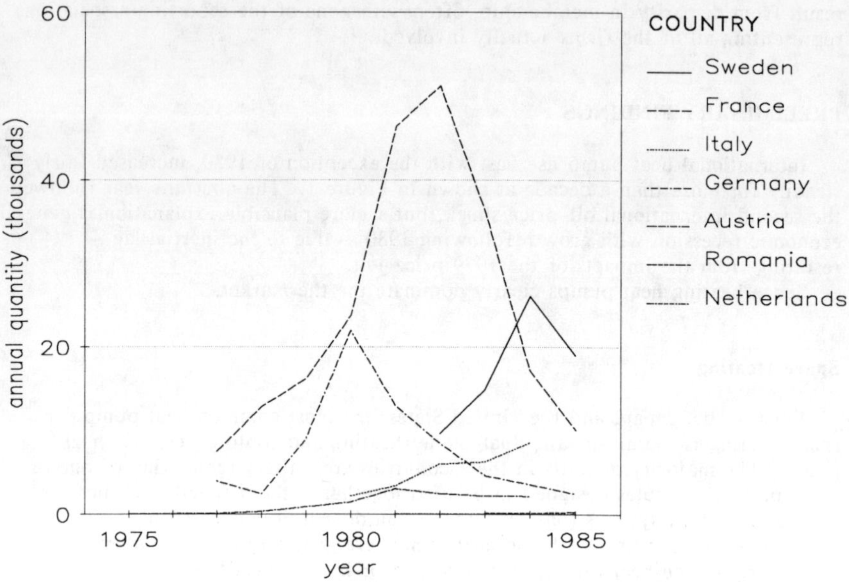

Figure 3. Heat pumps for space heating in Europe (preliminary data)

typical European heat pump is bivalent (using a combustion-based system, such as an oil- or gas-fired boiler or furnace, for peak and backup heating).

Figure 2 contrasts heat pump use for space heating in Japan, the United States, and Canada — sometimes referred to as *the Pacific countries* in international heat pump discussions — with that of the European countries. The sheer quantities of heat pumps in Japan and the U.S. overwhelm those in the rest of the world.

The European situation is elaborated, on an expanded ordinate, in Figure 3. Except for Austria, Italy, and Romania (discussed individually below), the characteristic market in the European countries with known data has been rapid growth followed by precipitous decline. Differences in the timing (for example, the rapid market increase in Sweden began long after the collapse of the German market and the respective peaks were separated by four years), strength, and duration of this pattern can be answered by differences in national energy prices and heat pump promotional incentives.

Water Heating

The heat pump market for service, including domestic, water heating is relatively small (see Figure 1) in comparison to that for space heating. It is, however, somewhat younger and — at least until 1986 — growing even in Europe (see Figure 4). The typical heat pump is an electric-powered air-to-water unit. High annual load factors result in better economic viability, in many cases, than for space heating. A limited number of integrated space and water heating heat pumps are sold, but their market is estimated to be relatively small; substantiating data have not been identified.

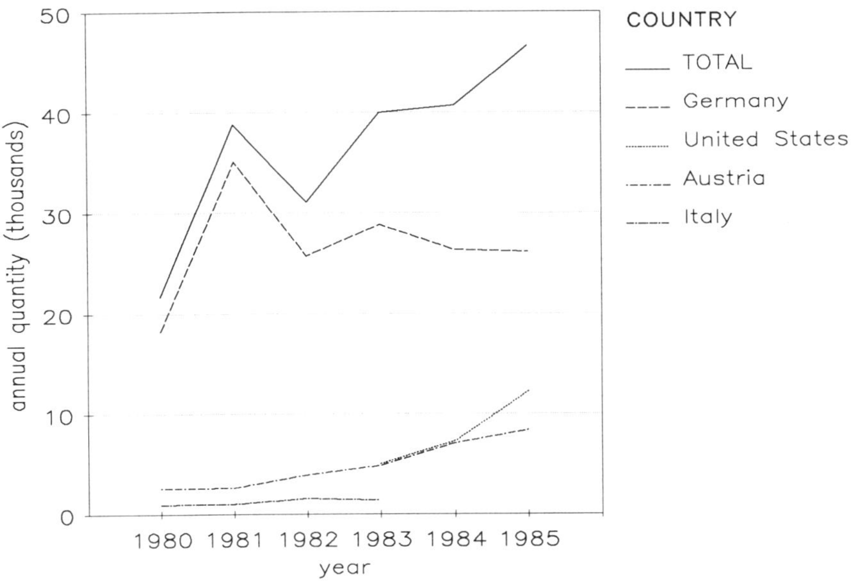

Figure 4. Heat pumps for water heating (preliminary data)

Pool Heating

Heating of swimming pools is a very favorable application of heat pumps, owing both to the high performance levels achieved at the moderate supply temperatures and, for indoor pools, to the inevitable need for concurrent dehumidification. The market for these devices is estimated to be growing and is already well established in Europe, where the first such application was completed in 1939.

Industrial Heating

The available data on the heat pump market for process heating is too limited for accurate depiction in Figure 1. Also, while the quantities involved are small compared to those for space heating, their average capacity is much larger. On bases of thermal capacity, financial value, or energy savings, the industrial heat pump market is quite significant.

Use of both open- and closed-cycle heat pumps is believed to be growing, although the former has a larger market size at present. Process heating is the only market in which absorption heat pumps (specifically excluding cooling-only devices) as well as heat transformers (sometimes referred to as type II heat pumps) have proven economic viability.

District Heating

Exploited most heavily in Sweden, heat pumps have also been applied to lesser extents for district heating in the other Nordic countries and Romania. Small numbers of installations have been made in France, the Federal Republic of Germany, the German Democratic Republic, Liechtenstein, the Netherlands, Switzerland, and the Soviet Union. Several installations of heat pumps for district heating and cooling also exist in Japan and the United States. The economics of these systems have generally been rather attractive.

DIVERSE NATIONAL MARKET PROFILES

The heat pump market complexion varies considerably among the countries for which data have been identified. The following profiles summarize the information obtained. Inclusion or exclusion of specific countries signifies only the availability of data and in no case implies any system of preferences or ranking.

Austria [1-5]

Austria claims the first heat pump installation, an open-cycle industrial system installed in a salt works in Ebensee in 1856 [3].

Heat pump sales for space heating have been nearly constant at approximately 2000 units/year, mostly air-source. Use of heat pump water heaters has grown steadily, reaching 8,400 units in 1985; two-thirds of these machines are produced domestically.

Interest is growing in ground-source systems, both direct expansion and ground-coupled, with a variety of heat exchanger designs. The feasibility of using large heat pumps in district heating is being evaluated for at least two systems.

Canada [6-7]

Cumulative sales of heat pumps in Canada for 1975-1985 are estimated at 100,000 units, of which approximately 60% are installed in the province of Ontario. Use of industrial heat pumps is limited, with approximately 15 existing medium and large installations. Heat pumps are accepted as the preferred means for and are used in approximately half of the hardwood dryers in Ontario.

Denmark [8]

The space heating market reached its maximum in 1981-1983 with approximately 2,000 small heat pumps. It has subsequently decreased to several hundred units per year. Total installations through 1985 are estimated at 8,000, of which nearly 50% are of the air-to-water type. Approximately 20,000 air-source heat pump water heaters are in use, with several domestic firms manufacturing exhaust-air heat reclaim heat pumps for single-family residences. Approximately 50 medium- and large-sized heat pumps are in use for district heating and other industrial uses.

Finland [9]

Still affected by an unsuccessful thrust into heat pump production in the 1970s, the market for heat pumps in Finland is rather limited. Approximately 5,000 units, mainly ground-source, were installed, but withdrawal or bankruptcy of most of the small manufacturers involved left these units without support for further maintenance. Six heat pumps, with a combined thermal capacity of 36 MW, are in use for district heating.

France [10-13]

The theoretical concept of a heat pump, thermodynamically a reverse heat engine cycle, originated in France in 1824 in a treatise by Sadi Carnot.

As shown in Figure 2, France has had the largest market for heat pumps in Europe, but this market has been declining since 1982 when it exceeded 50,000 units/year, far short of an initially targeted market of more than 100,000 heat pumps per year. 1985 use is estimated at 10,000 units.

Several hundred heat pumps are in use in the 50-500 kw_t range for multifamily-residential and commercial buildings.

Approximately 15 heat pumps are used in agricultural applications. In the industrial sector, use grew from approximately 500 heat pumps in 1979 to nearly 1,200 in 1982 (later data are unavailable, but the market is believed to have declined); many of these units were, however, for space heating of industrial facilities. Low-temperature drying remains an attractive market for heat pumps.

Germany [14-18]

The Federal Republic of Germany has the largest stock of gas-engine drive heat pumps with more than 700 operating units. The market for these machines has, however, sharply declined in the last two to three years.

The German heat pump market peaked in 1980 with nearly 22,000 electric-powered units for space heating and in 1981 with nearly 39,000 units for water heating. The latter application continues with more than 26,000 units in 1985, but the former has declined to approximately one-tenth of the cited peak (see Figures 3 and 4). German heat pump water heater sales are the highest in the world.

Italy [19-21]

Somewhat unique for Europe, Italian heat pumps derive from air conditioning designs. Most of the installed machines are, therefore, air-source and air is more common than hot water for distribution.

The market for space heating heat pumps nearly tripled from 3,100 units in 1980 to 8,600 in 1984, but more recent data have not been obtained. The water heating market peaked in 1982 with 1,500 heat pumps, but data are unavailable beyond 1983.

A small number of imported engine-driven and a few experimental absorption heat pumps are in operation.

Japan [22-27]

In quantities of heat pumps, Japan[1] has the largest market in the world as can be seen in Figure 2. Small ductless air-to-air units, referred to as *room air conditioner (RAC) heat pumps*, with capacities up to 3 kW_t dominate the market.

---

[1] Japanese annual data reflect the period from the beginning of October to end of the following September (e.g., 1985 data are for October 1984 through September 1985). The data for other countries are for the calendar year.

Available in window and through-the-wall configurations as well as in split systems, both single- and multizone, 1985 domestic sales were nearly 2.2 million heat pumps. Approximately 0.4 million additional heat pumps in larger sizes, termed *commercial air conditioner heat pumps* and *water chiller heat pumps*, were also sold. Virtually all of these units are dual mode (heating and cooling). Strong growth is expected in the near term market, influenced to a large measure by innovative room heat pumps incorporating inverters, scroll compressors, in-floor condensers, and even nonazeotropic mixture refrigerants with charge composition control for operation optimization.

Industrial heat pumps, including absorption type, also are used with 1985 installations amounting to an estimated 6 MW_e.

Netherlands [28-29]

The Netherlands are believed to employ the highest fraction of gas-engine-driven heat pumps in the world. Owing to large natural gas resources, a highly developed gas distribution infrastructure, and relatively high electric-to-gas price ratios, gas is the dominant heating fuel. The majority of heat pumps use air as the heat source with surface water a close second (nearly 40%). The total installed capacity of heat pumps exceeds 40 MW_t, including approximately 90 engine-driven (most with capacities exceeding 100 kW_t), 4 absorption, and 40-50 electric-motor powered heat pumps.

Norway [30]

The energy situation in Norway is envied by most other countries; in addition to very large, producing oil fields, the nation's electrical generation is — except in abnormally dry years for which thermal generating plants are in reserve — practically entirely by hydroelectric power. The consequently low prices for energy, compounded by relatively high interest rates, have resulted in less favorable economics for heat pumps than in other Scandinavian countries. Residential use, single- and multifamily, is very limited. However, approximately 120 installations have been made in greenhouses and another 100-150 for fish hatcheries, some with capacities exceeding 750 kW_t. 40 heat pumps are in use for fish drying and an estimated 200 for lumber drying. Open-cycle systems, with coefficients of performance as high as 30, have been used for concentration of liquids for more than 50 years. Domestic heat pump water heaters are rare, but one 6.5 MW_t water heater is in use in an alginate plant.

Romania [31]

Although small heat pumps are not used, Romania is the leading country in application of absorption heat pumps in district heating. The largest known resorption heat pump in the world, 8.7 MW_t using an ammonia/water solution, was installed in Bucharest in 1984. Steam-powered lithium bromide/water absorption cycles have been used since 1983, including 10 heat pumps of 2.2 MW_t and 15 of 5.8 MW_t.

Installations of smaller heat pumps, mainly for space and process heating, have increased from 90 units in 1983 to 200 in 1985 with capacities ranging from 30 to 230 kW_t. The heat sources used are primarily cooling water from transformers and compressors, industrial waste heat, and groundwater. A 7.3 MW_t vapor-compression heat pump, using a centrifugal compressor, is under construction for district heating.

Sweden [32-36]

Swedish use of large heat pumps in district heating is unmatched elsewhere, with at least four operating plants producing more than 100 MW_t each; the biggest, Stockholm Ropsten, uses six 25 MW_t machines. The largest individual heat pump, located in Göteborg, has a capacity of 45 MW_t,[2] employing a three-stage centrifugal compressor driven by an electric motor; it uses treated sewage effluent as its heat-source. Approximately 100 heat pumps with capacities exceeding 1 MW_t are in use. A single absorption machine of 7 MW_t capacity is also in operation for district heating.

The market for smaller machines peaked in 1984 with more than 26,000 units, mostly using outside-air and exhaust-air systems; nearly 8,000 of the latter were installed in 1984 and again in 1985. Ground-source (referred to as *bedrock*) heat pump systems, both ground-coupled and direct expansion, are also widely used.

United Kingdom [37-38]

Although both the first proposal for use of heat pumps for space heating, by Lord Kelvin in 1852, and the first such application, by T. G. N. Haldane in 1924, were in the UK, the market for such devices is limited. The present installation level is several hundred units per annum, equivalent to approximately 50 MW_e, with the majority being imported. Application in commercial buildings, particularly in stores, pubs, and facilities requiring air conditioning, is increasing, primarily with use of air-to-air packaged units. Heat pump use for swimming pools is also growing and represents a larger than anticipated market.

While the numbers of heat pumps used in various applications have not been identified, the financial value of installed equipment has increased rapidly with the both the highest share and gain being for air-to-air heat pumps. A 1984 survey projected a doubling of the heat pump market within 5 years.

United States [39-44]

The first commercialization of heat pumps for space heating began in the United States, which in aggregate heating capacities has the highest amounts of both existing and new heat pumps in the world. In terms of quantities, the U.S. follows Japan as shown in Figure 2.

Introduced in the early 1930's, heat pumps are now used to heat, or more commonly to heat and cool, approximately 30% of all new building in the United States. Their use is generally lower for domestic water and industrial process heating, and is still at the pioneering stage for district heating and cooling. Heat pump sales, as shown in Figures 2 and 4, have generally increased for the last decade and have established successive record levels in each of the last three years.

---

[2] Capacity ratings of large heat pumps are somewhat nebulous since each such machine is uniquely designed and operated with varying conditions. The unit cited, for example, has eight operating modes ranging from 25 to 45 MW_t capacity.

Continuation of this trend[3] is expected in 1986 for space heating devices, but a sharp decline is likely for heat pump water heaters.

Aggregate sales in 1985 are estimated to exceed 1.1 million units, including more than 12,000 for water heating and approximately 5,000 for process use. Note that a much larger market exists for cooling-only or cooling-controlled air-conditioning devices

CONTINUATION

As indicated above, findings reported herein should be regarded as preliminary. Data collection will continue, with emphasis on obtaining more complete information and resolving the conflicts cited. Any assistance that can be provided in this regard would be welcome; data or suggestions should be sent to: James M. Calm, IEA Heat Pump Center, FIZ-Karlsruhe, D-7514 Eggenstein-Leopoldshafen 2, Federal Republic of Germany.

SUMMARY

With the exception of 1980, ironically immediately following the second international oil shock, heat pump use has consistently expanded in each year of the last decade; it has increased nearly sevenfold in that period. The majority of use is for residential space heating or space heating and cooling. It appears that usage for commercial, institutional, and industrial space conditioning as well as for water heating, swimming pool heating and dehumidification, industrial process heating and drying, and district heating is also growing. Greatest use is in the highly industrialized and developed countries, particularly those where building air conditioning is common such as Japan and the United States.

The overall picture is radically different in Europe, where the much newer market is rapidly declining. Exceptions are for applications other than space conditioning and, to a limited extent, for building heating where air conditioning is emerging.

Uncertainties regarding future energy prices and electric generation, particularly nuclear, impact heat pump use. The market reaction, most noticeably in Europe, reflects a gamble on low oil prices in the future.

The availability, consistency, and completeness of market data for heat pumps and their use is insufficient, and the results presented in this continuing study should necessarily be regarded as preliminary.

ACKNOWLEDGEMENTS

This work is an extension of market data collection and analysis begun by A. Mitchell and J. A. Corfee, both formerly with the IEA Heat Pump Center (HPC),

---

[3] Although the quantity of new heat pump installations is increasing in single-family residential construction, the penetration rate is likely to have decreased in 1986 for the first time since 1979 (actual data will be unavailable until the summer of 1987). Both the absolute quantity and penetration rate decreased in multifamily-residential construction in 1985 and continuation is expected. For commercial and institutional construction, installation quantities have been increasing while penetration has declined slightly. Although highly speculative, penetration growth is likely to resume following 1987 when electricity/gas price ratios are projected to decline again.

under the leadership of K. O. Holzapfel [45]. Assisting with data gathering in the member countries of the HPC are the national teams including, to date, P. V. Gilli and R. Brandner (Austria), I. R. G. Lowe (Canada), K. O. Holzapfel (FRG), F. Trombetti (Italy), E. Miura (Japan), J. A. W. Oldenhof (The Netherlands), B. Lundqvist (Sweden), and F. A. Creswick and K. H. Zimmerman (USA).

REFERENCES

1. Arbeitsgemeinschaft Technik und Nutzung der Umweitenergie, "Marktstatistik," Vienna, Austria, 1980-1984.
2. R. Brandner et al., "National Position Paper for Austria," IEA Heat Pump Center November 1985 Workshop on Electric Heat Pumps for Retrofit in Small Residential Buildings with Hydronic Systems, Graz, Austria, October 1985.
3. K. Krenn, W. Ritter, and M. F. Schneeberger, "The World's First Industrial Heat Pump," *IEA Heat Pump Center Newsletter*, **4**(2), 14-16, June 1986.
4. "Marktentwicklung in Österreich," *Die Zeitung Clima Commerce International* (CCI), **3**, page 26, 2 March 1984.
5. M. F. Schneeberger, "Heat Pumps in Austria," 13th Congress of the World Energy Conference, Experts Meeting on Heat Pumps, Cannes, France, 72-76, October 1986.
6. J. M. Bell, "Heat Pumps in Canada," 13th Congress of the World Energy Conference, Experts Meeting on Heat Pumps, Cannes, France, 77-81, October 1986, and clarification by private communication, December 1986.
7. Ontario Research Foundation, "A Review of the Heat Pump Research and Development Program," National Research Council of Canada, DSS file number 55SZ.31025-5-5010, June 1986.
8. B. Nielsen, "Heat Pumps in Denmark," 13th Congress of the World Energy Conference, Experts Meeting on Heat Pumps, Cannes, France, 82-86, October 1986.
9. B. Landtman, "Heat Pumps in Finland," 13th Congress of the World Energy Conference, Experts Meeting on Heat Pumps, Cannes, France, 87-92, October 1986.
10. Électricité de France, SEPAC, "Situation and Prospects in France," Royal Swedish Academy of Engineering Sciences Symposium on International Heat Pump Promotion, October 1985.
11. P. Berhondo, "Heat Pumps for Space Heating in Buildings and Industrial Processes: The French Situation," Électricité de France Report AC/EEA NS 1524, Paris, France, undated (ca. 1986).
12. P. Berhondo, "Heat Pumps in France: The Present Situation," 13th Congress of the World Energy Conference, Experts Meeting on Heat Pumps, Cannes, France, 13-22, October 1986.
13. Groupe MAC, "Recent Developments in the French Heat Pump Market," *Minutes of the September 1986 International Workshop on Approaches to the European Heat Pump Market, Graz, Austria*, attachment 4, October 1986.
14. Gesellschaft für Wirtschaftsförderung und Marktplanung mbH, "National Position Paper for the Federal Republic of Germany," IEA Heat Pump Center

November 1985 Workshop on Electric Heat Pumps for Retrofit in Small Residential Buildings with Hydronic Systems, Graz, Austria, July 1985.

15. P. Göricke and P. Kalischer, "Heat Pumps in the Federal Republic of Germany (FRG)." 13th Congress of the World Energy Congress, Experts Meeting on Heat Pumps, Cannes, France, 3-12, October 1986.

16. I. R. Heinemann, "VDEW-Erhebung über elektrische Wärmepumpen-Heizungsanlagen," Vereinigung Deutscher Elektrizitätswerke Hauptgesellschäftsstelle, Frankfurt, FRG, July, 1985.

17. M. Raschka, "Ölspur im Wärmepumpenmarkt," Karlsruhe, FRG, *Die Zeitung Clima Commerce International* (CCI), **3**, pages 1 and 63, 2 March 1984.

18. Zentralverband der Elektrotechnischen Industrie e.V., "German Sales of German-Manufactured Heat Pumps," private communication, Frankfurt, FRG, 1986.

19. ANIMA and CO.AER, "Indagine Statistica sulle Pompe di Calore," Milan, Italy, 1980 - 1983.

20. M. Constantino, E. Piantoni, and P. Zuccala, "National Position Paper for Italy," IEA Heat Pump Center November 1986 Workshop on Electric Heat Pumps for Retrofit in Small Residential Buildings with Hydronic Systems, Graz, Austria, February 1985.

21. F. Trombetti, private communication, telex 1145/86, Milan, Italy, May 1986.

22. "Draft National Position Paper of Japan," IEA Heat Pump Center November 1985 Workshop on Electric Heat Pumps for Retrofit in Small Residential Buildings with Hydronic Systems, Graz, Austria, October 1984.

23. Japan Refrigeration and Air Conditioning Industry Association (JRAIA), "Heat Pumps in Japan," Tokyo, Japan, November 1983.

24. JRAIA, private communication from E. Miura, telex dated 7 August 1986.

25. K. Narita, "Bridging the Gap: Heat Pumps for Industrial Use in Japan," *IEA Heat Pump Center Newsletter*, **4**(2), 5-7, June 1986.

26. I. Takasawa, "The Present Status of the Gas Engine Heat Pump Market in Japan and the New Advanced Gas Engine Heat Pump," *Proceedings of the IEA Heat Pump Conference — Current Situation and Future Prospects*, Verlag für die Technische Universität Graz, 53-63, May 1984.

27. K. Tanaka and E. Miura, "Heat Pumps in Japan 1975-1990," *Proceedings of the IEA Heat Pump Conference — Current Situation and Future Prospects*, Verlag für die Technische Universität Graz, 195-205, May 1984.

28. P. J. Collet, "Heat Pumps in the Netherlands," 13th Congress of the World Energy Conference, Experts Meeting on Heat Pumps, Cannes, France, 93-98, October 1986.

29. J. A. W. Oldenhof, "Market Situation on Heat Pumps in the Netherlands," private communication, June 1986.

30. G. Lorentzen, "Heat Pumps in Norway," 13th Congress of the World Energy Conference, Experts Meeting on Heat Pumps, Cannes, France, 99-106, October 1986.

31. G. M. Pop et al., "Topical Activities Conducted in the Field of Heat Pumps in Romania," 13th Congress of the World Energy Conference, Experts Meeting on Heat Pumps, Cannes, France, October 1986.

32. EPRO, "Heat Pump Sales in Sweden," publication unknown.

33. C. Jacobsen, "Field Experience of Medium-Sized Heat Pumps for Residential and Commercial Areas and for Industrial Applications," *Proceedings of the IEA Heat Pump Conference — Current Situation and Future Prospects*, Verlag für die Technische Universität Graz, 139-148, May 1984.

34. B. Nordström, "Heat Pumps in Sweden," 13th Congress of the World Energy Conference, Experts Meeting on Heat Pumps, Cannes, France, 47-56, October 1986.

35. "Statistik Avseende Värmepumpe i Sverige," publication unknown, March 1984.

36. Svenska Värmepumpförningen, "Statistik," Vällingby, Sweden, SVEP Informations and Service AB.

37. Building Services Research and Information Association, "BSRIA Statistics Bulletin, Product Profile, Heat Pumps," **9**(1), Berkshire, UK, 1984.

38. A. T. Churchman, "Heat Pumps in the United Kingdom," 13th Congress of the World Energy Conference, Experts Meeting on Heat Pumps, Cannes, France, 107-108, October 1986.

39. Air-Conditioning and Refrigeration Institute (ARI), "Statistical Profile of the Air-Conditioning, Refrigeration, and Heating Industry," Section C, Arlington, VA, USA, July 1986.

40. J. M. Calm, "Electric Heat Pump Trends in the United States," *Proceedings of the IEA Heat Pump Conference — Current Situation and Future Prospects*, Verlag für die Technische Universität Graz, 15-26, May 1984.

41. J. M. Calm, "The Heat Pump," *ASHRAE Journal*, **26**(8), 40-47, August 1984.

42. J. M. Calm, "Heat Pumps in the United States," 13th Congress of the World Energy Conference, Experts Meeting on Heat Pumps, Cannes, France, 57-71, October 1986. Revision in publication, *International Journal of Refrigeration*, 1987.

43. J. G. Crawford, "Heat Pumps in the United States: A Status Report and View to the Future," *Proceedings of the IEA Heat Pump Conference — Current Situation and Future Prospects*, Verlag für die Technische Universität Graz, 27-38, May 1984.

44. G. C. Groff and R. E. Ertinger, "Heat Pumps in the USA — Projections for the Future," *Proceedings of the IEA Heat Pump Conference — Current Situation and Future Prospects*, Verlag für die Technische Universität Graz, 207-217, May 1984.

45. A. Mitchell, "Heat Pump Sales in Member Countries," *IEA Heat Pump Center Newsletter*, **3**(1/2), 18-20, July 1985.

Authors

Alefeld, Georg, Technical University of Munich, James-Franck-Strasse, D-8046 Garching bei Muenchen, Federal Republic of Germany

Astrand, Lars E., Uppsala Energy AB, P.O. Box 125, S-75104 Uppsala, Sweden

Azarniouch, M. K., Pulp & Paper Research Institute of Canada, 570 St. John's Boulevard, Pointe Claire Quebec, Canada H9R 3J9

Berghmans, Jan, Kathiolieke Universiteit Leuven, Celestijnenlaan 300 A, B-3030 Heverlee-Leuven, Belgium

Berntsson, Thore, Chalmers University of Technology, S-41296 Gothenberg, Sweden

Bokelmann, Horst, GEA GmbH, Reisingstrasse 9, D-4630 Bochum, Federal Republic of Germany

Bostroem, Torbjorn, Swedish Council for Building Research, St. Gorangatan 66 S-112 33 Stockholm, Sweden

Braswell, Arnold, Air Conditioning and Refrigeration Institute, 1501 Wilson Boulevard, 6th Floor, Arlington, VA 22209

Bullock, Charles E., United Technologies Carrier, Carrier Parkway, P.O. Box 4808, Syracuse, NY 13221

Calm, James M., IEA Heat Pump Center, Fachinformationszentrum-Karlsruhe, D-7514 Eggenstein-Leopoldshafen 2, Federal Republic of Germany

Chappell, Robert N., U.S. Department of Energy, Idaho National Engineering Laboratory, 785 DOE Place, Idaho Falls, ID 83402

Chaump, Louis E., Trane DPG, Inc., 6200 Troup Highway, Tyler, TX 75711

Chiba, Takanori, Takenaka Komuten Company, Ltd., Tokyo Main Office, 21-1, 8-chome, Ginza, Chuo-ku, Tokyo, Japan

DeVault, Robert C., Oak Ridge National Laboratory, Building 9102-2, P.O. Box Y, Oak Ridge, TN 37831

Enstroem, Henrik, Skandinavisk Termoekonomi AB, Ostermalmsgatan 7, S-11424 Stockholm, Sweden

Fairchild, Phillip D., Oak Ridge National Laboratory, Building 9102-2, P.O. Box Y, Oak Ridge, TN 37831

Frivik, Per E., Norges Tekniske Hoegskole, Institut F. Kuldetechnik, N-7034 Trondheim, Norway

Fukuda, Tadashi, Osaka Gas Company, Ltd., 2-95 Chiyozaki 3-chome, Nishi-ku, Osaka 550 Japan

Gilbert, Joel S., Dames and Moore Company, 7101 Wisconsin Avenue, Suite 700, Bethesda, MD 20814

Gilles, Ted C., Lennox Industries, Inc., P.O. Box 809000, Dallas, TX 75380

Glas, Lars-Olof, Skandinavisk Termoekonomi AB, Ostermalmsgatan, S-11424 Stockholm, Sweden

Gonzales, Jose A., International Energy Agency, 2 rue Andre-Pascal, 75775 Paris Cedex 16, France

Halozan, Hermann, University of Technology, Inffeldgassse 25, A-8010 Graz, Austria

Harris, Jean-Francis, Management Analysis Center, 11 Blvd. Latour Maubourg, 75007 Paris, France

Haskins, N., Management Analysis Center, 75007 Paris, France

Hiller, Carl C., Electric Power Research Institute, 3412 Hillview Avenue, Palo Alto, CA 94303

Hirschbichler, Franz, MAN Technologie GmbH, Dachauerstrasse 667, P.F. 500620, D-8000 Muenchen 50, Federal Republic of Germany

Hochegger, Werner, Energiesparhaus-Forschungsges mbH, Petersgasse 45, A-8010 Graz, Austria

Holzapfel, Karl O., IEA Heat Pump Center, Fachinformationszentrum-Karlsruhe, D-7514 Eggenstein-Leopoldshafen 2, Federal Republic of Germany

Hughes, H. Michael, Climate Master Corporation, 2007 Beechgrove Place, Utica, NY 13501

Kapus, Theodore E., U.S. Department of Energy, 1000 Independence Avenue, SW, Washington, DC 20585

Kawahara, Katsushige, Daikin Industries, Ltd., 1304 Kanaokacho, Sakai, Japan

Krauss, William E., Gas Research Institute, 8600 West Bryn Mawr Avenue, Chicago, IL 60631

Kruse, Horst H., University of Hannover, Welfengarten 1A, D-3000 Hannover 1, Federal Republic of Germany

Kubota, Kihachiro, Shimizu Construction Co., Ltd., Institute of Technology, No. 4-17 Etchujima 3-chome, Koto-ku, Tokyo, Japan

Kuno, Hiroaki, Shimizu Works, Hitachi, Ltd., 390 Muramatsu, Shimizu-shi, Shizuoka-ken 424 Japan

Kurosawa, Shigekichi, Tokyo Gas Company, Ltd., 5-20 Kaigan 1-chome, Minato-ku, Tokyo 105 Japan

Lannus, Arvo, Electric Power Research Institute, 3412 Hillview Avenue, P.O. Box 10412, Palo Alto, CA 94303

Letout, Philippe P., Alsthom-Rateau, 141 rue Rateau, F-93123 La Courneuve, France

Macriss, Robert A., Institute of Gas Technology, 4201 West 36th Street, Chicago, IL 60632

Maier, Richard G., Kulmbacher Klimageraete-Werke GmbH, Am Goldenen Feld 18, D-8650 Kulmbach, Federal Republic of Germany

Maret, Arthur R., Gas Research Institute, 8600 West Bryn Mawr Avenue, Chicago, IL 60631

Mashimo, Katsuyuki, Tokyo Sanyo Electric Co., Ltd., 180 Sakata, Oizumi-machi, Ora-gun, Gunma-ken, Japan

Matsubara, Katsumi, Hitachi, Ltd., 502 Kandatsu-machi, Tsuchiura-shi, Ibaraki-ken 300 Japan

Matsuoka, Fumio, Mitsubishi Electric Corporation, Consumer Products, 2-14-40, Ofuna, Kamakura 247 Japan

McGrath, Michael, Edison Electric Institute, 1111 19th Street, NW, Washington, DC 20036-3691

Minagawa, Tsunehiko, Tohoku Electric Power Co., Ltd., 7-2-1 Nakayama, Sendai 980 Japan

Monahan, Russell E., Stirling Power Systems Corporation, 7101 Jackson Road, Ann Arbor, MI 48103

Morino, Kimio, Shimizu Construction Co., Ltd., Institute of Technology, No. 4017, Etchujima 3-chome, Koto-ku, Tokyo, Japan

Narita, Katsuhiko, Tokyo Electric Power Co., 1-1-3 Uchisaiwai-cho, Chiyoda-ku, Tokyo 100 Japan

Niess, Richard C., Dames and Moore Company, 7101 Wisconsin Avenue, Suite 700, Bethesda, MD 20814

Nordstrom, Bengt F., Swedish State Power Board, Vattenfall, Fack Roecksta, S-16287 Vaellingby, Sweden

Nyman, Hans O., District Heating Development Co., 76 West Kellogg Boulevard, St. Paul, MN 55102

Paul, Joachim, A/S Thomas T. Sabroe and Co., P.O. Box 1810, DK-8270 Aarhus, Hojbjerg, Denmark

Reynaud, Jean-Francois, Ampere 82, Tour Atlantique, 92080 Paris de la Defense, Cedex 6, France

Richlen, Scott, U.S. Department of Energy, Office of Industrial Programs, CE-141/FORS, 1000 Independence Avenue, SW, Washington, DC 20585

Romagnino, J., Pulp & Paper Research Institute of Canada, 570 St. John's Boulevard, Pointe Claire Quebec, Canada

Rostek, Horst, Gaswaermepumpen-Gesellschaft, Halterner Str. 125, D-4270 Dorsten, Federal Republic of Germany

Ryan, John D., U.S. Department of Energy, 1000 Independence Avenue, SW, CE 112, Room GF-217/FORS, Washington, DC 20585

Saito, Takamoto, The University of Tokyo, 7-3-1 Hongo, Bunkyo, Tokyo 113, Japan

Sakashita, Shigeru, Mayekawa Manufacturing Co., Ltd., 13-1 Botan 2, Koto-ku, Tokyo 135 Japan

Schneeberger, Michael F., Oberoesterreichische Kraftwerke AG, Boehmerwaldstrasse 3, A-4021 Linz, Austria

Shimura, Takehiko, New Energy Development Organization, 28th Floor, Sunshine 60, No.1-1, Higashi-Ikebukuro, 3-chome, Tokyo 170, Japan

Struck, Wilhelm, MAN Technologie GmbH, Dachauerstrasse 667, P.f. 500620, D-8000 Muenchen 50, Federal Republic of Germany

Suefuji, Kazutaka, Hitachi, Ltd., 502 Kandatsu-machi, Tsuchiura-shi, Ibaraki-ken 300 Japan

Suma, Sei, Toshiba Corporation, Fuji Works, 336 Tadehara, Fuji-shi, Shizuoka 416 Japan

Tanaka, Kazuo, Daikin Industries, Ltd., 1304 Kanaokacho, Sakai, Japan

Ukaji, Masaaki, Kumagai Gumi Co., Ltd., 17-1 Tsukudo-cho, Shinjuku-ku, Tokyo 162 Japan

Vidal, Christian A., Electricity of France - SEPAC, 2 rue Louis Murat, 75008 Paris, France

Wakabayashi, Masahiro, Mayekawa Manufacturing Co., Ltd., 13-1 Botan 2, Koto-ku, Tokyo 135, Japan

Westerlund, Rolf, K-Konsult - Energisektionen, Liljeholmtorget 7, S-11780 Stockholm, Sweden

Yamaguchi, Zenkichi, Tohoku Electric Power Co., Ltd., 7-2-1 Nakayama, Sendai 980 Japan

Yoshii, Takeshi, Mitsubishi Heavy Industries, Ltd., 5-1 Marunouchi 2-chome, Chiyoda-ku, Tokyo, Japan

Young, David J., Ontario Hydro, Toronto, Ontario, Canada

Zegers, Peter, Commission of the European Communities, Wetstraat 200, B-1049 Brussels, Belgium

Zellhofer, Otto, Federal Ministry for Science & Research, Freyung 1, A-1014 Vienna, Austria

Index